THE GIFT OF
The Macmillan Company

PLANE AND SOLID ANALYTIC GEOMETRY

THE MACMILLAN COMPANY
NEW YORK · BOSTON · CHICAGO · DALLAS
ATLANTA · SAN FRANCISCO

MACMILLAN & CO., LIMITED
LONDON · BOMBAY · CALCUTTA
MELBOURNE

THE MACMILLAN CO. OF CANADA, LTD.
TORONTO

PLANE AND SOLID ANALYTIC GEOMETRY

BY

WILLIAM F. OSGOOD, Ph.D., LL.D.
PERKINS PROFESSOR OF MATHEMATICS
IN HARVARD UNIVERSITY

AND

WILLIAM C. GRAUSTEIN, Ph.D.
ASSISTANT PROFESSOR OF MATHEMATICS
IN HARVARD UNIVERSITY

New York
THE MACMILLAN COMPANY
1922

All rights reserved

PRINTED IN THE UNITED STATES OF AMERICA

Copyright, 1920, 1921,
By THE MACMILLAN COMPANY.

Set up and electrotyped. Published July, 1921.

Norwood Press
J. S. Cushing Co. — Berwick & Smith Co.
Norwood, Mass., U.S.A.

PREFACE

THE object of an elementary college course in Analytic Geometry is twofold: it is to acquaint the student with new and interesting and important geometrical material, and to provide him with powerful tools for the study, not only of geometry and pure mathematics, but in no less measure of physics in the broadest sense of the term, including engineering.

To attain this object, the geometrical material should be presented in the simplest and most concrete form, with emphasis on the geometrical content, and illustrated, whenever possible, by its relation to physics. This principle has been observed throughout the book. Thus, in treating the ellipse, the methods actually used in the drafting room for drawing an ellipse from the data commonly met in descriptive geometry are given a leading place. The theorem that the tangent makes equal angles with the focal radii is proved mechanically: a rope which passes through a pulley has its ends tied at the foci and is drawn taut by a line fastened to the pulley. Moreover, the meaning of foci in optics and acoustics is clearly set forth. Again, there is a chapter on the deformations of an elastic plane under stress, with indications as to the three-dimensional case (pure strain, etc.).

The methods of analytic geometry, even in their simplest forms, make severe demands on the student's ability to comprehend the reasoning of higher mathematics. Consequently, in presenting them for the first time, purely algebraic difficulties, such as are caused by literal coefficients and long formal computations, should be avoided. The authors have followed this principle consistently, beginning each new subject of the early chapters with the discussion of a simple special, but typical, case, and giving immediately at the close of the paragraph simple examples of the same sort. They have not, however, stopped here, but through carefully graded problems, both of geometric and of analytic character, have led the student to the more difficult applications of the methods, and collections of examples at the close of the chapters contain such as put to the test the initiative and originality of the best students.

As a result of this plan the presentation is extraordinarily elastic. It is possible to make the treatment of any given topic brief without rendering the treatment of later topics unintelligible, and thus the instructor can work out a course of any desired extent. For example, one freshman course at Harvard devotes about thirty periods to analytic geometry and the material covered consists of the essential parts of the first nine chapters. Another freshman course gives twice the time to analytic geometry (the students having already had trigonometry), taking up determinants and the descriptive properties of the quadric surfaces, and also devoting more time to the less elementary applications of the methods of analytic geometry. The advanced courses in the calculus and mechanics require the material of the later chapters. In fact, a thorough elementary treatment of the rudiments of Solid Analytic Geometry is indispensable for the understanding of standard texts on applied mathematics. It is true that these texts are chiefly Continental. But we shall never have American treatises which are up to the best scientific standards of the day until the subjects above mentioned are available in simply intelligible form for the undergraduate.

The subject of loci is brought in early through a brief introductory chapter, and problems in loci are spread throughout the book. A later chapter is devoted to a careful explanation of the method of auxiliary variables. There is a chapter on determinants, with applications both to analytic geometry and to linear equations. Diameters and poles and polars in the plane and in space receive a thorough treatment. Cylindrical and spherical coördinates and quadric surfaces are illumined by the concept of triply orthogonal systems of surfaces. The reduction of the general equation of the second degree in space to normal forms by translations and rotations is sketched and illustrated by numerical examples.

The question may be asked: In so extensive a treatment of analytic geometry should not, for example, homogeneous coordinates find a place? The authors believe that the student, before proceeding to the elaborate methods of modern geometry, should have a thorough knowledge both of the material and the methods which may fairly be called elementary, and they felt that a book which, avoiding the conciseness of some of the current texts and the looseness of others, is clear because it is rigorous will meet a real need.

This book is designed to be at once an introduction to the subject and a handbook of the elements. May it serve alike the needs of the future specialist in geometry, the analyst, the mathematical physicist, and the engineer.

Harvard University
April, 1921

CONTENTS

PLANE ANALYTIC GEOMETRY

INTRODUCTION

DIRECTED LINE-SEGMENTS. PROJECTIONS

CHAPTER I

COÖRDINATES. CURVES AND EQUATIONS

CHAPTER II

THE STRAIGHT LINE

CHAPTER VII

THE ELLIPSE

CHAPTER VIII

THE HYPERBOLA

CHAPTER IX

CERTAIN GENERAL METHODS

CHAPTER XIII

A SECOND CHAPTER ON LOCI. AUXILIARY VARIABLES. INEQUALITIES

CHAPTER XIV

DIAMETERS. POLES AND POLARS

CHAPTER XV

TRANSFORMATIONS OF THE PLANE. STRAIN

CHAPTER XVIII

DIRECTION COSINES. DIRECTION COMPONENTS

CHAPTER XIX

THE PLANE

CHAPTER XX

THE STRAIGHT LINE

CHAPTER XXI

THE PLANE AND THE STRAIGHT LINE. ADVANCED METHODS

CHAPTER XXII

SPHERES, CYLINDERS, CONES. SURFACES OF REVOLUTION

CHAPTER XXIII

QUADRIC SURFACES

CHAPTER XXIV

SPHERICAL AND CYLINDRICAL COÖRDINATES. TRANSFORMATION OF COÖRDINATES

PLANE ANALYTIC GEOMETRY

INTRODUCTION

DIRECTED LINE-SEGMENTS. PROJECTIONS

Elementary Geometry, as it is studied in the high school to-day, had attained its present development at the time when Greek culture was at its height. The first systematic treatment of the subject which has come down to us was written by Euclid about 300 B.C.

Algebra, on the other hand, was unknown to the Greeks. Its beginnings are found among the Hindus, to whom the so-called Arabic system of numerals may also be due. It came into Western Europe late, and not till the close of the middle ages was it carried to the point which is marked by any school book of to-day that treats this subject.

When scholars had once possessed themselves of these two subjects — Geometry and Algebra — the next step was quickly taken. The renowned philosopher and mathematician, René Descartes, in his Géometrie of 1637, showed how the methods of algebra could be applied to the study of geometry. He thus became the founder of Analytic Geometry.*

The "originals" and the locus problems of Elementary Geometry depend for their solution almost wholly on ingenuity. There are no general methods whereby one can be sure of solving a new problem of this class. Analytic Geometry,

*Also called Cartesian Geometry, from the Latinized form of his name, Cartesius.

on the other hand, furnishes universal methods for the treatment of such problems; moreover, these methods make possible the study of further problems not thought of by the ancients, but lying at the heart of modern mathematics and mathematical physics. Indeed, these two great subjects owe their very existence to the new geometry and the Calculus.

The question of how to make use in geometry of the negative, as well as the positive, numbers is among the first which must be answered in applying algebra to geometry. The solution of this problem will become clear in the following paragraphs.

1. Directed Line-Segments. Let an indefinite straight line, L, be given, and let two points, A and B, be marked on L. Then the portion of L which is bounded by A and B is what is called in Plane Geometry a *line-segment*, and is written as AB.

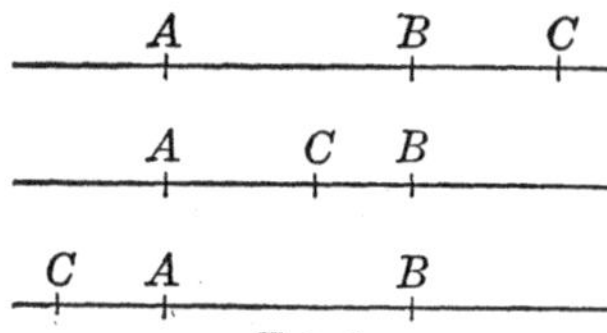

FIG. 1

Let a third point, C, be marked on L. Then three cases arise, as indicated in the figure. Corresponding to these three cases we have:

$$(a) \qquad AB + BC = AC;$$

$$(b) \qquad AB - CB = AC;$$

$$(c) \qquad CB - AB = CA.$$

Three other cases will arise if the original points A and B are taken in the opposite order on the line. Let the student write down the three corresponding equations.

A unification of all these cases can be effected by means of an extension of the concept of a line-segment. We no longer consider the line-segments AB and BA as identical, but we distinguish between them by giving each a *direction* or *sense*. Thus, AB shall be directed from A to B and BA shall be directed from B to A, *i.e.* oppositely to AB. These *directed line-segments* we denote by $\overline{AB}$ and $\overline{BA}$, to distinguish them from the ordinary, or undirected, line-segments.

We may, for the moment, interpret the directed line-segment $\overline{AB}$ as the act of walking from A to B; then $\overline{BA}$ represents the act of walking from B to A. With this in mind, let us return to Fig. 1 and consider the directed line-segments $\overline{AB}$, $\overline{BC}$, and $\overline{AC}$. We have, in all three cases represented by Fig. 1, and also in the other three:

$$\overline{AB} + \overline{BC} = \overline{AC},$$

since walking from A to B and then walking from B to C is equivalent, with reference to the point reached, to walking from A to C.

Accordingly, we unify all six cases by defining, as the sum of the directed line-segments $\overline{AB}$ and $\overline{BC}$, the directed line-segment $\overline{AC}$:

$$(1) \qquad \overline{AB} + \overline{BC} = \overline{AC}.$$

From this definition it follows that, if A, B, C, and D are any four points of L,

$$(2) \qquad \overline{AB} + \overline{BC} + \overline{CD} = \overline{AD}.$$

For, by (1), the sum of the first two terms in (2) is $\overline{AC}$, and, by the definition, the sum of $\overline{AC}$ and $\overline{CD}$ is $\overline{AD}$.

Similarly, if the points M, M_1, M_2, $\cdots$, M_{n-1}, N are any points of L, we have

$$(3) \qquad \overline{MM_1} + \overline{M_1M_2} + \cdots + \overline{M_{n-2}M_{n-1}} + \overline{M_{n-1}N} = \overline{MN}.$$

Given two directed line-segments on the same line or on two parallel lines, we say that these two directed line-segments are *equal*, if they have equal lengths and the same direction or sense.

2. Algebraic Representation of Directed Line-Segments. On the line L let one of the two opposite directions or senses be chosen arbitrarily and defined as the *positive direction* or *sense* of L; and let the other be called the *negative direction* or *sense.*

A directed line-segment $\overline{AB}$, which lies on L, is then called *positive*, if its sense is the same as the positive sense of L, and *negative*, if its sense is the same as the negative sense of L.

To such a directed line-segment $\overline{AB}$ we assign a number, which we shall also represent by $\overline{AB}$, as follows. If l is the length of the ordinary line-segment AB, then

$$\overline{AB} = l, \qquad \text{if } \overline{AB} \text{ is a positive line-segment;}$$

$$\overline{AB} = -l, \quad \text{if } \overline{AB} \text{ is a negative line-segment.}$$

If $\overline{AB} = l$, then $\overline{BA} = -l$; and if $\overline{AB} = -l$, then $\overline{BA} = l$. In either case

$$(1) \qquad \overline{AB} + \overline{BA} = 0 \qquad \text{or} \qquad \overline{AB} = -\overline{BA}.$$

Since the act of walking from A to B is nullified by the act of walking from B to A, we might have arrived at equations (1) from consideration of the line-segments themselves, instead of by use of the numbers which represent them.

It is easy to verify the fact that equations (1), (2), and (3) of the preceding paragraph, which relate to directed line-segments, hold for the corresponding numbers. Consequently, no error or confusion arises from using the same notation $\overline{AB}$ for both the directed line-segment and the number corresponding to it. We shall, however, adopt a still simpler notation, dropping the dash altogether and writing henceforth AB to denote, not merely the directed line-segment or the number corresponding to it, but also the line-segment itself, stating explicitly what is meant, unless the meaning is clear from the context.

Absolute Value. It is often convenient to be able to express merely the *length* of a directed line-segment, AB. The notation for this length is $|AB|$; read: "the absolute value of AB."

The numerical, or absolute, value of a number, a, is denoted in the same way: $|a|$. Thus, $|-3| = 3$. Of course, $|3| = 3$.

3. Projection of a Broken Line. By the *projection* of a point P on a line L is meant the foot, M, of the perpendicular dropped from P on L. If P lies on L, it is its own projection on L.

Let PQ be any directed line-segment, and let L be an arbitrary line. Let M and N be respectively the projections of P and Q on L. The *projection* of the directed line-segment PQ on L shall be defined as the directed line-segment MN, or the number which represents MN algebraically. Since $MN = -NM$, it follows that

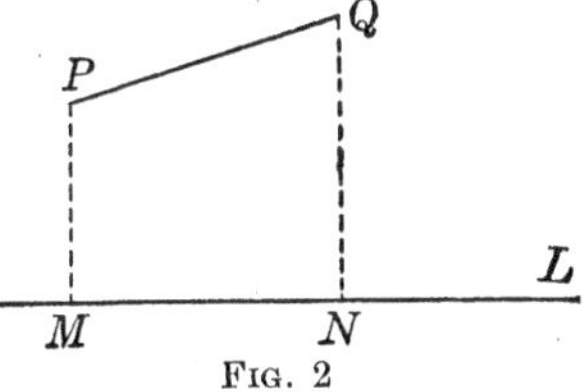

FIG. 2

$$\text{Proj. } PQ = -\text{Proj. } QP.$$

If PQ lies on a line perpendicular to L, the points M and N coincide, and we say that the projection MN of PQ on L is zero. Such a directed line-segment MN, whose end-points are identical, we may call a *nil-segment*; to it corresponds the number zero. It is evident that in taking the sum of a number of directed line-segments, any of them which are nil-segments may be disregarded, just as, in taking the sum of a set of numbers, any of them which are zero may be disregarded.

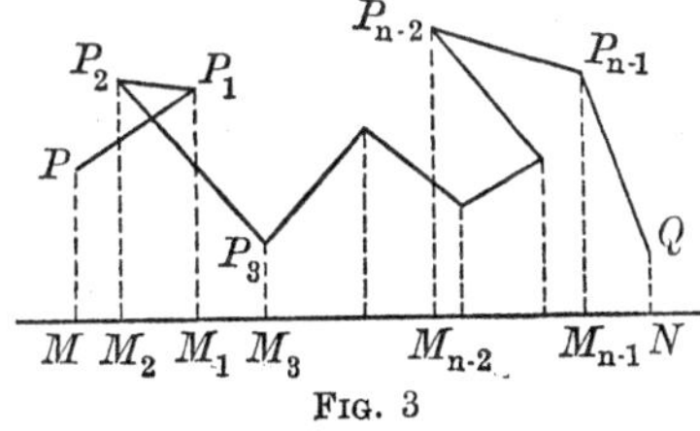

FIG. 3

Consider an arbitrary broken line $PP_1P_2 \cdots P_{n-1}Q$. By its *projection* on L is meant the sum of the projections of the directed line-segments PP_1, P_1P_2, $\cdots$, $P_{n-1}Q$, or

$$MM_1 + M_1M_2 + \cdots + M_{n-1}N.$$

This sum has the same value as MN, the projection on L of the directed line-segment PQ; cf. § 1, (3):

$$MM_1 + M_1M_2 + \cdots + M_{n-1}N = MN.$$

Hence the theorem:

THEOREM 1. *The sum of the projections on L of the segments PP_1, P_1P_2, $\cdots$, $P_{n-1}Q$ of a broken line joining P with Q is equal to the projection on L of the directed line-segment PQ.*

If, secondly, the same points P and Q be joined by another broken line, $PP'_1P'_2 \cdots P'_{m-1}Q$, the projection of the latter on L will also be equal to MN:

$$MM'_1 + M'_1M'_2 + \cdots + M'_{m-1}N = MN.$$

Hence the theorem:

THEOREM 2. *Given two broken lines having the same extremities,*

$$PP_1P_2 \cdots P_{n-1}Q \quad \textit{and} \quad PP'_1P'_2 \cdots P'_{m-1}Q.$$

Let L be an arbitrary straight line. Then the sum of the projections on L of the segments PP_1, P_1P_2, $\cdots$, $P_{n-1}Q$, of which the first broken line is made up, is equal to the corresponding sum for the second broken line.

CHAPTER I

COÖRDINATES. CURVES AND EQUATIONS

1. Definition of Rectangular Coördinates. Let a plane be given, in which it is desired to consider points and curves. Through a point O in this plane take two indefinite straight lines at right angles to each other, and choose on each line a positive sense.

Let P be any point of the plane. Consider the directed line-segment OP. Let its projections on the two directed lines through O be OM and ON. The numbers which represent algebraically these projections, that is, the lengths of OM and ON taken with the proper signs (cf. Introduction, § 2), are called the *coördinates* of P. We shall denote them by x and y:

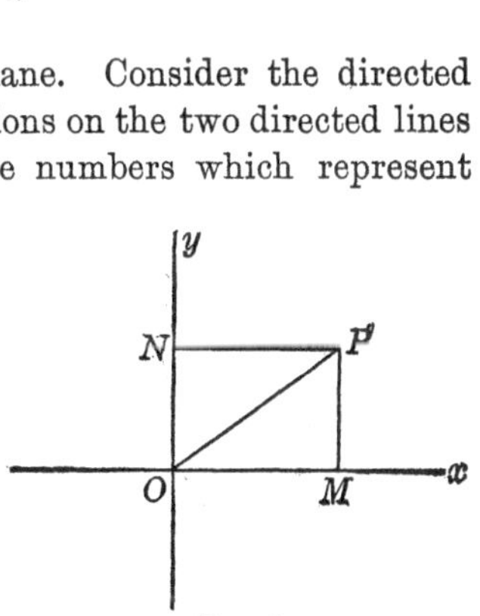

$$x = OM, \qquad y = ON,$$

and write them in parentheses: (x, y). The first number, x, is known as the *x-coördinate*, or *abscissa*, of P; the second, y, as the *y-coördinate*, or *ordinate*, of P.

FIG. 1

The point O is called the *origin* of coördinates. The directed lines through O are called the *axes of coördinates* or the *coördinate axes*; the one, the *axis of x*; the other, the *axis of y*. It is customary to take the coördinate axes as in Fig. 1, the axis of x being positive from left to right, and the axis of y, positive from below upward. But, of course, the opposite sense on one or both axes may be taken as positive, and an oblique position of the axes which conforms to the definition is legitimate, the essential thing being solely that the axes be taken *perpendicular* to each other.

Every point, P, in the plane has definite coördinates, (x, y). Conversely, to any pair of numbers, x and y, corresponds a point P whose coördinates are (x, y). This point can be constructed by laying off $OM = x$ on the axis of x, erecting a perpendicular at M to that axis, and then laying off $MP = y$. We might equally well have begun by laying off $ON = y$ on the axis of y (cf. Fig. 1), and then erected a perpendicular to that axis at N and laid off on it $NP = x$. It shall be understood that the positive sense on any line parallel to one of the coördinate axes, such as the perpendicular to the axis of x at M, shall be the same as the positive sense of that axis. For other lines of the plane there is no general principle governing the choice of the positive sense.

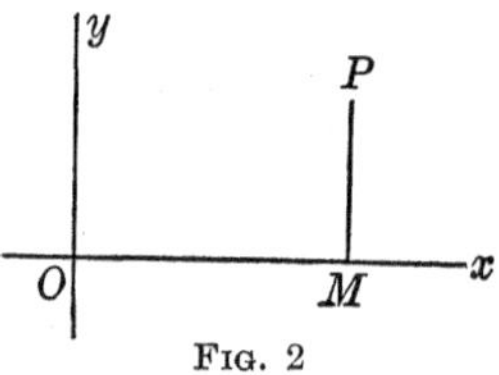

Fig. 2

The coördinates of the origin are (0, 0). Every point on the axis of x has 0 as its ordinate, and these are the only points of the plane for which this is true. *Hence the axis of x is represented by the equation*

$$y = 0, \qquad \text{(axis of } x\text{)}.$$

Similarly, the axis of y is represented by the equation

$$x = 0, \qquad \text{(axis of } y\text{)}.$$

The axes divide the plane into four regions, called *quadrants.* The *first quadrant* is the region included between the positive axis of x and the positive axis of y; the *second quadrant,* the region between the positive axis of y and the negative axis of x; etc. It is clear that the coördinates of a point in the first quadrant are both positive; that a point of the second quadrant has its abscissa negative and its ordinate positive; etc.

The system of coördinates just described is known as a *system of rectangular* or *Cartesian coördinates.*

EXERCISES

The student should provide himself with some squared paper for working these and many of the later exercises in this book. Paper ruled to centimeters and subdivided to millimeters is preferable.

1. Plot the following points, taking 1 cm. as the unit:

(*a*) $(0, 1)$;	(*b*) $(1, 0)$;	(*c*) $(1, 1)$;
(*d*) $(1, -1)$;	(*e*) $(-1, -1)$;	(*f*) $(2, -3)$;
(*g*) $(0, -2\frac{1}{2})$;	(*h*) $(-3.7, 0)$;	(*i*) $(-1\frac{1}{2}, -1\frac{3}{4})$;
(*j*) $(-4, 3.2)$;	(*k*) $(3.24, -0.87)$;	(*l*) $(-1, 1)$.

2. Determine the coördinates of the point P in Fig. 1 when 1 in. is taken as the unit of length; also when 1 cm. is the unit of length.

3. The same for the point marked by the period in "Fig. 1."

2. Projections of a Directed Line-Segment on the Axes. Let P_1, with the coördinates (x_1, y_1), and $P_2: (x_2, y_2)$* be any two points of the plane. Consider the directed line-segment P_1P_2. It is required to find its projections on the axes.

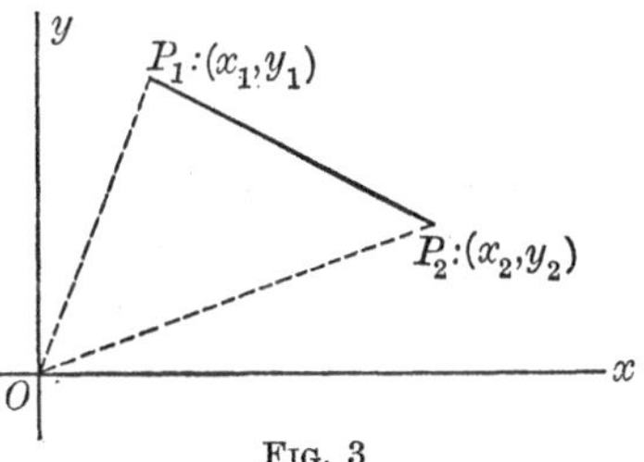

FIG. 3

To do this, draw the broken line P_1OP_2. By Introduction, § 3, Th. 1, the projections of this broken line on the axes are the same as those of the directed line-segment P_1P_2. Hence, taking first the projections on the axis of x, we have:

$$\begin{aligned}\text{Proj. } P_1P_2 &= \text{Proj. } P_1O + \text{Proj. } OP_2\\ &= -\text{Proj. } OP_1 + \text{Proj. } OP_2.\end{aligned}$$

* We shall frequently use this shorter notation, $P_2: (x_2, y_2)$, as an abbreviation for "P_2, with the coördinates (x_2, y_2)."

But the terms in the last expression are by definition $-x_1$ and x_2. So

$$\text{(1)} \qquad \text{Proj. } P_1P_2 \text{ on } x\text{-axis} = x_2 - x_1.$$

Similarly,

$$\text{(2)} \qquad \text{Proj. } P_1P_2 \text{ on } y\text{-axis} = y_2 - y_1.$$

The projections of P_1P_2 on two lines drawn parallel to the axes are obviously given by the same expressions.

EXERCISES

1. Plot P_1P_2 when P_1 is the point (a) of Ex. 1, § 1, and P_2 is (b). Determine the projections from the foregoing formulas, and verify directly from the figure.

2. The same, when

i) P_1 is (e) and P_2 is (f);
ii) P_1 is (c) and P_2 is (d);
iii) P_1 is (i) and P_2 is (l).

3. Distance between Two Points. Let the points be P_1, with the coördinates (x_1, y_1), and $P_2 : (x_2, y_2)$. Through P_1 draw a line parallel to the axis of x and through P_2, a line parallel to the axis of y; let Q denote the point of intersection of these lines. Then, by the Pythagorean Theorem,

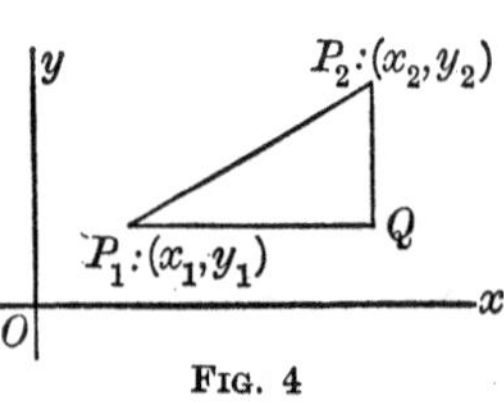

Fig. 4

$$\text{(1)} \qquad P_1P_2^2 = P_1Q^2 + QP_2^2,$$

or

$$\text{(2)} \qquad D^2 = (x_2 - x_1)^2 + (y_2 - y_1)^2,$$

where D denotes the distance between P_1 and P_2. Hence

$$\text{(3)} \qquad D = \sqrt{(x_2 - x_1)^2 + (y_2 - y_1)^2}.$$

In the foregoing analysis, we have used P_1Q (and similarly, QP_2) in two senses, namely, i) as the *length* of the ordinary line-segment P_1Q of Elementary Geometry; ii) as the algebraic expression $x_2 - x_1$ for the projection P_1Q of the directed line segment P_1P_2 on a parallel to the axis of x. Since, however, these two numbers differ at most in sign, their *squares* are equal, and hence equation (2) is equivalent to equation (1).

In particular, P_1P_2 may be parallel to an axis, *e.g.* the axis of x. Here, $y_2 = y_1$, and (3) becomes

$$D = \sqrt{(x_2 - x_1)^2}.$$

The student must not, however, hastily infer that

$$D = x_2 - x_1.$$

It may be that $x_2 - x_1$ is negative, and then *

$$D = -(x_2 - x_1).$$

A single formula which covers both cases can be written in terms of the *absolute value* (cf. Introduction, § 2) as follows:

$$D = |x_2 - x_1|. \tag{4}$$

EXERCISES

1. Find the distances between the following pairs of points, expressing the result correct to three significant figures. Draw a figure each time, showing the points and the line connecting them, and verify the result by actual measurement.

(*a*) (2, 1) and (– 2, – 2). (*b*) (– 7, 6) and (2, – 3).
(*c*) (13, 5) and (– 2, 5). (*d*) (7, 3) and (12, 3).
(*e*) (4, 8) and (4, – 8). (*f*) (– 1, 2) and (– 1, 6).

2. Find the lengths of the sides of the triangle whose vertices are the points (– 2, 3), (– 2, – 1), (4, – 1).

3. How far are the vertices of the triangle in question 2 from the origin?

* There is no contradiction here, or conflict with the ordinary laws of algebra. For, the $\sqrt{\ }$-sign always calls for the *positive* square root, — that being the definition of the symbol, — and we must see to it in any given case that we fulfill the contract.

4. Find the lengths of the diagonals of the convex quadrilateral whose vertices are the points (4, 1), (1, 3), (−3, 1), (−2, −1).

4. Slope of a Line. By the *slope*, λ, of a line is meant the trigonometric tangent of the angle, θ, which the line makes with the *positive* axis of x:

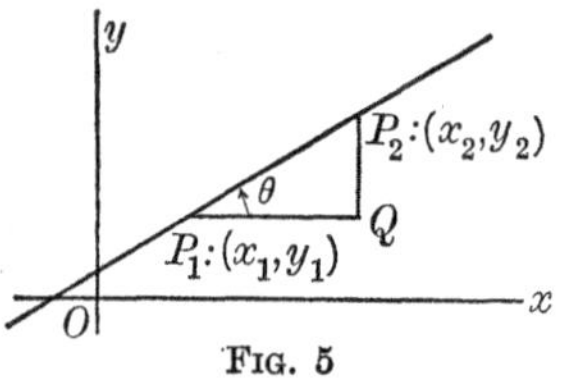

FIG. 5

(1) $$\lambda = \tan\theta.$$

To find the slope of the line, let P_1, with the coördinates (x_1, y_1), and $P_2:(x_2, y_2)$ be the extremities of any directed line-segment P_1P_2 on the line. Then

(2) $$\tan\theta = \frac{QP_2}{P_1Q} = \frac{y_2 - y_1}{x_2 - x_1},$$

or

(3) $$\lambda = \frac{y_2 - y_1}{x_2 - x_1}.$$

If, instead of P_1P_2, we had taken its opposite, P_2P_1, we should have obtained for λ the value $(y_1 - y_2)/(x_1 - x_2)$. But this is equal to the value of λ given by (3). Thus, λ is the same, whether the line is directed in the one sense or in the opposite sense. Hence we think of λ as the slope of the line without regard to sense.

Variation of the Slope. Consider the slopes, λ, of different lines, L, through a given point, P. When L is parallel to the axis of x, λ has the value zero. When L rotates as shown in the figure, λ becomes positive and increases steadily in value. As L approaches the vertical line L', λ becomes very large, increasing without limit.

When L passes beyond L', λ changes sign, being still numerically large. As L continues to rotate, λ increases *algebraically* through negative values. Finally, when L has again become parallel to the axis of x, λ has increased algebraically through all negative values and becomes again zero.

When L is in the position of L', θ is $90°$ and $\tan\theta = \lambda$ is undefined, that is, has no value. Hence L' *has no slope.* One often sees the expression: $\tan 90° = \infty$, and, in accordance with it, one might write here, $\lambda = \infty$. This does not mean that L' *has* a slope, which is infinite, for "infinity" is not a number. It is merely a brief and symbolic way of describing the behavior of λ for a line L, near to, *but not coincident with* L'; it says that for such a line λ is numerically very large; and further that, when the line L approaches L' as its limit, λ increases numerically without limit, — that is, increases numerically beyond any preassigned number, as 10,000,000 or 10,000,000!, and stays numerically above it.

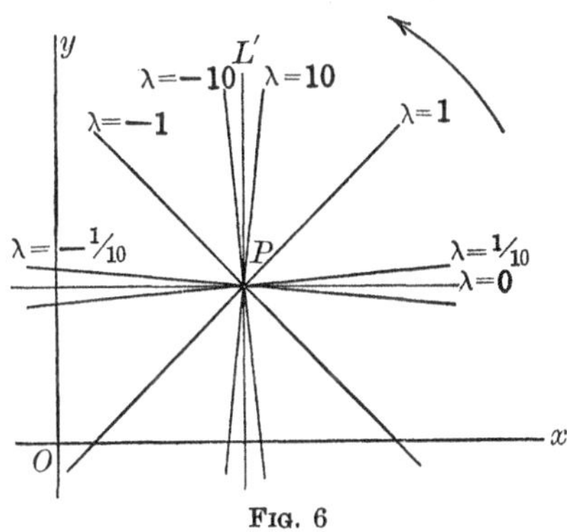

FIG. 6

The Angle θ. In measuring the angle *from* one line *to* another, it is essential, first of all, to agree on which direction of rotation shall be considered as *positive.* We shall take always as the *positive* direction of rotation that *from* the positive axis of x *to* the positive axis of y; so that the angle from the positive axis of x to the positive axis of y is $+90°$, and not $-90°$.

The complete definition of θ is, then, as follows: *The slope-angle θ of a line is the angle from the positive axis of x to the direction of the line.* There are in general two positive values for θ less than $360°$; if the smaller of them is denoted by θ, the other is $180° + \theta$. Which of these angles is chosen is immaterial, since $\tan(180° + \theta) = \tan\theta$; this result is in agreement with the previous one, to the effect that the slope pertains to the undirected line without regard to a sense on it.

The student should now draw a variety of lines, indicating for each the angle θ, and assure himself that the deduction of formula (3) holds, not merely when the quantities $x_2 - x_1$ and $y_2 - y_1$ are positive, but also when one or both are negative.

Right-Handed and Left-Handed Coördinate Systems. For the choice of axes in Fig. 1, the positive direction for angles is the counter-clockwise direction. But for such a choice as is indicated in the present figure, — a choice equally legitimate, — it is the clockwise sense which is positive.

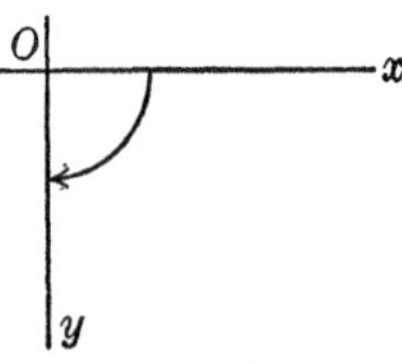

Fig. 7

The above formulas apply to either system of axes. The first system is called a *right-handed* system; the other, a *left-handed* system. We shall ordinarily use a right-handed system.

Problem. *To draw a line through a given point having a given slope.* In practice, this problem is usually to be solved on squared paper. The solution will be sufficiently clearly indicated by an example or two.

Example 1. To draw a line through the point $(-2, 3)$ having the slope -4.

Proceed along the parallel to the x-axis through the given point by any convenient distance, as 1 unit, toward the left.* Then go up the line through this point, parallel to the y-axis, by 4 times the former distance, — here, 4 units. Thus, a second point on the desired line is determined, and the line can now be drawn with a ruler.

If the given point lay near the edge of the paper, so that the above construction is inconvenient, it will do just as well to proceed from the first point toward the *right* by 1 unit, and then *down* by four units.

* The student will follow these constructions step by step on a piece of squared paper.

Example 2. To draw a line through the point (1.32, 2.78) having the slope .6541.

Here, it is clear that we cannot draw accurately enough to be able to use the last significant figure of the given slope. Open the compasses to span 10 cm. (if the squared paper is ruled to cm.) and lay off a distance of 10 cm. to the right on a parallel to the x-axis through the given point. This parallel need not actually be drawn. Its intersection, Q, with the circular arc is all that counts, and this point, Q, can be estimated and marked. Its distance above the axis of x will be 2 cm. and 7.8 mm. The error of drawing will be of the order of the last significant figure, namely, more than $\frac{1}{10}$ mm. and less than .5 mm.

Next, open the compasses to span 6 cm. and 5.4 mm. Put the point of the compasses on Q, and lay off the above distance, 6.54 cm., on a parallel through Q to the y-axis and above Q. The point R, thus found, will be a second point on the desired line, which now can be drawn.

EXERCISES

1. The points P_1, P_2, P_3, with the coördinates (2, 5), (7, 3), (– 3, 7) respectively, lie on a line. Show that the value for the slope of the line as given by equation (3) is the same, no matter which two of the three points are used in obtaining it.

2. Find the slopes of the sides of the triangle of Ex. 2, § 3.

3. Find the angles which the sides of that triangle make with the axes, and hence determine the angles of the triangle.

4. Show that the points (– 2, – 3), (5, – 4), (4, 1), (– 3, 2) are the vertices of a parallelogram.

5. Draw a line through the point (1, –2) having the slope 3.

6. Draw a line through the point (– 2, –1) having the slope $-1\frac{1}{2}$.

7. Draw a line through the point (–1.32, 0.14) having the slope – .2688.

5. Mid-Point of a Line-Segment. Let P_1, with the coördinates (x_1, y_1), and $P_2 : (x_2, y_2)$ be the extremities of a line-segment. It is desired to find the coördinates of the point P which bisects P_1P_2.

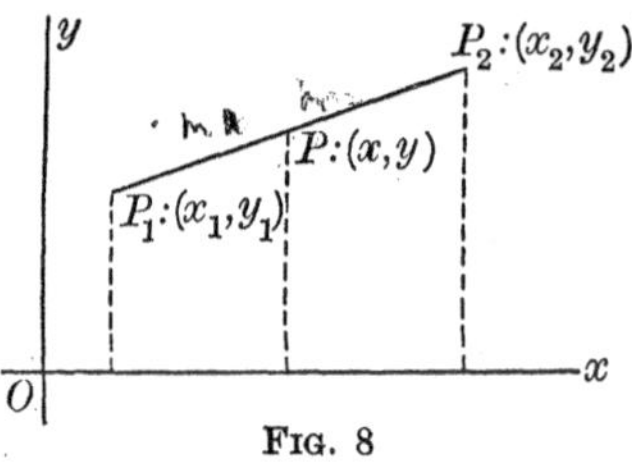

FIG. 8

Let the coördinates of P be (x, y). It is evident that the directed line-segment P_1P is equal to the directed line-segment PP_2. Hence the projection of P_1P on the axis of x, or $x - x_1$, must equal the projection of PP_2 on that axis, or $x_2 - x$:

$$x - x_1 = x_2 - x.$$

Hence

$$x = \frac{x_1 + x_2}{2}.$$

Similar considerations apply to the projections on the axis of y, and consequently

$$y = \frac{y_1 + y_2}{2}.$$

We have thus obtained the following result: *The coördinates (x, y) of the point P which bisects the line-segment P_1P_2 are given by the equations:*

(1) $$x = \frac{x_1 + x_2}{2}, \qquad y = \frac{y_1 + y_2}{2}.$$

EXERCISES

1. Determine the coördinates of the mid-point of each of the line-segments given by the pairs of points in Ex. 1, § 3. Draw figures and check your answers.

2. Find the mid-points of the sides of the triangle mentioned in Ex. 2, § 3, and check by a figure.

3. Determine the coördinates of the mid-point of the line joining the points $(a + b, a)$ and $(a - b, b)$.

4. Show that the diagonals of the parallelogram of Ex. 4, § 4 bisect each other.

6. Division of a Line-Segment in Any Ratio.* Let it be required to find the coördinates (x, y) of the point P which divides the line-segment P_1P_2 in an arbitrary ratio, m_1/m_2: †

$$\frac{P_1P}{PP_2} = \frac{m_1}{m_2}.$$

Obviously the projections of P_1P and PP_2 on the axis of x must be in the same ratio, m_1/m_2, and hence

$$\frac{x - x_1}{x_2 - x} = \frac{m_1}{m_2}.$$

On solving this equation for x, it is found that

$$x = \frac{m_2x_1 + m_1x_2}{m_2 + m_1}.$$

Similar considerations, applied to the projections on the axis of y, lead to the corresponding formula for y, and thus the coördinates of P are shown to be the following:

$$(1) \qquad x = \frac{m_2x_1 + m_1x_2}{m_2 + m_1}, \qquad y = \frac{m_2y_1 + m_1y_2}{m_2 + m_1}.$$

If m_1 and m_2 are equal, these formulas reduce to those of § 5.

External Division. It is also possible to find a point P on the indefinite straight line through P_1 and P_2 and lying outside the line-segment P_1P_2, which makes

$$\frac{P_1P}{P_2P} = \frac{m_1}{m_2},$$

where m_1 and m_2 are any two unequal positive numbers. Here,

$$\frac{x_1 - x}{x_2 - x} = \frac{m_1}{m_2}.$$

* This paragraph may well be omitted till the results are needed in later work.

† The given numbers m_1 and m_2 may be precisely the lengths P_1P and PP_2; but in general they are merely proportional respectively to them, *i.e.* they are these lengths, each multiplied by the same positive or negative number.

On solving this equation for x and the corresponding one for y, we find, as the coördinates of the point P, the following:

$$(2) \qquad x = \frac{m_2x_1 - m_1x_2}{m_2 - m_1}, \qquad y = \frac{m_2y_1 - m_1y_2}{m_2 - m_1}.$$

The point P is here said to divide the line P_1P_2 *externally* in the ratio m_1/m_2; and, in distinction, the division in the earlier case is called *internal* division. Both formulas, (1) and (2), can be written in the form (1) if one cares to consider external division as represented by a negative ratio, m_1/m_2, where, then, one of the numbers m_1, m_2 is positive, the other, negative.

EXERCISES

1. Find the coördinates of the point on the line-segment joining $(-1, 2)$ with $(5, -4)$ which is twice as far from the first point as from the second. Draw the figure accurately and verify.

2. Find the point on the line through the points given in the preceding problem, which is outside of the line-segment bounded by them and is twice as far from the first point as from the second.

3. Find the point which divides internally the line-segment bounded by the points $(3, 8)$ and $(-6, 2)$ in the ratio $1:5$, and lies nearer the first of these points.

4. The same question for external division.

7. Curve Plotting. Equation of a Curve. Since the subject of graphs is now very generally taught in the school course in Algebra, most students will already have met some of the topics taken up on the foregoing pages, and moreover they will have plotted numerous simple curves on squared paper from given equations. Thus, in particular, they will be familiar with the fact that all the points whose coördinates satisfy a *linear equation*, i.e. an *equation of the first degree*, like

$$(1) \qquad 2x - 3y - 1 = 0,$$

lie on a straight line, though they may never have seen a formal proof.

A number of points, whose coördinates satisfy equation (1), can be determined by giving to x simple values, computing the corresponding values of y from (1), and then plotting the points (x, y). Thus

if $x = 0$, $y = -\frac{1}{3}$, and the point is $(0, -\frac{1}{3})$,
if $x = 1$, $y = \frac{1}{3}$, and the point is $(1, \frac{1}{3})$;
if $x = 2$, $y = 1$, and the point is $(2, 1)$;
if $x = -1$, $y = -1$, and the point is $(-1, -1)$;
etc.

Of course, if it is known that (1) represents a straight line, — *i.e.* that all the points whose coördinates satisfy (1) lie on a straight line, — it is sufficient to determine *two* points as above, and then to draw the line through them.

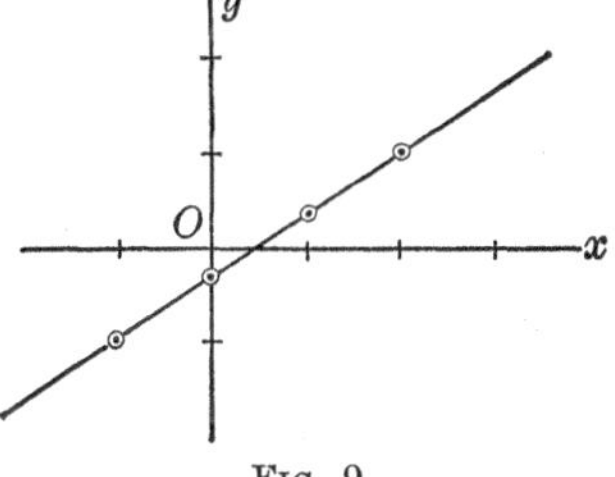

FIG. 9

This process of determining a large number of points whose coördinates satisfy a given equation and then passing a smooth curve through them is known as "plotting a curve* from its equation."

The mathematical curve† defined by an equation in x and y consists of all those points and only those points whose coördinates, when substituted for x and y in the equation, satisfy it.

Suppose, for example, that the equation is

(2) $$y = x^2.$$

The point (2, 4) lies on the curve defined by (2), because, when

* In Analytic Geometry the term *curve* includes straight lines as well as crooked curves.

† This curve is sometimes called the *locus of the equation.*

x is set equal to 2 and y is set equal to 4 in (2), the resulting equation,

$$4 = 4,$$

is true. We say, equation (2) is *satisfied* by the coördinates of the point (2, 4), or that the point (2, 4) lies on the curve (2)

On the other hand, the point $(-1, 2)$, for example, does not lie on the curve defined by (2). For, if we set $x = -1$ and $y = 2$, equation (2) becomes

$$2 = 1.$$

This is not a true equation; *i.e.* equation (2) is not satisfied by the coördinates of the point $(-1, 2)$, and so this point does not lie on the curve (2).

Equation of a Curve. A curve may be determined by simple geometric conditions; as, for example, that all of its points be at a distance of 2 units from the origin. This is a circle with its center at the origin and having a radius of length 2.

It is easy to state analytically the condition which the coördinates of any point (x, y) on the circle must satisfy. Since by § 3 the distance of any point (x, y) from the origin is

$$\sqrt{x^2 + y^2},$$

the condition that (x, y) be a point of the curve is clearly this, that

$$\sqrt{x^2 + y^2} = 2,$$

or that

(3) $$x^2 + y^2 = 4.$$

Equation (3) is called the *equation of the curve* in question.

The equation of a curve is an equation in x and y which is satisfied by the coördinates of every point of the curve, and by the coördinates of no other point.

In this book we shall be engaged for the most part in finding the equations which represent the simpler and more important curves, and in discovering and proving, from these equations, properties of the curves.

Nevertheless, the student should at the outset have clearly in mind the fact that any equation between x and y, like

$$y = x^3, \qquad y = \log x, \qquad y = \sin x,$$

represents a perfectly definite mathematical curve, which he can plot on paper. Moreover, he is in a position to determine whether, in the case of a chosen one of these curves, a given point lies on it. He will find it desirable to plot afresh a few simple curves, and to test his understanding of other matters taken up in this paragraph by answering the questions in the following exercises.

EXERCISES

1. What does each of the following equations represent? Draw a graph in each case.

(*a*) $x = 2$; (*c*) $x - y = 0$; (*e*) $2x - 3y + 6 = 0$;
(*b*) $2y + 3 = 0$; (*d*) $2x + 5y = 0$; (*f*) $5x + 8y - 4 = 0$.

Plot the following curves on squared paper.

2. $$y = x^2.$$

Take 2 cm. or 1 in. as the unit of length. Use a table of squares.

3. $$y^2 = x.$$

Take the same unit as in question 2 and use a table of square roots.

4. Show that, when one of the curves of Exs. 2 and 3 has been plotted from the tables, the other can be plotted from the first without the tables.

Work the corresponding exercises for the following curves.

5. $y = x^3$. 6. $y = \sqrt[3]{x}$. 7. $y^2 = x^3$. 8. $y^3 = x^2$.

9. Plot the curve

$$y = \log_{10} x$$

from a table of logarithms for values of x from 1 to 10, taking 1 cm. as the unit.

10. Which of the straight lines of Ex. 1 go through the origin?

11. Show that the curve

(a) $$y = \sin x$$

goes through the origin.

Do the curves

$$(b)\ y = \tan x, \qquad (c)\ y = \cos x,$$

go through the origin?

12. Do the following points lie on the curve

$$xy = 1?$$

(a) $(-1, -1)$; (b) $(-1, 1)$; (c) $(\frac{2}{3}, \frac{3}{2})$;
(d) $(-\frac{2}{3}, -\frac{3}{2})$; (e) $(\frac{1}{2}, -2)$; (f) $(0, 1)$.

13. Find the equations of the following curves.

(a) The line parallel to the axis of x and 8 units above it.

(b) The line parallel to the axis of y and $1\frac{2}{3}$ units to the left of it.

(c) The line bisecting the angle between the positive axis of y and the negative axis of x.

(d) The circle, center in the origin, radius ρ.

(e) The circle, center in the point (1, 2), radius 3.

Ans. $(x-1)^2 + (y-2)^2 = 9$.

8. Points of Intersection of Two Curves. Consider, for example, the problem of finding the point of intersection of the lines

$$L: \qquad 2x - 3y = 4,$$

$$L': \qquad 3x + 4y = -11.$$

Let (x_1, y_1) be the coördinates of this unknown point, P_1. Any point P, with the coördinates (x, y), which lies on L, has its x and y satisfying the *first* of the above equations. Hence, in particular, since P_1 lies on L, x_1 and y_1 must satisfy that equation, or

$$(1) \qquad 2x_1 - 3y_1 = 4.$$

Similarly, a point $P:(x, y)$, which lies on L', has its x and y satisfying the *second* of the above equations. Hence, in particular, since P_1 lies on L', x_1 and y_1 must satisfy that equation, or

(2) $$3x_1 + 4y_1 = -11.$$

Thus it appears that the two *unknown quantities*, x_1 and y_1, satisfy the two *simultaneous equations*, (1) and (2). Hence these equations are to be solved as simultaneous by the methods of Algebra.

$$\begin{array}{rl|l} 2x_1 - 3y_1 = & 4, & 4 \\ 3x_1 + 4y_1 = & -11, & 3 \end{array}$$

To do this, eliminate y_1 by multiplying the first equation through by 4, the second by 3, and then adding:

$$17x_1 = -17, \qquad \text{or} \qquad x_1 = -1.$$

On substituting this value of x_1 in either equation (1) or (2), the value of y_1 is found to be: $y_1 = -2$. Hence P_1 has the coördinates $(-1, -2)$.

The equations (1) and (2) are the same, except for the subscripts, as the equations of the given lines, L and L'. Hence we may say: *To find the coördinates of the point of intersection of two lines given by their equations, solve the latter as simultaneous equations in the unknown quantities, x and y, by the methods of Elementary Algebra.*

The generalization to the case of any two curves given by their equations is obvious. The equations are to be regarded as *simultaneous equations between the unknown quantities, x and y*, and solved as such.

The student should observe that the letters "x" and "y" have totally different meanings when they appear as the *coördinates of a variable point in the equation of a curve*, and when they represent *unknown quantities in a pair of simultaneous equations*. In the first case, they are *variables*, and a pair of values, (x, y), which satisfy equation L will *not*, in general, satisfy L' In the second case, x and y are *constants*, the coördinates of a *single point*, or of several points; but of *isolated* and *not variable* points.

EXERCISES

Determine the points of intersection of the following curves. Check your results by plotting the curves and reading off as accurately as possible the coördinates of the points of intersection.

1. The straight lines (*a*) and (*d*) of Ex. 1, § 7.

2. The straight lines (*c*) and (*e*) of Ex. 1, § 7.

3. The straight lines (*e*) and (*f*) of Ex. 1, § 7.

4. $\begin{cases} y^2 = 4x, \\ x + y = 3. \end{cases}$ *Ans.* (1, 2), (9, − 6).

5. $\begin{cases} x^2 + y^2 = 13, \\ xy = 6. \end{cases}$

6. $\begin{cases} x^2 + y^2 = a^2, \\ x + y = 0. \end{cases}$

7. $\begin{cases} x^2 + y^2 = 25, \\ 4x^2 + 36y^2 = 144. \end{cases}$

8. $\begin{cases} y^2 + 6x = 0, \\ 2x + y = 7. \end{cases}$

9. $\begin{cases} x^2 + y^2 = 2, \\ xy = 1. \end{cases}$ *Ans.* (1, 1), (− 1, − 1).

10. Show that the curves

$$y = \log_{10} x, \qquad x + y = 1,$$

intersect in the point (1, 0).

11. Show that the curves

$$x^2 + y^2 = 25, \qquad 3x - 4y = 0,$$

intersect in the point (4, 3), and also in (− 4, − 3).

EXERCISES ON CHAPTER I

1. Show that the points (2, 0), (0, 2), $(1 + \sqrt{3}, 1 + \sqrt{3})$ are the vertices of an equilateral triangle.

2. Prove that the triangle with vertices in the points (1, 8), (3, 2), (9, 4) is an isosceles right triangle.

3. Show that the points $(-1, 2)$, $(4, 10)$, $(2, 3)$, and $(-3, -5)$ are the vertices of a parallelogram.

4. Given the points A, B, C with coördinates $(-7, -2)$, $(-\frac{11}{5}, 0)$, $(5, 3)$. By proving that

$$AB + BC = AC,$$

show that the three points lie on a line.

5. Show that the three points of the previous problem lie on a line by proving that AB and AC have the same slope.

6. Prove that the two points $(5, 3)$ and $(-10, -6)$ lie on a line with the origin.

7. Prove that the two points (x_1, y_1), (x_2, y_2) lie on a line with the origin when, and only when, their coördinates are proportional:

$$x_1 : y_1 = x_2 : y_2.$$

8. Determine the point on the axis of x which is equidistant from the two points $(3, 4)$, $(-2, 6)$.

9. If $(3, 2)$ and $(-3, 2)$ are two vertices of an equilateral triangle which contains within it the origin, what are the coördinates of the third vertex?

10. If $(3, -1)$, $(-4, -3)$, $(1, 5)$ are three vertices of a parallelogram and the fourth lies in the first quadrant, find the coördinates of the fourth. *Ans.* $(8, 7)$.

11. If P is the mid-point of the segment P_1P_2, and P and P_1 have coördinates $(8, 17)$, $(-5, -3)$ respectively, what are the coördinates of P_2?

12. If P divides the segment P_1P_2 in the ratio $2:1$, and P_1 and P have coördinates $(3, 8)$ and $(1, 12)$ respectively, determine the coördinates of P_2. *Ans.* $(0, 14)$.

13. Find the ratio in which the point B of Ex. 4 divides the segment AC of that exercise. *Ans.* $2:3$.

14. A point with the abscissa 6 lies on the line joining the two points $(2, 5)$, $(8, 2)$. Find its ordinate.

Suggestion. Determine the ratio in which the point divides the line-segment between the two given points.

15. Prove that the sum of the squares of the distances of any point in the plane of a given rectangle to two opposite vertices equals the sum of the squares of the distances from it to the two other vertices.

Suggestion. Choose the axes of coördinates skillfully.

16. If D is the mid-point of the side BC of a triangle ABC, prove that

$$AB^2 + AC^2 = 2\,AD^2 + 2\,BD^2.$$

17. Show that the lines joining the mid-points of opposite sides of a quadrilateral bisect each other.

18. Prove that the lines joining the mid-points of adjacent sides of a quadrilateral form a parallelogram.

19. Prove that, if the diagonals of a parallelogram are equal, the parallelogram is a rectangle.

20. If two medians of a triangle are equal, show that the triangle is isosceles.

CHAPTER II

THE STRAIGHT LINE

1. Equation of Line through Two Points. Let $P_1 : (x_1, y_1)$ and $P_2 : (x_2, y_2)$ be two given points, and let it be required to find the equation of the line through them.

The slope of the line, by Ch. I, § 4, is

$$\frac{y_2 - y_1}{x_2 - x_1}.$$

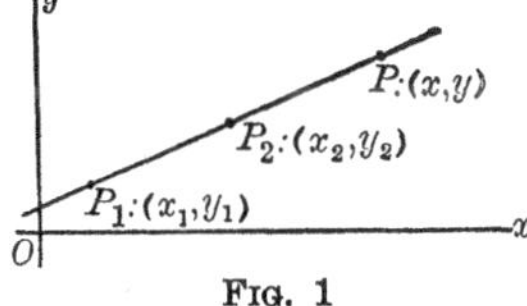

FIG. 1

Let P, with the coördinates (x, y), be any point on the line other than P_1. Then the slope of the line is also given by

$$\frac{y - y_1}{x - x_1}.$$

Hence

$$(1) \qquad \frac{y - y_1}{x - x_1} = \frac{y_2 - y_1}{x_2 - x_1}.$$

Conversely, if $P : (x, y)$ is any point whose coördinates satisfy equation (1), this equation then says that the slope of the line P_1P is the same as the slope of the line P_1P_2 and hence that P lies on the line P_1P_2.

A more desirable form of equation (1) is obtained by multiplying each side by $(x - x_1)/(y_2 - y_1)$. We then have:

$$(\text{I}) \qquad \frac{x - x_1}{x_2 - x_1} = \frac{y - y_1}{y_2 - y_1}.$$

Equation (I) is satisfied by the coördinates of those points and only those points which lie on the line P_1P_2. Consequently, by Ch. 1, § 7, (I) *is the equation of the line through the two given points.*

Example 1. Find the equation of the line which passes through the points $(1, -2)$ and $(-3, 4)$.

Here

$$x_1 = 1, \quad y_1 = -2 \qquad \text{and} \qquad x_2 = -3, \quad y_2 = 4.$$

By (I) the equation of the line is

$$\frac{x-1}{-3-1} = \frac{y-(-2)}{4-(-2)}, \qquad \text{or} \qquad \frac{x-1}{-4} = \frac{y+2}{6}.$$

On clearing of fractions and reducing, the equation becomes

$$3x + 2y + 1 = 0.$$

Let the student show that, if (x_1, y_1) had been taken as $(-3, 4)$ and (x_2, y_2) as $(1, -2)$, the same equation would have resulted.

Example 2. Find the equation of the line passing through the origin and the point (a, b).

Here, $(x_1, y_1) = (0, 0)$ and $(x_2, y_2) = (a, b)$, and (I) becomes

$$\frac{x}{a} = \frac{y}{b}, \qquad \text{or} \qquad bx - ay = 0.$$

Lines Parallel to the Axes. In deducing (I) we tacitly assumed that

$$y_2 - y_1 \neq 0 \qquad \text{and} \qquad x_2 - x_1 \neq 0;$$

for otherwise we could not have divided by these quantities.

If $y_2 - y_1 = 0$, the line is parallel to the axis of x. Its equation is, then, obviously

$$(2) \qquad y = y_1.$$

Similarly, if $x_2 - x_1 = 0$, the line is parallel to the axis of y and has the equation

$$(3) \qquad x = x_1.$$

These two special cases are not included in the result embodied in equation (I). We see, however, that they are so simple, that they can be dealt with directly.*

Example 3. Find the equation of the line passing through the two points $(-5, 1)$ and $(-5, 8)$.

It is clear from the figure that this line is parallel to the axis of y and 5 units distant from it to the left. Accordingly, the abscissa of every point on it is -5; conversely, every point whose abscissa is -5 lies on it. Therefore, its equation is

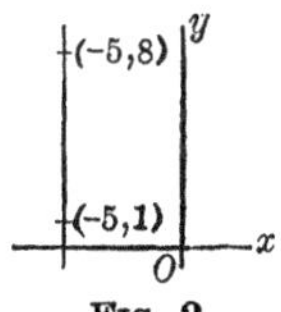

FIG. 2

$$x = -5, \quad \text{or} \quad x + 5 = 0.$$

EXERCISES †

Draw the following lines and find their equations.

1. Through $(1, 1)$ and $(3, 4)$. *Ans.* $3x - 2y - 1 = 0$.
2. Through $(5, 3)$ and $(-8, 6)$.
3. Through $(0, -5)$ and $(-2, 0)$. *Ans.* $5x + 2y + 10 = 0$.
4. Through the origin and $(-1, 2)$.
5. Through the origin and $(-2, -3)$.
6. Through $(2, -3)$ and $(-4, -3)$. *Ans.* $y + 3 = 0$.

* It is not difficult to replace (I) by an equation which holds in *all* cases, — namely, the following:

$$\text{(I')} \qquad (y_2 - y_1)(x - x_1) = (x_2 - x_1)(y - y_1).$$

We prefer, however, the original form (I). For (I) is more compact and easier to remember, and the special cases not included in it are best handled without a formula.

† In substituting numerical values for (x_1, y_1) and (x_2, y_2) in (I), the student will do well to begin with a framework of the form

$$\frac{x - \quad}{\quad - \quad} = \frac{y - \quad}{\quad - \quad},$$

and then fill in each place in which x_1 occurs; next, each place in which y_1 occurs; and so on. When x_1 or y_1 is negative, substitute it first in parentheses; thus, if $x_1 = -3$, begin by writing

$$\frac{x - (-3)}{\quad - (-3)} = \frac{y - \quad}{\quad - \quad}.$$

7. Through (0, 8) and (0, -56).

8. Through (5, 3) and parallel to the axis of y.

9. Through (5, 3) and parallel to the axis of x.

10. Through (a, b) and (b, a). *Ans.* $x + y = a + b$

11. Through $(a, 0)$ and $(0, b)$. *Ans.* $\frac{x}{a} + \frac{y}{b} = 1.$

2. One Point and the Slope Given. Let it be required to find the equation of the line which passes through a given point $P_1 : (x_1, y_1)$ and has a given slope, λ.

If $P : (x, y)$ be any second point on the line, the slope of the line will be, by Ch. I, § 4,

$$\frac{y - y_1}{x - x_1}.$$

But the slope of the line is given as λ. Hence

$$\frac{y - y_1}{x - x_1} = \lambda,$$

or

(II) $$y - y_1 = \lambda(x - x_1).$$

The student can now show, conversely, that any point, whose coördinates (x, y) satisfy (II), lies on the given line. Hence (II) is the equation of the line passing through the given point and having the given slope.

Example. Find the equation of the line which goes through the point $(2, -3)$ and makes an angle of 135° with the positive axis of x.

Here, $\lambda = -1$ and $(x_1, y_1) = (2, -3)$, and hence, by (II), the equation of the line is

$$y + 3 = -1(x - 2),$$

or

$$x + y + 1 = 0.$$

Slope-Intercept Form of Equation. It is frequently convenient to determine a line by its slope λ, and the y-coördinate of the point in which it cuts the axis of y.

Here, $x_1 = 0$; and, if we denote y_1 by the letter b, (II) becomes

(III) $$y = \lambda x + b.$$

This is known as the *slope-intercept form* of the equation of a straight line; b is known as the *intercept* of the line on the axis of y.

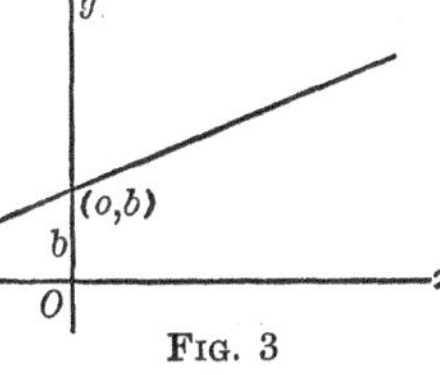

FIG. 3

Example. Find the equation of the line which makes an angle of 60° with the axis of x and whose intercept on the axis of y is -2.

Since $\lambda = \sqrt{3}$ and $b = -2$, the equation is

$$y = \sqrt{3}\,x - 2.$$

EXERCISES

Draw the following lines and find their equations.

1. Through $(-4, 5)$ and with slope -2. *Ans.* $2x + y + 3 = 0$.
2. Through $(3, 0)$ and with slope $\frac{8}{3}$.
3. Through $(\frac{2}{5}, -\frac{1}{2})$ and with slope $-\frac{5}{3}$.
4. Through the origin and making an angle of 60° with the axis of x.
5. Through $(-4, 0)$ and making an angle of 45° with the axis of y.
6. With intercept 1 on the axis of y and with slope $-\frac{3}{2}$. *Ans.* $3x + 2y - 2 = 0$.
7. With intercept $\frac{1}{2}$ on the axis of y and making an angle of 30° with the axis of x.
8. With slope -1 and intercept $-c$ on the axis of y.
9. With slope a/b and intercept b on the axis of y. *Ans.* $ax - by + b^2 = 0$.

3. The General Equation of the First Degree. Let there be given an arbitrary line of the plane. If the line is parallel to neither axis, its equation is of the form (I), § 1, — an equation of the first degree in x and y. If the line is parallel to the axis of x, its equation is of the form $y = y_1$, — a special equation of the first degree in x and y, in which it happens that the term in x is lacking. Similarly, if the line is parallel to the axis of y, its equation is of the form $x = x_1$, — an equation of the first degree which lacks the term in y. Consequently, we can say: *The equation of every straight line is of the first degree in x and y.*

Given, conversely, *the general equation of the first degree* in x and y, namely

$$(1) \qquad Ax + By + C = 0,$$

where A, B, C are any three constants, of which A and B are not both zero;* *this equation represents always a straight line.*

The Case $B \neq 0$. In general, B will not be zero and we can divide equation (1) through by it:

$$\frac{A}{B}x + y + \frac{C}{B} = 0,$$

and then solve for y:

$$y = -\frac{A}{B}x - \frac{C}{B}.$$

But this equation is precisely of the form (III), § 2, where

$$\lambda = -\frac{A}{B}, \qquad b = -\frac{C}{B}.$$

Therefore, it represents a straight line whose slope is $-A/B$ and whose intercept on the axis of y is $-C/B$.

The Case $B = 0$. If, however, B is zero, the equation (1) becomes

$$Ax + C = 0.$$

Now, A cannot be zero, since the case that both A and B are zero was excluded at the outset. We can, therefore, divide by A and then solve for x:

$$x = -\frac{C}{A}.$$

* In dealing with equation (1), now and henceforth, we shall always assume that A and B are not both zero.

This is the equation of a straight line parallel to the axis of y, if $C \neq 0$. If $C = 0$, it is the equation of this axis.

This completes the proof that every equation of the first degree represents a straight line. In accordance with this property, such an equation is frequently called a *linear* equation.

Example. What line is represented by the equation

$$6x + 3y + 1 = 0?$$

If we solve for y, we obtain

$$y = -2x - \tfrac{1}{3}.$$

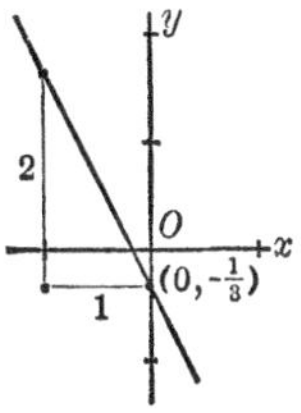

FIG. 4

Hence the equation represents the line of slope -2 with intercept $-\frac{1}{3}$ on the axis of y. From these data we may draw the line.

EXERCISES

Find the slopes and the intercepts on the axis of y of the lines represented by the following equations. Draw the lines.

1. $4x + 2y - 1 = 0$.
2. $7x + 8y + 5 = 0$.
3. $2x - 5y = 0$.
4. $2x - y = 1$.
5. $y = 0$.
6. $x = 3 - y$.

Find the slopes of each of the following lines.

7. $-x + 2y = 7$. *Ans.* $\frac{1}{2}$.
8. $x = y + 1$.
9. $3 - 2x = 5y$.
10. $2x - 3y = 4$.
11. $2y - 3 = 0$.
12. $2x = 3y$.
13. $x = 5y + 1$.
14. $bx + ay = ab$.

4. Intercepts. In the preceding paragraph we learned to plot the line represented by a given equation, from the values of its slope and its intercept on the axis of y, as found from the equation. It is often simpler, however, in the case of a line which cuts the axes in two distinct points, to determine from the equation the coördinates of these two points and then to plot the points and draw the line through them.

The point of intersection of a line, for example,

$$(1) \qquad 2x - 3y + 4 = 0,$$

with the axis of x has its y-coördinate equal to 0. Consequently, to find the x-coördinate of the point, we have but to set $y = 0$ in the equation of the line and solve for x. In this case we have, then,

$$2x + 4 = 0, \qquad \text{or} \qquad x = -2.$$

Similarly, the x-coördinate of the point of intersection of the line with the axis of y is 0, and its y-coördinate is obtained by setting $x = 0$ in the equation of the line and solving for y. In the present case this gives

$$-3y + 4 = 0, \qquad \text{or} \qquad y = \tfrac{4}{3}.$$

The points of intersection of the line (1) with the axes of coördinates are, then, $(-2, 0)$ and $(0, \frac{4}{3})$. We now plot these points and draw the line through them.

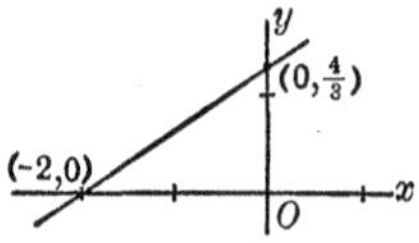

We recognize the number $\frac{4}{3}$ as the intercept of the line (1) on the axis of y; the number -2 we call the intercept on the axis of x. We have plotted the line (1), then, by finding its intercepts.

FIG. 5

In general, the *intercept* of a line on the axis of x is the x-coördinate of the point in which the line meets that axis. The intercept on the axis of y is similarly defined. These definitions admit of extension to any curve. Thus, the circle of Ch. I, § 7, has two intercepts on the axis of x, namely, $+2$ and -2.

An axis or a line parallel to an axis has no intercept on that axis. Every other line has definite intercepts on both axes, and these intercepts determine the position of the line unless they are both zero, that is, unless the line goes through the origin.

EXERCISES

Determine the intercepts of the following lines on each of the coördinate axes, so far as such intercepts exist, and draw the lines.

1. $2x + 3y - 6 = 0$. *Ans.* 3, 2.
2. $x - y + 1 = 0$.
3. $3x - 5y + 10 = 0$.
4. $5x + 7y + 13 = 0$.
5. $2x - 3y = 0$.
6. $2x + 3 = 0$. *Ans.* $-1\frac{1}{2}$, none.
7. $8 - 5y = 0$.
8. $x = 0$.
9. $x + y = a$.
10. $2ax - 3by = ab$.

5. The Intercept Form of the Equation of a Line. Given a line whose position is determined by its intercepts. Let the intercept on the axis of x be a, and let that on the axis of y be b. To find the equation of the line in terms of a and b.

Since one point on this line is $(a, 0)$ and a second is $(0, b)$, we have, by (I), § 1,

$$\frac{x-a}{0-a} = \frac{y-0}{b-0},$$

or

$$\frac{x}{a} + \frac{y}{b} = 1. \quad \text{(IV)}$$

Only lines which intersect the axes in two points that are distinct can have their equations written in this form. A line through the origin is an exception, because one or both its intercepts are zero and division by zero is impossible. Also a line parallel to an axis is an exception, since it has no intercept on that axis.

EXERCISES

Find the equations of the following lines.

1. With intercepts 5 and 3.
2. With intercepts $-2\frac{1}{2}$ and 8.
3. With intercepts $\frac{4}{5}$ and $-\frac{3}{4}$.

4. The diagonals of a square lie along the coördinate axes, and their length is 2 units. Find the equations of the four sides (produced).

Ans. $x + y = 1$; $x - y = 1$; $-x + y = 1$; $-x - y = 1$.

5. A triangle has its vertices at the points $(0, 1)$, $(-2, 0)$, $(1, 0)$. Draw the triangle and find the equations of its sides (produced). Use formula (IV), when possible.

6. A triangle has its vertices at the points $(a, 0)$, $(b, 0)$, $(0, c)$. Find the equations of the sides (produced).

7. A line goes through the origin and the mid-point of that side of the triangle of Ex. 5 which lies in the first quadrant. Find its equation.

8. Find the equations of the lines through the origin and the respective mid-points of the sides of the triangle of Ex. 6.

6. Parallel and Perpendicular Lines. *Parallels.* Given two lines oblique to the axis of y, so that both have slopes. The lines are parallel if, and only if, they have equal slopes. For, if they are parallel, their slope angles, and hence their slopes, are equal; and conversely.

Example 1. To find the equation of the line through the point $(1, 2)$ parallel to the line

(1) $$3x - 2y + 6 = 0.$$

The slope of the line (1) is $\frac{3}{2}$. The required line has the same slope and passes through the point $(1, 2)$. By (II), § 2, its equation is

$$y - 2 = \tfrac{3}{2}(x - 1),$$

or

$$3x - 2y + 1 = 0.$$

If the given line is parallel to the axis of y, it has no slope and hence the method of Example 1 is inapplicable. But then the required line must also be parallel to the axis of y and its equation can be written down directly. For example, if the given line is $3x + 8 = 0$, and there is required the line parallel to it passing through the point $(-8, 2)$, it is clear that the required line is parallel to the axis of y and 8 units to the left of it, and consequently has the equation $x = -8$, or $x + 8 = 0$.

Perpendiculars. Given two lines oblique to the axes, so that both have slopes, neither of which is zero. The lines are perpendicular if, and only if, their slopes, λ_1 and λ_2, are *negative reciprocals* of one another:

$$(2) \qquad \lambda_2 = -\frac{1}{\lambda_1}, \qquad \text{or} \qquad \lambda_1 = -\frac{1}{\lambda_2}, \qquad \lambda_1 \neq 0,\ \lambda_2 \neq 0.$$

For, if the lines are perpendicular, one of their slope angles, θ_1 and θ_2, may be taken as 90° greater than the other, viz.:

$$\theta_2 = \theta_1 + 90°,$$

and hence

$$\lambda_2 = \tan\theta_2 = \tan(\theta_1 + 90°) = -\cot\theta_1 = -\frac{1}{\tan\theta_1} = -\frac{1}{\lambda_1},$$

or

$$\lambda_2 = -\frac{1}{\lambda_1}.$$

Conversely, if this last equation is valid, the steps can be retraced and the lines shown to be perpendicular to each other.

Example 2. To find the equation of the line through the point (1, 2) perpendicular to the line (1).

The slope of (1) is $\frac{3}{2}$. Hence the required line has the slope $-\frac{2}{3}$. We have, then, to find the equation of the line through the point (1, 2) with slope $-\frac{2}{3}$. By (II), § 2, this equation is

$$y - 2 = -\tfrac{2}{3}(x - 1),$$

or

$$2x + 3y - 8 = 0.$$

If the given line is parallel to an axis, it has no slope or its slope is zero. In either case, equation (2) and the method of Example 2 are inapplicable. But then the required line must be parallel to the other axis and it is easy to write its equation. Suppose, for example, that the given line is $2y - 3 = 0$, — a line parallel to the axis of x, — and that the required line perpendicular to it is to go through the point (3, 5). Then this line must be parallel to the axis of y and at a distance of 3 units to the right of it. Consequently, its equation is $x - 3 = 0$.

The methods of this paragraph are applicable to all problems in which it is required to find the equation of a line which passes through a given point and is parallel, or perpendicular, to a given line.

EXERCISES

In each of the following exercises find the equations of the lines through the given point parallel and perpendicular to the given line.

	Line	*Point*
1.	$4x - 8y = 5$,	$(-1, -3)$. *Ans.* $x - 2y - 5 = 0$; $2x + y + 5 = 0$.
2.	$x - y = 1$,	$(0, 0)$.
3.	$5x + 13y - 3 = 0$,	$(2, -1)$.
4.	$3x + 5y = 0$,	$(5, 0)$.
5.	$2x = 3$,	$(5, -6)$.
6.	$\sqrt{2}y + \pi = 0$,	$(-2, 0)$. *Ans.* $y = 0$; $x + 2 = 0$.
7.	$1 - x = 0$,	$(0, \pi)$.

8. Find the equations of the altitudes of the triangle of § 5, Ex. 5.

9. Find the equations of the perpendicular bisectors of the sides of the triangle of § 5, Ex. 5.

10. Show that the equation of the line through the point (x_1, y_1) parallel to the line

$$Ax + By = C \tag{3}$$

is

$$Ax + By = Ax_1 + By_1.$$

11. Show that the equation of the line through the point (x_1, y_1) perpendicular to the line (3) of Ex. 10 is

$$Bx - Ay = Bx_1 - Ay_1.$$

7. Angle between Two Lines. Let L_1 and L_2 be two given lines, whose slopes are, respectively,

$$\lambda_1 = \tan\theta_1, \qquad \text{and} \qquad \lambda_2 = \tan\theta_2.$$

To find the angle, ϕ, *from* L_1 *to* L_2.

Since

$$\phi = \theta_2 - \theta_1,$$

it follows from Trigonometry that

$$\tan \phi = \frac{\tan \theta_2 - \tan \theta_1}{1 + \tan \theta_1 \tan \theta_2},$$

and hence that

$$(1) \qquad \tan \phi = \frac{\lambda_2 - \lambda_1}{1 + \lambda_1 \lambda_2}.$$

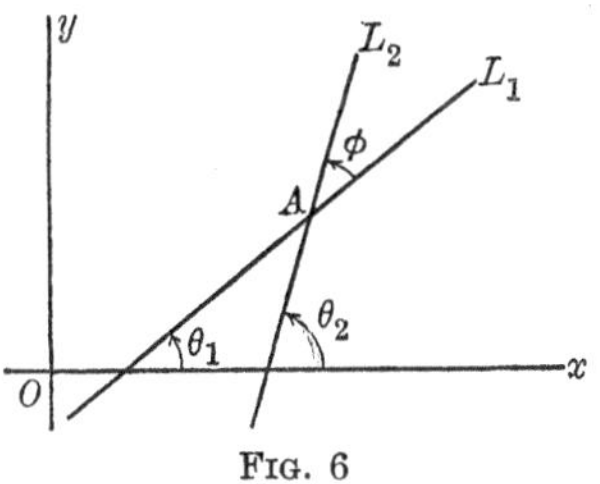

FIG. 6

The angle ϕ is the angle *from* L_1 *to* L_2. That is, it is the angle through which L_1 must be rotated *in the positive sense*, about the point A, in order that it coincide with L_2. In particular, we agree to take it as the smallest such angle, always less, then, than 180° : $0 \leq \phi < 180°$.*

If L_1 and L_2 are perpendicular, then, by (2), § 6, $\lambda_2 = -1/\lambda_1$ and $1 + \lambda_1 \lambda_2 = 0$. Consequently, cot ϕ, which is equal to the reciprocal of the right-hand side of (1), has the value zero, and so $\phi = 90°$.

Example. Let L_1 and L_2 be given by the equations,

$$L_1: \qquad 4x - 2y + 7 = 0,$$

$$L_2: \qquad 12x + 4y - 5 = 0.$$

Here $\lambda_1 = 2$ and $\lambda_2 = -3$, and (1) becomes

$$\tan \phi = \frac{-3 - 2}{1 - 6} = 1.$$

Hence the angle ϕ from L_1 to L_2 is 45°.

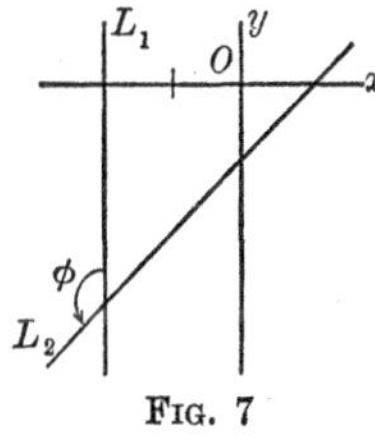

FIG. 7

In deducing (1) it was assumed that L_1 and L_2 both have slopes. If this is not the case, at least one of the lines is parallel to the axis of y and no formula is needed. The angle ϕ may be found directly. Suppose, for example, that L_1 and L_2 are, respectively,

$$x + 2 = 0 \qquad \text{and} \qquad x - y = 1.$$

* The figure shows L_1 and L_2 as intersecting lines, but formula (1) and the deduction of it are valid also in case L_1 and L_2 are parallel. In this

Then L_1 is parallel to the axis of y, and L_2 is inclined at an angle of $45°$ to the positive axis of x, since $\lambda_2 = 1$. Consequently, $\phi = 135°$.

EXERCISES

In each of the following exercises determine whether the given lines are mutually parallel or perpendicular, and in case they are neither, find the angle from the first line to the second.

1. $x + 2y = 3$, $\qquad x + 2y = 4$.
2. $2x - y + 5 = 0$, $\qquad 4x - 2y - 7 = 0$.
3. $x - y = 1$, $\qquad x + y = 2$.
4. $x + 2y + 11 = 0$, $\qquad 6x - 3y - 4 = 0$.
5. $3x - y = 0$, $\qquad 2x + y = 0$.
6. $x + 2y + 1 = 0$, $\qquad 2x + y - 1 = 0$.
7. $4x + 3y = 3$, $\qquad 9x - 3y = 5$.
8. $2x - 3y = 1$, $\qquad x - 3 = 0$.
9. $x + y = 0$, $\qquad y = 0$.
10. $2x - 3y + 1 = 0$, $\qquad 3x - 4y - 1 = 0$.

11. By the method of this paragraph determine each of the three angles of the triangle whose sides have the equations

$$x - 2y - 6 = 0, \qquad 2x + y - 4 = 0, \qquad 3x - y + 3 = 0.$$

Check your results by adding the angles.

12. Prove that if L_1 and L_2 are represented by the equations

$$L_1: \qquad A_1x + B_1y + C_1 = 0,$$
$$L_2: \qquad A_2x + B_2y + C_2 = 0,$$

then

$$\tan \phi = \frac{A_1B_2 - A_2B_1}{A_1A_2 + B_1B_2}.$$

What can you say of L_1 and L_2 if $A_1B_2 - A_2B_1 = 0$? If $A_1A_2 + B_1B_2 = 0$?

case, we take the angle from L_1 to L_2 as $0° -$, not as $180°$, as is conceivable. Hence arises the sign $\leq$ (less than or equal to) in the place in which it stands in the double inequality.

13. Show that the formula of Ex. 12 for $\tan\phi$ is valid even if one or both of the lines has no slope, *i.e.* is parallel to the axis of y.

8. Distance of a Point from a Line. Let $P:(x_1, y_1)$ be a given point and let

$$L: \qquad Ax + By + C = 0$$

be a given line. To find the distance, D, of P from L.

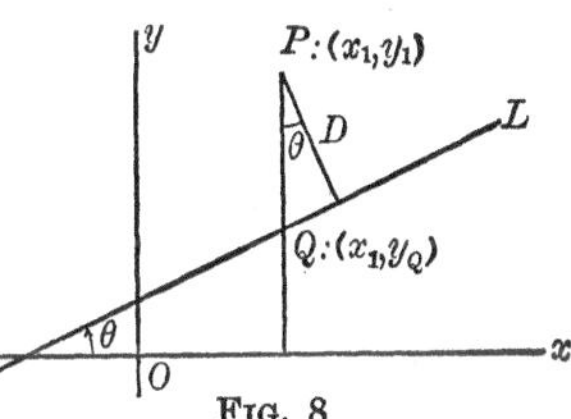

FIG. 8

Drop a perpendicular from P on the axis of x, and denote the point in which it cuts L by Q. The abscissa of Q is x_1. Denote its ordinate by y_Q. Then

$$QP = y_1 - y_Q.$$

Since $Q:(x_1, y_Q)$ lies on L, its coördinates satisfy the equation of L; thus

$$Ax_1 + By_Q + C = 0.$$

Solving this equation for y_Q, we find:

$$y_Q = -\frac{Ax_1 + C}{B}.$$

Hence

$$(1) \qquad QP = \frac{Ax_1 + By_1 + C}{B}.$$

Let θ be the slope-angle of L and form the product $QP\cos\theta$. One or both of the factors of this product may be negative, according to the positions of P and L.* But always the numerical value of the product is equal to the distance D:

$$(2) \qquad D = |QP\cos\theta|.$$

This is clear in case P and L are situated as in Fig. 8;

* There are four essentially different positions for P and L, for L may have a positive or a negative slope, and P may lie on the one or on the other side of L.

the student should draw the other typical figures and show that for them, also, (2) is valid.

Since the slope of L is

$$\lambda = \tan\theta = -\frac{A}{B},$$

we have

$$\sec^2\theta = 1 + \tan^2\theta = \frac{A^2 + B^2}{B^2}.$$

Consequently,

$$\cos\theta = \pm\frac{B}{\sqrt{A^2 + B^2}}. \tag{3}$$

It is immaterial to us which sign in (3) is the proper one. For, according to (2), we have now to multiply together the values of QP and $\cos\theta$, as given by (1) and (3), and take the *numerical* value of the product. The result is the desired formula:

$$(4)\qquad \begin{cases} D = \dfrac{|Ax_1 + By_1 + C|}{\sqrt{A^2 + B^2}}, \\ \text{or} \\ D = \pm\dfrac{Ax_1 + By_1 + C}{\sqrt{A^2 + B^2}}, \end{cases}$$

where, in the second formula, that sign is to be chosen which makes the right-hand side positive.

Example. The distance of the point $(3, -2)$ from the line

$$3x + 4y - 7 = 0$$

is

$$D = \frac{|3\cdot 3 + 4(-2) - 7|}{\sqrt{3^2 + 4^2}} = \frac{|-6|}{\sqrt{25}} = \frac{6}{5} = 1\tfrac{1}{5}.$$

The deduction of formula (4) involves division by B and hence tacitly assumes that $B \neq 0$, *i.e.* that L is not parallel to the axis of y. The formula holds, however, even when L is parallel to the axis of y. For, in this case it is clear from a figure that

$$D = \left|x_1 + \frac{C}{A}\right|,$$

and (4) reduces precisely to this when $B = 0$.

EXERCISES

In each of the first seven exercises find the distance of the given point from the given line.

	Point	*Line*	
1.	(5, 2),	$3x - 4y + 6 = 0.$	*Ans.* $2\frac{3}{5}$.
2.	(2, 3),	$5x + 12y + 2 = 0.$	
3.	(6, − 1),	$3x - y + 1 = 0.$	*Ans.* $2\sqrt{10}$, or 6.32.
4.	(3, 4),	$3x + 5 = 0.$	
5.	(− 2, − 5),	$y = 0.$	
6.	Origin,	$x + y - 1 = 0.$	
7.	Origin,	$3x + 2y - 6 = 0.$	

8. Find the lengths of the altitudes of the triangle with vertices in the points (2, 0), (3, 5), (− 1, 2).

9. Area of a Triangle. Let a triangle be given by means of its vertices (x_1, y_1), (x_2, y_2), (x_3, y_3). To find its area.

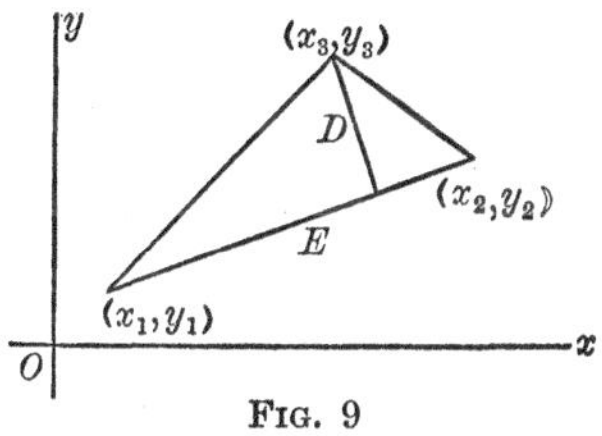

Drop a perpendicular from one of the vertices, as (x_3, y_3), on the opposite side. Then the required area is

$$A = \tfrac{1}{2} DE,$$

FIG. 9

where D denotes the length of the perpendicular and E, the length of the side in question.

By Ch. I, § 3, we have

$$E = \sqrt{(x_2 - x_1)^2 + (y_2 - y_1)^2}.$$

D is the distance of (x_3, y_3) from the line joining (x_1, y_1) and (x_2, y_2). The equation of this line, as given by (I) or (I′), § 1, may be put into the form:

$$(y_2 - y_1)x - (x_2 - x_1)y - x_1y_2 + x_2y_1 = 0.$$

Consequently, by (4), § 8, we find:

$$D = \pm \frac{(y_2 - y_1)x_3 - (x_2 - x_1)y_3 - x_1y_2 + x_2y_1}{\sqrt{(x_2 - x_1)^2 + (y_2 - y_1)^2}}.$$

Thus

$$A = \pm \tfrac{1}{2}[(y_2 - y_1)x_3 - (x_2 - x_1)y_3 - x_1y_2 + x_2y_1].$$

The result may be written more symmetrically in either of the forms

(1) $$A = \pm \tfrac{1}{2}[(x_1 - x_2)y_3 + (x_2 - x_3)y_1 + (x_3 - x_1)y_2],$$

or

(2) $$A = \pm \tfrac{1}{2}[(y_1 - y_2)x_3 + (y_2 - y_3)x_1 + (y_3 - y_1)x_2],$$

where in each case that sign is to be chosen which makes the right-hand side positive.

EXERCISES

Find the area of the triangle whose vertices are in the points

1. $(1, 2), (-1, 2), (-2, 1)$.
2. $(5, 3), (-3, 4), (-2, -1)$.
3. $(1, 2), (2, 1), (0, 0)$.

Find the area of the triangle whose sides lie along the lines

4. $x - y = 0, \quad x + y = 0, \quad 2x + y - 3 = 0.$
5. $2x + y - 6 = 0, \quad x - y + 3 = 0, \quad x - 2y - 8 = 0.$
6. Find the area of the convex quadrilateral whose vertices are in the points $(4, 2), (-1, 4), (-3, -2), (5, -8)$.
7. What do formulas (1) and (2) become when one of the vertices, say (x_3, y_3), is in the origin?

Ans. $A = \pm \tfrac{1}{2}(x_1y_2 - x_2y_1)$.

10. General Theory of Parallels and Perpendiculars. Identical Lines.* The line through the point (x_1, y_1) parallel to the line

(1) $$Ax + By = C,$$

has the equation, according to § 6, Ex. 10,

$$Ax + By = Ax_1 + By_1.$$

* The discussion in the class-room of the subjects treated in this and the following paragraph may well be postponed until the need for them arises.

This equation is of the form

$$(2) \qquad Ax + By = C',$$

since the constant $Ax_1 + By_1$ may be denoted by the single letter C'.

Conversely, equations (1) and (2), for $C' \neq C$, always represent parallel lines. For, if $B \neq 0$, the lines have the same slope, $-A/B$; if $B = 0$, A cannot be zero, and the lines are parallel to the axis of y and hence to each other.

THEOREM 1. *Two lines are parallel when and only when their equations can be written in the forms* (1) *and* (2), *where* $C \neq C'$.

The line through the point (x_1, y_1), perpendicular to the line (1), has the equation (§ 6, Ex. 11):

$$Bx - Ay = Bx_1 - Ay_1,$$

and this equation is of the form

$$(3) \qquad Bx - Ay = C'.$$

Let the student show, conversely, that equations (1) and (3) always represent perpendicular lines.

THEOREM 2. *Two lines are perpendicular when and only when their equations can be written in the forms* (1) *and* (3).

The equations of two parallel lines *can* always be written in the forms (1) and (2). But they need not be so written. Thus the lines,

$$2x - y = -1,$$
$$6x - 3y = 2,$$

are parallel, though the equations are not in the forms (1) and (2). The coefficients of the terms in x and y are not respectively equal. They are, however, proportional: $2:6 = -1:-3$.

This condition holds in all cases. For the two lines

$$L_1: \qquad A_1x + B_1y + C_1 = 0,$$
$$L_2: \qquad A_2x + B_2y + C_2 = 0,$$

we may state the theorem:

THEOREM 3. *The lines L_1 and L_2 are parallel* if and only if*

$$A_1 : A_2 = B_1 : B_2.$$

For, L_1 and L_2 are parallel if and only if the angle ϕ between them, as defined in § 7, is zero; but, according to § 7, Ex. 12, ϕ, or better, $\tan \phi$, is zero, when and only when $A_1B_2 - A_2B_1 = 0$. But this equation is equivalent to the proportion $A_1 : A_2 = B_1 : B_2$.

As a second consequence of § 7, Ex. 12, we obtain the following theorem.

THEOREM 4. *The lines L_1 and L_2 are perpendicular if and only if*

$$A_1A_2 + B_1B_2 = 0.$$

Identical Lines. Two equations do not have to be identically the same in order to represent the same line. For example, the equations,

$$2x - y + 1 = 0,$$
$$6x - 3y + 3 = 0,$$

represent the same line. The corresponding constants in them are not equal, but they are proportional. We have, namely,

$$2 : 6 = -1 : -3 = 1 : 3,$$

or, what amounts to the same thing,

$$2 : -1 : 1 = 6 : -3 : 3.$$

This condition is general. We formulate it as a theorem:

THEOREM 5. *The lines L_1 and L_2 are identical if and only if*

$$A_1 : A_2 = B_1 : B_2 = C_1 : C_2,$$

or

$$A_1 : B_1 : C_1 = A_2 : B_2 : C_2.$$

For, L_1 and L_2 are the same line when and only when they have the same slope and the same intercept on the axis of y, that is, when and only when

$$-\frac{A_1}{B_1} = -\frac{A_2}{B_2} \quad \text{and} \quad -\frac{C_1}{B_1} = -\frac{C_2}{B_2},$$

* Or, in a single case, identical. Cf. Th. 5.

or $\qquad A_1 : A_2 = B_1 : B_2 \qquad$ and $\qquad B_1 : B_2 = C_1 : C_2,$

or, finally, $\qquad A_1 : A_2 = B_1 : B_2 = C_1 : C_2.$

This proof assumes that $B_1 \neq 0$ and $B_2 \neq 0$. The proof, when this is not the case, is left to the student.

EXERCISES

1. Prove Th. 3 directly, without recourse to the results of § 7.

2. The same for Th. 4.

See also Exs. 15, 16, 17, 18 at the end of the chapter.

11. Second Method of Finding Parallels and Perpendiculars. *Problem* 1. To find the equation of a line parallel to the given line

(1) $$Ax + By = C,$$

and satisfying a further condition.

By § 10, Th. 1, the desired equation can be written in the form

(2) $$Ax + By = C',$$

where C' is to be determined by the further condition.

Example. Consider the first example treated in § 6. In this case the equation of the desired line can be written in the form

$$3x - 2y = k,$$

where we have replaced the C' of (2) by k. The "further condition," by means of which the value of k is to be determined, is that the line go through the point (1, 2). Hence $x = 1$, $y = 2$ must satisfy the equation of the line, or

$$3 \cdot 1 - 2 \cdot 2 = k.$$

Consequently, $k = -1$, and the equation of the line is

$$3x - 2y + 1 = 0.$$

Problem 2. To find the equation of a line perpendicular to the given line (1) and satisfying a further condition.

By § 10, Th. 2, the desired equation can be written in the form

$$Bx - Ay = C', \tag{3}$$

where C' is to be determined by applying the further condition.

This condition does not always have to be that the line should go through a given point. It may be any single condition, not affecting the slope of the line, which it seems desirable to apply. We give an example illustrating the method in such a case.

Example. To find the equation of the line perpendicular to

$$2x - y - 4 = 0$$

and cutting from the first quadrant a triangle whose area is 16.

Equation (3) may, in this case, be written as

$$x + 2y = k. \tag{4}$$

We are to determine k so that the line (4) cuts from the first quadrant a triangle of area 16. The intercepts of the line (4) are k and $\frac{1}{2}k$, and hence the area of the triangle in question is $\frac{1}{4}k^2$. Accordingly, $\frac{1}{4}k^2 = 16$, and $k = \pm 8$. But the line cuts the first quadrant only if k is positive, and so we must have $k = 8$. The equation of the desired line is, then,

$$x + 2y - 8 = 0.$$

EXERCISES

1. Work Exs. 1–4, 8, 9 of § 6 by this method.

2. There are two lines parallel to the line

$$x - 2y = 6$$

and forming with the coördinate axes triangles of area 9. Find their equations.

3. Find the equations of the lines parallel to the line of Ex. 2 and 3 units distant from it.

Suggestion. Write the equation of the required line in the form (2) and demand that the distance from it of a chosen point of the given line be 3.

4. Find the equations of the lines parallel to the line

$$5x + 12y - 3 = 0$$

and 2 units distant from the origin.

5. The same as Ex. 2, if the lines are to be perpendicular, instead of parallel, to the given line.

6. The same as Ex. 4, if the lines are to be perpendicular, instead of parallel, to the given line.

7. A line is parallel to the line $3x + 2y - 6 = 0$, and forms a triangle in the first quadrant with the lines,

$$x - 2y = 0 \quad \text{and} \quad 2x - y = 0,$$

whose area is 21. Find the equation of the line.

Ans. $3x + 2y - 28 = 0$.

EXERCISES ON CHAPTER II

1. Find the equation of the line whose intercepts are twice those of the line $2x - 3y - 6 = 0$.

2. Find the equation of the line having the same intercept on the axis of x as the line $\sqrt{3}x - y - 3 = 0$, but making with that axis half the angle.

3. Find the equation of the line joining the point $(7, -2)$ with that point of the line $2x - y = 8$ whose ordinate is 2.

4. A perpendicular from the origin meets a line in the point $(5, 2)$. What is the equation of the line?

5. The coördinates of the foot of the perpendicular dropped from the origin on a line are (a, b). Show that the equation of the line is

$$ax + by = a^2 + b^2.$$

6. The line through the point $(5, -3)$ perpendicular to a given line meets it in the point $(-3, 2)$. Find the equation of the given line.

7. Prove that the line with intercepts 6 and 3 is perpendicular to the line with intercepts 3 and -6. Is it also perpendicular to the line with intercepts -3 and 6?

8. Prove that the line with intercepts a and b is perpendicular to the line with intercepts b and $-a$.

9. Show that the two points (5, 2) and (6, − 15) subtend a right angle at the origin.

10. Prove that the two points, (x_1, y_1) and (x_2, y_2), subtend a right angle at the origin when, and only when, $x_1x_2 + y_1y_2 = 0$.

11. Do the points (6, − 1) and (− 3, 4) subtend a right angle at the point (4, 6)? At the point (− 4, − 2)?

12. Given the triangle whose sides lie along the lines,

$$x - 2y + 6 = 0, \qquad 2x - y = 3, \qquad x + y - 3 = 0.$$

Find the coördinates of the vertices and the equations of the lines through the vertices parallel to the opposite sides.

13. Two sides of a parallelogram lie along the lines,

$$2x + 3y - 6 = 0, \qquad 4x - y = 4.$$

A vertex is at the point (− 2, 1). Find the equations of the other two sides (produced).

14. One side of a rectangle lies along the line,

$$5x + 4y - 9 = 0.$$

A vertex on this side is at the point (1, 1) and a second vertex is at (2, − 1). Find the equations of the other three sides (produced).

15. For what value of λ will the two lines,

$$3x - 2y + 6 = 0, \qquad \lambda x - y + 2 = 0,$$

(*a*) be parallel? (*b*) be perpendicular?

16. For what value, or values, of m will the two lines,

$$4x - my + 6 = 0, \qquad x + my + 3 = 0,$$

(*a*) be parallel? (*b*) be perpendicular?

17. For what value of m will the two equations,

$$mx + y + 5 = 0, \qquad 4x + my + 10 = 0,$$

represent the same line?

18. For what pairs of values for k and l will the two equations,

$$12x + ky + l = 0, \qquad lx - 5y + 3 = 0,$$

represent the same line?

19. The equations of the sides of a convex quadrilateral are

$$x = 2, \qquad y = 4, \qquad y = x, \qquad 2y = x.$$

Find the coördinates of the vertices and the equations of the diagonals.

20. Find the equation of the line through the point of intersection of the lines,

$$3x - 5y - 11 = 0, \qquad 2x - 7y = 11,$$

and having the intercept -5 on the axis of y.

21. Find the equation of the line through the point of intersection of the lines,

$$2x + 5y = 4, \qquad 3x - 4y + 17 = 0,$$

and perpendicular to the first of these two lines.

22. Find the distance between the two parallel lines,

$$3x - 4y + 1 = 0, \qquad 6x - 8y + 9 = 0.$$

Suggestion. Find the distance of a chosen point of the first line from the second.

23. Let

$$Ax + By + C = 0 \qquad \text{and} \qquad Ax + By + C' = 0$$

be any two parallel lines. Show that the distance between them is

$$\frac{|C' - C|}{\sqrt{A^2 + B^2}}, \qquad \text{or} \qquad \pm\frac{C' - C}{\sqrt{A^2 + B^2}}.$$

24. There are two points on the axis of x which are at the distance 4 from the line $2x - 3y - 4 = 0$. What are their coördinates?

25. Find the coördinates of the point on the axis of y which is equidistant from the two points $(3, 8)$, $(-2, 5)$.

26. There are two lines through the point (1, 1), each cutting from the first quadrant a triangle whose area is $2\frac{1}{4}$. Find their slopes. *Ans.* $-\frac{1}{2}, -2$.

27. Find the equation of the line through the point (3, 7) such that this point bisects the portion of the line between the axes. *Ans.* $7x + 3y - 42 = 0$.

28. The origin lies on a certain line and is the mid-point of that portion of the line intercepted between the two lines,

$$3x - 5y = 6, \qquad 4x + y + 6 = 0.$$

Find the equation of the line. *Ans.* $x + 6y = 0$.

29. The line

$$(1) \qquad 3x - 8y + 5 = 0$$

goes through the point (1, 1). Find the equation of the line (2) through this same point, if the angle from the line (1) to the line (2) is 45°. *Ans.* $11x - 5y - 6 = 0$.

30. Find the equations of the two lines through the origin making with the line $2x - 3y = 0$ angles of 60°.

CHAPTER III

APPLICATIONS

1. Certain General Methods. *Lines through a Point.* In many theorems and problems of Plane Geometry the question is to show that *three lines pass through a point.* Plane Geometry affords, however, no general method for dealing with this question. Each new problem must be discussed as if it were the first of this class to be considered.

Analytic Geometry, on the other hand, affords a universal method, whereby in any given case the question can be settled. For, from the data of the problem, the equation of each of the lines can be found. These will all be linear, and can be written in the form

L_1: $$A_1x + B_1y + C_1 = 0,$$

L_2: $$A_2x + B_2y + C_2 = 0,$$

L_3: $$A_3x + B_3y + C_3 = 0.$$

The coördinates of the point of intersection of two of these lines, as L_1 and L_2, can be found by solving the corresponding equations, regarded as simultaneous, for the unknown quantities x and y. Let the solution be written as

$$x = x', \qquad y = y'.$$

The third line, L_3, will pass through this point (x', y'), if and only if the coördinates of the latter satisfy the equation of L_3; i.e. *if and only if*

$$A_3x' + B_3y' + C_3 = 0.$$

Points on a Line. A second question which presents itself in problems of Plane Geometry is to determine when *three points lie on a straight line.* Here, again, the reply of Analytic Geometry is methodical and universal. From the data of the problem it will be possible in any given case to obtain the coördinates of the three points. Call them

$$(x_1, y_1), \qquad (x_2, y_2), \qquad (x_3, y_3).$$

Now, we know how to write down the equation of a line through two of them, as (x_1, y_1) and (x_2, y_2). This equation will always be linear, and can be written in the form

$$Ax + By + C = 0.$$

The third point, (x_3, y_3), *will lie on this line if and only if its coördinates satisfy the equation of the line;* i.e. *if and only if*

$$Ax_3 + By_3 + C = 0.$$

The student should test his understanding of the foregoing theory by working Exs. 1–6 at the end of the chapter.

2. The Medians of a Triangle. We recall the proposition from Plane Geometry, that *the medians of a triangle meet in a point.* The proof there given is simple, provided one remembers the construction lines it is desirable to draw. By means, however, of Analytic Geometry we can establish the proposition, not by artifices, but by the natural and direct application of the general principle enunciated in the preceding paragraph.

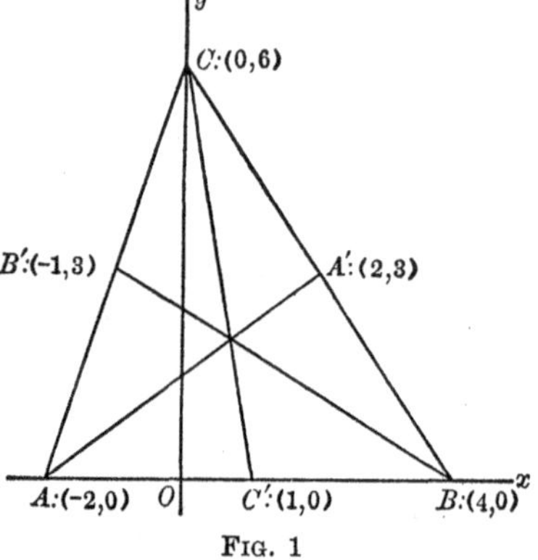

FIG. 1

The first step consists in the choice of the coördinate axes. This choice is wholly in our hands, and we make it in such a way as to simplify the coördinates of the given points. Thus, clearly, it will be well to take one of the axes along a side of the triangle. Let this be the axis of x.

A good choice for the axis of y will be one in which this axis passes through a vertex. Let this be the vertex not on the axis of x.

We begin with a numerical case, choosing the vertices A, B, C at the points indicated in the figure.

The Equations of the Medians. Consider the median AA'. One point on this line is given, namely $A:(-2, 0)$. A second point is the mid-point A' of the line-segment BC. By Ch. I, § 5, the coördinates of A' are $(2, 3)$.

The student can now solve for himself the problem of finding the equation of the line L_1 through $A:(-2, 0)$ and $A':(2, 3)$. The answer is,

$$L_1: \qquad 3x - 4y + 6 = 0.$$

In a precisely similar way the coördinates of B' are found to be $(-1, 3)$, and the equation of the median BB' is

$$L_2: \qquad 3x + 5y - 12 = 0.$$

Finally, the coördinates of C' are $(1, 0)$, and the equation of the median CC' is

$$L_3: \qquad 6x + y - 6 = 0.$$

The Point of Intersection of the Medians. The next step consists in finding the point in which two of the medians, as L_1 and L_2, intersect. The coördinates of this point will be given by solving as simultaneous the equations of these lines:

$$3x - 4y + 6 = 0,$$
$$3x + 5y - 12 = 0.$$

The solution is found to be:

$$x = \tfrac{2}{3}, \qquad y = 2.$$

And now the third median, L_3, will go through this point, $(\frac{2}{3}, 2)$, if the coördinates of the point satisfy the equation of L_3,

$$6x + y - 6 = 0.$$

On substituting for x in this equation the value $\frac{2}{3}$ and for y the value 2, we are led to the equation

$$6 \cdot \tfrac{2}{3} + 2 - 6 = 0.$$

This is a true equation, and hence the three lines L_1, L_2, and L_3 pass through the same point.

Remark. It can be shown by the formulas of Ch. I, § 6, that the above point $(\frac{2}{3}, 2)$ trisects each of the medians AA', BB', and CC'.

EXERCISES

1. Taking the same triangle as before, choose the axis of x along the side AB, but take the axis of y through A. The coördinates of the vertices will then be:

$$A: (0, 0); \qquad B: (6, 0); \qquad C: (2, 6).$$

Prove the theorem for this triangle.

2. The vertices of a triangle lie at the points (0, 0), (3, 0), (0, 9). Prove that the medians meet in a point.

3. Continuation. The General Case. We now proceed to prove the theorem of the medians for any triangle, ABC. Let

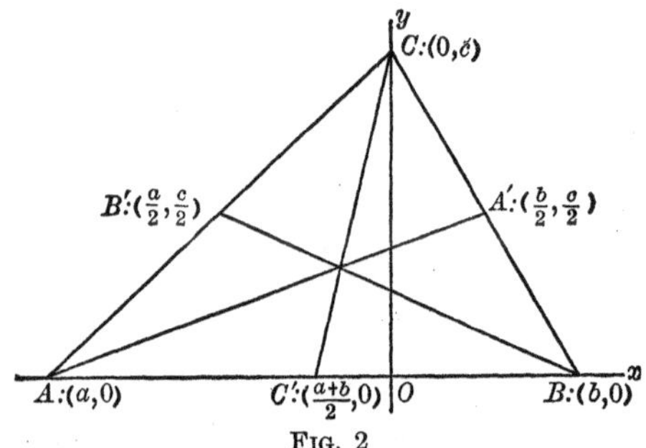

FIG. 2

the axes be chosen as in the text of § 2. Then the coördinates of A will be $(a, 0)$, where a may be any number whatever, positive, negative, or zero. The coördinates of B will be $(b, 0)$, where b may be any number distinct from a:

$$b \neq a, \qquad \text{or} \qquad a - b \neq 0.^*$$

Finally, the coördinates of C can be written as $(0, c)$, where c is any positive number.

Next, find the coördinates of A', B', C'. They are as shown in the figure.

The equation of L_1 is given by Ch. II, (I), where

$$(x_1, y_1) = (a, 0); \qquad (x_2, y_2) = \left(\frac{b}{2}, \frac{c}{2}\right).$$

It is:

$$\frac{x - a}{\frac{b}{2} - a} = \frac{y - 0}{\frac{c}{2} - 0},$$

or

L_1: $$cx + (2a - b)y = ac.$$

The equation of L_2 can be worked out in a similar manner. But it is not necessary to repeat the steps, since interchanging the letters a and b interchanges the points A and B, and also A' and B'. Thus L_1 passes over into L_2. Hence the equation of L_2 is:

L_2: $$cx + (2b - a)y = bc.$$

The line L_3 is determined by its intercepts, $\frac{1}{2}(a + b)$ and c; by Ch. II, (IV), its equation is found to be:

L_3: $$2cx + (a + b)y = (a + b)c.$$

To find the coördinates of the point in which L_1 and L_2 intersect, solve as simultaneous the equations of L_1 and L_2:

$$\begin{cases} cx + (2a - b)y = ac, \\ cx + (2b - a)y = bc. \end{cases}$$

The result is:

$$x = \frac{a + b}{3}, \qquad y = \frac{c}{3}.$$

* The figure has been drawn for the case in which a is negative and b positive.

Finally, to show that this point, $\left(\frac{a+b}{3}, \frac{c}{3}\right)$, lies on L_3, substitute its coördinates in the equation of L_3:

$$2c\frac{a+b}{3}+(a+b)\frac{c}{3}=(a+b)c.$$

Since this is a true equation, the point lies on the line, and we have proved the theorem that *the medians of a triangle pass through a point.*

That this point trisects each median can be proved as in the special case of the preceding paragraph, by means of Ch. I, § 6. The details are left to the student.

EXERCISE

Prove the theorem of the medians by taking the coördinate axes as in the first exercise of the preceding paragraph. Here, the vertices are

$$A:(0,0); \qquad B:(a,0); \qquad C:(b,c),$$

where a may be any number not 0, b any number whatever, and c any positive number. Draw the figure, and write in the coördinates of each point used.

4. The Altitudes of a Triangle. Another proposition of Plane Geometry is, that *the perpendiculars dropped from the vertices of a triangle on the opposite sides meet in a point.*

The proof of the proposition by Analytic Geometry is direct and simple. Let us begin with a numerical case, taking the triangle of Fig. 1. One of the perpendiculars is, then, the axis of y, and so all that is necessary to show is that the other two meet on this axis, or that the x-coördinate of their point of intersection is 0.

The equation of the line BC can be written down at once in terms of its intercepts:

$$\frac{x}{4}+\frac{y}{6}=1, \qquad \text{or} \qquad 3x+2y=12.$$

The slope of this line is $\lambda = -\frac{3}{2}$. The slope of any line perpendicular to it is $\lambda' = \frac{2}{3}$. Hence the equation of L_1, the perpendicular which passes through the point $A:(-2, 0)$, is

$$y - 0 = \tfrac{2}{3}(x + 2),$$

or

L_1: $$2x - 3y + 4 = 0.$$

In a similar manner the student can obtain the equation of the perpendicular L_2 from B on the side AC. It is,

L_2: $$x + 3y - 4 = 0.$$

On computing the x-coördinate of the point in which L_1 and L_2 intersect, it is found that $x = 0$, and hence the proposition is established for this triangle.

Remark. For use in a later problem it is necessary to know the exact point in which the perpendiculars meet. It is readily shown that this point is $(0, \frac{4}{3})$.

EXERCISES

1. Prove the above proposition for the special triangle considered, choosing the coördinate axes as in Ex. 1 of § 2.

2. Prove the proposition for the triangle of Ex. 2, § 2.

3. Prove the proposition for the general case, choosing the axes as in Fig. 2. First show that the equation of the perpendicular L_1 from A on BC is

L_1: $$bx - cy = ab,$$

and that the equation of the perpendicular L_2 from B on AC is

L_2: $$ax - cy = ab.$$

Then show that these lines intersect each other on the axis of y.

4. Show that the point in which the perpendiculars in the preceding question meet is $\left(0, -\dfrac{ab}{c}\right)$.

5. Prove the theorem of the altitudes, when the axes of coördinates are taken as in the exercise of § 3.

5. The Perpendicular Bisectors of the Sides of a Triangle. It is shown in Plane Geometry that these lines meet in a point. Since the student is now in full possession of the method employed in Analytic Geometry for the proof of this theorem, he will find it altogether possible to work out that proof without further suggestion. Let him begin with the special triangle of Fig. 1. He will find that the equations of the perpendicular bisectors of the sides are the following:

L_1: $$2x - 3y + 5 = 0;$$
L_2: $$x + 3y - 8 = 0;$$
L_3: $$x - 1 = 0.$$

These lines are then shown to meet in the point $(1, \frac{7}{3})$.

He can work further special examples corresponding to the exercises at the end of § 2 if this seems desirable.

Finally, let him work out the proof for the general case, taking the coördinate axes as in Fig. 2. The three lines will be found to have the equations

L_1: $$bx - cy = \tfrac{1}{2}(b^2 - c^2),$$
L_2: $$ax - cy = \tfrac{1}{2}(a^2 - c^2),$$
L_3: $$x = \tfrac{1}{2}(a + b).$$

They meet in the point

$$\left(\frac{a+b}{2}, \frac{ab+c^2}{2c}\right).$$

EXERCISE

Give the proof when the axes of coördinates are taken as in the exercise of § 3.

6. Three Points on a Line. The foregoing three propositions about triangles have led to three points, namely, the three points of intersection of the three lines in the various cases. In the case of the special triangle of Fig. 1, these points are

$$(\tfrac{2}{3}, 2); \qquad (0, \tfrac{4}{3}); \qquad (1, \tfrac{7}{3}).$$

These points lie on a straight line. Let the student try to prove this theorem by Plane Geometry.

The proof by Analytic Geometry is given immediately as a direct application of the second of the general principles enunciated in the opening paragraph of the chapter.

Write down the equation of the line through two of these points, — say, through the first and third. It is found to be:

$$3x - 3y + 4 = 0.$$

The coördinates of the second point,

$$x = 0, \qquad y = \tfrac{4}{3},$$

are seen to satisfy this equation, and the proposition is proved.

EXERCISES

1. Prove the proposition for the general case (Fig. 2). The points have been found to be:

$$\left(\frac{a+b}{3}, \frac{c}{3}\right); \quad \left(0, -\frac{ab}{c}\right); \quad \left(\frac{a+b}{2}, \frac{ab+c^2}{2c}\right).$$

2. On plotting the three points obtained in the special case discussed in the text it is observed that the line-segment determined by the extreme points is divided by the intermediate point in the ratio of $1:2$. Prove this analytically. Is it true in general?

EXERCISES ON CHAPTER III

1. Prove that the three lines,

$$2x - 3y - 5 = 0, \quad 3x + 4y - 16 = 0, \quad 4x - 23y + 7 = 0,$$

go through a point.

2. Prove that the three lines,

$$ax + by = 1, \quad bx + ay = 1, \quad x - y = 0,$$

go through a point.

3. Prove that the three points $(4, 1)$, $(-1, -9)$, and $(2, -3)$ lie on a line.

4. Prove that the three points (a, b), (b, a), and $(-a, 2a+b)$ lie on a line.

5. Find the condition that the three lines,

$$bx + ay = 2ab, \quad ax + by = a^2 + b^2, \quad 3x - 2y = 0,$$

where a^2 is not equal to b^2, meet in a point.

6. Find the condition that the three points (a, b), (b, a), and $(2a, -b)$, where a is not equal to b, lie on a line.

Lines through a Point

7. Show that the line drawn through the mid-points of the parallel sides of a trapezoid passes through the point of intersection of the non-parallel sides.

8. Show that, in a trapezoid, the diagonals and the line drawn through the mid-points of the parallel sides meet in a point.

9. A right triangle has its vertices A, B, and O in the points $(4, 0)$, $(0, 3)$, and $(0, 0)$. The points $A' : (4, -4)$ and $B' : (-3, 3)$ are marked. Prove that the lines AB', BA', and the perpendicular from O on the hypothenuse meet in a point.

10. (Generalization of Ex. 9.) Given a right triangle ABO with the right angle at O. On the perpendicular to OA in the point A measure off the distance AA', equal to OA, in the direction away from the hypothenuse. In a similar fashion mark the point B' on the perpendicular to OB in B, so that $BB' = OB$. Prove that the lines AB', BA', and the perpendicular from O on the hypothenuse meet in a point.

11. Let P be any point (a, a) of the line $x - y = 0$, other than the origin. Through P draw two lines, of arbitrary slopes λ_1 and λ_2, intersecting the x-axis in A_1 and A_2 and the y-axis in B_1 and B_2 respectively. Prove that the lines A_1B_2 and A_2B_1 will, in general, meet on the line $x + y = 0$.

12. If on the three sides of a triangle as diagonals parallelograms, having their sides parallel to two given lines, are

described, the other diagonals of the parallelograms meet in a point.

Prove this theorem, when the given lines are the coördinate axes, and the triangle has as its vertices the points (1, 6), (4, 11), (9, 3).

13. Prove the theorem of the preceding exercise, when the given lines are the axes, and the triangle has its vertices in the points (0, 0), (a, a), (b, c).

Points on a Line

14. Show that in the parallelogram $ABCD$ the vertex D, the mid-point of the side AB, and a point of trisection of the diagonal AC lie on a line.

15. Prove that the feet of the perpendiculars from the point $(2, -1)$ on the sides of the triangle with vertices in the points (0, 0), (3, 0), and (0, 1) lie on a line.

16. Prove that the feet of the perpendiculars from the point $(-1, 4)$ on the sides of the triangle with vertices in the points (2, 0), $(-3, 0)$, and (0, 4) lie on a line.

17. Show that the feet of the perpendiculars from the point $\left(0, \dfrac{ab}{c}\right)$ on the sides of the triangle with vertices in the points $(a, 0)$, $(b, 0)$, and $(0, c)$ lie on a line.

18. Let M be the point of intersection of two opposite sides of a quadrilateral, and N, the point of intersection of the other two sides. The mid-point of MN and the mid-points of the diagonals lie on a right line.

Prove this proposition for the special case that the vertices of the quadrilateral are situated at the points (0, 0), (8, 0), (6, 4), (1, 6).

19. Prove the proposition of Ex. 18 for the general case.

Suggestion. Take the axis of x through M and N, the origin being at the mid-point. The equations of the sides can then be written in the form

$$y = \lambda_1(x - h), \qquad y = \lambda_2(x - h),$$
$$y = \lambda_3(x + h), \qquad y = \lambda_4(x + h).$$

20. Let O be the foot of the altitude from the vertex C of the triangle ABC on the side AB. Then the feet of the perpendiculars from O on the sides BC and AC and on the other two altitudes lie on a line.

Prove this theorem for the triangle ABC with vertices in the points (1, 0), (– 4, 0), (0, 2).

21. Prove the theorem of the preceding exercise for the triangle with vertices in the points $(a, 0)$, $(b, 0)$, $(0, c)$. It will be found that

$$\left(\frac{ac^2}{a^2+c^2}, \frac{a^2c}{a^2+c^2}\right), \left(\frac{bc^2}{b^2+c^2}, \frac{b^2c}{b^2+c^2}\right),$$
$$\left(\frac{a^2b}{a^2+c^2}, \frac{-abc}{a^2+c^2}\right), \left(\frac{ab^2}{b^2+c^2}, \frac{-abc}{b^2+c^2}\right),$$

are the coördinates of the four points which are to lie on a line, and that

$$c(a + b)x + (ab - c^2)y = abc$$

is the equation of the line.

CHAPTER IV

THE CIRCLE

1. Equation of the Circle. According to Ch. I, § 7, the equation of the circle whose center is at the origin, and whose radius is ρ, is

$$(1) \qquad x^2 + y^2 = \rho^2.$$

In a precisely similar manner, the equation of a circle with its center at an arbitrary point $C: (\alpha, \beta)$ of the plane, the length of the radius being denoted by ρ, is found to be:

$$(2) \qquad (x - \alpha)^2 + (y - \beta)^2 = \rho^2.$$

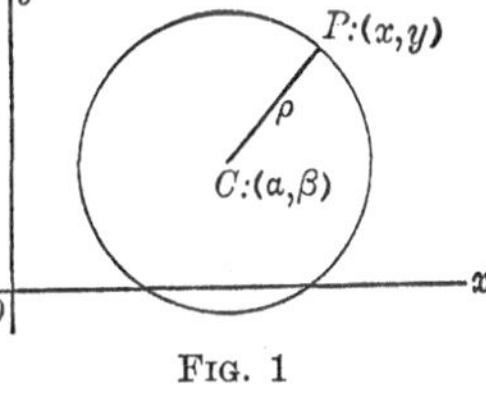

FIG. 1

Example. Find the equation of the circle whose center is at the point $(-\frac{4}{3}, 0)$, and whose radius is $\frac{2}{3}$.

Here, $\alpha = -\frac{4}{3}$, $\beta = 0$, and $\rho = \frac{2}{3}$. Hence, from (2):

$$(x + \tfrac{4}{3})^2 + y^2 = \tfrac{4}{9}.$$

This equation can be simplified as follows:

$$x^2 + \tfrac{8}{3}x + \tfrac{16}{9} + y^2 = \tfrac{4}{9},$$

or, finally,

$$3x^2 + 3y^2 + 8x + 4 = 0.$$

EXERCISES

Find the equations of the following circles, and reduce the results to their simplest form. Draw the figure each time.

1. Center at (4, 6); radius, 3.

Ans. $x^2 + y^2 - 8x - 12y + 43 = 0.$

2. Center at (0, -2); radius, 2. *Ans.* $x^2 + y^2 + 4y = 0.$

3. Center at $(-3, 0)$; radius, 3.
4. Center at $(2, -4)$; radius, 8.
5. Center at $(0, \frac{4}{3})$; radius, $\frac{2}{3}$.
6. Center at $(3, -4)$; radius, 5.
7. Center at $(-5, 12)$; radius, 13.
8. Center at $(\frac{7}{5}, -\frac{2}{5})$; radius, 2.
9. Center at $(-\frac{5}{4}, \frac{8}{3})$; radius, $\frac{11}{6}$.
10. Center at $(a, 0)$; radius, a.
11. Center at $(0, a)$; radius, a.
12. Center at (a, a); radius, $a\sqrt{2}$.

2. A Second Form of the Equation. Equation (2) of § 1 can be expanded as follows:

$$x^2 + y^2 - 2\alpha x - 2\beta y + \alpha^2 + \beta^2 - \rho^2 = 0.$$

This equation is of the form

$$x^2 + y^2 + Ax + By + C = 0. \tag{1}$$

Let us see whether, conversely, equation (1) always represents a circle.

Example 1. Determine the curve represented by the equation

$$x^2 + y^2 + 2x - 6y + 6 = 0. \tag{2}$$

We can rewrite this equation as follows:

$$(x^2 + 2x \qquad) + (y^2 - 6y \qquad) = -6.$$

The first parenthesis becomes a perfect square if 1 is added; the second, if 9 is added. To keep the equation true, these numbers must be added also to the right-hand side. Thus

$$(x^2 + 2x + 1) + (y^2 - 6y + 9) = -6 + 1 + 9,$$

or

$$(x + 1)^2 + (y - 3)^2 = 4.$$

This equation is precisely of the form (2), § 1, where $\alpha = -1$, $\beta = 3$, $\rho = 2$. It therefore represents a circle whose center is at $(-1, 3)$, and whose radius is 2.

Example 2. What curve is represented by the equation

$$(3) \qquad x^2 + y^2 + 1 = 0\,?$$

It is clear that *no* point exists whose coördinates satisfy this equation. For, x^2 and y^2 can never be negative. Their least values are 0, — namely, for the origin, (0, 0), — and even for this point, the left-hand side of the equation has the value $+1$. *Hence, there is no curve corresponding to equation* (3).

Example 3. Discuss the equation

$$(4) \qquad x^2 + y^2 + 2x - 4y + 5 = 0.$$

Evidently, this equation can be written in the form:

$$(5) \qquad (x+1)^2 + (y-2)^2 = 0.$$

The coördinates of the point $(-1, 2)$ satisfy the equation. But, for any other point (x, y), at least one of the quantities, $x+1$ and $y-2$, is not zero, and the left-hand side of the equation is positive. Thus the point $(-1, 2)$ is the only point whose coördinates satisfy the equation. *Hence equation* (4) *represents a single point* $(-1, 2)$.

Remark. Equation (5) can be regarded as the limiting case of the equation

$$(x+1)^2 + (y-2)^2 = \rho^2,$$

when ρ approaches the limit 0. This equation represents a circle of radius ρ for all positive values of ρ. When ρ approaches 0, the circle shrinks down toward the point $(-1, 2)$ as its limit. Accordingly, equation (5) is sometimes spoken of as representing a circle of zero radius or a *null circle.*

The General Case. It is now clear how to proceed in the general case, in order to determine what curve equation (1) represents. The equation can be written in the form:

$$(x^2 + Ax + \tfrac{1}{4}A^2) + (y^2 + By + \tfrac{1}{4}B^2) = -C + \tfrac{1}{4}A^2 + \tfrac{1}{4}B^2,$$

or

$$\left(x + \frac{A}{2}\right)^2 + \left(y + \frac{B}{2}\right)^2 = \frac{A^2 + B^2 - 4C}{4}.$$

If the right-hand side is positive, *i.e.* if

$$A^2 + B^2 - 4C > 0,$$

then equation (1) represents a circle, whose center is at the point $(-\frac{1}{2}A, -\frac{1}{2}B)$ and whose radius is

$$\rho = \tfrac{1}{2}\sqrt{A^2 + B^2 - 4C}.$$

If, however, $A^2 + B^2 - 4C = 0$, then equation (1) represents just one point, namely, $(-\frac{1}{2}A, -\frac{1}{2}B)$, — or, if one prefers, a circle of zero radius or a null circle.

Finally, when $A^2 + B^2 - 4C < 0$, there are no points whose coördinates satisfy (1). To sum up, then:

Equation (1) *represents a circle, a single point, or there is no point whose coördinates satisfy* (1), *according as the expression*

$$A^2 + B^2 - 4C$$

is positive, zero, or negative.

Consider, more generally, the equation

$$a(x^2 + y^2) + bx + cy + d = 0. \tag{6}$$

If $a = 0$, but b and c are not both 0, the equation represents a straight line.

If, however, $a \neq 0$, the equation can be divided through by a, and it thus takes on the form:

$$x^2 + y^2 + \frac{b}{a}x + \frac{c}{a}y + \frac{d}{a} = 0.$$

This is precisely the form of equation (1), and hence the above discussion is applicable to it.

EXERCISES

Determine what the following equations represent. Apply each time the *method* of completing the square and examining the right-hand side of the new equation. Do *not* merely substitute numerical values in the formulas developed in the text.

1. $x^2+y^2+6x-8y=0.$

Ans. A circle, radius 5, with center at $(-3, 4)$.

2. $x^2+y^2-6x+4y+13=0.$ *Ans.* The point $(3, -2)$.

3. $x^2+y^2+2x+4y+6=0.$ *Ans.* No point whatever.

4. $x^2+y^2-10x+24y=0.$

5. $x^2+y^2-7x=5.$

6. $x^2+y^2-6x+8y+25=0.$

7. $49x^2+49y^2-14x+28y+5=0.$

Ans. The point $(\frac{7}{49}, -\frac{14}{49})$.

8. $x^2+y^2+8y=10.$

9. $x^2+y^2=2ax.$

10. $x^2+y^2=2ay.$

11. $x^2+y^2-6ax-2by+9a^2=0.$

12. $x^2+y^2+4ax-8by+16b^2=0.$

13. $x^2+y^2+3=0.$

14. $x^2+y^2-2x+4y+10=0.$

15. $3x^2+3y^2-4x+2y+7=0.$

16. $5x^2+5y^2-6x+8y=12.$

17. $3x^2+3y^2-x+y=6.$

3. Tangents. Let the circle

$$x^2+y^2=\rho^2 \tag{1}$$

be given, and let $P_1:(x_1, y_1)$ be any point of this circle. To find the equation of the tangent at P_1.

The tangent at P_1 is, by Elementary Geometry, perpendicular to the radius, OP_1. Hence its slope, λ', is the negative reciprocal of the slope, y_1/x_1, of OP_1; or

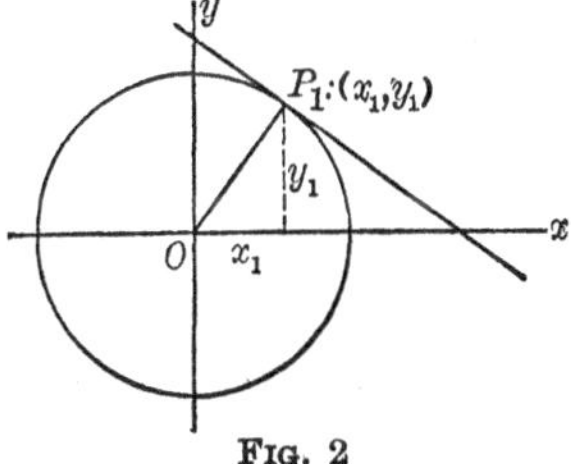

FIG. 2

$$\lambda' = -\frac{x_1}{y_1}.$$

We wish, therefore, to find the equation of the line which passes through the point (x_1, y_1) and has the slope $\lambda' = -x_1/y_1$.

By Ch. II, § 2, (II), the equation of this line is

$$(2) \qquad y - y_1 = -\frac{x_1}{y_1}(x - x_1).$$

This equation can be simplified by multiplying through by y_1 and transposing:

$$(3) \qquad x_1x + y_1y = x_1^2 + y_1^2.$$

Now, the point (x_1, y_1) is, by hypothesis, on the circle; hence its coördinates satisfy the equation (1) of the circle:

$$x_1^2 + y_1^2 = \rho^2.$$

The right-hand side of equation (3) can, therefore, be replaced by the simpler expression, ρ^2.

We thus obtain, as the final form of the equation of the tangent, the following:

$$(4) \qquad x_1x + y_1y = \rho^2.$$

In deducing this equation it was tacitly assumed that $y_1 \neq 0$, since otherwise we could not have divided by it in obtaining λ'. The final formula, (4), is true, however, even when $y_1 = 0$, as can be directly verified. For, if $y_1 = 0$ and $x_1 = \rho$, then (4) becomes

$$\rho x = \rho^2 \qquad \text{or} \qquad x = \rho,$$

and this is the equation of the tangent in the point $(\rho, 0)$. Similarly, when $y_1 = 0$ and $x_1 = -\rho$.

Any Circle. If the given circle is represented by the equation

$$(5) \qquad (x - \alpha)^2 + (y - \beta)^2 = \rho^2,$$

precisely the same reasoning can be applied. The equation of the tangent to (5) at the point $P_1 : (x_1, y_1)$ of that circle is thus found to be:

$$(6) \qquad (x_1 - \alpha)(x - \alpha) + (y_1 - \beta)(y - \beta) = \rho^2.$$

The proof is left to the student as an exercise.

If the equation of the circle is given in the form

(7) $$x^2 + y^2 + Ax + By + C = 0,$$

or in the form (6), § 2, the equation can first be thrown into the form (5), and then the equation of the tangent is given by (6).

Example. To find the equation of the tangent to the circle

(8) $$3x^2 + 3y^2 + 8x - 5y = 0$$

at the origin.

First, reduce the coefficients of the terms in x^2 and y^2 to unity:

$$x^2 + y^2 + \tfrac{8}{3}x - \tfrac{5}{3}y = 0.$$

Next, complete the squares:

$$x^2 + \tfrac{8}{3}x + (\tfrac{4}{3})^2 + y^2 - \tfrac{5}{3}y + (\tfrac{5}{6})^2 = \tfrac{16}{9} + \tfrac{25}{36} = \tfrac{89}{36},$$

or $$(x + \tfrac{4}{3})^2 + (y - \tfrac{5}{6})^2 = \tfrac{89}{36},$$

Now, apply the theorem embodied in formula (6). Since

$$x_1 = 0, \quad y_1 = 0, \quad \alpha = -\tfrac{4}{3}, \quad \beta = \tfrac{5}{6},$$

we have $$\tfrac{4}{3}(x + \tfrac{4}{3}) - \tfrac{5}{6}(y - \tfrac{5}{6}) = \tfrac{89}{36},$$

or $$8x - 5y = 0,$$

as the equation of the tangent to (8) at the origin.

EXERCISES

Find the equation of the tangent to each of the following circles at the given point.

1. $x^2 + y^2 = 25$ at $(-3, 4)$. *Ans.* $3x - 4y + 25 = 0$.

2. $x^2 + y^2 = a^2$ at $(0, a)$. *Ans.* $y = a$.

3. $x^2 + y^2 = 49$ at $(-7, 0)$.

4. $(x - 1)^2 + (y + 2)^2 = 25$ at $(4, 2)$. *Ans.* $3x + 4y = 20$.

5. $(x + 5)^2 + (y - 3)^2 = 49$ at $(2, 3)$.

6. $x^2 + y^2 - 9x + 11y = 0$ at the origin.

7. $2x^2 + 2y^2 - 3x - y = 11$ at $(-1, 2)$.

8. Find the intercepts on the axis of x made by the tangent at $(-5, 12)$ to

$$x^2 + y^2 = 169. \qquad \textit{Ans.} \ -33\tfrac{4}{5}.$$

9. Find the area of the triangle cut from the first quadrant by the tangent at $(1, 1)$ to

$$3x^2 + 3y^2 + 8x + 16y = 30.$$

10. If the equation

$$x^2 + y^2 + Ax + By + C = 0$$

represents a circle, and if the point (x_1, y_1) lies on the circle, show that the equation of the tangent at this point can be written in the form:

$$(9) \qquad x_1x + y_1y + \frac{A}{2}(x + x_1) + \frac{B}{2}(y + y_1) + C = 0.$$

Suggestion. Find the values of α, β, and ρ for the circle, substitute them in (6), and simplify the result.

11. Do Exs. 6 and 7, using formula (9), Ex. 10.

12. The same for the tangent to the circle in Ex. 9.

13. Show that, if $P_1 : (x_1, y_1)$ is any point of the circle

$$x^2 + y^2 + Ax + By + C = 0,$$

at which the tangent is not parallel to the axis of y, then the slope of the tangent at P_1 is

$$-\frac{2x_1 + A}{2y_1 + B}.$$

4. Circle through Three Points. It is shown in Elementary Geometry that a circle can be passed through any three points not lying in a straight line.

If the points are (x_1, y_1), (x_2, y_2), and (x_3, y_3), and if the equation of the circle through them is written in the form

$$x^2 + y^2 + Ax + By + C = 0,$$

then clearly the following three equations must hold:

$$x_1^2 + y_1^2 + Ax_1 + By_1 + C = 0,$$
$$x_2^2 + y_2^2 + Ax_2 + By_2 + C = 0,$$
$$x_3^2 + y_3^2 + Ax_3 + By_3 + C = 0.$$

We thus have three simultaneous linear equations for determining the three unknown coefficients A, B, C.

Suppose, for example, that the given points are the following:

$$(1, 1), \qquad (1, -1), \qquad (-2, 1).$$

The equations can be thrown at once into the form

$$A + B + C = -2,$$
$$A - B + C = -2,$$
$$-2A + B + C = -5.$$

Solve two of these equations for two of the unknowns in terms of the third. Then substitute the values thus found in the third equation. Thus the third unknown is completely determined, and hence the other two unknowns can be found.

Here, it is easy to solve the first two equations for A and B in terms of C. On subtracting the second equation from the first, we find:

$$2B = 0; \qquad \text{hence} \qquad B = 0.$$

Then either of the first two equations gives for A the value:

$$A = -C - 2.$$

Next, set for A and B in the third equation the values just found:

$$2C + 4 + C = -5, \qquad C = -3.$$

Hence, finally,

$$A = 1, \qquad B = 0, \qquad C = -3,$$

and the equation of the desired circle is:

$$x^2 + y^2 + x - 3 = 0.$$

Check the result by substituting the coördinates of the given points successively in this last equation. They are found each time to satisfy the equation.

The circle through the three given points has its center in the point $(-\frac{1}{2}, 0)$. Its radius is of length $\sqrt{3.25} = 1.803$.

EXERCISES

Find the equations of the circles through the following triples of points. Plot the points and draw the circles.

1. $(1, 0)$, $(0, 1)$, the origin. *Ans.* $x^2 + y^2 - x - y = 0$.

2. $(1, 1)$, $(-1, -1)$, $(1, -1)$.

3. $(5, 10)$, $(6, 9)$, $(-2, 3)$.

4. The vertices of the triangle of Ex. 15 at the end of Ch. III, p. 63. Show that the point $(2, -1)$ of that exercise lies on the circle.

5. The same question for Ex. 17, p. 63. Show that the point $\left(0, \dfrac{ab}{c}\right)$ of that exercise lies on the circle.

6. The vertices of the triangle of Ch. III, Fig. 1. Find the coördinates of the center and check by comparing them with those of the point of intersection of the perpendicular bisectors of the sides of the triangle, as determined in Ch. III, § 5.

7. The same question for the triangle of Ch. III, Fig. 2. Check.

8. The vertices of the triangle formed by the coördinate axes and the line $2x - 3y = 6$.

9. The vertices of the triangle whose sides are:

$$x - y - 1 = 0, \quad x + y + 2 = 0, \quad 2x - y + 3 = 0.$$

Ans. $3x^2 + 3y^2 + 17x + 16y + 25 = 0$.

EXERCISES ON CHAPTER IV

1. Find the equation of the circle with the line-segment joining the two points $(3, 0)$ and $(5, 2)$ as a diameter.

2. A circle goes through the origin and has intercepts -5 and 3 on the axes of x and y respectively. Find its equation.

3. A circle goes through the origin and has intercepts a and b. Find its equation.

4. Find the equation of the circle which has its center in the point $(-3, 4)$ and is tangent to the line $3x + 8y - 6 = 0$.

5. A circle has its center on the line $2x - 3y = 0$ and passes through the points $(4, 3)$, $(-2, 5)$. Find its equation.

6. Find the equation of the circle which passes through the point $(5, -2)$ and is tangent to the line $3x - y - 1 = 0$ at the point $(1, 2)$.

7. There are two circles passing through the points $(3, 2)$, $(-1, 0)$ and having 6 as their radius. Find their equations.

8. There are two circles with their centers on the line, $5x - 3y = 8$, and tangent to the coördinate axes. Find their equations.

9. Find the equations of the circles tangent to the axes and passing through the point $(1, 2)$.

10. Find the equations of the circles passing through the points $(3, 1)$, $(1, 0)$ and tangent to the line $x - y = 0$.

Suggestion. Demand that the center (α, β) be equally distant from the two points and the line.

11. Find the equations of the circles passing through the origin, tangent to the line $x + y - 8 = 0$, and having their centers on the line $x = 2$.

12. Find the equations of the circles of the preceding exercise, if their centers lie on the line $2x - y - 2 = 0$.

13. Find the equation of the circle inscribed in the triangle formed by the axes and the line $3x - 4y - 12 = 0$.

14. Find the equation of an arbitrary circle, referred to two perpendicular tangents as axes.

15. Do the four points $(0, 0)$, $(6, 0)$, $(0, -4)$, $(5, 1)$ lie on a circle?

16. Find the coördinates of the points of intersection of the circles

$$x^2 + y^2 - x + 2y = 0,$$
$$x^2 + y^2 + 2x - y = 9.$$

17. Find the coördinates of the points of intersection of the circles

$$x^2 + y^2 + ax + by = 0,$$
$$x^2 + y^2 + bx - ay = 0.$$

Orthogonality

18. A circle and a line intersect in a point P. The acute angle between the line and the tangent to the circle at P is known as the angle of intersection of the line and the circle at P. If the line meets the circle in two points, the angles of intersection at the two points are equal. Determine the angle in the case of the circle

$$x^2 + y^2 = 25,$$

and the line

$$2x - y - 5 = 0.$$

19. A circle and a line are said to intersect orthogonally if their angle of intersection is a right angle. Prove that the circle,

$$x^2 + y^2 - 4x + 6y + 3 = 0,$$

is intersected by the line, $5x + y = 7$, orthogonally.

Suggestion. First answer geometrically the question: What lines cut a given circle orthogonally?

20. Show that the circle,

$$x^2 + y^2 + Ax + By + C = 0,$$

intersects the line,

$$ax + by + c = 0,$$

orthogonally when and only when

$$aA + bB = 2c.$$

21. If two circles intersect in a point P, the acute angle between their tangents at P is known as their angle of intersection. If the circles intersect in two points, their angles of intersection at these points are equal. Find this angle in the case of the circles,

$$x^2 + y^2 = 25,$$
$$x^2 + y^2 - 7x + y = 0.$$

22. Prove geometrically that two circles intersect orthogonally, that is, at right angles, when and only when the sum of the squares of their radii equals the square of the distance between their centers. Then show that the circles

$$x^2 + y^2 - 4x + 5y - 2 = 0,$$
$$2x^2 + 2y^2 + 4x - 6y - 19 = 0,$$

intersect orthogonally.

23. Prove that the two circles,

$$x^2 + y^2 + A_1x + B_1y + C_1 = 0,$$
$$x^2 + y^2 + A_2x + B_2y + C_2 = 0,$$

intersect orthogonally when and only when

$$A_1A_2 + B_1B_2 = 2C_1 + 2C_2.$$

24. Find the equation of the circle which cuts the circle

$$x^2 + y^2 + 2x = 0$$

at right angles and passes through the points (1, 0) and (0, 1).

25. There are an infinite number of circles cutting each of the two circles,

$$x^2 + y^2 - 4y + 2 = 0,$$
$$x^2 + y^2 + 4y + 2 = 0,$$

orthogonally. Show that they are all given by the equation

$$x^2 + y^2 + ax - 2 = 0,$$

where a is an arbitrary constant. Where are their centers? Draw a figure.

26. Find the equation of the circle cutting orthogonally the three circles,

$$x^2 + y^2 = 9,$$
$$x^2 + y^2 + 3x - 5y + 6 = 0,$$
$$x^2 + y^2 - 2x + 3y - 19 = 0.$$

Ans. $x^2 + y^2 + 10x + 9 = 0.$

Miscellaneous Theorems

27. Prove analytically that every angle inscribed in a semicircle is a right angle.

28. Prove analytically that the perpendicular dropped from a point of a circle on a diameter is a mean proportional between the segments in which it divides the diameter.

29. The tangents to a circle at two points P, Q meet in the point T. The lines joining P and Q to one extremity of the diameter parallel to PQ meet the perpendicular diameter in the points R and S. Prove that $RT = ST$.

30. In a triangle the circle through the mid-points of the sides passes through the feet of the altitudes and also through the points halfway between the vertices and the point of intersection of the altitudes. This circle is known as the *Nine-Point Circle* of the triangle.

For the triangle with vertices in the points $(-4, 0)$, $(2, 0)$, $(0, 6)$ construct the circle and mark the nine points through which it passes.

31. For the triangle in the preceding exercise find the equation of the nine-point circle, as the circle through the mid-points of the sides. *Ans.* $3x^2 + 3y^2 + 3x - 11y = 0$.

32. Show that this circle goes through the other six points.

33. For the triangle with vertices in the points $(a, 0)$, $(b, 0)$, $(0, c)$ find the equation of the nine-point circle, as the circle through the mid-points of the sides.

Ans. $2c(x^2 + y^2) - (a + b)cx + (ab - c^2)y = 0$.

34. Show that this circle goes through the other six points.

CHAPTER V

INTRODUCTORY PROBLEMS IN LOCI. SYMMETRY OF CURVES

1. Locus Problems.* *A point is moving under given conditions; its locus is required.* This type of problem the student studied in Plane Geometry. But he found there no general method, by means of which he could always determine a locus; for each problem he had to devise a method, depending on the particular conditions of the problem.

Analytic Geometry, however, provides a general method for the determination of loci. Some simple examples of the method have already been given. Thus, in finding the equation of a circle, we determined the locus of a point whose distance from a fixed point is constant. Again, in deducing the equation of a line through two points, we found the locus of a point moving so that the line joining it to a given point has a given direction.

The method in each of these cases consisted merely in expressing in analytic terms — *i.e.* in the form of an equation involving the variable coördinates, x and y, of the moving point — the given geometric condition under which the point moved. We proceed to show how this method applies in less simple cases.

Example 1. The base of a triangle is fixed, and the distance from one end of the base to the mid-point of the opposite side is given. Find the locus of the vertex.

* The locus problems in this chapter may be supplemented, if it is desired, by §§ 6–8 of the second chapter on loci, Ch. XIII, in which the loci of inequalities and the bisectors of the angles between two lines, together with related subjects, are considered.

Let the triangle be OAP, with M as the mid-point of AP. Let a be the length of the base OA, and let l be the given distance. It is required to find the locus of P, so that always

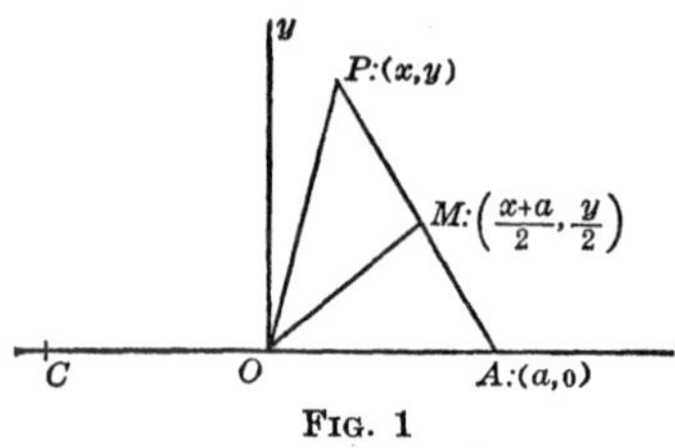

Fig. 1

(1) $$OM = l.$$

It is convenient to take the origin of coördinates in O and the positive axis of x along the base. The coördinates of A are then $(a, 0)$. The coördinates of the moving point P we denote by (x, y). The coördinates of the point M are

$$\left(\frac{x+a}{2}, \frac{y}{2}\right).$$

The distance OM is

$$\sqrt{\left(\frac{x+a}{2}\right)^2+\left(\frac{y}{2}\right)^2}.$$

Thus condition (1), expressed analytically, is

$$\sqrt{\left(\frac{x+a}{2}\right)^2+\left(\frac{y}{2}\right)^2}=l.$$

Squaring both sides of this equation and simplifying, we have

(2) $$(x+a)^2+y^2=(2\,l)^2.$$

This equation represents the circle whose center is at $(-a, 0)$ and whose radius is $2\,l$. We have shown, therefore, that, if (1) is always satisfied, the coördinates (x, y) of P satisfy (2), and P lies on the circle. The locus of P appears, then, to be the circle.

How do we know, though, that P traces the entire circle? To prove this, we must show, conversely, that, if the coördinates (x, y) of P satisfy (2), condition (1) is valid. If (x, y) satisfy (2), then, on dividing both sides of (2) by 4 and extracting the square root of each side, we obtain *two* equations:

i) $$\sqrt{\left(\frac{x+a}{2}\right)^2+\left(\frac{y}{2}\right)^2}=l,$$

ii) $$\sqrt{\left(\frac{x+a}{2}\right)^2+\left(\frac{y}{2}\right)^2}=-l.$$

Equation ii) says that a positive or zero quantity equals a negative quantity, and is therefore impossible. Thus only equation i) remains. This equation says that $OM = l$. Hence condition (1) is satisfied by every point of the circle,* and so the circle is the locus of P.

We have yet to describe the locus, independently of the coördinate system, with reference merely to the original triangle. Produce the base, in the direction from A to O, to the point C, doubling its length. Then the locus of P is a circle, whose center is at C and whose radius is twice the given distance.

Example 2. Determine the locus of a point P which moves so that the difference of the squares of its distances from two fixed points P_1, P_2 is constant, and equal to c:

$$(3)\qquad \begin{cases} PP_1^2 - PP_2^2 = c, \\ \qquad\text{or} \\ PP_2^2 - PP_1^2 = c. \end{cases}$$

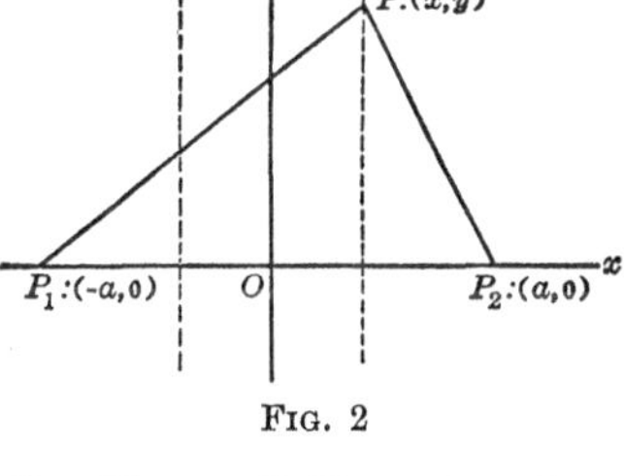

FIG. 2

Take the mid-point of the segment P_1P_2 as origin and the axis of x along P_1P_2. The coördinates of P_1 and P_2 can be written as $(-a, 0)$, $(a, 0)$; those of P, as (x, y).

By Ch. I, § 3,

$$PP_1^2 = (x+a)^2 + y^2, \qquad PP_2^2 = (x-a)^2 + y^2.$$

Then the equations (3), expressed analytically, are

* The two points in which the circle cuts the axis of x are exceptions, since these do not lead to a triangle, OAP.

$$(x+a)^2+y^2-(x-a)^2-y^2=c,$$
$$(x-a)^2+y^2-(x+a)^2-y^2=c.$$

These reduce to

$$(4) \qquad 4\,ax=c, \qquad 4\,ax=-\,c.$$

Hence, if condition (3) is satisfied, P lies on one or the other of the lines

$$(5) \qquad x=\frac{c}{4\,a}, \qquad x=-\frac{c}{4\,a}.$$

Conversely, if P lies on one or the other of the lines (5), then (4) holds, and from (4) we show by retracing the steps that one or the other of the equations (3) is valid.

Consequently, the locus of P consists of *two* straight lines, perpendicular to the line P_1P_2, and symmetrically situated with reference to the mid-point of P_1P_2, the distance of either line from the mid-point being $c/4\,a$. Thus the locus consists of two entirely unconnected pieces, one corresponding to each of the equations (3). If $c=0$, these equations are the same, and the two lines forming the locus coincide in the perpendicular bisector of the segment P_1P_2.

EXERCISES

In solving the following problems, the first step is to find the equation of a curve, — or the equations of curves, — on which points of the locus lie. The student must then take care (*a*) to show, conversely, that every point lying on the curve or curves obtained satisfies the given conditions; and (*b*) to describe the locus, finally, without reference to the coördinate system used.

1. A point P moves so that the sum of the squares of its distances to two fixed points P_1, P_2 is a constant, c, greater than $\frac{1}{2}\,\overline{P_1P_2}^2$. Show that the locus of P is a circle, with its center at the mid-point of P_1P_2.

What is the locus if $c=\frac{1}{2}\,\overline{P_1P_2}^2$? If $c<\frac{1}{2}\,\overline{P_1P_2}^2$?

2. Find the locus of the mid-point of a line of fixed length which moves so that its end points always lie on two mutually perpendicular lines.

3. Determine the locus of a point which moves so that the sum of the squares of its distances to the sides, or the sides produced, of a given square is constant. Is there any restriction necessary on the value of the constant?

4. Determine the locus of a point which moves so that the square of its distance to the origin equals the sum of its coördinates. *Ans.* A circle, center at $(\frac{1}{2}, \frac{1}{2})$, radius $= \frac{1}{2}\sqrt{2}$.

5. Show that the locus of a point which moves so that the sum of its distances to two mutually perpendicular lines equals the square of its distance to their point of intersection consists of the arcs of four circles, forming a continuous curve. Where are the circles, and which of their arcs belong to the locus?

6. The base of a triangle is fixed, and the trigonometric tangent of one base angle is a constant multiple, not -1, of the trigonometric tangent of the other. Find the locus of the vertex.

2. Symmetry. In the problems of the preceding paragraph, the equations of the loci were familiar and the curves they represented were easily identified. In subsequent chapters, however, we shall have locus problems to consider in which the resulting equations will be new to us. In drawing the curves which these equations represent, it will be useful to have at hand the salient facts concerning the symmetry of curves.

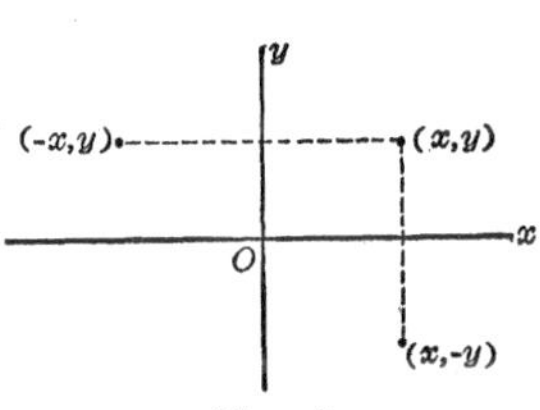

$(-x,y)$ (x,y) y x O $(x,-y)$

FIG. 3

Symmetry in a Line. Two points, P and P', are said to be symmetric in a line L, if L is the perpendicular bisector of PP'.

If L is the axis of x and (x, y) are the coördinates of P, then it is clear that $(x, -y)$ are the coördinates of P'.

Similarly, if L is the axis of y and P has the coördinates (x, y), then P' has the coördinates $(-x, y)$.

Example 1. Given the curve

$$(1) \qquad y^2 = x.$$

Let $P:(x_1, y_1)$ be any point on it, *i.e.* let

$$(2) \qquad {y_1}^2 = x_1$$

be a true equation. Then the point $P':(x_1, -y_1)$, symmetric to P in the axis of x, also lies on the curve. For, if we substitute the coördinates of P' into (1), the result is $(-y_1)^2 = x_1$, or (2), and (2) we know is a true equation. We say, then, that the curve (1) is symmetric in the axis of x.

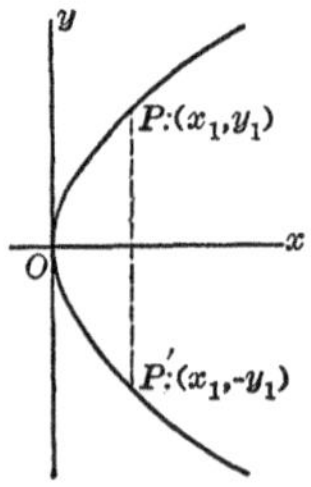

Fig. 4

The test for symmetry in the axis of x, employed in this example, is general in application. We state it, and the corresponding test for symmetry in the axis of y, in the form of theorems.

Theorem 1. *A curve is symmetric in the axis of x if the substitution of $-y$ for y in its equation leaves the equation unchanged.*

Theorem 2. *A curve is symmetric in the axis of y if the substitution of $-x$ for x in its equation leaves the equation unchanged.*

Symmetry in a Point. Two points, P and P', are symmetric in a given point, if the given point is the mid-point of PP'.

If the given point is the origin of coördinates and P has the coördinates (x, y), then the coördinates of P' are evidently $(-x, -y)$.

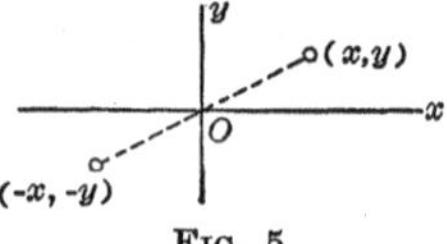

Fig. 5

Example 2. Consider the curve

$$(3) \qquad y = x^3.$$

If $P:(x_1, y_1)$ is any point on this curve, then the point $P':(-x_1, -y_1)$, symmetric to P in the origin, is also on the curve. For, the condition that P' lies on the curve, namely,

$$-y_1 = (-x_1)^3 \qquad \text{or} \qquad -y_1 = -{x_1}^3,$$

is equivalent to the condition: $y_1 = x_1^3$, that P lie on the curve. We say, then, that the curve (3) is symmetric in the origin.

This test, too, is general in application; we formulate it as a theorem.

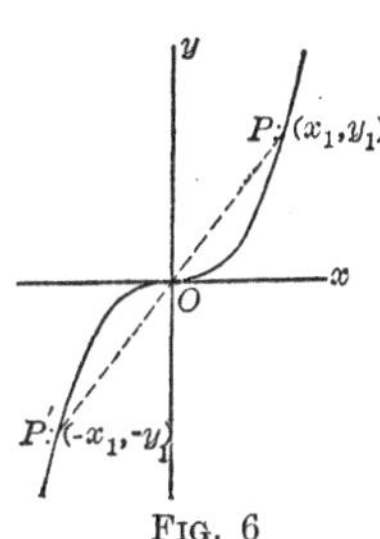

Fig. 6

Theorem 3. *A curve is symmetric in the origin of coördinates, if the substitution of* $-x$ *for* x, *and of* $-y$ *for* y, *in its equation leaves the equation essentially unchanged.*

A case in which the test leaves the equation wholly unchanged is that of the circle, $x^2 + y^2 = \rho^2$, or the curve $xy = a^2$ (Fig. 7).

Now the circle in question is symmetric in both axes. It follows then, without further investigation, that it is symmetric in the origin, the point of intersection of the axes. This conclusion holds always; in fact, we may state the theorem.

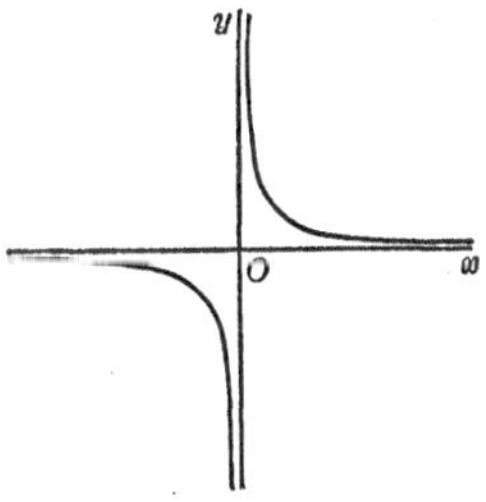

Fig. 7

Theorem 4. *If a curve is symmetric in both axes of coördinates, it is symmetric in the origin.*

The details of the proof are left to the student as an exercise. It is to be noted that the converse of the theorem, namely, that if a curve is symmetric in the origin, it is symmetric in the axes, is not true. For, the curve of Example 2 is symmetric in the origin, but not symmetric in either axis; this is true also of the curve $xy = a^2$ of Fig. 7.

EXERCISES

1. Prove Theorem 4.

2. Test, for symmetry in each axis and in the origin, the curves given in the following exercises of Ch. I, § 7:

(a) Exercise 2; (c) Exercise 7;
(b) Exercise 6; (d) Exercise 8.

In each of the following exercises test the given curve for symmetry in each axis and in the origin. Plot the curve.

3. $xy + 1 = 0$.
4. $10y = x^4$.
5. $20x = y^5$.
6. $y^2 + 4x = 0$.
7. $x^2 - y^2 = 4$.
8. $x^2 + 2y^2 = 16$.

EXERCISES ON CHAPTER V

1. The base of a triangle is fixed and the ratio of the lengths of the two sides is constant. Find the locus of the vertex. *Ans.* A circle, except for one value of the constant.

2. A point P moves so that its distance from a given line L is proportional to the square of its distance to a given point K, not on L. If P remains always on the same side of L as K, show that its locus is a circle.

3. Find the locus of P in the preceding exercise, if it remains always on the opposite side of L from K. Does your answer cover all cases?

4. If, in Ex. 2, K lies on L and P may be on either side of L, what is the locus of P?

5. Three vertices of a quadrilateral are fixed. Find the locus of the fourth, if the area of the quadrilateral is constant.

6. Find the locus of a point moving so that the sum of the squares of its distances from the sides of an equilateral triangle is constant. Discuss all cases.

Ans. A circle, center at the point of intersection of the medians; this point; or no locus.

7. The feet of the perpendiculars from the point $P: (X, Y)$ on the sides of the triangle with vertices in the points (0, 0), (3, 0), (0, 1) lie on a line. Find the locus of P.

Ans. The circle circumscribing the triangle.

8. The preceding problem, if the triangle has the points (2, 0), (− 3, 0), (0, 4) as vertices.

9. Problem 7, for the general triangle, with vertices at $(a, 0)$, $(b, 0)$, $(0, c)$.

10. Show that the equation of the circle described on the line-segment joining the points (x_1, y_1), (x_2, y_2) as a diameter may be written in the form

$$(x - x_1)(x - x_2) + (y - y_1)(y - y_2) = 0.$$

Suggestion. Find the locus of a point P moving so that the two given points always subtend at P a right angle.

11. The two points, P and P', are symmetric in the line, $x - y = 0$, bisecting the angle between the positive axes of x and y. Show that, if (x, y) are the coördinates of P, then (y, x) are the coördinates of P'.

12. Prove that a curve is symmetric in the line $x - y = 0$ if the interchange of x and y in its equation leaves the equation unchanged.

13. If P and P' are symmetric in the line $x + y = 0$ and P has the coördinates (x, y), show that the coördinates of P' are $(-y, -x)$.

14. Give a test for the symmetry of a curve in the line $x + y = 0$.

15. Test each of the following curves for symmetry in the lines $x - y = 0$ and $x + y = 0$.

(*a*) $xy = a^2$; (*c*) $x^2 - y^2 = a^2$;
(*b*) $xy = -a^2$; (*d*) $(x - y)^2 - 2x - 2y = 0$.

16. Plot the curve of Ex. 15, (*d*).

In each of the following exercises find the equation of the locus of the point P. Plot the locus from the equation, making all the use possible of the theory of symmetry.

17. The distance of P from the line $x + 2 = 0$ equals its distance from the point $(2, 0)$.

18. The sum of the distances of P from the points $(3, 0)$ and $(-3, 0)$ is 10.

19. The difference of the distances of P from the points $(5, 0)$ and $(-5, 0)$ is 8.

CHAPTER VI

THE PARABOLA

1. Definition. A *parabola* is defined as the locus of a point P, whose distance from a fixed line D is always equal to its distance from a fixed point F, not on the line. It is understood, of course, that P is restricted to the plane determined by D and F.

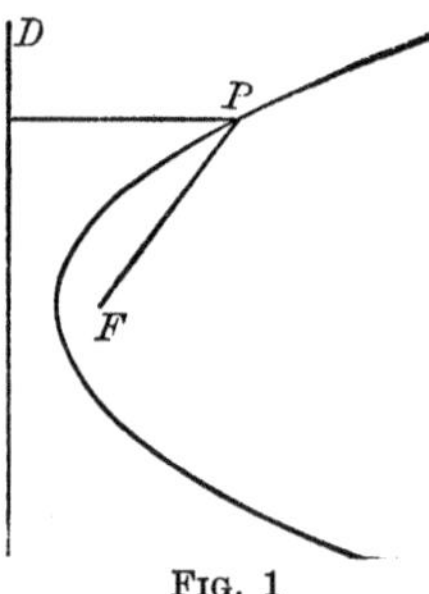

FIG. 1

One point of the locus is the mid-point A, Fig. 2, of the perpendicular FE dropped from F on D. Through A draw T parallel to D. Then no other point on T, or to the left of T, can belong to the locus, for all such points are clearly nearer to D than they are to F.

Further points of the locus can be obtained as follows. To the right of T draw L parallel to D, cutting AF, produced if necessary, in S. With ES as radius and F as center describe a circle, cutting L in P and Q. Then P and Q lie on the locus.

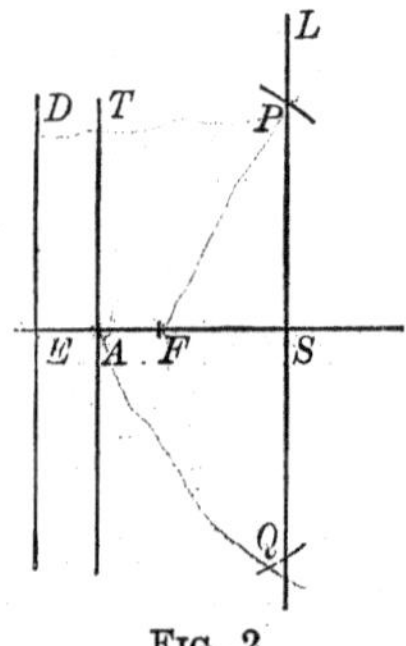

FIG. 2

A large number of points having been obtained in this way, a smooth curve can be passed through them. The curve is symmetric in the line AF, and evidently has T as a tangent.

The line D is called the *directrix*, and the point F, the *focus*, of the parabola; A is the *vertex*, and the indefinite line AF, the *axis*; FP is a *focal radius*.

The student is familiar with the fact that all circles are *similar*; *i.e.* have the same shape, and differ only in size. A like relation holds for any two parabolas. Think of them as lying in different planes, and choose in each plane as the unit length the distance between the focus and the directrix. Then the one parabola, in its plane, is the replica of the other, in its plane. Consequently, the two parabolas differ only in the scale to which they are drawn, and are, therefore, similar.

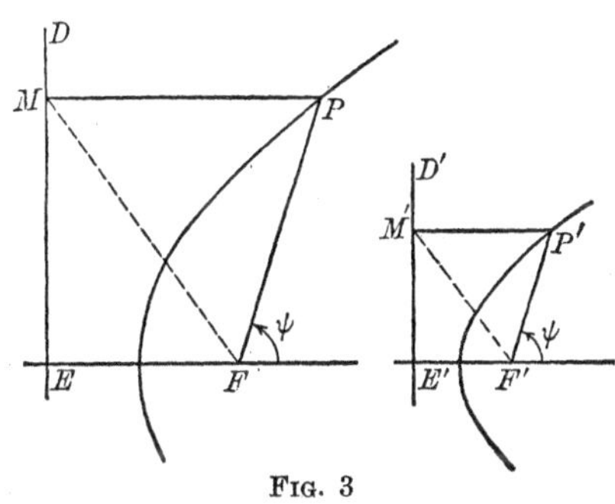

FIG. 3

The details of the proof just outlined can be supplied at once by showing that the triangles FPM and MFE are similar, respectively, to $F'P'M'$ and $M'F'E'$, the angles ψ in Fig. 3 being equal by construction. Hence

$$\frac{FP}{F'P'}=\frac{EF}{E'F'},$$

i.e. focal radii, FP and $F'P'$, which make the same angle with the axes always bear to each other the same fixed ratio.

EXERCISES

1. Take a sheet of squared paper and mark D along one of the vertical rulings near the edge of the paper. Choose F at a distance of 1 cm. from D. Then the points of the locus on the vertical rulings — or on as many of them as one desires — can be marked off rapidly with the compasses. Make a clean, neat figure.

2. Place a card under the curve of Ex. 1 and, with a needle, prick numerous points of the curve through on the card, and mark, also, the focus and axis in this way. Cut the card along

the curve with sharp scissors. The piece whose edge is convex forms a convenient parabolic ruler, or templet, to be used whenever an accurate drawing is desired.

A small hole at the focus and a second hole farther along the axis make it possible, in using the templet, to mark the focus and draw the axis.

A second templet, to twice the above scale, will also be found useful.

3. The focus of a parabola is distant 5 units from the directrix. In a second parabola, this distance is 2 units. How much larger is the first parabola than the second, *i.e.*, how do their scales compare with each other?

2. Equation of the Parabola. The first step is to choose the axes of coördinates in a convenient manner. Evidently, one good choice would be to take the axis of x perpendicular to D and passing through F. Let us do this, choosing the positive sense from A toward F.

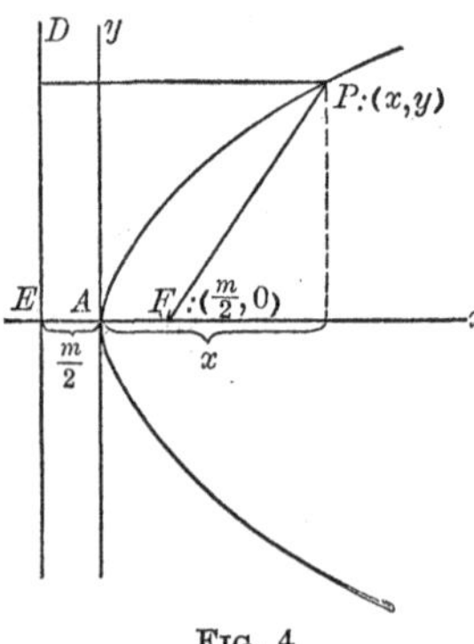

Fig. 4

For the axis of y three simple choices present themselves, namely:

(*a*) through A;
(*b*) along D;
(*c*) through F.

Perhaps (*b*) seems most natural; but (*a*) has the advantage that the curve then passes through the origin, and this choice turns out in practice to be the most useful one. We will begin with it.

Let $P:(x, y)$ be any point on the curve. Denote the distance of F from D by m. Then

$$EF = m, \qquad AF = \frac{m}{2}, \qquad \text{and} \qquad EA = \frac{m}{2}.$$

By Ch. I, § 3,

$$FP = \sqrt{\left(x - \frac{m}{2}\right)^2 + y^2}.$$

On the other hand, the distance of P from D is

$$x + \frac{m}{2}.$$

By definition, these two distances are equal, or:

$$(1) \qquad \sqrt{\left(x - \frac{m}{2}\right)^2 + y^2} = x + \frac{m}{2}.$$

Square each side of the equation, so as to remove the radical, and expand the binomials:

$$(2) \qquad x^2 - mx + \frac{m^2}{4} + y^2 = x^2 + mx + \frac{m^2}{4}.$$

The result can be reduced at once to the form

$$(3) \qquad y^2 = 2mx,$$

and this is the equation of the parabola, referred to its vertex as origin and to its axis as the axis of x.

The proof of this last statement is not yet, however, complete; for it remains to show conversely that, if (x, y) be any point whose coördinates satisfy (3), it is a point of the parabola. From (3) we can pass to (2). On extracting the square root of each side of (2), we have two equations:

$$\text{i)} \qquad \sqrt{\left(x - \frac{m}{2}\right)^2 + y^2} = x + \frac{m}{2},$$

$$\text{ii)} \qquad \sqrt{\left(x - \frac{m}{2}\right)^2 + y^2} = -\left(x + \frac{m}{2}\right),$$

one of which *must* be true, and both of which *may conceivably* be true. Now, x is a positive quantity or zero; for, by hypothesis, the coördinates of the point (x, y) satisfy equation (3). Hence ii) is impossible, for it says that a positive or zero quantity is equal to a negative quantity. Thus only i) remains, and this equation is precisely the condition that the distance of (x, y) from D be equal to its distance from F. Hence the point (x, y) lies on the parabola, q. e. d.

EXERCISES

1. Show that the choice (*b*) leads to the equation

$$y^2 = 2mx - m^2. \tag{4}$$

This is the equation of the parabola referred to its directrix and axis as the axes of y and x respectively, with the positive axis of x in the direction in which the curve opens.

2. Show that the choice (*c*) leads to the equation

$$y^2 = 2mx + m^2. \tag{5}$$

This is the equation of the parabola when the focus is the origin and the positive axis of x is along the axis of the curve in the direction in which the curve opens.

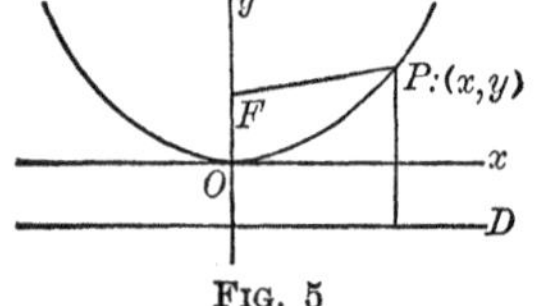

Fig. 5

3. Taking the axes as indicated in Fig. 5, show that the equation of the parabola is

$$x^2 = 2my.$$

4. Choosing the axis of y as in the foregoing question, show that the equation of the parabola is

$$x^2 = 2my - m^2,$$

in case the axis of x is along D, and is

$$x^2 = 2my + m^2,$$

in case F is taken as the origin.

5. If the axis of x is taken along the axis of the parabola, but positively in the direction from F toward D, and if the origin is taken at the vertex, show that the equation of the curve is

$$y^2 = -2mx.$$

6. If the axis of y is taken along the axis of the parabola, but positively in the direction from F toward D, and if the origin is taken at the vertex, show that the equation of the curve is

$$x^2 = -2my.$$

7. Determine the focus and directrix of each of the following parabolas:

(a) $y^2 = 4x$. *Ans.* $(1, 0)$; $x + 1 = 0$.

(b) $y = x^2$. *Ans.* $(0, \frac{1}{4})$; $4y + 1 = 0$.

(c) $3y^2 - 5x = 0$. (d) $3y^2 + 22x = 0$.

(e) $y = -2x^2$. (f) $5x^2 + 12y = 0$

(g) $y^2 = px$. (h) $x^2 = 4ay$.

8. It appears from the foregoing that any equation of the form

$$y^2 = \pm Ax, \qquad \text{or} \qquad x^2 = \pm Ay,$$

where A is any positive constant, represents a parabola with its vertex at the origin. Formulate a general rule for ascertaining the distance of the focus of such a parabola from the vertex.

9. Find the equations of the following parabolas:

(a) Vertex at $(0, 0)$ and focus at $(2, 0)$.
(b) Vertex at $(0, 0)$ and $2x + 5 = 0$ as directrix.
(c) Vertex at $(0, 0)$ and focus at $(0, -\frac{3}{5})$.
(d) Vertex at $(0, 0)$ and $2y - 1 = 0$ as directrix.
(e) Focus at $(0, 0)$ and vertex at $(-3, 0)$.
(f) Focus at $(0, 0)$ and $3y + 4 = 0$ as directrix.
(g) Focus at $(6, 0)$ and axis of y as directrix.
(h) Focus at $(0, -7)$ and axis of x as directrix.

3. Tangents. The student will next turn to Chapter IX and study §§ 1, 2. It is there shown that the slope of the parabola

(1) $$y^2 = 2mx$$

at any one of its points (x_1, y_1) is, in general, given by the formula

$$\lambda = \frac{m}{y_1}; \tag{2}$$

and that the equation of the tangent line at any point (x_1, y_1) can, without exception, be written in the form

$$y_1 y = m(x + x_1). \tag{3}$$

Latus Rectum. The chord, PP', of a parabola which passes through the focus and is perpendicular to the axis is called the *latus rectum* (plural, *latera recta*).

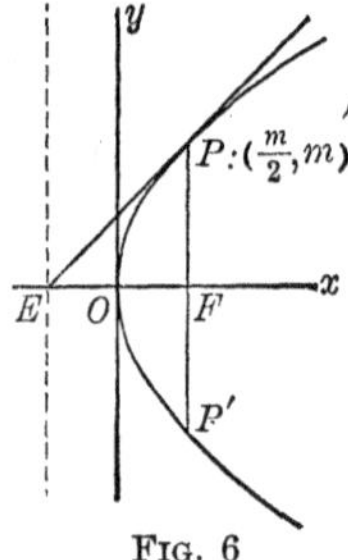

Fig. 6

Its half-length is found by setting $x = m/2$ in the equation of the parabola, and solving for the positive y:

$$y^2 = 2m\left(\frac{m}{2}\right) = m^2, \qquad y = m.$$

Thus the length, PP', of the latus rectum is $2m$.

The tangent at either P or P' makes an angle of 45° with the axis of x. For, the slope of the tangent at P is, from (2):

$$\frac{m}{y_1} = \frac{m}{m} = 1.$$

Let E be the point in which the tangent at P meets the axis of x. Since $FP = m$, and $\angle FEP = 45°$, $EF = m$ and so E lies on the directrix. Consequently, the tangents at P and P' cut the axis of x at the point of intersection of the directrix with that axis.

This theorem can also be proved by writing down the equation of the tangent at P,

$$y = x + \frac{m}{2},$$

and finding the intercept of this line on the axis of x.

EXERCISES

1. Find the equation of the tangent to the parabola $y^2 = 3x$ at the point (12, 6). *Ans.* $x - 4y + 12 = 0$.

2. Find the equation of the normal to the same parabola at the given point. *Ans.* $4x + y = 54$.

3. Find the length of the latus rectum of the parabola of Ex. 1.

4. Show that the tangents to any parabola at the extremities of the latus rectum are perpendicular to each other.

5. Show that the tangent to the parabola $y^2 = 4x$ at the point (36, 12) cuts the negative axis of x at a point whose distance from the origin is 36.

6. At what point of the parabola of Ex. 5 is the tangent perpendicular to the tangent mentioned in that exercise?

Ans. $(\frac{1}{36}, -\frac{1}{3})$.

7. Show that the two tangents mentioned in Exs. 5 and 6 intersect on the directrix, and that the *chord of contact* of these tangents, *i.e.* the right line drawn through the two points of tangency, passes through the focus.

8. Show that the tangent to the parabola (1) at any point P cuts the negative axis of x at a point M whose distance from the origin is the same as the distance of P from the axis of y.

9. Prove that the two parabolas,

$$y^2 = 4x + 4 \quad \text{and} \quad y^2 = -6x + 9,$$

intersect at right angles. Assume that the slope of the parabola of Ex. 2, § 2, at the point (x_1, y_1) is m/y_1.

10. If two parabolas have a common focus and their axes lie along the same straight line, their vertices, however, being on opposite sides of the focus, show that the curves cut each other at right angles.

4. Optical Property of the Parabola. If a polished reflector, like the reflector of the headlight of a locomotive or a searchlight, be made in the form of a paraboloid of revolution, *i.e.* the surface generated by a parabola which is revolved about its axis, and if a source of light be placed at the focus, the reflected rays will all be parallel.

This phenomenon is due to the fact that the focal radius FP drawn to any point P of the parabola makes the same angle with the tangent at P as does the line through P parallel to the axis.

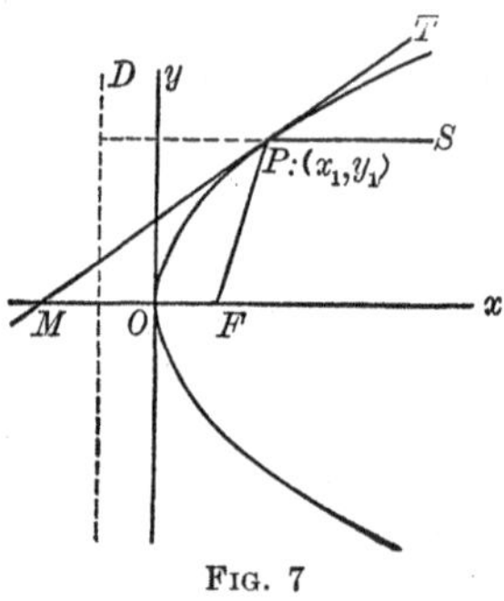

Fig. 7

The proof of this property can be given as follows. Let the tangent at $P:(x_1, y_1)$ cut the axis of x in M. Then the length of OM is equal to x_1, by § 3, Ex. 8. Furthermore, $OF = m/2$. Hence the distance from M to F is

$$MF = x_1 + \frac{m}{2}.$$

But this is precisely the distance of P from D, § 2, and hence, by the definition of the parabola, it is also equal to FP. We have, then, that $MF = FP$. Consequently, the triangle MFP is isosceles, and

$$\measuredangle FMP = \measuredangle MPF.$$

But
$$\measuredangle FMP = \measuredangle SPT,$$

and the proposition is proved.

The result can be restated in the following

Theorem. *The focal radius FP of a parabola at any point P of the curve and the parallel to the axis at P make equal angles with the tangent at P.*

Heat. If such a parabolic reflector as the one described above were turned toward the sun, the latter's rays, being practically parallel to each other and to the axis of the reflector, would, after impinging on the polished surface, proceed along lines, all of which would pass through F. Thus, in particular, the heat rays would be collected at F, and if a minute charge of gunpowder were placed at F, it might easily be fired.

It is to this property that the *focus* (German, *Brennpunkt*) owes its name. The Latin word means *hearth*, or *fireplace*. The term was introduced into the science by the astronomer Kepler in 1604.

EXERCISES ON CHAPTER VI

1. A parabola opens out along the positive axis of y as axis. Its focus is in the point $(0, 3)$ and the length of its latus rectum is 12. Find its equation. *Ans.* $x^2 = 12y$.

2. A parabola has its vertex in the origin and its axis along the axis of x. If it goes through the point $(2, -3)$, what is its equation? *Ans.* $2y^2 - 9x = 0$.

3. Show that the equation of a parabola with the line $x = c$ as directrix and with the point $(c + m, 0)$ or $(c - m, 0)$ as focus is

$$y^2 = 2m(x - c) - m^2, \qquad \text{or} \qquad y^2 = -2m(x - c) - m^2.$$

Hence prove that every parabola with the axis of x as axis has an equation of the form: $x = ay^2 + b$, where a and b are constants, $a \neq 0$.

4. Find the equation of the parabola which has its axis along the axis of x and goes through the two points $(3, 2)$, $(-2, -1)$. *Ans.* $3x = 5y^2 - 11$.

5. Prove that every parabola with an axis parallel to the axis of y has an equation of the form

$$y = ax^2 + bx + c,$$

where a, b, c are constants, $a \neq 0$.

Suggestion. Find the equation of the parabola which has the line $y = k$ as directrix and the point $(l, k + m)$ or $(l, k - m)$ as focus.

6. Find the equation of the parabola which has a vertical axis and goes through the points $(0, 0)$, $(1, 0)$, and $(3, 6)$.
Ans. $y = x^2 - x$.

7. A circle is tangent to the parabola $y^2 = x$ at the point (4, 2) and goes through the vertex of the parabola. Find its equation.

8. What is the equation of the circle which is tangent to the parabola $y^2 = 2mx$ at both extremities of the latus rectum?

Ans. $4x^2 + 4y^2 - 12mx + m^2 = 0.$

9. Find the coördinates of the points of tangency of the tangents to the parabola $y^2 = 2mx$ which make the angles 60°, 45°, and 30° with the axis of the parabola. Show that the abscissæ of the three points are in geometric progression, and that this is true also of the ordinates.

10. Show that the common chord of a parabola, and the circle whose center is in the vertex of the parabola and whose radius is equal to three halves the distance from the vertex to the focus, bisects the line-segment joining the vertex with the focus.

11. Let N be the point in which the normal to a parabola at a point P, not the vertex, meets the axis. Prove that the projection on the axis of the line-segment PN is equal to one half the length of the latus rectum.

12. On a parabola, P is any point other than the vertex, and N is the point in which the normal at P meets the axis. Show that P and N are equally distant from the focus.

13. The tangent to a parabola at a point P, not the vertex, meets the directrix in the point L. Prove that the segment LP subtends a right angle at the focus.

14. Show that the length of a focal chord of the parabola $y^2 = 2mx$ is equal to $x_1 + x_2 + m$, where x_1, x_2 are the abscissæ of the end-points of the chord. Hence show that the mid-point of a focal chord is at the same distance from the directrix as it is from the end-points of the chord.

Exercises 15–26. The following exercises express properties of the parabola which involve an arbitrary point on the parabola. In order to prove these properties, it will, in general, be necessary to make actual use of the equation which expresses analytically the fact that the point lies on the parabola.

15. An arbitrary point P of a parabola, not the vertex, is joined with the vertex A, and a second line is drawn through P, perpendicular to AP, meeting the axis in Q. Prove that the projection on the axis of PQ is equal to the length of the latus rectum.

16. The tangent to a parabola at a point P, not the vertex, meets the tangent at the vertex in the point K. Show that the line joining K to the focus is perpendicular to the tangent at P.

17. The tangent to a parabola at a point P, not the vertex, meets the directrix and the latus rectum produced in points which are equally distant from the focus. Prove this theorem.

18. Prove that the coördinates of the point of intersection of the tangents to the parabola $y^2 = 2mx$ at the points (x_1, y_1), (x_2, y_2) may be put in the form

$$\left(\frac{y_1 y_2}{2m}, \quad \frac{y_1 + y_2}{2}\right).$$

Suggestion. To reduce the coördinates to the desired form, use the equations which express analytically the fact that the two points lie on the parabola.

19. Show that the intercept on the axis of x of the line joining the points (x_1, y_1), (x_2, y_2) of the parabola $y^2 = 2mx$ may be expressed as

$$-\frac{y_1 y_2}{2m}.$$

By means of the results of the two preceding exercises prove the following theorems.

20. The point of intersection of two tangents to a parabola and the point of intersection with the axis of the line joining their points of contact are equally distant from the tangent at the vertex, and are either on it or on opposite sides of it.

21. Tangents to a parabola at the end-points of a focal chord meet at right angles on the directrix.

22. If the points of contact of two tangents to a parabola are on the same side of the axis and at distances from the axis whose product is the square of half the length of the latus rectum, the tangents intersect on the latus rectum produced.

23. The end-points of a chord of a parabola, which subtends a right angle at the vertex, are on opposite sides of the axis and at distances from the axis, whose product is the square of the length of the latus rectum.

24. The chords of a parabola, which subtend a right angle at the vertex, pass through a common point on the axis; this point is at a distance from the vertex equal to the latus rectum.

25. The distance from the focus of a parabola to the point of intersection of two tangents is a mean proportional between the focal radii to the points of tangency.

26. The tangents to a parabola at the points P and Q intersect in T, and the normals at P and Q meet in N. Then the segment TM, where M is the mid-point of TN, subtends a right angle at the focus.

Locus Problems

27. Show that the locus of a point which moves so that the difference of the slopes of the lines joining it to two fixed points is constant is a parabola through the two fixed points. What are its axis and vertex?

28. Determine the locus of a point which moves so that its distance from a fixed circle equals its distance from a fixed line passing through the center of the circle.

Ans. Two equal parabolas, with foci at the center of the circle and axes perpendicular to the fixed line.

29. The base of a triangle is fixed and the sum of the trigonometric tangents of the base angles is constant. Find the locus of the vertex.

CHAPTER VII

THE ELLIPSE

1. Definition. An ellipse is defined as the locus of a point P, the sum of whose distances from two given points, F and F', is constant. It is found convenient to denote this constant by $2a$. Then

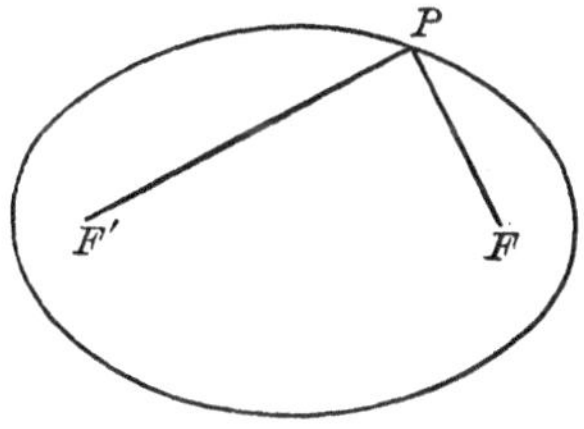

(1) $$FP + F'P = 2a.$$

It is understood, of course, that P always lies in a fixed plane passing through F and F'.

FIG. 1

The points F and F' are called the *foci* of the ellipse. It is clear that $2a$ must be greater than the distance between them.

Mechanical Construction. From the definition of the ellipse a simple mechanical construction readily presents itself. Let a string, of length $2a$, have its ends fastened at F and F', and let the string be kept taut by a pencil point at P. As the pencil moves, its point obviously traces out on the paper the ellipse.

The student will find it convenient to use two thumb tacks partially inserted at F and F'. A silk thread can be tied to one of the thumb tacks and wound round the other so that it will not slip. Thus a variety of ellipses with different foci and different values of a can be drawn.

Let the student make finally *one* ellipse in this manner, and draw it neatly.

Center, Vertices, Axes. It is obvious from the definition,—and the fact becomes more striking from the mechanical construction,—that the ellipse is symmetric in the line through the foci. It is also symmetric in the perpendicular bisector of FF'. Hence it is symmetric, furthermore, in the mid-point, O, of the line FF'.

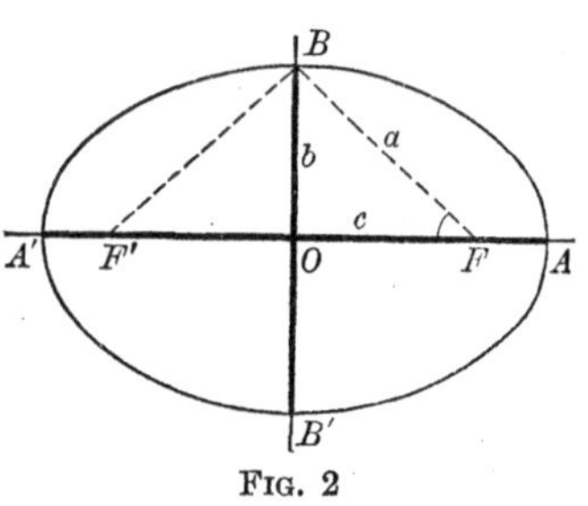

Fig. 2

The indefinite line through the foci, F and F', is called the *transverse axis* of the ellipse; the perpendicular bisector of FF', the *conjugate axis.* The point O is called the *center* of the ellipse; the points A, A', its *vertices.*

The line-segments AA' and BB', which measure the length and breadth of the ellipse, are known respectively as the *major axis* and the *minor axis* of the ellipse. The word "axes" refers sometimes to the transverse and conjugate axes, and sometimes to the major and minor axes, or their lengths, the context making clear in any case the meaning.

When P is at A, equation (1) becomes

$$FA + F'A = 2a.$$

But $$FA = A'F'.$$

Hence $$AA' = 2a \qquad \text{and} \qquad OA = a.$$

Thus it appears that the length of the *semi-axis major*, OA, is a. Let the length of the minor axis be denoted by $2b$, and the distance between the foci by $2c$. Then, from the triangle FOB, we have:

$$a^2 = b^2 + c^2. \tag{2}$$

Note that, of the three quantities a, b, and c, the quantity a is always the largest.

Eccentricity. All circles have the same shape, *i.e.* are similar; and the same is true of parabolas. But it is not true of ellipses. As a measure of the roundness or flatness of an ellipse a number, called the *eccentricity*, has been chosen; this number is defined as the radio c/a and is denoted by e:

$$(3) \qquad e = \frac{c}{a}.$$

Since c is always less than a, it is seen that the eccentricity of an ellipse is always *less* than unity:

$$e < 1.$$

In terms of a and b, e has the value:

$$(4) \qquad e = \frac{\sqrt{a^2 - b^2}}{a}.$$

All ellipses with the same eccentricity are similar, and conversely. For the shape of an ellipse depends only on b/a, the ratio of its breadth to its length, and since from (4)

$$e = \sqrt{1 - \left(\frac{b}{a}\right)^2},$$

all ellipses for which the ratio b/a is the same have the same eccentricity, and conversely.

A circle is the limiting case of an ellipse whose foci approach each other, the length $2a$ remaining constant. The eccentricity approaches 0, and a circle is often spoken of as an ellipse of eccentricity 0.

EXERCISES

1. The semi-axes of an ellipse are of lengths 3 cm. and 5 cm. Find the distance between the foci, and the eccentricity.

Ans. 8; $\frac{4}{5}$.

2. The eccentricity of an ellipse is $\frac{3}{5}$ and the semi-axis minor is 4 in. long. How long is the major axis?

3. The major axis of an ellipse is twice as great as the minor axis. What is the eccentricity of the ellipse?

4. The major axis of an ellipse is 39 yards, and the eccentricity, $\frac{5}{13}$. Find the minor axis.

5. Express the eccentricity of an ellipse in terms of b and c.

6. Show, from Fig. 2, that the eccentricity is given by the formula

$$e = \cos OFB.$$

7. Give a proof, based on similar triangles, that two ellipses having the same eccentricity are similar.

2. Geometrical Construction. Points on the ellipse may be obtained with speed and accuracy by a simple geometrical construction. Draw the major axis and mark the points A, F, F', A' on it. Mark an arbitrary point Q between F and F'. With F as center and AQ as radius describe a circle, and with F' as center and $A'Q$ as radius describe a second circle. The points of intersection of these two circles will lie on the ellipse, since the sum of the radii is

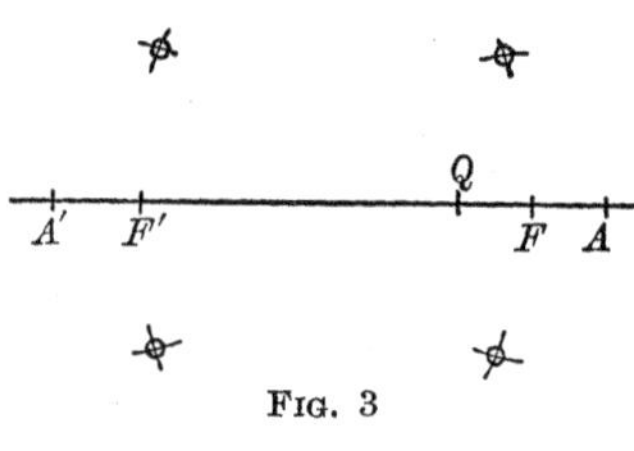

FIG. 3

$$AQ + A'Q = 2a.$$

It is, of course, not necessary to draw the complete circles, but only so much of them as to determine their points of intersection. Moreover, *four* points, instead of *two*, can be obtained from each pair of settings of the compasses by simply reversing the rôles of F and F'.

EXERCISES

1. Construct the ellipse for which $c = 2\frac{1}{2}$ cm., $a = 4$ cm.

2. From the ellipse just constructed make a templet, with holes at the foci and with the axes properly drawn.

3. Construct the ellipse whose axes are 4 cm. and 6 cm.

3. Equation of the Ellipse. It is natural to choose the axes of the ellipse as the coördinate axes (Fig. 4). Let the foci lie on the axis of x, and let P: (x, y) be any point of the ellipse. Then, from (1), § 1,

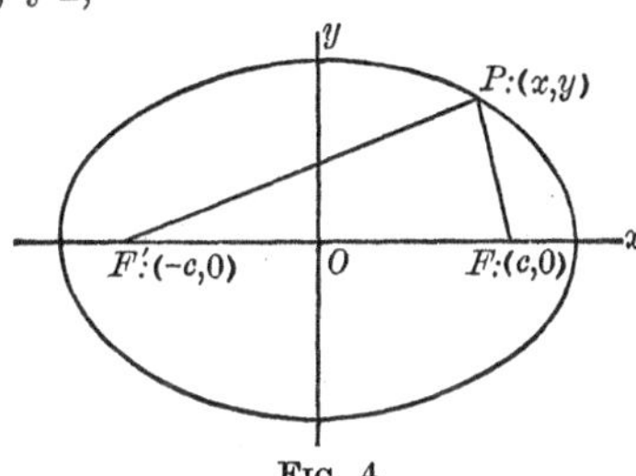

FIG. 4

(1) $$\sqrt{(x-c)^2+y^2}+\sqrt{(x+c)^2+y^2}=2a.$$

Transpose one of the radicals and square:

$$(x-c)^2+y^2=(x+c)^2+y^2-4a\sqrt{(x+c)^2+y^2}+4a^2.$$

Hence

(2) $$a\sqrt{(x+c)^2+y^2}=a^2+cx.$$

To remove this radical, square again:

(3) $$a^2x^2+2a^2cx+a^2c^2+a^2y^2=a^4+2a^2cx+c^2x^2,$$

or $$(a^2-c^2)x^2+a^2y^2=a^2(a^2-c^2).$$

But, by (2), § 1, $$a^2-c^2=b^2,$$

and hence

(4) $$b^2x^2+a^2y^2=a^2b^2,$$

or

(5) $$\frac{x^2}{a^2}+\frac{y^2}{b^2}=1.$$

This is the standard form of the equation of the ellipse, referred to its axes as the axes of coördinates. The proof, however, is not as yet complete, for it remains to show, conversely, that any point (x, y) whose coördinates satisfy equation (5) is a point of the ellipse. To do this, it is sufficient to show that x, y satisfy (1). From (5) we mount up to (4) and thence to (3), since all of these are equivalent equations. When, however, we extract a square root we obtain *two* equations each time, and so we are led, finally, to the *four* equations

$$\pm\sqrt{(x-c)^2+y^2}\pm\sqrt{(x+c)^2+y^2}=2a,$$

the ambiguous signs being chosen in all possible ways. The four equations can be characterized as follows:

i)	$+$	$+$;	ii)	$+$	$-$;
iii)	$-$	$+$;	iv)	$-$	$-$.

We wish to show that i) is the only possible one of the four equations. This is done as follows.

Equation iv) is satisfied by no pair of values for x and y, since the left-hand side is always negative and so can never be equal to the positive quantity $2a$.

Equations ii) and iii) say that the difference of the distances of (x, y) from F and F' is equal to $2a$, and hence greater than the line $FF'=2c$. Thus, in the triangle FPF' the difference of two sides is greater than the third side, and this is absurd.* Hence equations ii) and iii) are impossible and equation i) alone remains, q. e. d.

Consequently, if we start with equation (5) as given and require that $a > b$, then (5) represents an ellipse with semi-axes a and b and foci in the points $(\pm c, 0)$, where $c=\sqrt{a^2-b^2}$.

The Focal Radii. From equation (2) we obtain a simple expression for the length of the *focal radius*, $F'P$. Dividing (2) by a and remembering that $c/a=e$, we have:

$$\sqrt{(x+c)^2+y^2}=a+ex.$$

But the value of the left-hand side of this equation is precisely $F'P$. Hence

$$F'P=a+ex. \tag{6}$$

* If, in particular, the point (x, y) lay on FF', we should not, it is true, have a triangle. But it is at once obvious that in this case, too, equations ii) and iii) are impossible.

If, in transforming (1), the other radical had been transposed to the right-hand side and we had then proceeded as before, we should have found the equation:

$$a\sqrt{(x-c)^2+y^2}=a^2-cx.$$

From this we infer that

$$\sqrt{(x-c)^2+y^2}=a-ex,$$

or

$$FP=a-ex. \tag{7}$$

EXERCISES

1. What is the equation of the ellipse whose axes are of lengths 6 cm. and 10 cm.? *Ans.* $\frac{x^2}{25}+\frac{y^2}{9}=1.$

2. Find the coördinates of the foci of the ellipse of Ex. 1.

3. The foci of an ellipse are at the points (1, 0) and (– 1, 0), and the minor axis is of length 2. Find the equation of the ellipse. *Ans.* $x^2+2y^2=2.$

4. Find the lengths of the axes, the coördinates of the foci, and the eccentricity of the ellipse

$$25x^2+169y^2=4225.$$

5. An ellipse, whose axes are of lengths 8 and 10, has its center at the origin and its foci on the axis of y. Obtain its equation.

6. Show that, if $B>A$, the equation

$$\frac{x^2}{A^2}+\frac{y^2}{B^2}=1$$

still represents an ellipse with its axes lying along the axes of coördinates; but the foci lie on the axis of y at the points $(0, C)$ and $(0, -C)$, where

$$B^2=A^2+C^2.$$

The eccentricity is

$$e=\frac{C}{B}.$$

7. Find the lengths of the axes, the coördinates of the foci, and the value of the eccentricity for each of the following ellipses:

(*a*) $9x^2 + 4y^2 = 36$;　　(*d*) $5x^2 + 3y^2 = 45$;

(*b*) $3x^2 + 2y^2 = 12$;　　(*e*) $2x^2 + 7y^2 = 10$;

(*c*) $x^2 + 2y^2 = 4$;　　(*f*) $11x^2 + y^2 = 3$.

4. Tangents. The ellipse has the remarkable property that the *tangent to the curve at any point makes equal angles with the focal radii drawn to that point:*

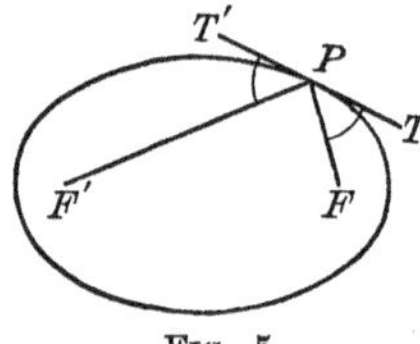

FIG. 5

$$\measuredangle FPT = \measuredangle F'PT'.$$

i) *Mechanical Proof.* The simplest proof of this theorem is a mechanical one. Think of a flexible, inelastic string of length $2a$ with its ends fastened at the foci, F and F'. Suppose a small, smooth bead to be threaded on this string. Let a cord be fastened to the bead and then pulled taut, so that the cord and the two portions of the string will be under tension. Evidently, the bead can be held in this manner at any point. (No force of gravity is supposed to act. The strings and bead may be thought of as resting on a smooth horizontal table.)

The forces that act on the bead are:

(*a*) the tension S in the cord;

(*b*) two equal tensions, R, in the string, directed respectively toward the foci.*

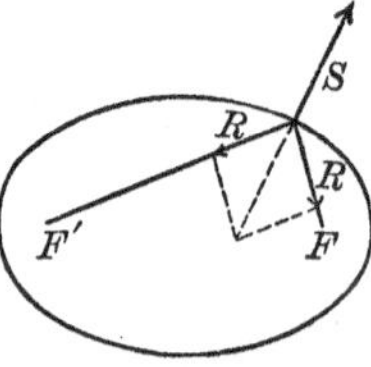

FIG. 6

Draw the parallelogram of forces for the forces R. It will be a rhombus, and so the resultant of these forces will bisect the angle between the focal radii.

On the other hand, the force S, equal and opposite to this resultant, is perpendicular to the tangent at P. In fact, if

* Since the bead is *smooth*, the tension in the string is the same at all its points, and so, in particular, is the same on the two sides of the bead.

instead of the flexible string we had a *smooth rigid wire,* in the form of the ellipse, for the bead to slide on, the bead would be held at P by the cord exactly as before. But the reaction of a smooth wire is at right angles to its tangent. This is the very conception of a *smooth* wire. For otherwise, if S were oblique, it could be resolved into a normal and a tangential component. But the smooth wire could not yield a reaction, part of which is along the tangent.

It follows, then, that the normal at P bisects the angle between the focal radii, and hence these make equal angles with the tangent at P, q. e. d.

ii) *Proof by Means of Minimum Distances. A Lemma.* A barnyard is bounded on one side by a straight river. The cows, as they come from the pasture, enter the barnyard by a gate at A, go to the river to drink, and then keep on to the door of the barn at B. What point, P, of the river should a cow select, in order to save her steps so far as possible?

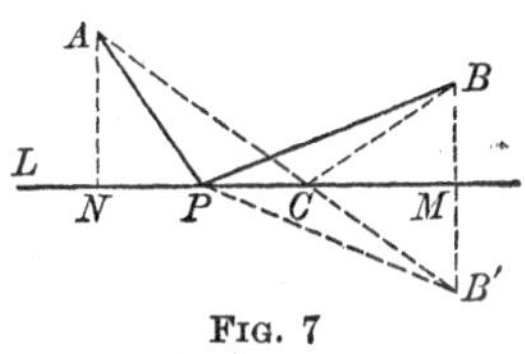

Fig. 7

It is easy to answer this question by means of a simple construction. From B drop a perpendicular BM on the line of the river bank, L, and produce it to B', making $MB' = BM$. Join A with B', and let AB' cut L at C. Then C is the position of P, for which the distance under consideration,

$$AP + PB,$$

is least.

For, the straight line AB' is shorter than any broken line APB':

$$AB' < APB'.$$

But $\qquad PB = PB' \qquad$ and $\qquad CB = CB'.$

Hence

$$AB' = AC + CB \qquad \text{and} \qquad APB' = AP + PB.$$

It follows, then, that

$$AC + CB < AP + PB,$$

if P is any point of L distinct from C. Hence C is the point for which APB is a minimum.

The point C is evidently characterized by the fact that

$$\measuredangle ACN = \measuredangle BCM.$$

We can state the result, then, by saying that *the point P, for which the distance APB is least, is the point for which*

$$\measuredangle APN = \measuredangle BPM.$$

Optical Interpretation. We have used a homely example of cows and a barnyard. The problem we have solved is, however, identical with the optical problem of finding the point at which a ray of light, emanating from A, will strike a plane mirror L, if the reflected ray is to pass through B. For, the law of light is, that it will travel the distance in the shortest possible time, and hence it will choose the shortest path.

Application to the Ellipse. The application of this result to the ellipse is as follows. The tangent to any smooth, closed, convex curve evidently is characterized by the fact that it meets the curve in one, and only one, point.

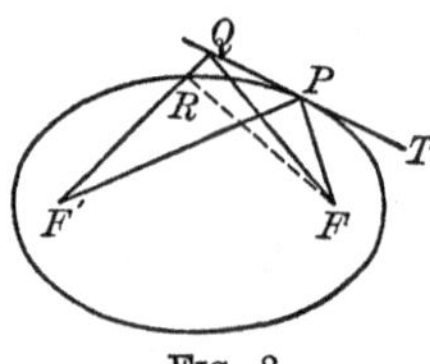

FIG. 8

Let P be any point of the ellipse. Draw the tangent, T, at P. Let Q be any point of T distinct from P. Now

$$F'P + FP = F'R + FR,$$

since the sum of the focal radii is the same for all points of an ellipse. But

$$FR < RQ + FQ$$

and so

$$F'R + FR < F'R + RQ + FQ = F'Q + FQ.$$

Therefore

$$F'P + FP < F'Q + FQ.$$

Hence P is that point of T for which the distance $F'QF$ is least, and consequently the lines $F'P$ and FP make equal angles with T, q.e.d.

EXERCISE

Show that the normal of an ellipse at any point distinct from the vertices A, A' cuts the major axis at a point which lies between the foci.

5. Optical and Acoustical Meaning of the Foci. Let a thin strip of metal, — say, a strip of brass a yard long and a quarter of an inch wide, — be bent into the form of an ellipse and polished on the concave side. Let a light be placed at one of the foci. Then the rays, after impinging on the metal, will be reflected and will come together again at the other focus, which will, therefore, be brilliantly illuminated.*

The same is true of heat, since heat rays are reflected from a polished surface by the same law as that of light rays. If, then, a candle is placed at one focus and some gunpowder at the other, the powder can be ignited by the heat from the candle.

Sound waves behave in a similar manner. The story is told of the Ratskeller in Bremen, the walls of which are shaped somewhat like an ellipse, that the city fathers were remarkably well informed concerning the feelings and views of the populace. For, the former drank their wine at a table which was situated at a focus, and thus could hear distinctly the conversation at a distant table, which stood at the other focus and about which the Bürger congregated.

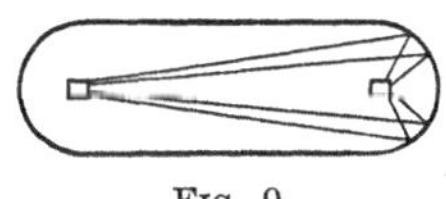

Fig. 9

6. Slope and Equation of the Tangent. The student will next turn to Ch. IX, § 2, where the slope of the ellipse

$$\frac{x^2}{a^2}+\frac{y^2}{b^2}=1$$

* The statement is, of course, strictly true only for such rays as travel in the plane through the foci, which is perpendicular to the elements of the cylinder formed by the polished band. Since, however, only a narrow strip of this cylinder is used, other rays will pass very near to the second focus and contribute to the illumination there.

at the point (x_1, y_1) is found to be

$$(1) \qquad \lambda = -\frac{b^2 x_1}{a^2 y_1}.$$

The equation of the tangent line at this point is

$$(2) \qquad \frac{x_1 x}{a^2} + \frac{y_1 y}{b^2} = 1.$$

Latus Rectum. The latus rectum of an ellipse is defined as a chord perpendicular to the major axis and passing through a focus. The term is also used to mean the *length* of such a chord.

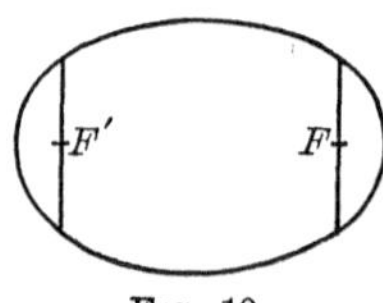

FIG. 10

Thus, in the ellipse

$$\frac{x^2}{25} + \frac{y^2}{16} = 1,$$

one focus is at the point (3, 0). The length of the latus rectum is twice that of the positive ordinate corresponding to this point. Setting, then, $x = 3$ in the equation of the curve and solving for that ordinate, we have

$$\frac{y^2}{16} = 1 - \frac{9}{25} = \frac{16}{25}, \qquad y = \frac{16}{5} = 3\tfrac{1}{5}.$$

Hence the length of the latus rectum is $6\tfrac{2}{5}$.

EXERCISES

1. Find the equation of the tangent to the ellipse

$$\frac{x^2}{225} + \frac{y^2}{25} = 1$$

at the point (9, 4). *Ans.* $x + 4y = 25$.

2. Find the equation of the normal to the ellipse of Ex. 1 at the same point. *Ans.* $4x - y = 32$.

3. At what point does the tangent to the ellipse

$$2x^2 + 3y^2 = 14$$

at the point $(-1, 2)$ cut the axis of y?

4. At what angle does the straight line through the origin, which bisects the angle between the positive axes of coördinates, cut the ellipse $3x^2 + 4y^2 = 7$? *Ans.* $81° 53'$.

5. Find the area of the triangle cut off from the first quadrant by the tangent to the ellipse of Ex. 3 at the point (1, 2). *Ans.* $8\frac{1}{6}$.

6. Find the length of the latus rectum of the ellipse of Ex. 1. *Ans.* $3\frac{1}{3}$.

7. The same for the ellipse of Ex. 3.

8. Show that the length of the latus rectum of the ellipse

$$\frac{x^2}{a^2} + \frac{y^2}{b^2} = 1, \qquad b < a,$$

is given by any one of the expressions

$$\frac{2b^2}{a}; \qquad 2b\sqrt{1-e^2}; \qquad 2a(1-e^2).$$

Find its value in terms of c and e.

9. Find the length of the latus rectum of the ellipse

$$25x^2 + 16y^2 = 400. \qquad \textit{Ans. } 6\tfrac{2}{5}.$$

10. Prove that the minor axis of an ellipse is a mean proportional between the major axis and the latus rectum.

7. A New Locus Problem. Given a line D and a point F distant m from D. To find the locus of a point P such that the ratio of its distance FP from F to its distance MP from D is always equal to a given number, ϵ:

$$(1) \qquad \frac{FP}{MP} = \epsilon, \qquad \text{or} \qquad FP = \epsilon MP.$$

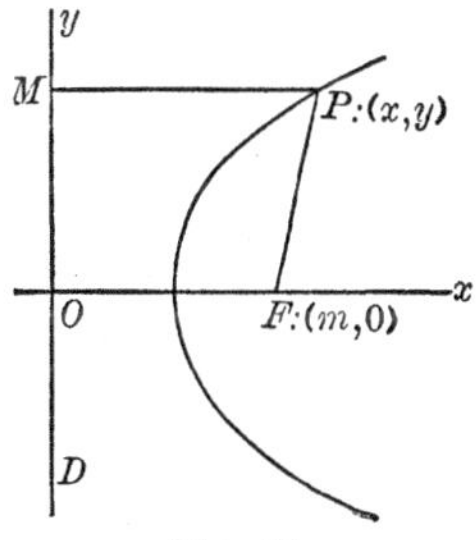

FIG. 11

It is understood that P shall be restricted to the plane determined by F and D.

If, in particular, $\epsilon = 1$, the locus

is a parabola with D as directrix and F as focus; Ch. VI, § 1.

To treat the general case, let D be taken as the axis of y and let the positive axis of x pass through F. Then

$$FP = \sqrt{(x-m)^2 + y^2}, \qquad MP = \pm\, x,$$

the lower sign holding only when x is negative, and (1) becomes

$$(2) \qquad \sqrt{(x-m)^2 + y^2} = \pm\, \epsilon x.$$

On squaring and transposing we obtain the equation:

$$(3) \qquad (1-\epsilon^2)x^2 - 2\,mx + y^2 + m^2 = 0.$$

This is the equation of the proposed locus.

The student will now turn to Ch. XI and study carefully § 1.

EXERCISES

1. Take $\epsilon = \frac{1}{2}$ and $m = 3$, the unit of length being 1 cm. With ruler and compasses construct a generous number of points of the locus,* and then draw in the locus with a clean, firm line.

2. Work out the equation of the locus of Ex. 1 directly, using the method of the foregoing text, but not looking at the formulas. *Ans.* $3\,x^2 + 4\,y^2 - 24\,x + 36 = 0.$

3. Take $\epsilon = \frac{3}{5}$ and $m = 4$, the unit of length being 1 cm. Draw the locus accurately, as in Ex. 1.

4. Work out directly the equation of the locus of Ex. 3.
Ans. $16\,x^2 + 25\,y^2 - 200\,x = -\,400.$

5. By means of a transformation to parallel axes show that the curve of Ex. 2 is an ellipse whose center is at the point (4, 0) and whose axes are of lengths 4 and $2\sqrt{3}$. What is its eccentricity?

* The details of the construction are an obvious modification of the corresponding construction for the parabola in Ch. VI, § 1. A circle of arbitrary radius is drawn with its center at F, and this circle is cut by a parallel to D, whose distance from D is *twice* the radius of the circle.

6. Show that the curve of Ex. 4 is an ellipse whose axes are $7\frac{1}{2}$ and 6. What is its eccentricity?

8. Discussion of the Case $\epsilon < 1$. The Directrices. From equation (3) of § 7 follows:

$$(1)\qquad x^2 - \frac{2m}{1-\epsilon^2}x + \frac{y^2}{1-\epsilon^2} = -\frac{m^2}{1-\epsilon^2}.$$

The first two terms on the left-hand side are also the first two in the expansion of

$$\left(x - \frac{m}{1-\epsilon^2}\right)^2 = x^2 - \frac{2m}{1-\epsilon^2}x + \frac{m^2}{(1-\epsilon^2)^2}.$$

If, then, we add the third term of the last expression to both sides of (1), we shall have:

$$x^2 - \frac{2m}{1-\epsilon^2}x + \frac{m^2}{(1-\epsilon^2)^2} + \frac{y^2}{1-\epsilon^2} = \frac{m^2}{(1-\epsilon^2)^2} - \frac{m^2}{1-\epsilon^2},$$

or

$$(2)\qquad \left(x - \frac{m}{1-\epsilon^2}\right)^2 + \frac{y^2}{1-\epsilon^2} = \frac{\epsilon^2 m^2}{(1-\epsilon^2)^2}.$$

This equation reminds us strongly of the equation of an ellipse. In fact, if we transform to parallel axes with the new origin, O', at the point

$$x_0 = \frac{m}{1-\epsilon^2}, \qquad y_0 = 0,$$

the equations of transformation are

$$(3)\qquad x' = x - \frac{m}{1-\epsilon^2}, \qquad y' = y,$$

and (2) then takes on the form

$$(4)\qquad x'^2 + \frac{y'^2}{1-\epsilon^2} = \frac{\epsilon^2 m^2}{(1-\epsilon^2)^2},$$

or

$$(5)\qquad \frac{x'^2}{a^2} + \frac{y'^2}{b^2} = 1,$$

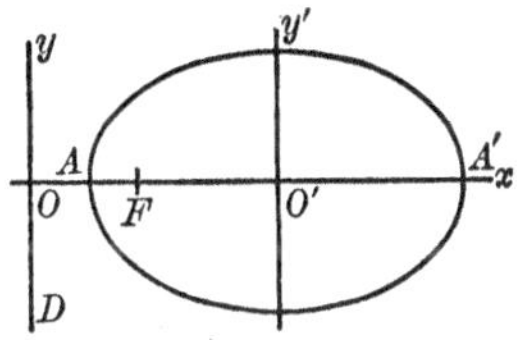

Fig. 12

where

$$(6)\qquad a = \frac{\epsilon m}{1-\epsilon^2}, \qquad b = \frac{\epsilon m}{\sqrt{1-\epsilon^2}}.$$

Thus the locus is seen to be an ellipse with its center, O', at the point

(7) $$\left(\frac{m}{1-\epsilon^2}, 0\right),$$

the semi-axes being given by (6).

The value of c is given by the equation $c^2 = a^2 - b^2$. Hence

(8) $$c = \frac{\epsilon^2 m}{1-\epsilon^2}.$$

The eccentricity, $e = c/a$, is now seen to be precisely ϵ:

$$e = \epsilon;$$

i.e. *the given constant,* ϵ, *turns out to be the eccentricity of the ellipse.*

Finally, F *is one of the foci.* For, the distance from F to O' is

$$OO' - OF = \frac{m}{1-\epsilon^2} - m = \frac{\epsilon^2 m}{1-\epsilon^2},$$

and this, by (8), is precisely c.

The line D is called a *directrix* of the ellipse. Its distance from the center is

$$OO' = \frac{m}{1-\epsilon^2} = \frac{m\epsilon}{1-\epsilon^2}\frac{1}{\epsilon} = \frac{a}{\epsilon}.$$

The Directrices. From the symmetry of the ellipse it is clear that there is a second directrix, D', on the other side of the conjugate axis, parallel to that axis, and at the same distance from it as D. This line D' and the focus F' stand in the same relation to the ellipse as the first line, D, and the focus F. Thus the ellipse is the locus of a point so moving that its distance from a focus always bears to its distance from the corresponding directrix the same ratio, e, the eccentricity.

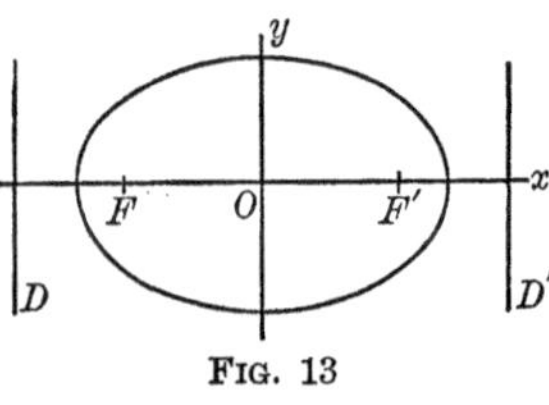

FIG. 13

Since the distance of D from the center of the ellipse is a/e, the equations of the directrices of the ellipse

$$\frac{x^2}{a^2}+\frac{y^2}{b^2}=1, \qquad a>b,$$

are

$$x=-\frac{a}{e}, \qquad x=\frac{a}{e}.$$

EXERCISES

1. Show that the distances of the vertices, A and A', from O are:

$$OA=\frac{m}{1+e}, \qquad OA'=\frac{m}{1-e}.$$

2. Collect the foregoing results in a syllabus, arranged in tabular form, giving each of the quantities a, b, c, OO', OA, OA', OF, OF' in terms of m and e.

3. Work out each of the quantities of Ex. 2 directly for the ellipse of § 7, Ex. 4, and verify the result by substituting the values $e=\frac{3}{5}$, $m=4$ in the formulas of the syllabus.

4. Between the *five* constants of the ellipse, a, b, c, e, m, there exist *three* relations, which may be written in a variety of ways; as, for example,

$$\text{i) } a^2=b^2+c^2; \qquad \text{ii) } e=\frac{c}{a}; \qquad \text{iii) } m=\frac{1-e^2}{e}a.$$

By means of these relations, any three of the five quantities can be expressed in terms of the other two. Thus, in Ex. 2, m and e are chosen as the quantities in terms of which all others shall be expressed.

Taking the semi-axes, a and b, $(a>b)$, as the preferred pair, express the other quantities in terms of them.

5. Show that the tangent to the ellipse

$$\frac{x^2}{25}+\frac{y^2}{16}=1$$

at an extremity of a latus rectum cuts the transverse axis in the same point in which this axis is cut by a directrix.

6. The same for any ellipse.

7. Prove directly that, if P is any point of the ellipse

$$\frac{x^2}{a^2}+\frac{y^2}{b^2}=1, \qquad b<a,$$

the ratio of its distance from a focus to its distance from the corresponding directrix is equal to the eccentricity.

8. Show that in an ellipse the major axis is a mean proportional between the distance between the foci and the distance between the directrices.

9. Show that the distances from the center and a focus of an ellipse to the directrix corresponding to the focus are in the same ratio as the squares of the semi-axis major and the semi-axis minor.

9. The Parabola as the Limit of Ellipses. We have proved that, when $\epsilon<1$, equation (3), § 7, represents an ellipse with eccentricity $e=\epsilon$. We know that, if $\epsilon=1$, the equation represents a parabola. If, then, in the equation we allow ϵ to approach 1 through values <1, the ellipse which the equation defines approaches a parabola as its limit.

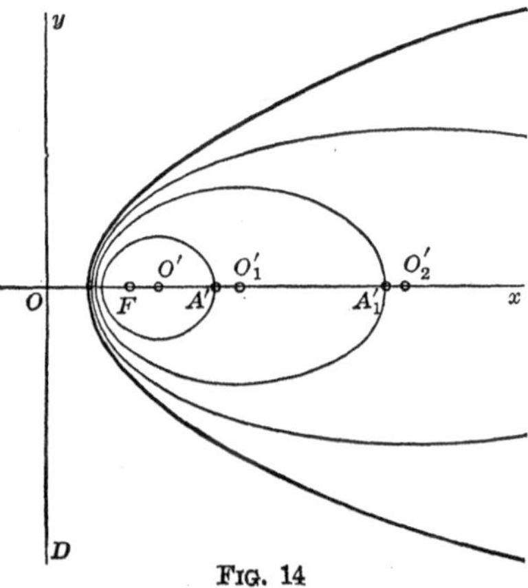

FIG. 14

We can visualize the ellipse, going over into a parabola, by drawing a number of ellipses having the same value of m, but having values for ϵ which are increasing toward 1 as their limit, viz. $\epsilon=\frac{1}{2}$, $\epsilon=\frac{3}{4}$, $\epsilon=\frac{7}{8}, \cdots$. The directrix D, along the axis of y, and the focus $F:(m,\ 0)$ are the same for all the ellipses. But the center O' and the right-hand vertex

A' of each successive ellipse are farther away from O, and their distances from O, namely,

$$OO' = \frac{m}{1-\epsilon^2}, \qquad OA' = \frac{m}{1-\epsilon},$$

increase without limit. Thus, as ϵ approaches 1, the ellipse approaches as its limit the parabola whose directrix is D and whose focus is F.

10. New Geometrical Construction for the Ellipse. Parametric Representation. Let it be required to draw an ellipse when its axes, AA' and BB', are given. Describe circles of radii $a = OA$ and $b = OB$, with the origin O as the common center. Draw any ray from O, making an angle ϕ with the positive axis of x, as shown in the figure. Through the points Q and R draw the parallels indicated. Their point of intersection, P, will lie on the ellipse. For, if the coördinates of P be denoted by (x, y), it is clear that

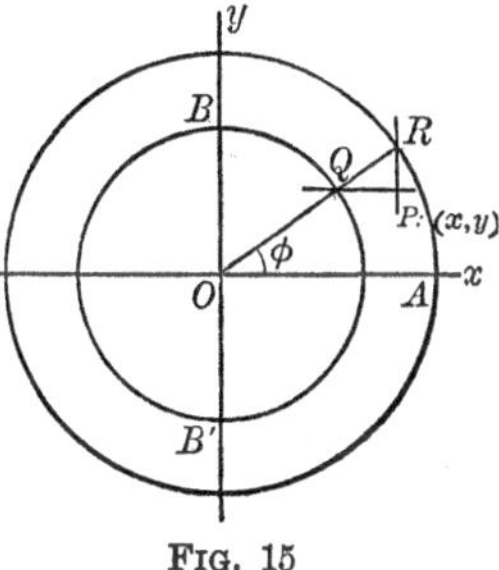

FIG. 15

(1) $$x = a\cos\phi, \qquad y = b\sin\phi.$$

From these equations ϕ can be eliminated by means of the trigonometric identity

$$\sin^2\phi + \cos^2\phi = 1.$$

Hence

(2) $$\frac{x^2}{a^2} + \frac{y^2}{b^2} = 1.$$

Conversely, any point (x, y) on the ellipse (2) has corresponding to it an angle ϕ, for which equations (1) are true.

Equations (1) afford what is known as a *parametric representation* of the coördinates of a variable point (x, y) of the ellipse in terms of the *parameter* ϕ. When $b = a$, the ellipse becomes a circle, and the equations (1) become

(3) $$x = a\cos\phi, \qquad y = a\sin\phi.$$

These parametric representations, though little used in Analytic Geometry, are an important aid in the Calculus.

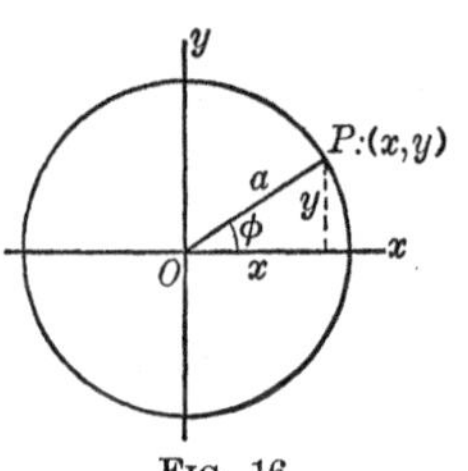

Fig. 16

The larger of the two circles in Fig. 15 is commonly called the *auxiliary circle* of the ellipse, and the points R and P are known as *corresponding points.* The angle ϕ is called the *eccentric* angle.

EXERCISE

By means of the foregoing method, draw on squared paper an ellipse whose axes are of length 4 cm. and 6 cm.

EXERCISES ON CHAPTER VII

1. The earth moves about the sun in an elliptic orbit.* The shortest and longest distances from it to the sun are in the ratio 29 : 30. What is the eccentricity of the orbit?

2. Show that the slopes of the tangents to an ellipse at the extremities of the latera recta are $\pm e$.

3. The axes of an ellipse which goes through the points (4, 1), (2, 2) are the axes of coördinates. Find its equation.

4. The center of an ellipse is in the origin and the foci are on the axis of x. The ellipse has an eccentricity of $\frac{3}{5}$ and goes through the point (12, 4). What is its equation?

$$\textit{Ans.}\quad \frac{x^2}{25}+\frac{y^2}{16}=\frac{169}{25}.$$

5. Solve the preceding problem if the foci may lie on either axis of coördinates.

6. Find the equations of the ellipses which have the axes of coördinates as axes, go through the point (3, 4), and have their major and minor axes in the ratio 3 : 2.

7. Show that the ellipses represented by the equation

$$2x^2+3y^2=c^2,$$

* The planets describe ellipses about the sun as a focus, and the comets usually describe parabolas with the sun as the focus.

where c^2 is an arbitrary positive constant, are similar. What is the common value of the eccentricity?

8. How many ellipses are there with eccentricity $\frac{1}{2}$, having their centers in the origin and their foci on the axis of x? Deduce an equation which represents them all.

Ans. $3x^2 + 4y^2 = c^2$.

9. The foci of an ellipse lie midway between the center and the vertices. What is the eccentricity? How many such ellipses are there, with centers in the origin and foci on the axis of x? Write an equation which represents them all.

10. The line joining the left-hand vertex of an ellipse with the upper extremity of the minor axis is parallel to the line joining the center with the upper extremity of the right-hand latus rectum. Answer the questions of the preceding exercise.

11. The foci of an ellipse subtend a right angle at either extremity of the minor axis. What is the eccentricity? Find the equation of all such ellipses with centers in the origin and foci on the axis of y.

12. Prove that the ratio of the distance from a focus of an ellipse to the intersection with the transverse axis of the normal at a point P, and the distance from this focus to P equals the eccentricity of the ellipse.

13. The projections of a point P of an ellipse on the transverse and conjugate axes are P_1 and P_2. The tangent at P meets these axes in T_1 and T_2. Prove that $OP_1 \cdot OT_1 = a^2$ and $OP_2 \cdot OT_2 = b^2$, where O is the center and a and b are the semi-axes of the ellipse.

14. Prove that the segment of a tangent to an ellipse between the point of contact and a directrix subtends a right angle at the corresponding focus.

15. Determine the points of an ellipse at which the tangents have intercepts on the axes whose absolute values are proportional to the lengths of the axes.

16. Through a point M of the major axis of an ellipse a line is drawn parallel to the conjugate axis, meeting the ellipse in P and the tangent at an extremity of the latus rectum in Q. Show that the distance MQ equals the distance of P from the focus corresponding to the latus rectum taken.

17. Prove that the line joining a point P of an ellipse with the center and the line through a focus perpendicular to the tangent at P meet on a directrix.

18. Prove that the distance from a focus F to a point P of an ellipse equals the distance from F to the tangent to the auxiliary circle at the point corresponding to P.

19. Find the equation of a circle which is tangent to the ellipse

$$\frac{x^2}{a^2}+\frac{y^2}{b^2}=1$$

at both ends of a latus rectum.

20. In an ellipse whose major axis is twice the minor axis, a line of length equal to the minor axis has one end on the ellipse, the other on the conjugate axis. The two ends are always on opposite sides of the transverse axis. Prove that the mid-point of the line lies always on the transverse axis.

21. A number of ellipses have the same major axis both in length and position. A tangent is drawn to each ellipse at the upper extremity of the right-hand latus rectum. Prove that these tangents all pass through a point.

Exercises 22–28. In these exercises, in which properties involving an arbitrary point P of an ellipse are to be proved, it will, in general, be necessary to make actual use of the equation expressing the fact that the point P lies on the ellipse.

22. The tangent to an ellipse at a point P meets the tangent at one vertex in Q. Prove that the line joining the other vertex to P is parallel to the line joining the center to Q.

23. The lines joining the extremities of the minor axis with a point P of an ellipse meet the transverse axis in the points

M and N. Prove that the semi-axis major is a mean proportional between the distances from the center to M and N.

24. Prove the theorem of the preceding exercise when the major and minor axes, and the transverse and conjugate axes, are interchanged.

25. Show that the segment of a directrix, between the points of intersection of the lines joining the vertices with a point on an ellipse, subtends a right angle at the corresponding focus.

26. Prove that the product of the distances of the foci of an ellipse from a tangent is a constant, independent of the choice of the tangent.

27. Let F' and F be the foci of an ellipse and P any point on it. Prove that $b^2 : FK^2 = F'P : FP$, where FK is the distance from F to the tangent at P.

28. The normal to an ellipse at a point P meets the axes in N_1 and N_2. Show that $PN_1 \cdot PN_2$ is equal to the product of the focal radii to P.

Loci

29. A point moves so that the product of the slopes of the two lines joining it to two fixed points is a negative constant. What is its locus?

30. A circle whose diameter is 10 cm. is drawn, center at O. On a radius OA a point B is marked distant 4 cm. from O. If OQ is any second radius, show how to construct, with ruler and compasses, a point P on OQ, whose distance from the circle equals its distance from B. In this way plot a number of points on the locus of P.

31. Find the equation of the locus of the point P of the preceding exercise. Take the origin of coördinates at the mid-point of OB.

32. The base of a triangle is fixed and the product of the tangents of the base angles is a positive constant. Find the locus of the vertex.

CHAPTER VIII

THE HYPERBOLA

1. Definition. A *hyperbola* is defined as the locus of a point P, the difference of whose distances from two given points, F and F', is constant. It is found convenient to denote this constant by $2a$. Then

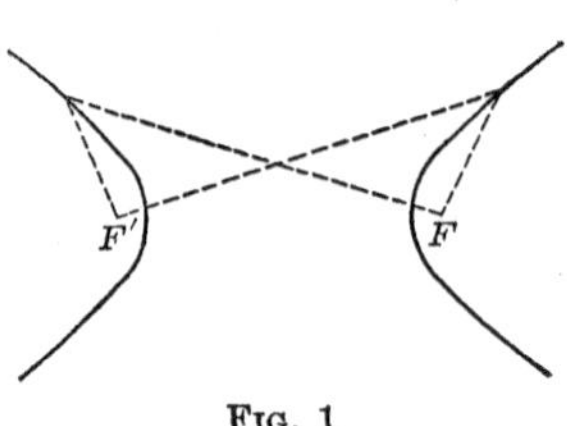

Fig. 1

$$FP - F'P = 2a,$$
or
$$F'P - FP = 2a.$$

It is understood, of course, that P is restricted to a particular plane through F and F'.

The points F and F' are called the *foci* of the hyperbola. It is clear that $2a$ must be *less* than the distance between them. Denote this distance by $2c$.

Geometrical Construction. Draw the indefinite line FF', mark the mid-point, O, of the segment FF', and the points A and A' each at a distance a from O:

$$OA = OA' = a; \qquad OF = OF' = c.$$

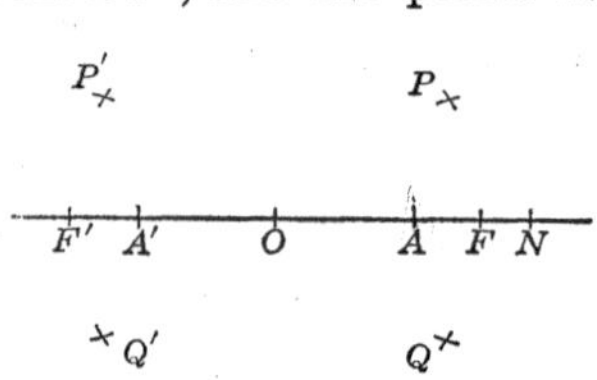

The point A lies on the locus; for,

$$FA = c - a, \qquad F'A = c + a,$$

Fig. 2

and hence
$$F'A - FA = 2a.$$

Likewise, A' lies on the curve.

Mark any point, N, to the right of F. With radius AN and center F, describe a circle. Next, with radius $A'N$ and center

F', describe a second circle. The points P and Q in which these circles intersect are points of the locus. For,

$$F'P - FP = A'N - AN = A'A = 2a.$$

Two more points, P' and Q', can be obtained from the same pair of settings by interchanging the centers, F and F', of the circles.

By repeating the construction a number of times, a goodly array of points of the hyperbola can be obtained. These points will lie on two distinct arcs, symmetric to each other in the perpendicular bisector BOB' of FF'. Thus it will be seen that the hyperbola consists of two parts, or *branches*, as they are called. These branches, besides being the images of each other in BB', are each the image of itself in FF'. It is natural to speak of the indefinite straight lines FF' and BB' as the *axes* of the hyperbola. FF' is called the *transverse*, BB' the *conjugate* axis; O is the *center*, and A, A' are the *vertices*.

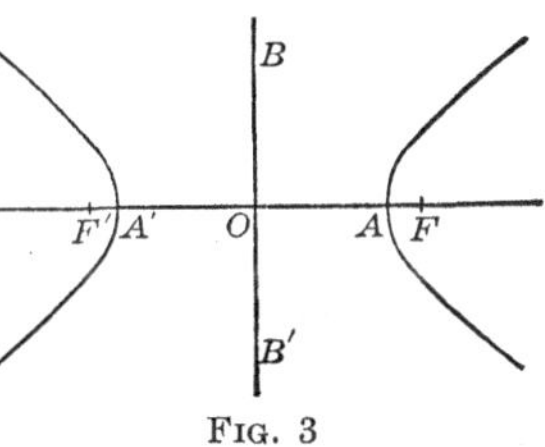

Fig. 3

EXERCISES

1. Taking $c = 3$ cm. and $a = 2$ cm., make a clean drawing of the corresponding hyperbola.

2. Reproduce the drawing on a rectangular card and, with a sharp knife or a small pair of scissors, cut out the center of the card along the hyperbola and two parallels to the transverse axis. On the templet which remains make holes at the foci and draw the two axes.

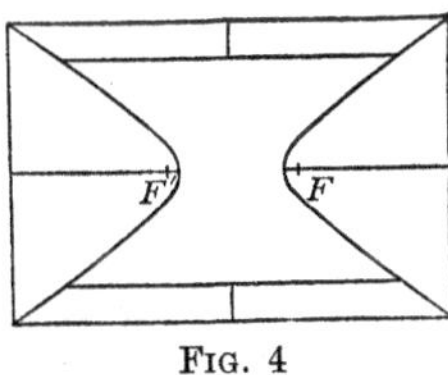

Fig. 4

2. Equation of the Hyperbola. The treatment here is parallel to that of the ellipse, Ch. VII, § 3. Let the transverse axis

be chosen as the axis of x; the conjugate axis, as the axis of y. Then the equation of the right-hand branch of the hyperbola can be written in the form

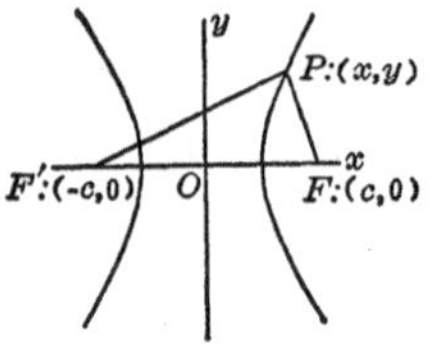

$$(1)\quad \sqrt{(x+c)^2+y^2}-\sqrt{(x-c)^2+y^2}=2a.$$

Fig. 5

Transpose the first radical and square:

$$(x-c)^2+y^2=(x+c)^2+y^2$$
$$-4a\sqrt{(x+c)^2+y^2}+4a^2.$$

Hence

$$(2)\quad a\sqrt{(x+c)^2+y^2}=a^2+cx.$$

Square again:

$$a^2x^2+2a^2cx+a^2c^2+a^2y^2=a^4+2a^2cx+c^2x^2,$$

or

$$(3)\quad (a^2-c^2)x^2+a^2y^2=a^2(a^2-c^2).$$

This is precisely the same equation that presented itself in the case of the ellipse; but the locus is a curve of wholly different nature. The reason is, that a and c have different *relative values*. In the ellipse, a was *greater* than c, and hence a^2-c^2 was positive. It could be denoted by b^2. Here, a is *less* than c; a^2-c^2 is negative, and it cannot be set equal to b^2. It can, however, be set equal to $-b^2$. This we will do:

$$(4)\quad a^2-c^2=-b^2, \qquad \text{or} \qquad c^2=a^2+b^2,$$

thus *defining* the quantity b in the case of the hyperbola by the equation:

$$b=\sqrt{c^2-a^2}.$$

The final equation between x and y can now be written in the form

$$(5)\quad \frac{x^2}{a^2}-\frac{y^2}{b^2}=1.$$

This equation is satisfied by the coördinates of all points on the right-hand branch, as is seen from the way in which it was deduced. It is, however, also satisfied by the coördinates of all points on the left-hand branch. For such a point, the signs of both radicals in (1) will be reversed. Starting, now, with the new equation and proceeding as before, we find the same equation (3), which we may again write in the form (5), and thus the truth of the statement is established.

Is (5) satisfied by the coördinates of still other points? To answer this question, let (x, y) be *any* point whose coördinates satisfy (5). Then, starting from (5), we retrace our steps, admitting, each time that we extract a square root, both signs of the radical as conceivably possible. Thus we can be sure that (x, y) will satisfy *one* of the four equations

$$\pm\sqrt{(x+c)^2+y^2}\pm\sqrt{(x-c)^2+y^2}=2a,$$

corresponding to the four conceivable choices of the signs of the radicals:

i)	$-$	$+$;	iii)	$-$	$-$;
ii)	$+$	$-$;	iv)	$+$	$+$.

If (x, y) satisfies i) or ii), the point lies on the hyperbola. The other two cases are impossible. For, case iii) says that a negative quantity is equal to a positive quantity, and case iv) says that $F'P+FP=2a$. Now $F'P+FP$, being the sum of two sides of the triangle FPF', is *greater* than the third side, FF', or $2c$. But $2a$ is actually *less* than $2c$. Hence we have a contradiction, and this case cannot arise.

We have shown then, finally, that (5) is the equation of the hyperbola.

EXERCISE

Plot the hyperbola

$$\frac{x^2}{25}-\frac{y^2}{16}=1$$

directly from its equation, taking 1 cm. as the unit of length.

3. Axes, Eccentricity, Focal Radii. The *transverse* and the *conjugate axis* have already been defined in § 1. The segment AA' of the transverse axis is called the *major axis*, and this term is also applied to its length, $2a$. The segment BB' of the conjugate axis, whose center is at O and whose length is

$2b$, is called the *minor axis*, and this term is also applied to its length, $2b$.

The major axis of an ellipse is always longer than the minor axis. In the case of the hyperbola, however, this is not always true. For example, if $2c$ and $2a$ are taken as 10 and 6 respectively, then $2b = 8$. Thus the major axis of the hyperbola is to be understood as the principal axis, but not necessarily as the longer axis.

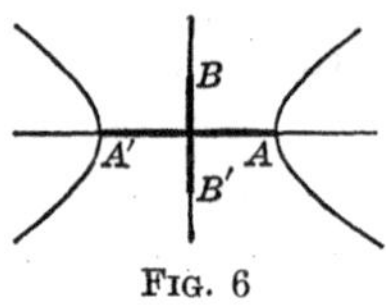

FIG. 6

The *eccentricity* of the hyperbola is defined as the number

$$e = \frac{c}{a}.$$

Since c is greater than a, the eccentricity of a hyperbola is always *greater* than unity.

The eccentricity characterizes the *shape* of the hyperbola. All hyperbolas having the same eccentricity are *similar*, differing only in the scale to which they are drawn, and conversely; cf. Exercise 8.

The *focal radii* FP, $F'P$ can be represented by simple expressions, similar to those which presented themselves in the case of the ellipse. On dividing equation (2), § 2, through by a, we have:

$$\sqrt{(x+c)^2+y^2} = a + ex.$$

Hence, when P is a point of the right-hand branch,

$$(1) \qquad F'P = ex + a.$$

The evaluation,

$$(2) \qquad FP = ex - a,$$

is obtained in a similar manner.*

If P is a point of the left-hand branch, these formulas become:

$$(3) \qquad F'P = -(ex + a); \qquad FP = -(ex - a).$$

* P being a point of the right-hand branch, x is positive and greater than or equal to a; also, $e > 1$. Hence $ex > a$, and $ex - a$ is positive, as it should be.

EXERCISES

1. Find the lengths of the axes, the coördinates of the foci, and the value of the eccentricity for each of the following hyperbolas.

(*a*) $\frac{x^2}{16}-\frac{y^2}{9}=1$. *Ans.* 8, 6; (5, 0), (– 5, 0); $1\frac{1}{4}$.

(*b*) $x^2-y^2=a^2$. *Ans.* $2a$, $2a$; $(a\sqrt{2}, 0)$, $(-a\sqrt{2}, 0)$; $\sqrt{2}$.

(*c*) $4x^2-3y^2=24$.

(*d*) $2x^2-y^2=4$.

(*e*) $5x^2-6y^2=8$.

(*f*) $6x^2-9y^2=4$.

2. If the eccentricity of a hyperbola is 2 and its major axis is 3, what is the length of its minor axis? *Ans.* $3\sqrt{3}$.

3. How far apart are the foci of the hyperbola in Ex. 2? *Ans.* 6.

4. What is the equation of the hyperbola whose eccentricity is $\sqrt{2}$ and whose foci are distant 4 from each other? *Ans.* $x^2-y^2=2$.

5. The extremities of the minor axis of a hyperbola are in the points $(0, \pm 3)$ and the eccentricity is 2. Find the equation of the hyperbola.

6. Show that, in terms of a and b, e has the value

$$e=\frac{\sqrt{a^2+b^2}}{a}.$$

7. Express b in terms of a and e.

8. Prove that two hyperbolas which have the same eccentricity are similar, and conversely.

9. Establish formulas (3).

4. The Asymptotes. Two lines, called the *asymptotes*, stand in a peculiar and important relation to the hyperbola. They are the lines

$$y=\frac{bx}{a} \quad \text{and} \quad y=-\frac{bx}{a}.$$

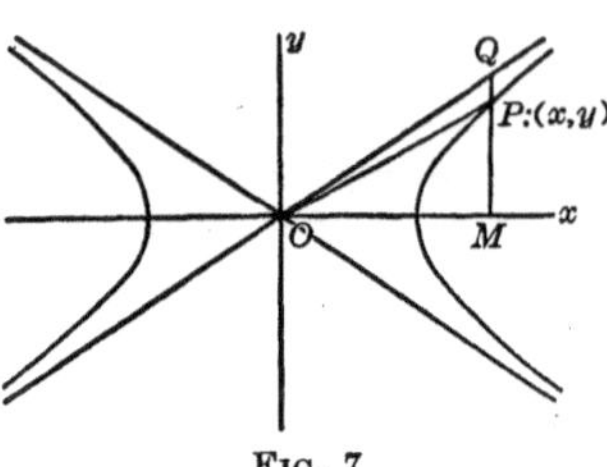

FIG. 7

Let a point $P:(x, y)$ move off along a branch of the hyperbola

$$(1) \qquad \frac{x^2}{a^2}-\frac{y^2}{b^2}=1,$$

and let this take place, for definiteness, in the first quadrant. The slope of the line OP is

$$\frac{MP}{OM}=\frac{y}{x}.$$

Since the coördinates (x, y) of P satisfy (1), it follows that

$$(2) \qquad y=\frac{b}{a}\sqrt{x^2-a^2},$$

and hence

$$(3) \qquad \frac{y}{x}=\frac{b}{a}\sqrt{1-\frac{a^2}{x^2}}.$$

When P recedes indefinitely, x increases without limit, and the right-hand side of this equation approaches the limit b/a. Thus we see that the slope of OP approaches that of the line OQ,

$$(4) \qquad y=\frac{b}{a}x,$$

as its limit, always remaining, however, less than the latter slope, so that P is always below OQ.

It seems likely that P will come indefinitely near to this line; but this fact does not follow from the foregoing, since P might approach a line parallel to (4) and lying below it. In that case, all that has been said would still be true.

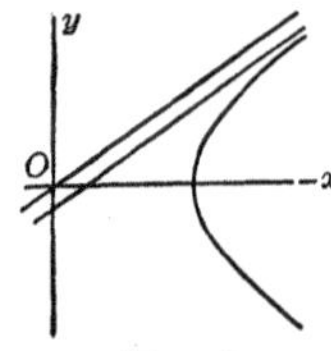

FIG. 8

That P does, however, actually approach (4) can be shown by proving that the distance PQ approaches 0 as its limit. Now,

$$PQ=MQ-MP,$$

and, from (4),

$$MQ=\frac{b}{a}x.$$

Furthermore, MP is the y-coördinate of the point P on the hyperbola:

$$MP = \frac{b}{a}\sqrt{x^2 - a^2}.$$

Hence
$$PQ = \frac{b}{a}[x - \sqrt{x^2 - a^2}].$$

To find the limit approached by the square bracket, we resort to an algebraic device. The value of the bracket will clearly not be changed if we multiply and divide it by the expression $x + \sqrt{x^2 - a^2}$:

$$x - \sqrt{x^2 - a^2} = \frac{(x - \sqrt{x^2 - a^2})(x + \sqrt{x^2 - a^2})}{x + \sqrt{x^2 - a^2}}.$$

But the numerator of the last expression reduces at once to a^2. Hence

$$x - \sqrt{x^2 - a^2} = \frac{a^2}{x + \sqrt{x^2 - a^2}}.$$

From this form it is evident that the bracket approaches 0 when x increases indefinitely; and hence the limit of PQ is zero,* q. e. d.

Similar reasoning, or considerations of symmetry, applied in the other quadrants, show that in the second and fourth quadrants P approaches the line

(5)
$$y = -\frac{b}{a}x,$$

while in the third quadrant, as in the first, P approaches (4).

The equations (4) and (5), of the asymptotes, can also be written in the form

$$\frac{x}{a} - \frac{y}{b} = 0, \qquad \frac{x}{a} + \frac{y}{b} = 0.$$

* The limit approached by the variable $x - \sqrt{x^2 - a^2}$ can be found geometrically as follows. Construct a variable right triangle, one leg of which is fixed and of length a, the hypothenuse being variable and of length x. Then the above variable, $x - \sqrt{x^2 - a^2}$, is equal to the difference in length between the hypothenuse and the variable leg. This difference obviously approaches 0 as x increases indefinitely.

a x $\sqrt{x^2-a^2}$

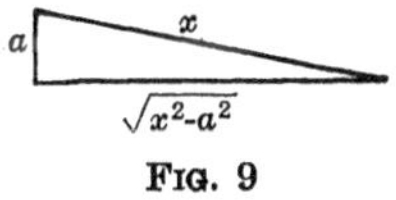

FIG. 9

It is easy to remember these equations, since they can be written down by replacing the right-hand side of (1) by 0, factoring the left-hand side:

$$\left(\frac{x}{a}-\frac{y}{b}\right)\left(\frac{x}{a}+\frac{y}{b}\right)=0,$$

and putting the individual factors equal to zero.

The slopes of the asymptotes are b/a and $-b/a$. Consequently, the asymptotes make equal angles with the transverse axis.

Since the ratio of b to a is unrestricted, the asymptotes can make any arbitrarily assigned angle with each other. If, in particular, $b = a$, this angle is a right angle, and the curve is called a *rectangular*, or *equilateral, hyperbola.* Its equation can be written in the form:

$$x^2 - y^2 = a^2. \tag{6}$$

Its eccentricity is $e = \sqrt{2}$.

Construction of the Asymptotes. Mark with heavy lines the major and minor axes, and through the extremities of each draw lines parallel to the other, thus obtaining a rectangle. The diagonals of this rectangle, produced, are the asymptotes, since their slopes are clearly $\pm b/a$.

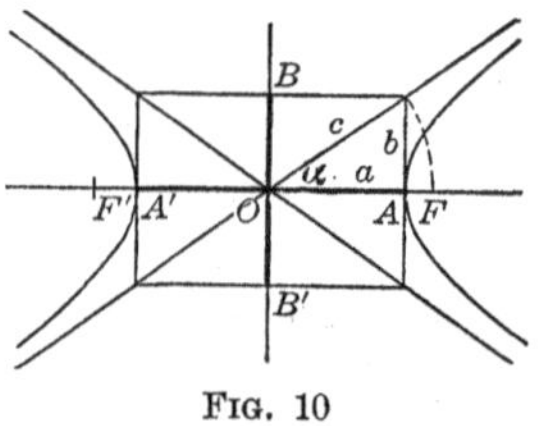

FIG. 10

The diagonals of the rectangle have lengths equal to the distance $2c$ between the foci, for, $c^2 = a^2 + b^2$ and the lengths of the sides of the rectangle are $2a$ and $2b$. If the acute angle between an asymptote and the transverse axis is denoted by α, then

$$e = \sec \alpha.$$

EXERCISES

1. Find the equations and slopes of the asymptotes of the hyperbolas of Exercise 1, § 3. Draw the hyperbolas.

2. Show that the asymptotes of the hyperbola

$$Ax^2 - By^2 = C,$$

where A, B, and C are any three positive quantities, are given by the equations

$$\sqrt{A}x + \sqrt{B}y = 0, \qquad \sqrt{A}x - \sqrt{B}y = 0.$$

3. Find the equation of the hyperbola whose asymptotes make angles of 60° with the axis of x and whose vertices are situated at the points (1, 0), and (– 1, 0). *Ans.* $3x^2 - y^2 = 3$.

4. Show that the slopes of the asymptotes are given by the expression $\pm\sqrt{e^2 - 1}$.

5. The slope of one asymptote of a hyperbola is $\frac{3}{4}$. Find the eccentricity. *Ans.* $e = 1\frac{1}{4}$.

6. The distance of a focus of a certain hyperbola from the center is 10 cm., and the distance of a vertex from the focus is 2 cm. What angle do the asymptotes make with the conjugate axis? *Ans.* 53° 8′.

7. Show that the circle circumscribed about the rectangle of the text passes through the foci.

8. A perpendicular dropped from a focus F on an asymptote meets the latter at E. Show that $OE = a$, and $EF = b$.

9. Find the equation of the equilateral hyperbola whose foci are at unit distance from the center.

10. Find the equation of the equilateral hyperbola which passes through the point (– 5, 4).

5. Tangents. The method of finding the slope of an ellipse, Ch. IX, § 2, can be applied to the hyperbola, and it is thus shown that the slope of this curve,

$$\frac{x^2}{a^2} - \frac{y^2}{b^2} = 1,$$

at the point (x_1, y_1) is

$$\lambda = \frac{b^2 x_1}{a^2 y_1}.$$

The equation of the tangent of the hyperbola at this point is

(1) $$\frac{x_1x}{a^2} - \frac{y_1y}{b^2} = 1.$$

THEOREM. *The tangent of a hyperbola at any point bisects the angle between the focal radii.*

To prove this proposition we recall the theorem of Plane Geometry which says that the bisector of an angle of a triangle divides the opposite side into segments which are proportional to the adjacent sides. It is easily seen that the converse * of this proposition is also true, and hence it is sufficient for our proof to show that

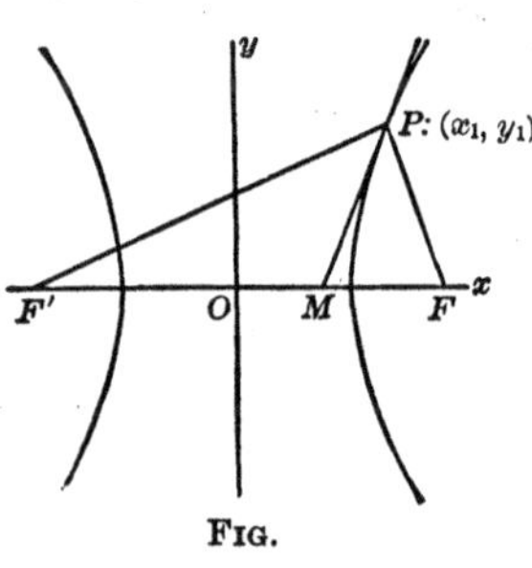

FIG.

(2) $$\frac{FP}{FM} = \frac{F'P}{F'M}.$$

We already have simple expressions for the numerators. If P: (x_1, y_1) be a point of the right-hand branch of the curve, then, by § 3,

$$FP = ex_1 - a; \qquad F'P = ex_1 + a.$$

To compute the denominators, find where the tangent at P, whose equation is given by (1), cuts the axis of x. Denoting the abscissa of M by x', we have:

$$x' = \frac{a^2}{x_1}.$$

Now, $$FM = OF - OM = c - x',$$

and $$c - x' = c - \frac{a^2}{x_1} = \frac{cx_1 - a^2}{x_1}.$$

But $c = ae$, and so

$$cx_1 - a^2 = a(ex_1 - a).$$

Thus $$c - x' = \frac{a}{x_1}(ex_1 - a),$$

* Let the student prove this proposition as an exercise.

and we arrive finally at the desired expression for FM:

$$FM = \frac{a}{x_1}(ex_1 - a).$$

In a similar manner it is shown that

$$F'M = \frac{a}{x_1}(ex_1 + a).$$

From these evaluations it appears that

$$\frac{FP}{FM} = \frac{x_1}{a} \quad \text{and} \quad \frac{F'P}{F'M} = \frac{x_1}{a}.$$

Hence (2) is a true equation, and the proof is complete for the case that P lies on the right-hand branch. Since, however, the curve is symmetric in the conjugate axis, the theorem is true for the left-hand branch also.

Latus Rectum. The latus rectum of a hyperbola is defined as a chord passing through a focus and perpendicular to the transverse axis. The term is also applied to the length of such a chord.

EXERCISES

1. Find the slope of the hyperbola $4x^2 - y^2 = 15$ at the point $(2, -1)$. *Ans.* -8.

2. Find the equation of the tangent of the hyperbola of Ex. 1 at the point there mentioned. *Ans.* $8x + y = 15$.

3. Find the angle at which the line through the origin bisecting the angle between the positive axes of coördinates cuts the hyperbola of Ex. 1. *Ans.* $30° 58'$.

4. Find the length of the latus rectum of the hyperbola

$$\frac{x^2}{16} - \frac{y^2}{9} = 1. \qquad \textit{Ans. } 4\tfrac{1}{2}.$$

5. Find the length of the latus rectum of the hyperbola of Ex. 1. *Ans.* 15.49.

6. Find the equation of the normal of the hyperbola

$$\frac{x^2}{25} - \frac{y^2}{144} = 1$$

at the extremity of the latus rectum which lies in the first quadrant. *Ans.* $25x + 65y = 2197$.

7. Show that the length of the latus rectum of the hyperbola

$$\frac{x^2}{a^2} - \frac{y^2}{b^2} = 1$$

is $\frac{2b^2}{a}$.

8. Prove that the tangents at the extremities of the latera recta have slopes $\pm e$.

9. In an ellipse, the focal radii make equal angles with the tangent. Prove this theorem by the method employed in this paragraph to prove the corresponding theorem relating to the hyperbola.

6. New Definition. The Directrices. The locus defined in Ch. VII, § 7, can now be shown to be a hyperbola when $\epsilon > 1$. The analytic treatment given there and in § 8 down to equation (2) and the transformation (3) holds unaltered for the present case.

When, however, $\epsilon > 1$, the new origin, O', lies to the left of O, in the point $\left(-\frac{m}{\epsilon^2 - 1}, 0\right)$, and it is more natural to write (3) in the form

$$(1) \qquad x' = x + \frac{m}{\epsilon^2 - 1}, \qquad y' = y,$$

and likewise (4) as

$$(2) \qquad x'^2 - \frac{y'^2}{\epsilon^2 - 1} = \frac{\epsilon^2 m^2}{(\epsilon^2 - 1)^2}.$$

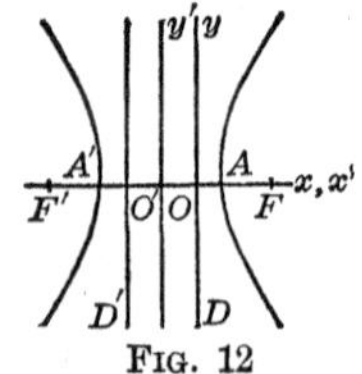

Fig. 12

This equation passes over into the form

$$(3) \qquad \frac{x'^2}{a^2} - \frac{y'^2}{b^2} = 1,$$

on setting

$$(4) \qquad a = \frac{\epsilon m}{\epsilon^2 - 1}, \qquad b = \frac{\epsilon m}{\sqrt{\epsilon^2 - 1}}.$$

Thus the locus is seen to be a hyperbola with its center, O', at the point $\left(-\frac{m}{\epsilon^2-1}, 0\right)$, the semi-axes being given by (4).

The value of c is given by the equation $c^2 = a^2 + b^2$. Hence

$$c = \frac{\epsilon^2 m}{\epsilon^2 - 1}. \tag{5}$$

The eccentricity, $e = c/a$, is seen to be precisely ϵ:

$$e = \epsilon,$$

and thus the given constant, ϵ, turns out to be the eccentricity of the hyperbola.

Finally, *F is one of the foci.* For, the distance from O' to F is

$$O'O + OF = \frac{m}{\epsilon^2 - 1} + m = \frac{\epsilon^2 m}{\epsilon^2 - 1},$$

and this, by (5), is precisely c.

The line D is called a *directrix* of the hyperbola. Its distance from the center is

$$O'O = \frac{m}{\epsilon^2 - 1} = \frac{\epsilon m}{\epsilon^2 - 1} \cdot \frac{1}{\epsilon} = \frac{a}{\epsilon}.$$

The Directrices. There is a second directrix, namely, the line D' symmetric to D in the conjugate axis. It is clear from the symmetry of the figure that what is true of the hyperbola with respect to the focus F and the corresponding directrix D is equally true with respect to the focus F' and the directrix D'. Accordingly, the hyperbola is the locus of a point whose distance from a focus bears to its distance from the corresponding directrix a fixed ratio, the eccentricity.

The equations of the directrices of the hyperbola,

$$\frac{x^2}{a^2} - \frac{y^2}{b^2} = 1,$$

are $\qquad x = \frac{a}{e} \qquad$ and $\qquad x = -\frac{a}{e}.$

EXERCISES

1. Take $\epsilon = 2$ and $m = 3$, the unit of length being 1 cm. With ruler and compasses construct a generous number of points of the locus, and then draw in the locus with a clean, firm line.*

2. Work out the equation of the locus of Ex. 1 directly, using the method of Ch. VII, § 7, but not looking at the formulas. *Ans.* $3x^2 - y^2 + 6x = 9$.

3. By means of a transformation to parallel axes show that the curve of Ex. 2 is a hyperbola whose center is at the point $(-1, 0)$ and whose axes are of lengths 4 and $4\sqrt{3}$.

4. Show that in the general case the distances of the vertices, A and A', from O are:

$$OA = \frac{m}{\epsilon + 1}, \qquad A'O = \frac{m}{\epsilon - 1}.$$

5. Collect the results of this paragraph in a syllabus, arranged in tabular form, giving each of the quantities, a, b, c, $O'O$, OA, $A'O$, OF, and $F'O$, in terms of m and ϵ.

6. Work out each of the quantities of Ex. 5 directly for the curve of Ex. 2 and verify the result by substituting the values $\epsilon = 2$, $m = 3$ in the formulas of the syllabus.

7. Show that the tangent to the hyperbola

$$\frac{x^2}{16} - \frac{y^2}{9} = 1$$

at an extremity of a latus rectum cuts the transverse axis in the same point in which this axis is cut by a directrix.

8. The same for any hyperbola.

* The footnote of p. 114 applies in the present case with the obvious modification that the distance of the parallel from D must now be *half* the radius of the circle. Moreover, *two* parallels to D must now be drawn, the second one, as soon as the radius has increased sufficiently, giving points on the left-hand branch.

9. Prove directly that, if P is any point of the hyperbola

$$\frac{x^2}{a^2}-\frac{y^2}{b^2}=1,$$

the ratio of its distance from a focus to its distance from the corresponding directrix equals the eccentricity.

10. Prove that the ratio of the distance between the foci of a hyperbola to the distance between the directrices equals the square of the eccentricity.

7. The Parabola as the Limit of Hyperbolas. Summary. Equation (3) of Ch. VII, § 7, namely,

$$(1) \qquad (1-\epsilon^2)x^2+y^2-2\,mx+m^2=0,$$

represents a hyperbola when $\epsilon>1$ and a parabola when $\epsilon=1$. If, then, we let ϵ approach 1 through values greater than 1, the hyperbola which (1) represents will approach a parabola as its limiting position.

Suppose, for example, that we take $m=2$ and let ϵ take on successively the values 2, $1\frac{1}{2}$, $1\frac{1}{4}$, $1\frac{1}{8}$, $\cdots$. Drawing the corresponding hyperbolas, we find that, whereas the directrix D and the right-hand focus F are always fixed, the center and the left-hand vertex keep receding to the left, and that their distances from O, namely,

$$O'O=\frac{m}{\epsilon^2-1}, \qquad A'O=\frac{m}{\epsilon-1},$$

increase without limit. Thus, when ϵ approaches 1, the left-hand branch of the hyperbola recedes indefinitely to the left and disappears in the limit, whereas, meanwhile, the right-hand branch gradually changes shape and in the limit becomes the parabola whose directrix is D and whose focus is F.

Summary. Let us now combine the results of § 6 with those of § 8, Ch. VII. We have proved that equation (1) represents an ellipse, a parabola, or a hyperbola, according as $\epsilon<1$, $\epsilon=1$, or $\epsilon>1$. In case of the ellipse and the hyperbola the constant ϵ turned out to be the eccentricity e. We are led then to give to the parabola an eccentricity, namely, $\epsilon = e = 1$.

THEOREM. *The locus of a point which moves so that its distance from a fixed point bears to its distance from a fixed line, not passing through the fixed point, a given ratio ϵ is an ellipse, a parabola, or a hyperbola, according as ϵ is less than, equal to, or greater than unity. In every case the constant ϵ equals the eccentricity.*

Since always $\epsilon = e$, we may suppress ϵ in future work, and use e exclusively. Thus equation (1) becomes

$$(1 - e^2)x^2 + y^2 - 2\,mx + m^2 = 0. \tag{2}$$

The theorem furnishes a blanket definition for the ellipse, parabola, and hyperbola, which might have been used instead of the separate definitions which we have given. It should be noted, however, that this blanket definition does not include the circle. For, if we set $e = 0$ in (2), the equation reduces to

$$(x - m)^2 + y^2 = 0,$$

which represents merely the focus $F: (m, 0)$.

The fact that the blanket definition does not yield a circle as a special case in no way discredits the circle as the limiting form of an ellipse when the eccentricity approaches zero, Ch. VII, § 1. The reason that a circle cannot be defined in the new manner is because it has no directrices. When the eccentricity of an ellipse approaches zero, the major axis remaining constant, the distance a/e of the directrices from the center increases indefinitely, so that in the limit, when the ellipse becomes a circle, the directrices have disappeared.*

* It is, of course, possible to obtain the circle as a *limiting* curve approached by ellipses defined in the new way. If the points F and A of Fig. 12, Ch. VII, are held fast and m is allowed to increase indefinitely, then it can be shown that ϵ approaches zero and that a and b both approach the fixed distance AF. Thus the variable ellipse approaches a circle as its limit.

8. Hyperbolas with Foci on the Axis of y. Conjugate Hyperbolas. Let the student show that the equation of the hyperbola whose foci are at the points $(0, \pm C)$ on the axis of y and the difference of whose focal radii is $2B$ is

$$\frac{x^2}{A^2} - \frac{y^2}{B^2} = -1,$$

where

$$C^2 = A^2 + B^2.$$

The transverse axis of this hyperbola is the axis of y; the conjugate axis, the axis of x. The length of the major axis is $2B$; that of the minor axis, $2A$. The eccentricity is C/B and the asymptotes have the equations,

$$\frac{x}{A} - \frac{y}{B} = 0 \quad \text{and} \quad \frac{x}{A} + \frac{y}{B} = 0.$$

Conjugate Hyperbolas. The two hyperbolas,

$$\frac{x^2}{a^2} - \frac{y^2}{b^2} = 1 \quad \text{and} \quad \frac{x^2}{a^2} - \frac{y^2}{b^2} = -1,$$

have the same asymptotes. The transverse axis of each is the conjugate axis of the other, and the major axis of each is the minor axis of the other.

Taken together, the two hyperbolas form what is called a *pair of conjugate hyperbolas*. The relationship between them is perfect in its duality. We say, then, that each is the *conjugate* of the other.

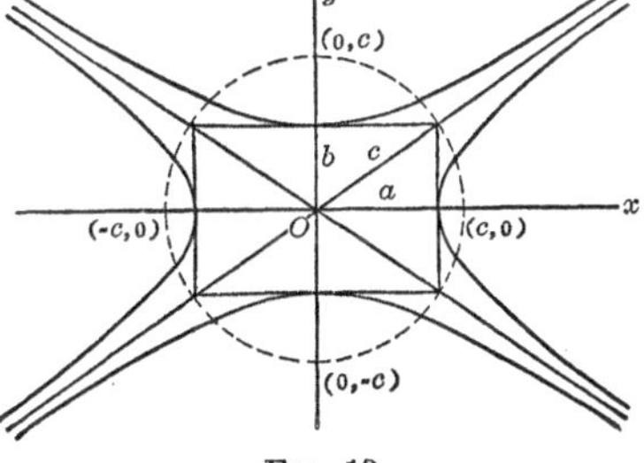

FIG. 13

The two hyperbolas together are tangent externally at their vertices to the rectangle of § 4 at the mid-points of its sides. Moreover, all straight lines through the common center O, except two, meet one hyperbola or the other in two points, and the segment thus terminated is bisected at O.

The student should compare these facts with the corresponding ones concerning a single ellipse and the circumscribed rectangle.

EXERCISES

1. Find the coördinates of the foci, the lengths of the axes, the slopes of the asymptotes, and the value of the eccentricity for each of the hyperbolas:

(*a*) $\frac{x^2}{9}-\frac{y^2}{16}=-1$; (*c*) $y^2-x^2=4$;

(*b*) $5x^2-4y^2+20=0$; (*d*) $3x^2-2y^2+6=0$.

Draw an accurate figure in each case.

2. What are the equations of the hyperbolas conjugate to the hyperbolas of Ex. 1?

3. Find the equation of the hyperbola whose vertices are in the points $(0, \pm 4)$ and whose eccentricity is $\frac{3}{2}$.

Ans. $4x^2-5y^2+80=0$.

4. Find the equation of the hyperbola the extremities of whose minor axis are in the points $(\pm 3, 0)$ and whose eccentricity is $\frac{5}{4}$.

5. Prove that the sum of the squares of the reciprocals of the eccentricities of the two conjugate hyperbolas

$$\frac{x^2}{9}-\frac{y^2}{16}=1, \qquad \frac{x^2}{9}-\frac{y^2}{16}=-1,$$

is equal to unity.

6. Prove the theorem of Ex. 5 for the general pair of conjugate hyperbolas.

7. Show that the foci of a pair of conjugate hyperbolas lie on a circle.

9. Parametric Representation. It is possible to construct a hyperbola, given its axes, AA' and BB', by a method much like that of Ch. VII, § 10, for the ellipse.

Let the two circles, C and C', and the ray from O, be drawn as before. At the point L draw the tangent to C', and mark the point Q where the ray cuts this line. At R draw the tangent to C and mark the point S where this tangent cuts the axis of x.

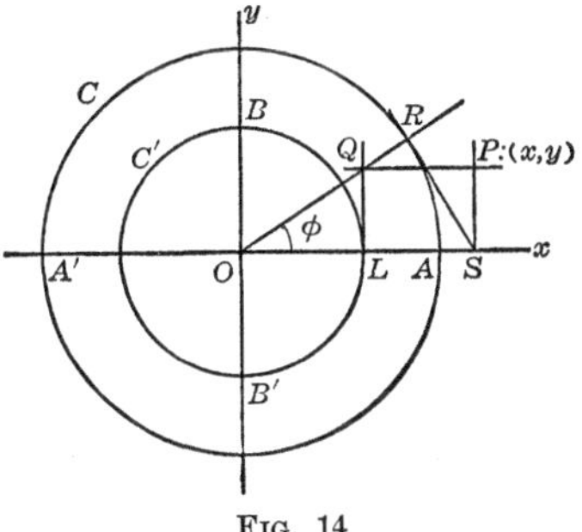

FIG. 14

The locus of the point $P:(x, y)$, in which the parallel to the axis of x through Q and the parallel to the axis of y through S intersect, is the hyperbola.

For, $\qquad OR = a, \qquad OL = b,$

and $\qquad x = OS = a \sec \phi, \qquad y = LQ = b \tan \phi.$

Hence $\qquad \dfrac{x}{a} = \sec \phi, \qquad \dfrac{y}{b} = \tan \phi,$

and since $\qquad \sec^2 \phi - \tan^2 \phi = 1,$

it follows that $\qquad \dfrac{x^2}{a^2} - \dfrac{y^2}{b^2} = 1.$

Conversely, any point (x, y) whose coördinates satisfy this equation is seen to lead to an angle ϕ, for which the above formulas hold.

We thus obtain the following parametric representation of the hyperbola:

$$x = a \sec \phi, \qquad y = b \tan \phi.$$

The circle C, constructed on the major axis of the hyperbola as a diameter, is known as the *auxiliary circle* of the hyperbola, and the angle ϕ is called the *eccentric angle.*

EXERCISES

1. Carry out the construction described above for the cases:

$$(a) \qquad a = 3 \text{ cm.}, \qquad b = 2 \text{ cm.}$$

(b) $a = 3$ cm., $b = 3$ cm.

(c) $a = 2$ cm., $b = 3$ cm.

2. Obtain a parametric representation of the hyperbola

$$\frac{x^2}{A^2} - \frac{y^2}{B^2} = -1.$$

10. Conic Sections. The ellipse (inclusive of the circle), the hyperbola, and the parabola are often called *conic sections,* because they are the curves in which a cone of revolution is cut by planes.

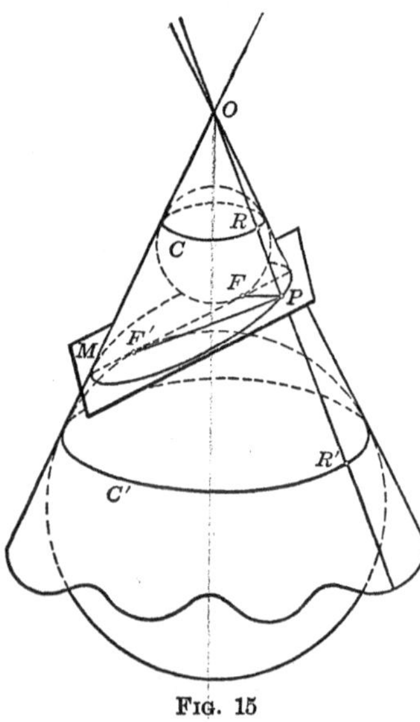

FIG. 15

Suppose a plane M cuts only one nappe of the cone, as is shown in the accompanying drawing. Let a small sphere be placed in the cone near O, tangent to this nappe along a circle. It will not be large enough to reach to the plane M. Now let the sphere grow, always remaining tangent to the cone along a circle. It will finally just reach the plane. Mark the point of tangency, F, of the plane M with the sphere, and also the circle of contact, C, of the sphere with the cone.

As the sphere grows still larger, it cuts the plane M, but finally passes beyond on the other side. In its last position, in which it still meets M, it will be tangent to M. Let the point of tangency be denoted by F', and the circle of contact of the sphere with the cone by C'.

Through an arbitrary point P of the curve of intersection of M with the cone passes a generator OP of the cone; let it cut C in R and C' in R'. Then RR', being the slant height of the frustum * cut from the cone by the planes of C and C', is of the same length, $2a$, for all points P.

Join P with F. Then PF and PR, being tangents from P to the same sphere, are equal. Similarly, PF' and PR' are equal. Hence

$$FP + F'P = RP + R'P = RR',$$

or

$$FP + F'P = 2a.$$

But this locus is by definition an ellipse with its foci at F and F', and hence the proposition is proved for the case that M cuts only one nappe, the intersection being a closed curve.

If the plane M cuts both nappes, but does not pass through O, it is a little harder to draw the figure, one sphere being inscribed in the one nappe, the other, in the other nappe. A similar study shows that here the *difference* between FP and $F'P$ is equal to RR', and hence the locus is a hyperbola.

The parabola corresponds to the case that M meets only one nappe, but does not cut it in a closed curve. This case is realized when M does not pass through O and is parallel to a generator of the cone.

Let L be a line which is perpendicular to the axis of the cone in a point of the axis distinct from the vertex. As a plane, M, rotates about L, it will cut from the cone all three kinds of conics. This will still be true if we take, as L, any line of space which does not pass through the vertex and is not parallel to a generator.

11. Confocal Conics. Two conics are said to be *confocal* if they have the same foci; in the case of two parabolas, we demand, further, that they have the same axis.

* No technical knowledge of Solid Geometry beyond the definitions of the terms used (which can be found in any dictionary) is here needed. On visualizing the figure, the truth of the statements regarding the space relations becomes evident.

Consider an ellipse and a hyperbola which are confocal. They evidently intersect in four points.*

Let P be one of these points. Join P with F and F'. Then FP and $F'P$ are focal radii both of the ellipse and of the hyperbola. Now, the tangent to a hyperbola at any point not a vertex bisects the angle between the focal radii drawn to that point, § 5; and the normal to an ellipse at any point not on the transverse axis bisects the angle between the focal radii drawn to that point, Ch. VII, § 4.

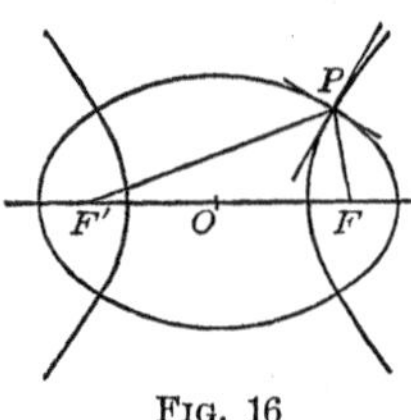

Fig. 16

It follows, then, that the tangent to the hyperbola at P and the normal to the ellipse at this point coincide. Hence the two curves intersect at right angles, or *orthogonally*, as we say. We have thus proved the following

Theorem. *A pair of confocal conics, one of which is an ellipse and the other a hyperbola, cut each other orthogonally.*

Confocal Parabolas. Consider two parabolas having the same focus and the same axis. If both open out in the same direction, they have no point in common. If, however, they open out in opposite directions, they intersect in two points which are symmetrically situated with respect to the axis.

In the latter case, the parabolas intersect orthogonally, as has already been proved analytically; cf. Ch. VI, § 3, Ex. 10.

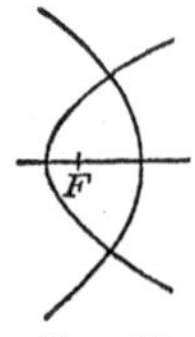

Fig. 17

This result could have been forecast, as a consequence of the relations established in § 7. For, if one focus, F, and the two corresponding directrices of a pair of confocal conics, consisting of an ellipse and a hyperbola, are held fast, and if the other focus is made to recede indefinitely, each of the conics approaches a parabola. But the

* Let the student satisfy himself that two confocal ellipses do not intersect, and that the same is true of two confocal hyperbolas.

conics always intersect orthogonally, and so the same will be true of the limiting curves, the parabolas.

To obtain a prescribed pair of parabolas, like those described above, as limiting curves, it is necessary merely to choose the two confocal conics so that the directrices corresponding to F are at the proper distances from F.

Mechanical Constructions. It is possible to draw with ease a large number of confocal ellipses by the method set forth in Ch. VII, § 1. Let thumb tacks be inserted at F and F', but not pushed clear down. Let a thread be tied to the tack at F, passed round the tack at F', and held fast at M. Then an ellipse can be drawn with F and F' as foci. Now let the thread be unwound at F' and drawn in or paid out slightly, so that the length of the free thread between F and F' is changed. On repeating the above construction, a second ellipse with its foci at F and F' is obtained; and so on.

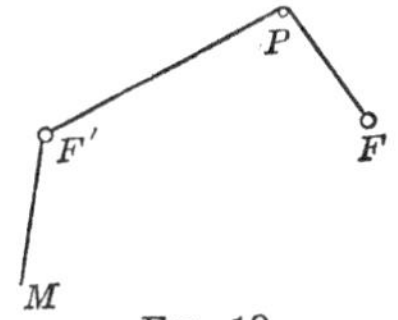

Fig. 18

There is an analogous construction for a hyperbola, which has not yet been mentioned. Tie a thread to a pencil point,* pass the thread round the pegs at F and F' as shown, hold the free ends firmly together at M, and, keeping the thread taut by pressing on the pencil, allow M to move. The pencil then obviously traces out a hyperbola.

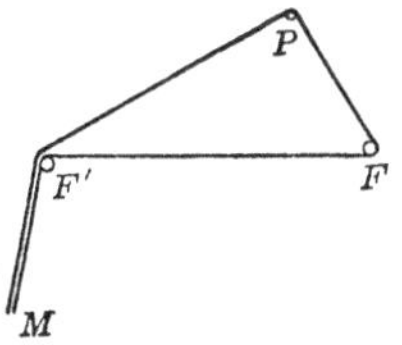

Fig. 19

By pulling one end of the thread in slightly at M, or by paying it out, and then repeating the construction, a new hyperbola with the same foci is obtained; and so on.

Parabolas. The accompanying figure suggests a means for drawing a parabola mechanically.

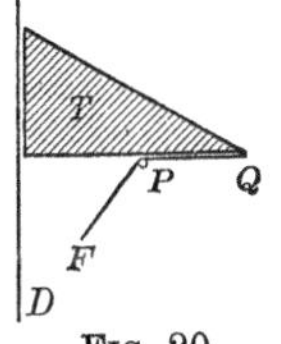

Fig. 20

* To keep the thread from slipping off, cut a groove in the lead, such as would be obtained if the pencil were turned about its axis in a lathe and the point of a chisel were held against the lead close to the wood.

A ruler, D, is held fast and a triangle, T, is allowed to slide along the ruler. A thread is tied at F and Q, and a pencil point, P, keeps the thread taut and pressed against the triangle.

EXERCISES

1. Show that the conics,

$$\frac{x^2}{24}+\frac{y^2}{8}=1 \quad \text{and} \quad \frac{x^2}{4}-\frac{y^2}{12}=1,$$

are confocal.

2. Prove that the equation,

$$\frac{x^2}{9+\lambda}+\frac{y^2}{5+\lambda}=1,$$

represents an ellipse for each value of λ greater than -5 and represents a hyperbola for each value of λ between -9 and -5. Show that all these ellipses and hyperbolas are confocal, with the points $(\pm 2, 0)$ as foci.

3. For what values of λ does the equation

$$\frac{x^2}{a^2+\lambda}+\frac{y^2}{b^2+\lambda}=1,$$

where a and b are given positive constants such that $a > b$, represent i) ellipses? ii) hyperbolas? Show that all these conics are confocal.

4. Draw a set of confocal ellipses and hyperbolas.

5. Draw a set of confocal parabolas, all having the same transverse axis, some opening in one direction, some in the other.

EXERCISES ON CHAPTER VIII

1. The axes of a hyperbola which goes through the points $(1, 4)$, $(-2, 7)$ are the axes of coördinates. Find the equation of the hyperbola. *Ans.* $y^2 - 11x^2 = 5$.

2. Show that the hyperbolas defined by the equation

$$4x^2 - 5y^2 = c,$$

where c is an arbitrary constant, not zero, all have the same asymptotes.

3. How many hyperbolas are there with the lines

$$3x^2 - 16y^2 = 0$$

as asymptotes? Find an equation which represents them all.

Ans. $3x^2 - 16y^2 = c, \quad c \neq 0.$

4. What is the equation of all the rectangular hyperbolas with the axes of coördinates as axes?

5. A hyperbola with the lines $4x^2 - y^2 = 0$ as asymptotes goes through the point (1, 1). What is its equation?

Ans. $4x - y^2 = 3.$

6. The asymptotes of a hyperbola go through the origin and have slopes ± 2. The hyperbola goes through the point (1, 3). Find its equation. *Ans.* $4x^2 - y^2 = -5.$

7. The two hyperbolas of Exs. 5 and 6 have the same asymptotes, but lie in the opposite pairs of regions into which the plane is divided by the asymptotes. Show that the sum of the squares of the reciprocals of their eccentricities equals unity.

8. Prove that of the hyperbolas of Ex. 2 those for which c is positive are all similar, and that this is true also of those for which c is negative. If e is the common value of the eccentricity of the hyperbolas of the first set and e' is that of the hyperbolas of the second set, show that

$$\frac{1}{e^2} + \frac{1}{e'^2} = 1. \tag{1}$$

9. Prove that the relation (1) is valid for the eccentricities of any two hyperbolas which have the same asymptotes but lie in the opposite regions between the asymptotes.

10. Show that two hyperbolas which are related as those described in the previous exercise have the same eccentricity if and only if they are rectangular hyperbolas.

11. A hyperbola with its center in the origin has the eccentricity 2. Find the equations of the asymptotes, (*a*) if the foci lie on the axis of x; (*b*) if the foci lie on the axis of y.

Ans. (*a*) $3x^2 - y^2 = 0$; (*b*) $x^2 - 3y^2 = 0$.

12. What is the equation representing all the hyperbolas which have their centers in the origin and eccentricity 2, (*a*) if the foci lie on the axis of x? (*b*) if the foci lie on the axis of y? Show that in either case the vertices lie midway between the center and the foci.

13. Prove that the vertices of the hyperbola

$$\frac{x^2}{a^2} - \frac{y^2}{b^2} = 1$$

subtend a right angle at each of the points $(0, \pm b)$ when and only when the hyperbola is rectangular. What is the corresponding theorem in the case of the ellipse?

14. The projections of a point P of a hyperbola on the transverse and conjugate axes are P_1 and P_2. The tangent at P meets these axes in T_1 and T_2. Show that $OP_1 \cdot OT_1 = a^2$ and $OP_2 \cdot OT_2 = -b^2$, where O is the center of the hyperbola and a and b are the semi-axes.

15. Prove that the segment of a tangent to a hyperbola between the point of contact and a directrix subtends a right angle at the corresponding focus.

16. The projection of a point P of a hyperbola on the transverse axis is P_1 and the normal at P meets this axis at N_1. Show that the ratio of the distances of the center from N_1 and P_1 equals the square of the eccentricity.

17. Prove that the line joining a point P of a hyperbola with the center and the line through a focus perpendicular to the tangent at P meet on a directrix.

18. Find the equation of the circle which is tangent to a hyperbola at the upper ends of the two latera recta.

19. Let O be the center, A a vertex, and F the adjacent focus of a hyperbola. The tangent at a point P meets the transverse axis at T and the tangent at A meets OP at V. Show that TV is parallel to AP.

20. Show that an asymptote, a directrix, and the line through the corresponding focus perpendicular to the asymptote go through a point.

21. A line through a focus F parallel to an asymptote meets the hyperbola at P. Show that the tangent at P, the other asymptote, and the line of the latus rectum through F meet in a point.

22. Let F be a focus and D the corresponding directrix of a hyperbola. A line through a point P of the hyperbola parallel to an asymptote meets D in the point K. Prove that the triangle FPK is isosceles.

Exercises 23–33. In proving the theorems in these exercises it will, in general, be necessary to make actual use of the equation expressing the fact that a certain point lies on the hyperbola.

23. The tangent to a hyperbola at a point P meets the tangent at one vertex in Q. Prove that the line joining the other vertex to P is parallel to the line joining the center to Q.

24. Let F be a focus and D the corresponding directrix of a hyperbola. Prove that the segment cut from D by the lines joining the vertices with an arbitrary point on the hyperbola subtends a right angle at F.

25. Prove that the product of the distances of the foci of a hyperbola from a tangent is constant, *i.e.* independent of the choice of the tangent.

26. Let A and A' be the vertices of a rectangular hyperbola and let P and P' be two points of the hyperbola symmetric in the transverse axis. Prove that AP is perpendicular to $A'P'$ and that AP' is perpendicular to $A'P$.

27. Show that the product of the focal radii to a point on a rectangular hyperbola is equal to the square of the distance of the point from the center.

28. Prove that the angles subtended at the vertices of a rectangular hyperbola by a chord parallel to the conjugate axis are supplementary.

29. Prove that the product of the distances of an arbitrary point on a hyperbola from the asymptotes is constant, *i.e.* the same for every choice of the point.

30. A line through an arbitrary point P on a hyperbola parallel to the conjugate axis meets the asymptotes in M and N. Show that the product of the segments in which P divides MN is constant.

31. Prove that the segment of a tangent to a hyperbola cut out by the asymptotes is bisected by the point of contact of the tangent.

32. Show that the tangent to a hyperbola at an arbitrary point forms with the asymptotes a triangle which has a constant area.

33. The tangent to a hyperbola at a point P meets the tangents at the vertices in M and N. Prove that the circle on MN as a diameter passes through the foci.

Loci

34. Find the locus of a point whose distance from a given circle always equals its distance from a given point without the circle. First give a geometric construction, with ruler and compass, for points on the locus. Then find the equation of the locus.

35. The base of a triangle is fixed and the product of the tangents of the base angles is a negative constant. What is the locus of the vertex?

36. A line moves so that the area of the triangle which it forms with two given perpendicular lines is constant. Find the locus of the mid-point of the segment cut from it by these lines.

Ans. Two conjugate rectangular hyperbolas, with the given lines as asymptotes.

37. Given a fixed line L and a fixed point A, not on L. A point P moves so that its distance from L always equals the distance AQ, where Q is the foot of the perpendicular dropped from P on L. What is the locus of P?

38. What is the locus of the point P of the preceding exercise, if the ratio of its distance from L to the distance AQ is constant?

CHAPTER IX

CERTAIN GENERAL METHODS

1. Tangents. Let it be required to find the tangent line to a given curve at an arbitrary point.

In the case of the circle the tangent is perpendicular to the radius drawn to the point of tangency. But this solution is of so special a nature that it suggests no general method of attack. A general method must be based on a general property of tangents, irrespective of the special curve considered. Such a method is the following. Let P be an arbitrary point of a given curve, C, at which it is desired to draw the tangent, T. Let a second point, P', be chosen on C, and draw the secant, PP'. As P' moves along C and approaches the fixed point P as its limit, the secant rotates about P as a pivot and approaches the tangent, T, as its limiting position. *Thus the tangent appears as the limit of the secant.*

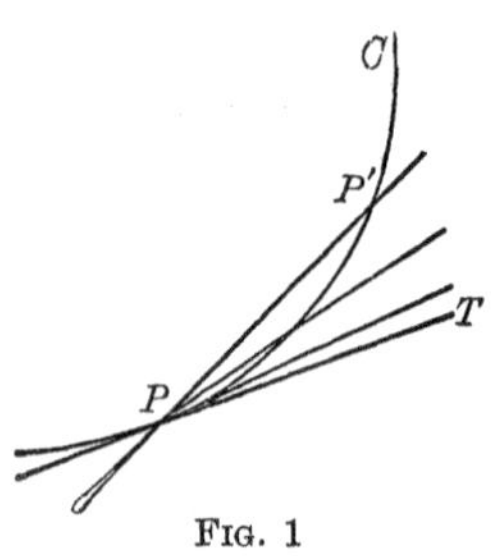

Fig. 1

If, now, in a given case we can find an expression for the *slope* of the secant, the limit approached by this expression will give us the *slope of the tangent.* The slope of the tangent to the curve at P we shall call, for the sake of brevity, the *slope of the curve* at P.

Example 1. Find the slope of the curve

$$(1) \qquad y = x^2$$

at a given point, P.

Let the coördinates of P be (x_1, y_1); those of P', (x', y'), or $(x_1 + h, y_1 + k)$. Then

$$PQ = h, \qquad QP' = k,$$

and we have, for the slope of the secant PP', the expression:

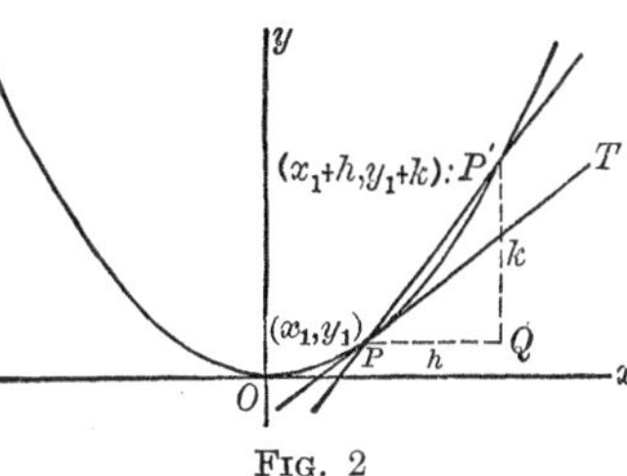

FIG. 2

$$(2) \qquad \tan \tau' = \frac{k}{h},$$

where $\tau' = \measuredangle\, QPP'$. The slope of the tangent line, T, at P is, then,

$$(3) \qquad \tan \tau = \lim_{P' \doteq P} \tan \tau' = \lim_{h \doteq 0} \frac{k}{h},$$

where $\tau = \measuredangle\, QPT$. The sign $\doteq$ is used to mean "approaches as its limit," and the expression: $\lim\limits_{P' \doteq P} \tan \tau'$, is read: "the limit of $\tan \tau'$, as P' approaches P."

Suppose, for example, that P is the point (1, 1). Let us compute k and $\tan \tau'$ for a few values of h. Here, $x_1 = 1$ and $y_1 = 1$. If $h = .1$, then

$$x' = x_1 + h = 1.1,$$
$$y' = y_1 + k = (1.1)^2 = 1.21,$$
$$k = .21,$$

and hence $$\tan \tau' = \frac{.21}{.1} = 2.1.$$

Next, let P' be the point for which

$$x' = 1.01.$$

Then $$y' = 1.0201,$$
$$h = .01, \qquad k = .0201,$$

and hence $$\tan \tau' = \frac{.0201}{.01} = 2.01.$$

Let the student work out one more case, taking $x' = 1.001$. He will find that here $k = .002001$ and

$$\tan \tau' = 2.001.$$

These results can be presented conveniently in the form of a table:

h	k	$\tan \tau' = \frac{k}{h}$
.1	.21	2.1
.01	.0201	2.01
.001	.002001	2.001

The numbers in the last column appear to be approaching nearer and nearer to the limit 2; in other words, the slope of the curve in the point (1, 1) appears to be 2. Let us prove that this is actually the case. Since the proof is just as simple for an arbitrary point P, we will return to the general case.

The point P being a point of the curve (1), its coördinates (x_1, y_1) must satisfy that equation. Hence

(4) $$y_1 = x_1^2.$$

Similarly, for the point P' whose coördinates are $(x_1 + h, y_1 + k)$:

$$y_1 + k = (x_1 + h)^2,$$

or

(5) $$y_1 + k = x_1^2 + 2\, x_1 h + h^2.$$

Subtracting (4) from (5), we get:

$$k = 2\, x_1 h + h^2.$$

Consequently, $$\tan \tau' = \frac{k}{h} = 2\, x_1 + h.$$

Now let P' approach P; h will then approach 0, and we shall have

$$\lim_{P' \doteq P} \tan \tau' = \lim_{h \doteq 0} \frac{k}{h} = \lim_{h \doteq 0} (2\, x_1 + h).$$

But

$$\lim_{P' \doteq P} \tan \tau' = \tan \tau, \qquad \text{and} \qquad \lim_{h \doteq 0} (2\, x_1 + h) = 2\, x_1.$$

Hence $$\tan \tau = 2\, x_1.$$

We can say, then, that *the slope of the curve* (1), *at an arbitrary point* $P:(x_1, y_1)$ *on it, is*

$$\lambda = 2x_1.$$

If, in particular, P is the point (1, 1), the slope of the tangent there is $\lambda = 2 \cdot 1 = 2$, and thus the indication given by the above table is seen to be borne out.

Example 2. Find the slope of the curve

$$y = \frac{a^2}{x} \tag{6}$$

at an arbitrary point $P:(x_1, y_1)$ of the curve.

Denote, as before, the coördinates of a second point, P', by

$$x' = x_1 + h, \qquad y' = y_1 + k.$$

Then, since P and P' lie on the curve,

$$y_1 = \frac{a^2}{x_1}$$

and

$$y_1 + k = \frac{a^2}{x_1 + h}.$$

Hence

$$k = \frac{a^2}{x_1 + h} - \frac{a^2}{x_1}.$$

Nothing is more natural than to reduce the right-hand side of this equation to a common denominator. Thus

$$k = \frac{-a^2 h}{x_1(x_1 + h)}.$$

Consequently,

$$\tan \tau' = \frac{k}{h} = \frac{-a^2}{x_1(x_1 + h)}.$$

We are now ready to let P' approach P:

$$\lim_{P' \doteq P} \tan \tau' = \lim_{h \doteq 0} \frac{-a^2}{x_1(x_1 + h)}.$$

The limit approached by the right-hand side is obviously $-a^2/x_1^2$, and so

$$\tan \tau = -\frac{a^2}{x_1^2}.$$

We have, then, as the final result: *The slope of the curve* (6), *at an arbitary point* (x_1, y_1) *on it, is*

$$\lambda = -\frac{a^2}{x_1^2}.$$

Equation of the Tangent. Since the tangent to the curve (1),

$$y = x^2,$$

at the point (1, 1) has the slope 2, its equation is

$$y - 1 = 2(x - 1), \qquad \text{or} \qquad 2x - y - 1 = 0.$$

Similarly, the equation of the tangent to the curve (1) at an arbitrary point $P:(x_1, y_1)$ is

$$y - y_1 = 2x_1(x - x_1),$$

or

$$y - y_1 = 2x_1x - 2x_1^2.$$

This equation may be simplified by use of the equality,

$$y_1 = x_1^2,$$

which says that the point P lies on the curve. For, if we replace the term $2x_1^2$ by its equal, $2y_1$, and then combine the terms in y_1, the equation becomes

$$y + y_1 = 2x_1x.$$

This equation of the tangent is of the first degree in x and y, as it should be. The quantities x_1 and y_1 are the arbitrary, but in any given case *fixed*, coördinates of P and are not variables.

Equation of the Normal. The line through a point P of a curve perpendicular to the tangent at P is known as the *normal* to the curve at P.

Since the tangent to the curve $y = x^2$ at the point (1, 1) has the slope 2, the normal at this point has the slope $-\frac{1}{2}$. Consequently, the equation of the normal is

$$y - 1 = -\tfrac{1}{2}(x - 1), \qquad \text{or} \qquad x + 2y - 3 = 0.$$

EXERCISES

1. Determine the slope of the curve $y = x^2 - x$ at the point (3, 6). First make out a table like that under Example 1, and hence infer the probable slope. Then take an arbitrary point (x_1, y_1) on the curve and determine the actual slope at this point by finding

$$\lim_{h \doteq 0} \frac{k}{h}.$$

2. The same for the curve $8y = 3x^3$ at the point (2, 3).

3. The same for the curve $y = 2x^2 - 3x + 1$ at the point (1, 0).

Find the slope of each of the following curves at an arbitrary point $P: (x_1, y_1)$. No preliminary study of a numerical case, like that which gave rise to the table under Example 1, is here required.

4. $y = x^2 - 3x + 1$. *Ans.* $\lambda = 2x_1 - 3$.

5. $y = 2x^2 - x - 4$.

6. $y = x^3 - x$.

7. $y = 4x^3 - 2x^2 + 5$.

8. $y = x^3 + x^2 + x + 1$.

9. $y = x^3 + px + q$. *Ans.* $\lambda = 3x_1^2 + p$.

10. $y = x^4 - a^4$. *Ans.* $\lambda = 4x_1^3$.

11. $y = \dfrac{1}{x^2}$. *Ans.* $\lambda = -\dfrac{2}{x_1^3}$.

12. $y = \dfrac{a^4}{x^3}$.

13. $y = \dfrac{1}{1 - x}$. *Ans.* $\lambda = \dfrac{1}{(1 - x_1)^2}$.

14. $y = \dfrac{2x}{3x - 4}$. *Ans.* $\lambda = -\dfrac{8}{(3x_1 - 4)^2}$.

15. $y = ax^2 + bx + c$. *Ans.* $\lambda = 2ax_1 + b$.

16. $y = ax^3 + bx^2 + cx + d$.

17. $y = x^n$, (n, a positive integer) *Ans.* $\lambda = nx_1^{n-1}$.

18. $y = cx^n$.

Find the equations of the tangents to the following curves at the points specified. In each case reduce the equation obtained to the simplest form.

19. The curve of Ex. 1 at the points (3, 6); (x_1, y_1).

Ans. $5x - y - 9 = 0$; $(2x_1 - 1)x - y - x_1^2 = 0$.

20. The curve of Ex. 3 at the points (x_1, y_1); (1, 0).

21. The curve of Ex. 4 at the points (x_1, y_1); $(-1, 5)$.

22. The curve of Ex. 11 at the points (1, 1); (x_1, y_1).

23. The curve of Ex. 17 at the point (x_1, y_1).

Ans. $nx_1^{n-1}x - y - (n-1)y_1 = 0$.

24. The curve of Ex. 13 at the point whose abscissa is 2.

25. The curve of Ex. 14 at the point whose abscissa is 4.

26. Find the equations of the normals to the curves of Exs. 21, 22 at the designated points.

2. Continuation. Implicit Equations. We have applied the general method to curves whose equations are given in the form: $y =$ a simple expression in x. More precisely, this "simple expression" has each time been a polynomial (or even a monomial), or the ratio of two such expressions.

But even the simplest forms of the equations of the conics are, as a rule, such that, if the equation be solved for y, radicals will appear. In such cases, the following method of treatment can be used with advantage.

The Parabola. Let it be required to find the slope of the parabola

$$y^2 = 2mx \tag{1}$$

at any point $P: (x_1, y_1)$ on the curve.

We will treat first a numerical case, setting $m = 2$:

$$y^2 = 4x. \tag{2}$$

Since P is on the curve, we have

$$y_1^2 = 4x_1. \tag{3}$$

Since $P' : (x_1 + h,\ y_1 + k)$ is also on the curve, we have:

$$(y_1 + k)^2 = 4(x_1 + h),$$

or

(4) $$y_1^2 + 2y_1k + k^2 = 4x_1 + 4h.$$

Subtract (3) from (4):

$$2y_1k + k^2 = 4h.$$

Divide this equation through by h, to obtain an equation for $\tan\tau' = k/h$:

$$2y_1\frac{k}{h} + k\frac{k}{h} = 4, \qquad \text{or} \qquad 2y_1\tan\tau' + k\tan\tau' = 4.$$

Solve the latter equation for $\tan\tau'$:

$$\tan\tau' = \frac{4}{2y_1 + k}.$$

We are now ready to let P' approach P as its limit. This means that h and k both approach 0. We have, then,

$$\lim_{P' \doteq P} \tan\tau' = \lim_{h \doteq 0} \frac{4}{2y_1 + k},$$

or

$$\tan\tau = \frac{4}{2y_1} = \frac{2}{y_1}.$$

It has been tacitly assumed that $y_1 \neq 0$. If $y_1 = 0$, then $\tan\tau'$ increases indefinitely as h, and with it k, approaches zero. Thus the tangent line is seen to be perpendicular to the axis of x at this point, as obviously is, in fact, the case, since the point is the vertex of the parabola.

The student will now carry through by himself the corresponding solution in the general case of equation (1). He will arrive at the result: *The slope* λ *of the parabola*

$$y^2 = 2mx$$

at an arbitrary point (x_1, y_1) *of the curve is*

(5) $$\lambda = \frac{m}{y_1}.$$

The Ellipse. The treatment in the case of the ellipse,

$$\frac{x^2}{a^2}+\frac{y^2}{b^2}=1, \tag{6}$$

is precisely similar. Writing (6), for convenience, in the form

$$b^2x^2+a^2y^2=a^2b^2, \tag{7}$$

we are led to the following equations*:

$$b^2x_1^2+a^2y_1^2=a^2b^2; \tag{8}$$

$$b^2(x_1+h)^2+a^2(y_1+k)^2=a^2b^2,$$

or

$$b^2x_1^2+a^2y_1^2+2\,b^2x_1h+2\,a^2y_1k+b^2h^2+a^2k^2=a^2b^2. \tag{9}$$

Subtract (8) from (9):

$$2\,b^2x_1h+2\,a^2y_1k+b^2h^2+a^2k^2=0.$$

Divide by h:

$$2\,b^2x_1+2\,a^2y_1\frac{k}{h}+b^2h+a^2k\frac{k}{h}=0,$$

or $$2\,b^2x_1+2\,a^2y_1\tan\tau'+b^2h+a^2k\tan\tau'=0.$$

Solve this equation for $\tan\tau'$:

$$\tan\tau'=-\frac{2\,b^2x_1+b^2h}{2\,a^2y_1+a^2k}.$$

Now let P' approach P as its limit:

$$\lim_{P'\doteq P}\tan\tau'=\lim_{h\doteq 0}-\frac{2\,b^2x_1+b^2h}{2a^2y_1+a^2k}.$$

Hence $$\tan\tau=-\frac{2\,b^2x_1}{2\,a^2y_1}=-\frac{b^2x_1}{a^2y_1}.$$

We have thus obtained the result: *The slope λ of the ellipse*

$$\frac{x^2}{a^2}+\frac{y^2}{b^2}=1$$

at an arbitrary one of its points (x_1, y_1) is

$$\lambda=-\frac{b^2x_1}{a^2y_1}. \tag{10}$$

* The student will do well to paraphrase the text at this point with a numerical case, — say, $4\,x^2+9\,y^2=36$.

The Hyperbola. The treatment is left to the student. The result is as follows.

The slope λ *of the hyperbola*

$$\frac{x^2}{a^2}-\frac{y^2}{b^2}=1$$

at an arbitrary one of its points (x_1, y_1) *is*

$$\lambda=\frac{b^2x_1}{a^2y_1}. \tag{11}$$

Equation of the Tangent. Since the slope of the ellipse at the point (x_1, y_1) is $-b^2x_1/a^2y_1$, the equation of the tangent at (x_1, y_1) is

$$y-y_1=-\frac{b^2x_1}{a^2y_1}(x-x_1)$$

or, after clearing of fractions and rearranging terms,

$$b^2x_1x+a^2y_1y=b^2x_1^2+a^2y_1^2.$$

If we divide both sides of this equation by a^2b^2, we have

$$\frac{x_1x}{a^2}+\frac{y_1y}{b^2}=\frac{x_1^2}{a^2}+\frac{y_1^2}{b^2}.$$

But, since the point (x_1, y_1) lies on the ellipse, it follows that

$$\frac{x_1^2}{a^2}+\frac{y_1^2}{b^2}=1,$$

and the equation of the tangent becomes

$$\frac{x_1x}{a^2}+\frac{y_1y}{b^2}=1.$$

The equation of the tangent to the ellipse

$$\frac{x^2}{a^2}+\frac{y^2}{b^2}=1$$

at the point (x_1, y_1) *is*

$$\frac{x_1x}{a^2}+\frac{y_1y}{b^2}=1. \tag{12}$$

In a similar manner let the student establish the equations of the tangents to the hyperbola and the parabola.

The tangent to the hyperbola

$$\frac{x^2}{a^2} - \frac{y^2}{b^2} = 1$$

at the point (x_1, y_1) *has the equation*

$$\frac{x_1 x}{a^2} - \frac{y_1 y}{b^2} = 1. \qquad (13)$$

The tangent to the parabola

$$y^2 = 2mx$$

at the point (x_1, y_1) *has the equation*

$$y_1 y = m(x + x_1). \qquad (14)$$

EXERCISES

Find the slope of each of the following six curves at an arbitrary one of its points, applying each time the method set forth in the text.

1. $2x^2 + 3y^2 = 12.$
2. $x^2 - 4y^2 = 4.$
3. $y^2 = 12x.$
4. $x^2 - y^2 = a^2.$
5. $Ax^2 + By^2 = C$, where A, B, C are all positive.
6. $y^2 = Ax + B$, where $A \neq O$.

7. Find the slope of the parabola $y^2 + 2y = 6x$ at the point (x_1, y_1). *Ans.* $\lambda = \dfrac{3}{y_1 + 1}$.

8. What is the slope of the parabola of Ex. 7 at the origin? *Ans.* 3.

9. Find the slope of the curve

$$x^3 - y^2 - 3x + 4y = 0$$

at the origin. *Ans.* $\lambda = \frac{3}{4}$.

Suggestion. First find the slope at an arbitrary point (x_1, y_1). Then substitute in the result the coördinates of the origin.

10. What angle does the curve

$$2x^3 - 3y^2 + x - y + 1 = 0$$

make with a parallel to the axis of x at the point (1, 1)? *Ans.* 45°.

11. Find the slope of the curve $xy = a^2$ at any point (x_1, y_1) by the method of the present paragraph, and show that your result agrees with that of § 1, Example 2.

Find the equation of the tangent to each of the following curves at the point designated, applying each time the method of the text. Reduce the equation to its simplest form.

12. The curve of Ex. 1 at the point (x_1, y_1).

Ans. $2x_1x + 3y_1y = 12.$

13. The curve of Ex. 3 at the points (x_1, y_1); $(3, -6)$.

14. The curve of Ex. 5 at the point (x_1, y_1).

Ans. $Ax_1x + By_1y = C.$

15. The curve of Ex. 6 at the point (x_1, y_1).

16. The curve of Ex. 7 at the points (x_1, y_1); $(\frac{1}{2}, 1)$.

17. The curve of Ex. 9 at the origin.

18. Find the equations of the normals to the curves of Exs. 12, 13 at the points specified.

3. The Equation $u + kv = 0$. Consider the following example.

The equations

(1) $$x + y - 2 = 0,$$

(2) $$x - y = 0,$$

represent two straight lines intersecting in the point (1, 1), as shown in Fig. 3. What can we say concerning the curve*

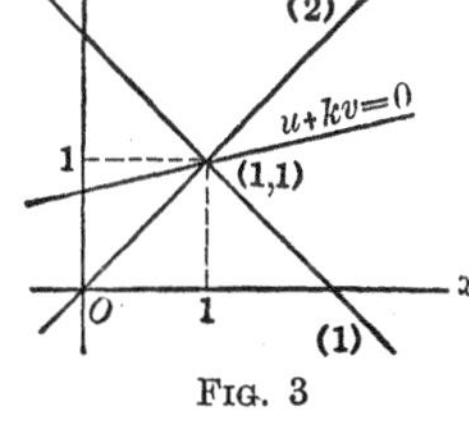

Fig. 3

(3) $$(x + y - 2) + k(x - y) = 0,$$

where k denotes a constant number?

This curve is a straight line, since (3) is an equation of the first degree in x and y. Suppose, now, that various different values are given to k. Then (3) represents various straight lines in turn. What do all these lines have in common?

* The word "curve" is used here in the sense common in analytic geometry, to denote merely the "locus of the equation." Consequently a *curve* in this sense is not necessarily crooked; it may be a straight line.

Since the lines (1) and (2) intersect in the point (1, 1), the coördinates of this point make the left-hand sides of equations (1) and (2), namely, the expressions,

$$x + y - 2 \qquad \text{and} \qquad x - y,$$

vanish. Consequently, they always make the left-hand side of equation (3) vanish. In other words, *equation* (3) *is satisfied by the coördinates of the point of intersection of the lines* (1) *and* (2), NO MATTER WHAT VALUE k HAS. This means that *all* the straight lines represented by (3) go through the point of intersection of the lines (1) and (2).

The result can be restated in the following form. Let the single letter u stand for the whole expression $x + y - 2$:

$$u \equiv x + y - 2,$$

the sign $\equiv$ meaning *identically equal, i.e.* equal, no matter what values x and y have. Similarly, let v stand for $x - y$:

$$v \equiv x - y.$$

Then (3) takes on the form:

$$u + kv = 0. \tag{4}$$

We now restate our result.

If $u = 0$ *and* $v = 0$ *are the equations of two intersecting straight lines, then the equation*

$$u + kv = 0$$

represents a straight line which goes through the point of intersection of the two given lines.

By giving to k a suitable value, $u + kv = 0$ can be made to represent any desired line through the point of intersection (x_1, y_1) of the given lines, with the sole exception of the line $v = 0$. For, let L be the desired line, and let (x_2, y_2) be a point of L distinct from (x_1, y_1). Then, on substituting for x and y the values x_2 and y_2 in the equation $u + kv = 0$, we obtain an equation, in which k is the unknown. This equation can be solved for k, since v does not vanish for the point (x_2, y_2).

Example. Find the equation of the line L which goes through the point of intersection of the lines (1) and (2) and cuts the axis of y in the point $(0, -4)$.

The required line, L, is one of the lines (3); *i.e.* for a suitable value of k, (3) will represent L. To find this value of k, we demand that (3) contain the given point $(0, -4)$ of L. We have, then, setting $x = 0$ and $y = -4$ in (3):

$$(0 - 4 - 2) + k(0 + 4) = 0 \qquad \text{or} \qquad k = \tfrac{3}{2}.$$

Consequently, the equation of the line L is

$$x + y - 2 + \tfrac{3}{2}(x - y) = 0 \qquad \text{or} \qquad 5x - y - 4 = 0.$$

That the line represented by the latter equation does actually go through the points $(1, 1)$ and $(0, -4)$ can be verified directly.

The principle which has been set forth for two straight lines evidently applies to any two intersecting curves whatever, so that we are now in a position to state the following general theorem.

THEOREM 1. *Let $u = 0$ and $v = 0$ be the equations of any two intersecting curves. Then the equation*

$$u + kv = 0, \qquad k \neq 0,$$

represents, in general, a curve which passes through all the points of intersection of the two given curves, and has no other point in common with either of them.*

The last statement in the theorem is new. To prove it, we have but to note that, if the coördinates of a point P satisfy the equation $u + kv = 0$ and also, for example, $v = 0$, they must satisfy the equation $u = 0$; that is, if P is a point on the curve $u + kv = 0$, which lies on one of the given curves, it lies also on the other and so is a point of intersection of the two.

* It may happen in special cases that the locus $u + kv = 0$ reduces to a point, as when, for example,

$$u = 2x^2 + 2y^2 - x, \qquad v = x^2 + y^2 - x, \qquad k = -1.$$

Suppose, now, that the equations $u = 0$ and $v = 0$ represent two curves which have no point of intersection. It follows, then, from the argument just given, that the curve

$$u + kv = 0, \qquad k \neq 0,$$

has no point in common with either of the given curves. But it may happen, in this case, that there are no points at all whose coördinates satisfy the equation $u + kv = 0$. Thus, if

$$u = x^2 + y^2 - 1,$$
$$v = x^2 + y^2 - 4,$$

and $k = -1$, we have

$$u + kv \equiv 3,$$

and there are no points whose coördinates satisfy the equation $3 = 0$.

The general result can be stated as

THEOREM 2. *Let $u = 0$ and $v = 0$ be the equations of two non-intersecting curves. Then the equation*

$$u + kv = 0, \qquad k \neq 0,$$

*represents, in general, a curve not meeting either of the two given curves. In particular, it may happen that the equation has no locus.**

In the special case that u and v are *linear* expressions in x and y, it is possible to say more.

If $u = 0$ and $v = 0$ are the equations of two parallel straight lines, the equation

$$u + kv = 0, \qquad k \neq 0,$$

represents, in general, a straight line parallel to the given lines. For a single value of k, the equation has no locus.

Thus, if the parallel lines are

$$u \equiv x + y = 0, \qquad v \equiv x + y + 1 = 0,$$

the equation

$$u + kv \equiv (1 + k)x + (1 + k)y + k = 0 \tag{5}$$

* It may happen, also, that the equation represents just one point, as when, for example,

$$u = x^2 + y^2 - 2, \qquad v = x^2 + y^2 - 1. \qquad k = -2,$$

has no locus when $k = -1$, but otherwise it represents a line, of slope -1, parallel to the given lines. In fact, it yields all the lines of slope -1, except the line $v = 0$, since, if we rewrite it in the form,

$$x + y + \frac{k}{1+k} = 0, \qquad k \neq -1,$$

the quantity $k/(1+k)$ may be made to take on any value, except 1, by suitably choosing k.

Pencils of Curves. All the lines through a point, or all the parallel lines with a given slope, form what is called a *pencil of lines.* Equation (5) represents, when k is considered as an arbitrary constant, all the lines of slope -1, except the line

$$v \equiv x + y + 1 = 0;$$

in this case, then, $u + kv = 0$ and $v = 0$ together represent all the lines of slope -1, that is, *a pencil of parallel lines.*

Similarly, $u + kv = 0$, when $u = 0$ and $v = 0$ are the lines (1) and (2), yields all the lines through the point (1, 1), except the line (2); hence $u + kv = 0$ and $v = 0$ together represent all the lines through the point (1, 1), — *a pencil of intersecting lines.*

Thus, if $u = 0$ and $v = 0$ are any two lines, the equations

(6) $$u + kv = 0 \qquad \text{and} \qquad v = 0$$

together represent a pencil of lines.

If we set $k = m/l$ in $u + kv = 0$ and multiply by l, the resulting equation

(7) $$lu + mv = 0$$

is equivalent to the equation $u + kv = 0$ when $l \neq 0$, and when $l = 0$ $(m \neq 0)$, it becomes $v = 0$. Consequently, the two equations (6) may be replaced by the single equation (7).

The pencil of lines through the point (1, 1), for example, may now be given by the single equation

$$l(x + y - 2) + m(x - y) = 0,$$

where l and m have arbitrary values, not both zero.

In general, if $u = 0$ and $v = 0$ are any two curves, all the curves represented by the equation

$$lu + mv = 0,$$

where l and m have arbitrary values, not both zero, form what is called *a pencil of curves.*

Applications. Example 1. Let

$$u \equiv x^2 + y^2 + ax + by + c = 0,$$
$$v \equiv x^2 + y^2 + a'x + b'y + c' = 0,$$

be the equations of any two circles which cut each other. Then the equation

$$u - v \equiv (a - a')x + (b - b')y + (c - c') = 0$$

represents a curve which passes through the two points of intersection of the circles. But this equation, being linear, represents a straight line, and is, therefore, the equation of the *common chord* of the circles.

The foregoing proof is open to the criticism that conceivably we might have

$$a - a' = 0, \qquad b - b' = 0,$$

and then the equation $u - v = 0$ would not represent a straight line. But in that case the circles would be concentric, and we have demanded that they cut each other.

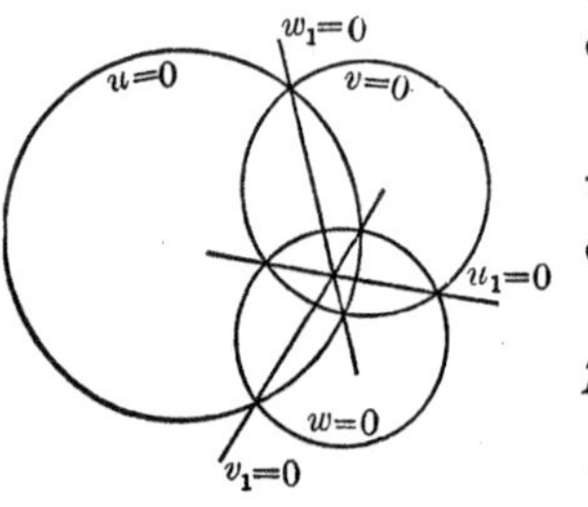

FIG. 4

Example 2. We can now prove the following theorem: *Given three circles, each pair of which intersect. Then their three common chords pass through a point, or are parallel.*

Let two of the three given circles be those of Example 1, and let the equation of the third circle be

$$w \equiv x^2 + y^2 + a''x + b''y + c'' = 0.$$

Then the equations of the three common chords can be written in the form:

$$u - v = 0, \qquad v - w = 0, \qquad w - u = 0.$$

Let

$$u_1 \equiv v - w, \qquad v_1 \equiv w - u, \qquad w_1 \equiv u - v.$$

We observe that the equation,

$$(8) \qquad u_1 + v_1 + w_1 \equiv 0 \qquad \text{or} \qquad - w_1 \equiv u_1 + v_1,$$

holds identically for all values of x and y. Consequently, the line $w_1 = 0$ is the same line as

$$u_1 + v_1 = 0,$$

and therefore it passes through the point of intersection of $u_1 = 0$ and $v_1 = 0$, or, if these lines are parallel, is parallel to them. Hence the theorem is proved.

The above proof is a striking example of a powerful method of Modern Geometry known as the *Method of Abridged Notation.** By means of this method many theorems, the proofs of which would otherwise be intricate, or for whose proof no method of attack is readily discerned, can be established with great ease.

EXERCISES

1. Find the equation of the straight line which passes through the origin and the point of intersection of the lines

$$2x - 3y - 2 = 0, \qquad 5x + 2y + 1 = 0.$$

Ans. $12x + y = 0$.

2. Find the equation of the straight line which passes through the point $(-1, 2)$ and meets the lines

$$x + y = 0, \qquad x + y + 3 = 0$$

at their point of intersection.

* The first general development of this method was given by the geometer, Julius Plücker, in his *Analytisch-geometrische Entwicklungen* of 1828 and 1831.

3. Find the equation of the straight line which passes through the point of intersection of the lines

$$5x - 2y - 3 = 0, \qquad 4x + 7y - 11 = 0$$

and is parallel to the axis of y.

4. Find the equation of the straight line which passes through the point of intersection of the lines given in Ex. 3 and makes an angle of 45° with the axis of x.

5. Find the equation of the straight line which passes through the point of intersection of the lines of Ex. 1 and is perpendicular to the first of the lines given in Ex. 3.

Ans. $38x + 95y + 58 = 0.$

6. The same, if the line is to be parallel instead of perpendicular.

7. Find the equation of the common chord of the parabolas

$$y^2 - 2y + x = 0, \qquad y^2 + 2x - y = 0.$$

Ans. $x + y = 0.$

8. The same for the parabolas

$$2x^2 - 5x + 2y = 3,$$
$$3x^2 + 7x - 9y = 4.$$

9. Write the equation of the pencil of curves determined by the two curves (*a*) of Ex. 1; (*b*) of Ex. 3; (*c*) of Ex. 7.

10. What is the equation of the pencil of circles determined by the two circles

$$x^2 + y^2 - 2x - 1 = 0,$$
$$x^2 + y^2 + 4x - 1 = 0\,?$$

Draw a figure showing the pencil. Find the equation of that circle of the pencil which goes through the point (2, 4).

11. Find the equation of the pencil of parallel lines (*a*) of slope 1; (*b*) of slope -3; (*c*) of slope λ_0.

Ans. (*a*) $y = x + k.$

12. Find the equation of the pencil of lines through (*a*) the point (0, 0); (*b*) the point (3, 2); (*c*) the point (0, b); (*d*) the point (x_0, y_0). *Ans.* (*a*) $lx + my = 0.$

4. The Equation $uv = 0$. Consider, for example, the equation

(1) $$x^2 - y^2 = 0.$$

Since $$x^2 - y^2 \equiv (x - y)(x + y),$$

it is clear that equation (1) will be satisfied

(*a*) if (x, y) lies on the line

(2) $$x - y = 0;$$

(*b*) if (x, y) lies on the line

(3) $$x + y = 0;$$

and in no other case. Equation (1), therefore, is equivalent to the two equations (2) and (3) taken together, and it represents, therefore, the two right lines (2) and (3).

It is clear from this example that we can generalize and say:

THEOREM. *The equation*

$$uv = 0$$

represents those points (x, y) *which lie on each of the two curves,*

$$u = 0, \qquad v = 0,$$

and no others.

It follows as an immediate consequence of the theorem that the equation

$$uvw \cdots = 0,$$

whose left-hand member is the product of any number of factors, represents the totality of curves corresponding to the individual factors, when these are successively set equal to zero.

Example. Consider the equation,

$$x^4 - y^4 = 0.$$

Here,*

$$x^4 - y^4 = (x^2 - y^2)(x^2 + y^2) = (x - y)(x + y)(x^2 + y^2).$$

* It is true that the following equation is an identity, and so the sign $\equiv$ instead of $=$ might be expected. The use of the sign $\equiv$ for an identical equation is not, however, considered obligatory, the sign $=$ being used when it is clear that the equation is an identity, so that the fact does not require special emphasis.

The given equation is, therefore, equivalent to the three equations:

$$x - y = 0, \qquad x + y = 0, \qquad x^2 + y^2 = 0.$$

The first two of these equations represent right lines. The third is satisfied by the coördinates of a single point, the origin. Since this point lies on the right lines, the third equation contributes nothing new to the locus.

EXERCISES

What are the loci of the following equations?

1. $\dfrac{x^2}{a^2} - \dfrac{y^2}{b^2} = 0.$ **2.** $x^2 + 3x + 2 = 0.$

3. $2x^2 + 3xy - 2y^2 = 0.$ **4.** $xy + x + 2y + 2 = 0.$

5. $x^2 + xy - 2x - 2y = 0.$ **6.** $x^3 + xy^2 = x.$

7. $3x^2y - 2xy = 0.$ **8.** $x^4 - y^4 - 2x^2 + 2y^2 = 0.$

9. $(x + y - 1)(x^2 + y^2) = 0.$ *Ans.* The line whose intercepts on the axes are both 1, and the origin.

10. $(x + y)(x^2 + y^2 + 1) = 0.$

11. $(x + y)[(x - 1)^2 + y^2] = 0.$

12. $x^3 + x^2y - xy^2 - y^3 = 0.$

Find, in each of the following exercises, a single equation whose locus is the same as that of the given systems of equations.

13. $x - 2 = 0, \qquad y - 4 = 0.$

14. $x = 2, \qquad y = 4.$

15. $x + y - 2 = 0, \qquad x - y + 2 = 0.$

16. $x - 3y = 5, \qquad 4x + 3 = 0.$

17. $\dfrac{x}{a} = \dfrac{y}{b}, \qquad \dfrac{x}{a} = -\dfrac{y}{b}.$

5. Tangents with a Given Slope. Discriminant of a Quadratic Equation. From elementary algebra we know that the roots of the quadratic equation

$$(1) \qquad Ax^2 + Bx + C = 0, \qquad A \neq 0,$$

are

$$x_1 = -\frac{B}{2A} + \frac{1}{2A}\sqrt{B^2 - 4AC},$$

$$x_2 = -\frac{B}{2A} - \frac{1}{2A}\sqrt{B^2 - 4AC}.$$

From these formulas the truth of the following theorem at once becomes apparent.

THEOREM 1. *The roots of the quadratic equation* (1) *are equal if and only if*

$$B^2 - 4AC = 0.$$

The quantity $B^2 - 4AC$ is known as the *discriminant* of the quadratic equation (1).

By means of the theorem we shall solve the following problem.

Problem. Let it be required to find the equation of the tangent to the parabola

$$(2) \qquad y^2 = 6x,$$

which is of slope $\frac{1}{2}$.

Let L be a line of slope $\frac{1}{2}$ which meets the parabola in two points, P_1 and P_2. If we allow L to move parallel to itself toward the tangent, T, the points P_1 and P_2 will move along the curve toward P, the point of tangency of T; and if L approach T as its limit, the points P_1 and P_2 will approach the one point P as their limit.

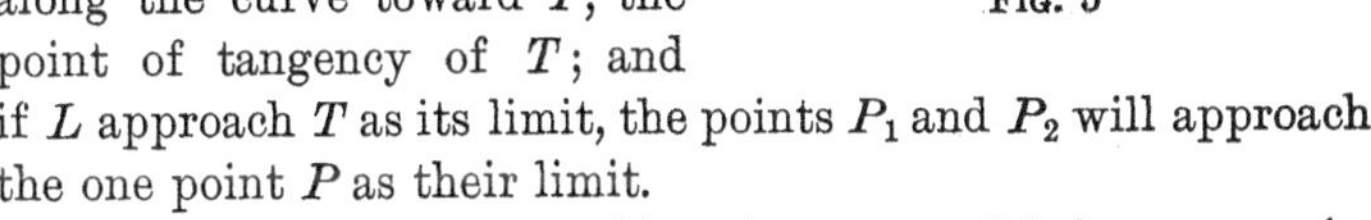

FIG. 5

It is clear that these considerations are valid for any conic. Accordingly, we may state the following theorem.

THEOREM 2. *A line which meets a conic intersects it in general in two points. If these two points approach coincidence* *in a single point, the limiting position of the line is a tangent to the conic.**

In applying Theorem 2 to the problem in hand, let us denote the intercept of the tangent T on the axis of y by β. The equation of T is, then,

$$(3)\qquad y = \tfrac{1}{2}x + \beta.$$

The coördinates of the point P, in which T is tangent to the parabola, are obtained by solving equations (2) and (3) simultaneously. Substituting in (2) the value of y given by (3), we have

$$(\tfrac{1}{2}x + \beta)^2 = 6x,$$

or

$$(4)\qquad x^2 + 4(\beta - 6)x + 4\beta^2 = 0.$$

The roots of equation (4) are equal, since they are both the abscissa of P. Accordingly, by Theorem 1, the discriminant of (4) is zero. Hence

$$16(\beta - 6)^2 - 16\beta^2 = 0, \qquad \text{or} \qquad -12\beta + 36 = 0.$$

Thus $\beta = 3$, and the tangent to the parabola (2) whose slope is $\frac{1}{2}$ has the equation

$$(5)\qquad x - 2y + 6 = 0.$$

If in (4) we set $\beta = 3$, the resulting equation,

$$x^2 - 12x + 36 = 0,$$

has equal roots, as it should. The common value is $x = 6$, and the corresponding value of y, from (2), is $y = 6$. The coördinates of the point of tangency, P, are, then, (6, 6).

Second Method. We proceed now to give a second method of solution for the type of problem just discussed. Let the conic be the ellipse

$$(6)\qquad 4x^2 + y^2 = 5,$$

and let the given slope be 4.

* A tangent to a conic might then be defined as the limiting position of a line having two points of intersection with the conic, when these points approach coincidence in a single point; this is a generalization of

It is evident from the figure that there are two tangents of slope 4 to the ellipse. Let the intercept on the axis of y of one of the tangents be β. The equation of this tangent is then

(7) $$y = 4x + \beta.$$

Our problem now is to determine the value of β. To this end, let the coördinates of the point of contact of the tangent be (x_1, y_1). Then a second equation of the tangent is, by (12), § 2,

(8) $$4x_1x + y_1y = 5.$$

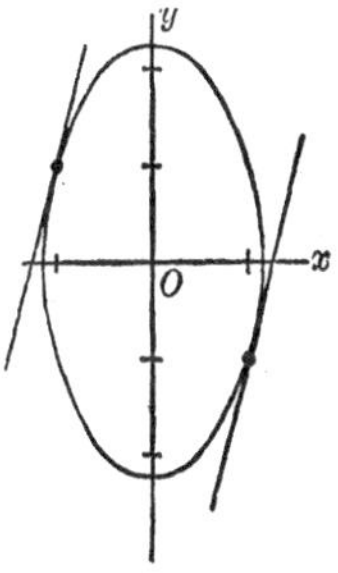

FIG. 6

Since equations (7) and (8), which we rewrite as

$$4x - y + \beta = 0,$$
$$4x_1x + y_1y - 5 = 0,$$

represent the same line, it follows, from Ch. II, § 10, Th. 5, that

$$\frac{4x_1}{4} = \frac{y_1}{-1} = \frac{-5}{\beta}.$$

From the equality of the first and third ratios we have

(9) $$x_1 = -\frac{5}{\beta}.$$

Since the second and third ratios are equal,

(10) $$y_1 = \frac{5}{\beta}.$$

Furthermore, the point (x_1, y_1) lies on the ellipse and so the values of x_1 and y_1, given by (9) and (10), satisfy equation (6). Accordingly,

$$\frac{100}{\beta^2} + \frac{25}{\beta^2} = 5, \qquad \text{or} \qquad \frac{25}{\beta^2} = 1.$$

Hence β has the value 5 or -5.

the definition of § 1. A tangent cannot be defined as a line meeting the conic in a single point, for there are lines of this character which are not tangents, viz., a line parallel to the axis of a parabola, or to an asymptote of a hyperbola.

Substituting these values of β in turn in (7), we obtain

$$4x - y + 5 = 0, \qquad 4x - y - 5 = 0,$$

as the equations of the two tangents of slope 4 to the ellipse (6). From equations (9) and (10) it follows that the points of contact of these tangents are, respectively, $(-1, 1)$ and $(1, -1)$.

Both the methods described in this paragraph are general in application. For the usual type of problem met with in a first course in Analytic Geometry either method may be used with facility. It is, however, to be noted that the second method presupposes that the equation of the tangent to the curve at an arbitrary point on the curve is known, whereas the first does not. Accordingly, in case a curve is given, for which the general equation of the tangent is not known, — for example, the parabola, $y = 3x^2 - 2x + 1$, — the first method will be shorter to apply.

EXERCISES

Determine in each of the following cases how many tangents there are to the given conic with the given slope. Find the equations of the tangents and the coördinates of the points of tangency. Use both methods in Exs. 1, 2, 3, checking the results of one by those of the other.

	Conic	*Slope*
1.	$x^2 + y^2 = 5,$	2.

Ans. $\begin{cases} 2x - y - 5 = 0, \text{ tangent at } (2, -1), \\ 2x - y + 5 = 0, \text{ tangent at } (-2, 1). \end{cases}$

2.	$y^2 = 3x,$	$\frac{3}{2}$.
3.	$2x^2 + y^2 = 11,$	$-\frac{2}{3}$.
4.	$x^2 + 8y = 0,$	2.
5.	$4x^2 - y^2 = 20,$	3.
6.	$x^2 + y^2 + 2x = 0,$	$\frac{1}{2}$.
7.	$6y^2 - 5x = 0,$	$1\frac{2}{3}$.

8. What are the equations of the tangents to the circle

$$x^2 + y^2 = 10,$$

which are parallel to the line $3x - y + 5 = 0$?

9. What is the equation of the tangent to the ellipse

$$4x^2 + 5y^2 = 20,$$

which is perpendicular to the line $x + 3y - 3 = 0$ and has a positive intercept on the axis of y?

10. Find the equation of the tangent to the parabola

$$y = 3x^2 - 2x + 1,$$

which is perpendicular to the line $x + 4y + 3 = 0$.

Ans. $4x - y - 2 = 0$.

11. Make clear geometrically that, no matter what direction is chosen, there are always two tangents to a given ellipse, which have that direction.

12. How many tangents are there to the parabola $y^2 = 2mx$, which have the slope 0? State a general theorem relating to the number of tangents to a parabola which have a given slope.

13. Are there any tangents of slope 3 to the hyperbola

$$4x^2 - y^2 = 5?$$

If so, what are their equations?

14. The preceding exercise, if the given slope is (*a*) 1; (*b*) 2. Give reasons for your answers.

6. General Formulas for Tangents with a Given Slope. Consider first the hyperbola

$$\frac{x^2}{a^2} - \frac{y^2}{b^2} = 1. \tag{1}$$

Before attempting to find a general formula for the equations of the tangents to the hyperbola, which have a given slope, λ, we shall do well to ask if such tangents exist. In answer to this question we state the following theorem.

THEOREM. *All the tangents to the hyperbola* (1) *are steeper than the asymptotes. Their slopes* λ *all satisfy the inequality*

$$(2) \qquad |\lambda| > \frac{b}{a} \qquad \text{or} \qquad \lambda^2 > \frac{b^2}{a^2}.$$

Conversely, if λ *satisfies* (2), *there are two tangents of slope* λ *to* (1). *If, however,* $\lambda^2 \leq b^2/a^2$, *there are no tangents of slope* λ *to* (1).

To prove the theorem, let a point P, starting from the vertex A, trace the upper half of the right-hand branch of (1). Then the tangent, T, at P, starting from the vertical position at A, turns continuously in one direction, and, as P recedes indefinitely, approaches the asymptote S as its limit. In other words, the slope, λ, of T decreases continuously through all positive values greater than the slope, b/a, of S, and approaches b/a as its limit.* Consequently, λ is always greater than b/a:

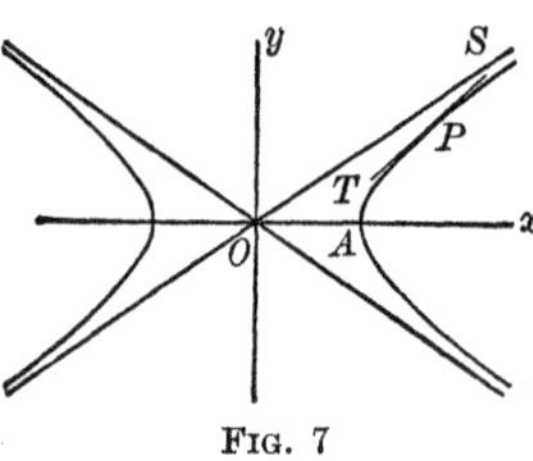

FIG. 7

$$\lambda > \frac{b}{a}.$$

* The geometrical evidence of this is convincing, but not conclusive. To clinch it, we give the following analytical proof: If the coördinates of P are (x, y), the slope λ of T is, by (11), § 2,

$$\lambda = \frac{b^2 x}{a^2 y}.$$

According to Ch. VIII, § 4, eq. (3), $\dfrac{x}{y} = \dfrac{a}{b}\dfrac{1}{\sqrt{1 - \dfrac{a^2}{x^2}}}.$

Hence $$\lambda = \frac{b}{a}\frac{1}{\sqrt{1 - \dfrac{a^2}{x^2}}}.$$

When P traces the upper half of the right-hand branch of (1) and recedes indefinitely, x increases continuously from the value a through all values greater than a. Then a^2/x^2 decreases continuously from 1 and approaches 0 as its limit; and $1 - a^2/x^2$, and hence $\sqrt{1 - a^2/x^2}$, in-

If P now traces the lower half of the right-hand branch, λ is negative, and always:

$$-\lambda > \frac{b}{a}.$$

These two inequalities can be combined into the single inequality (2). Thus (2) is satisfied by the slope λ of every tangent to the right-hand branch of (1), and hence also, because of the symmetry of the curve, by the slope λ of every tangent to the left-hand branch.

From the reasoning given in the first case, when P traces the upper half of the right-hand branch of (1), it follows, not only that $\lambda > b/a$, but also that λ takes on *every* value greater than b/a. Hence, if a value of λ, greater than b/a, is arbitrarily chosen, there is surely at least one tangent of this slope λ to (1), and consequently, because of the symmetry of the curve, there are actually two. Similarly, if a value of λ less than $-b/a$ is given.

To find the equations of the two tangents of slope λ to (1), in the case that λ does satisfy (2), we apply the first of the two methods of § 5. Let the equation of one of the tangents be

$$(3) \qquad y = \lambda x + \beta,$$

where β is to be determined. Proceeding to solve (1) and (3) simultaneously, we substitute for y in (1) its value as given by (3) and obtain the equation,

$$b^2x^2 - a^2(\lambda x + \beta)^2 = a^2b^2,$$

or

$$(4) \qquad (b^2 - a^2\lambda^2)x^2 - 2a^2\beta\lambda x - a^2(b^2 + \beta^2) = 0.$$

The roots of equation (4) are both equal to the abscissa of the

creases continuously from 0 and approaches 1 as its limit. Consequently, the reciprocal, $1/\sqrt{1 - a^2/x^2}$, of $\sqrt{1 - a^2/x^2}$ decreases continuously through all positive values greater than 1 and approaches 1 as its limit. Hence, finally, λ decreases continuously through all positive values greater than b/a and approaches b/a as its limit, q. e. d.

point of contact of the tangent (3), and hence the discriminant of (4) must vanish. We have, then,

$$4a^4\beta^2\lambda^2 + 4a^2(b^2 + \beta^2)(b^2 - a^2\lambda^2) = 0,$$

or, simplifying,

$$\beta^2 = a^2\lambda^2 - b^2. \tag{5}$$

Hence β has either of the values

$$\pm\sqrt{a^2\lambda^2 - b^2},$$

and the equations of the two tangents, written together, are

$$y = \lambda x \pm \sqrt{a^2\lambda^2 - b^2}. \tag{6}$$

Since λ satisfies (2), or the equivalent inequality $a^2\lambda^2 - b^2 > 0$, the quantity under the radical is positive and so has a square root.* We have thus obtained the following result.

The equations of the tangents to the hyperbola (1), *which have the given slope* λ, *where* λ *satisfies the inequality* (2), *are given by* (6).

Let the student deduce the following results, using either of the two methods of § 5.

The equations of the tangents to the ellipse

$$\frac{x^2}{a^2} + \frac{y^2}{b^2} = 1, \tag{7}$$

which have an arbitrarily given slope λ, *are*

$$y = \lambda x \pm \sqrt{a^2\lambda^2 + b^2}. \tag{8}$$

The equation of the tangent to the parabola

$$y^2 = 2mx, \tag{9}$$

which has a given slope λ, *not* 0, *is*

$$y = \lambda x + \frac{m}{2\lambda}. \tag{10}$$

* If we take a value of λ, for which $\lambda^2 < b^2/a^2$, then $a^2\lambda^2 - b^2$ is negative and has no square root. Consequently, there are no tangents with this slope, as the theorem states. Finally, if $\lambda = \pm b/a$, then $a^2\lambda^2 - b^2 = 0$, and (4) is not a quadratic equation.

Condition that a Line be Tangent to a Conic. The two methods used to find the tangent to a conic with a given slope apply equally well to the problem of determining the condition that an arbitrary line be tangent to a given conic. In fact, in finding the equations of the tangents of slope λ to the hyperbola (1), we have at the same time shown that *the condition that the line*

$$y = \lambda x + \beta, \tag{11}$$

where we now consider λ and β both arbitrary, be tangent to the hyperbola (1), *is that λ and β satisfy the equation* (5):

$$\beta^2 = a^2\lambda^2 - b^2.$$

Similarly, the work of deriving formula (8) or (10) involves finding the condition that the line (11) be tangent to the ellipse (7) or the parabola (9).

Example. Is the line $3x - 2y + 5 = 0$ tangent to the hyperbola $x^2 - 4y^2 = 4$?

It is, if, when we write the equations of the line and the hyperbola in the forms (11) and (1), the values which we obtain for λ, β, a^2, and b^2, namely, $\frac{3}{2}$, $\frac{5}{2}$, 4, and 1, satisfy (5). It is seen that they do not, and hence the line is not tangent to the hyperbola.

EXERCISES

1. Derive formula (8) and at the same time show that the condition that the line (11), where now λ and β are both arbitrary, be tangent to the ellipse (7) is that λ and β satisfy the equation

$$\beta^2 = a^2\lambda^2 + b^2. \tag{12}$$

2. Show that the line (11) is tangent to the parabola (9) if and only if

$$2\lambda\beta = m. \tag{13}$$

Hence prove the validity of formula (10).

3. By direct application of the methods of the text, show that the condition that the line (11) be tangent to the circle

$$x^2 + y^2 = a^2 \tag{14}$$

is that

$$\beta^2 = a^2(1 + \lambda^2). \tag{15}$$

4. Using formulas (6), (8), and (10), find the equations of the tangents which are required in Exs. 1, 2, 3, 5, and 7 of § 5.

5. Has the hyperbola $9x^2 - 4y^2 = 36$ any tangents whose inclination to the axis of x is 60°? Whose inclination is 45°? If so, find their equations.

6. Find the equations of the tangents to the parabola $y^2 = 8x$, one of which is parallel to and the other perpendicular to the line $3x - 2y + 5 = 0$. Show that these tangents intersect on the directrix.

7. Prove that any two perpendicular tangents to a parabola intersect on the directrix.

In each of the following exercises determine whether the given line is tangent to the given conic. If it is, find the coördinates of the point of contact.

	Conic	*Line*
8.	$2x^2 + 3y^2 = 5,$	$2x - 3y - 5 = 0.$
9.	$y^2 = 2x,$	$x + 4y + 8 = 0.$
10.	$3x^2 - 5y^2 = 7,$	$6x - 5y - 8 = 0.$

In each of the following cases the equation of the given line contains an arbitrary constant. Find the value or values of this constant, if any exist, for which the line is tangent to the given conic.

	Conic	*Line*
11.	$x^2 + 3y^2 = 4,$	$x - 3y + c = 0.$ *Ans.* $c = \pm 4.$
12.	$x^2 - y^2 = 3,$	$2x + dy - 3 = 0.$
13.	$5y^2 = 3x,$	$kx - 10y + 15 = 0.$
14.	$4x^2 - 3y^2 = 1,$	$x + 2y + k = 0.$

15. Is the line $x + y = 1$ tangent to the parabola $y = x - x^2$?

16. Show that the lines $3x \pm y + 10 = 0$ are common tangents of the circle $x^2 + y^2 = 10$ and the parabola $y^2 = 120x$.

17. Find the equations of the common tangents of the parabola $y^2 = 4\sqrt{2}x$ and the ellipse $x^2 + 2y^2 = 4$.

Ans. $\sqrt{2}x \pm 2y + 4 = 0$.

7. Tangents to a Conic from an External Point. Given a point P external to a conic, that is, lying on the convex side of the curve. From P it is possible, in general, to draw two tangents to the conic. It is required to find the equations of these tangents.

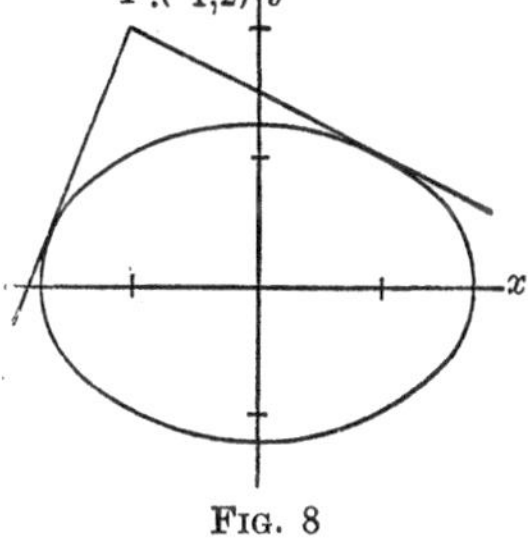

FIG. 8

Let the conic be the ellipse

$$x^2 + 2y^2 = 3 \tag{1}$$

and let P be the point $(-1, 2)$. We find the equations of the two tangents drawn from P to the ellipse by finding first the coördinates of the points of tangency. Let P_1 be the point of tangency of one of the tangents, and let the coördinates of P_1, which are as yet unknown, be (x_1, y_1). The equation of this tangent is then, by (12), § 2,

$$x_1x + 2y_1y = 3. \tag{2}$$

There are two conditions on the point P_1, to serve as a means of determining the values of x_1 and y_1. In the first place, the tangent (2) at P_1 must go through the point $(-1, 2)$; hence

$$-x_1 + 4y_1 = 3. \tag{3}$$

Secondly, the point P_1 lies on the ellipse (1); that is,

$$x_1^2 + 2y_1^2 = 3. \tag{4}$$

Equations (3) and (4) are two simultaneous equations in the unknowns x_1, y_1. If we solve (3) for x_1:

$$x_1 = 4y_1 - 3, \tag{5}$$

and substitute its value in (4), we obtain, on simplification, the following equation for y_1:

(6) $$3y_1^2 - 4y_1 + 1 = 0.$$

The roots of this equation are $y_1 = 1$ and $y_1 = \frac{1}{3}$; the corresponding values of x_1 are, from (5), 1 and $-\frac{5}{3}$. Hence $(x_1, y_1) = (1, 1)$ and $(x_1, y_1) = (-\frac{5}{3}, \frac{1}{3})$ are the solutions of (3) and (4).

The coördinates of the points of tangency are, therefore, (1, 1) and $(-\frac{5}{3}, \frac{1}{3})$. Substituting the coördinates of each point in turn for $x_1\, y_1$ in (2) and simplifying the results, we obtain, as the equations of the two tangents,

(7) $$x + 2y - 3 = 0 \quad \text{and} \quad 5x - 2y + 9 = 0.$$

The method used in this example is universal in its application, not only to conics, but to other curves as well. It should be noted, however, that the equation corresponding to (6) does not, in general, have rational, that is, fractional or integral, roots. Usually its roots involve radicals and hence so do the final equations of the tangents. If one were dealing with an arbitrary point P external to an arbitrary conic, for example, the ellipse

$$\frac{x^2}{a^2} + \frac{y^2}{b^2} = 1,$$

these radicals would be complicated. Accordingly, we make no attempt to set up general formulas for the tangents to a given conic from an external point. We have expounded a method which is applicable in all cases, and this is the purpose we set out to achieve.

Second Method. We give briefly an alternative method of finding the equations of the tangents from the point $(-1, 2)$ to the ellipse (1).

Suppose one of the tangents is the line

(8) $$y = \lambda x + \beta.$$

Since it is a tangent to (1), we have, according to § 6, Ex. 1,

$$2\beta^2 = 6\lambda^2 + 3.$$

Since it contains the point $(-1, 2)$,

$$2 = -\lambda + \beta.$$

If we solve these equations in λ and β simultaneously, we find that $\lambda = -\frac{1}{2}$ or $\frac{5}{2}$ and that $\beta = \frac{3}{2}$ or $\frac{9}{2}$. Substituting these pairs of values for λ and β in turn in (8) and simplifying the results, we obtain the equations (7).

EXERCISES

1. Make clear geometrically that from a point external to an ellipse or a parabola there can always be drawn just two tangents to the curve.

2. How many tangents can be drawn to a hyperbola from its center? From a point on an asymptote, not the center? From any other external point? Summarize your answers in the form of a theorem.

3. Let P be a point external to a hyperbola from which two tangents can be drawn to the curve. How must the position of P be restricted, if the two tangents are drawn to the same branch of the hyperbola? To different branches?

4. The point $(2, 0)$ is a point internal to the hyperbola $x^2 - 2y^2 = 2$. Prove analytically that no tangent can be drawn from it to the curve.

In each of the following exercises determine how many tangents there are from the point to the conic, and when there are tangents, find their equations. Use the first method.

	Conic	*Point*	
5.	$x^2 + y^2 = 5$,	$(3, 1)$.	*Ans.* $\begin{cases} x + 2y - 5 = 0, \\ 2x - y - 5 = 0. \end{cases}$
6.	$x^2 - 3y^2 = 4$,	$(\frac{1}{2}, -1)$.	
7.	$x^2 - 2y^2 = 2$,	$(1, -2)$.	
8.	$4x^2 - 9y^2 = 36$,	$(4, 1)$.	
9.	$y^2 - 4x = 0$,	$(4, 5)$.	
10.	$x^2 - 4y^2 = 4$,	$(2, 1)$.	

11. $x^2 - 8y = 0$, $(3, 2)$.

12. $2x^2 - 3y^2 = -10$, $(-2, 1)$.

13. $x^2 + y^2 - 4x - y = 0$, $(5, 2)$.

14. $x^2 + y^2 = 25$, $(-1, 7)$.

15. Work Exercises 5–10 by the second method.

16. Show, by use of the second method, that the tangents from the point $(2, 3)$ to the ellipse $4x^2 + 9y^2 = 36$ are perpendicular.

EXERCISES ON CHAPTER IX

1. Prove that the slope of the conic

$$(1 - e^2)x^2 + y^2 - 2mx + m^2 = 0$$

at the point (x_1, y_1) is

$$\lambda = -\frac{(1 - e^2)x_1 - m}{y_1}.$$

Hence show that the equation of the tangent at (x_1, y_1) is

$$(1 - e^2)x_1x + y_1y - m(x + x_1) + m^2 = 0.$$

2. Show that the slope of the curve

$$Ax^2 + Bxy + Cy^2 + Dx + Ey + F = 0$$

at the point (x_1, y_1) is

$$\lambda = -\frac{2Ax_1 + By_1 + D}{Bx_1 + 2Cy_1 + E}.$$

Then prove that the equation of the tangent at (x_1, y_1) is

$$Ax_1x + \frac{B}{2}(y_1x + x_1y) + Cy_1y + \frac{D}{2}(x + x_1) + \frac{E}{2}(y + y_1) + F = 0.$$

3. The following equations contain arbitrary constants. What does each represent?

(*a*) $y = \lambda x + 3$;

Ans. All the lines through $(0, 3)$ except $x = 0$.

(*b*) $3ax + 2y + a - 3 = 0$;

(*c*) $7x + 5y - c + 3 = 0$;

(*d*) $(2a + 5)x + (7a - 3)y = 9a + 2$;

(*e*) $lx + (2l + m)y - 3m = 0$;

(*f*) $(2l + 3m)x - (4l + 6m)y + 5l = 0$.

4. A line moves so that the sum of the reciprocals of its intercepts is constant. Show that it always passes through a fixed point.

5 A line with positive intercepts moves so that the excess of the intercept on the axis of x over the intercept on the axis of y is equal to the area of the triangle which the line forms with the axes. Show that it always passes through a fixed point.

6. Prove that the straight lines,

$$5x - 2y + 6 = 0,$$
$$2x - 4y + 3 = 0,$$
$$3x + 2y + 3 = 0,$$

meet in a point, by showing that the equation of one of them can be written in the form $lu + mv = 0$, where $u = 0$ and $v = 0$ are the equations of the others.

7. Show that the three lines,

$$x + 3y - 4 = 0,$$
$$5x - 3y + 6 = 0,$$
$$3x - 9y + 14 = 0,$$

meet in a point.

8. Prove that the three lines

$$k\alpha - l\beta = 0, \qquad l\beta - m\gamma = 0, \qquad m\gamma - k\alpha = 0,$$

where $\alpha=0$, $\beta=0$, and $\gamma=0$ are themselves equations of straight lines and k, l, and m are constants, meet in a point.

9. Find the equation of the common chord of the two intersecting circles

$$x^2 + y^2 + 6x - 8y + 3 = 0,$$
$$2x^2 + 2y^2 - 3x + 4y - 12 = 0.$$

10. Show that the two circles,

$$x^2 + y^2 - 4x - 4y - 10 = 0,$$
$$x^2 + y^2 + 6x + 6y + 10 = 0,$$

are tangent to one another. Find the equation of the common tangent and the coördinates of the common point.

11. Find the equation of the circle which goes through the points of intersection of the two circles of Ex. 9 and through the origin.

12. Find the equation of the circle which is tangent to the circles of Ex. 10 at their common point and meets the axis of x in the point $x = 2$.

13. What is the equation of the circle which passes through the points of intersection of the line

$$2x - y + 4 = 0$$

and the circle

$$x^2 + y^2 + 2x - 4y + 1 = 0,$$

and goes through the point (1, 1)?

14. Determine the equation of the ellipse which passes through the points of intersection of the ellipse

$$x^2 + 4y^2 = 4$$

and the line

$$3x - 4y - 3 = 0,$$

and goes through the point (2, 1). By a transformation to parallel axes (cf. Ch. XI, § 1), prove that this ellipse has axes parallel to those of the given ellipse and has the same eccentricity.

15. Find a single equation representing both diagonals of the rectangle whose center is at the origin and one of whose vertices is at the point (a, b).

16. What is the condition that the equation

$$a^2x^2 - b^2y^2 = 0$$

represent two perpendicular lines?

17. Find the locus of each of the following equations:

$$(a) \quad 6x^2 + 5xy - 4y^2 = 0;$$
$$(b) \quad 4x^2 - 20xy + 25y^2 = 0;$$
$$(c) \quad x^2 + xy + 6y^2 = 0;$$

18. Prove that the equation

(1) $$Ax^2 + Bxy + Cy^2 = 0$$

represents the origin, a single straight line, or two straight lines, according as the discriminant, $B^2 - 4AC$, is negative, zero, or positive.

19. Show that, if equation (1), Ex. 18, represents two straight lines, the slopes of these lines are the roots of the equation

$$C\lambda^2 + B\lambda + A = 0.$$

20. Prove that the equation

$$14x^2 - 45xy - 14y^2 = 0$$

represents two perpendicular straight lines.

21. Show that equation (1), Ex. 18, represents two perpendicular straight lines if and only if $A + C = 0$.

22. Prove that the equation

$$y^2 - 2xy \sec\theta + x^2 = 0$$

represents two straight lines which form with one another the angle θ.

23. A regular hexagon has its center at the origin and two vertices on the axis of x. Find a single equation which represents all three diagonals. *Ans.* $y^3 - 3x^2y = 0$.

24. Determine the points of contact of the tangents drawn to an ellipse from the points on the conjugate axis which are at a distance from the center equal to the semi-axis major.

25. Find the equations of the common tangents of each of the following pairs of conics:

(a) $x^2 + y^2 = 16$, $y^2 = 6x$;

(b) $\dfrac{x^2}{25} + \dfrac{y^2}{9} = 1$, $\dfrac{x^2}{16} + \dfrac{y^2}{25} = 1$;

(c) $\dfrac{x^2}{16} + \dfrac{y^2}{9} = 1$, $\dfrac{x^2}{25} - \dfrac{y^2}{16} = 1$.

Draw a good figure in each case, showing the common tangents.

26. Show that the line

$$(2) \qquad \frac{x}{A}+\frac{y}{B}=1$$

is tangent to the circle

$$x^2+y^2=a^2$$

if and only if

$$\frac{1}{A^2}+\frac{1}{B^2}=\frac{1}{a^2}.$$

27. Find the condition that the line (2), Ex. 26, be tangent to the ellipse

$$\frac{x^2}{a^2}+\frac{y^2}{b^2}=1. \qquad \textit{Ans.}\ \frac{a^2}{A^2}+\frac{b^2}{B^2}=1.$$

28. What will the condition obtained in Ex. 27 become in the case of the hyperbola

$$\frac{x^2}{a^2}-\frac{y^2}{b^2}=1\,?$$

29. Prove that the line (2), Ex. 26, is tangent to the parabola $y^2=2\,mx$, if and only if $2\,B^2+Am=0$.

30. Find the condition that the line $y=\lambda x+\beta$ be tangent to the conic

$$(1-e^2)x^2+y^2-2\,mx+m^2=0.$$

Ans. $(\beta+m\lambda)^2-e^2(\beta^2+m^2)=0.$

31. In an ellipse there is inscribed a rectangle with sides parallel to the axes. In this rectangle there is inscribed a second ellipse, with axes along the axes of the first. Show that a line joining extremities of the major and minor axes of the first ellipse is tangent to the second.

CHAPTER X

POLAR COÖRDINATES

1. Definition. It is possible to describe completely the position of a point in a plane by telling its *distance* and its *direction* from a given point. This idea forms the basis of the system of polar coördinates.

Let O be the given point, and draw from O a ray, OA, from which to measure angles. Let P be any point of the plane. Denote its distance from O by r, and the angle AOP by θ. Then (r, θ) form the *polar coördinates* of the point P. O is called the *pole* or *origin*; OA, the *prime direction* or *initial ray*; and r, the *radius vector* (pl. *radii vectores*).

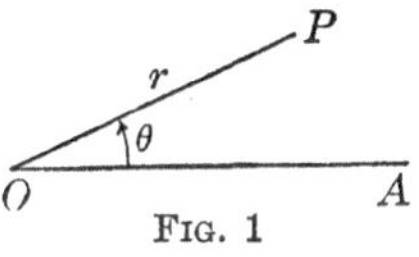

FIG. 1

When r and θ are given, one, and only one, point is determined. When, on the other hand, a point is given, r is completely determined, but θ may have any one of an infinite set of values differing from one another by multiples of 360° (or 2π). Thus, if θ' is one value of θ, the others will all be comprised in the formula

$$\theta = \theta' \pm 360\,n \qquad (\text{or} \quad \theta' \pm 2\pi n),$$

where n is a whole number.

For the point O, $r = 0$; but there is no more reason for assigning to θ one value rather than another. As the coördinates of O, therefore, we take $(0, \theta)$, where θ may be any number whatever.

It is possible to define polar coördinates so that r can be negative. Thus the point $(-2, 30°)$ would be obtained by drawing the ray which makes an angle of 30° with OA and then laying off on the opposite ray a distance of 2 units. This system of polar coördinates is not widely used in later work in mathematics, and even in analytic geometry it is a matter of custom rather than of any logical necessity. We shall, therefore, adhere to the original definition and exclude negative values of r, unless an explicit statement to the contrary is made.

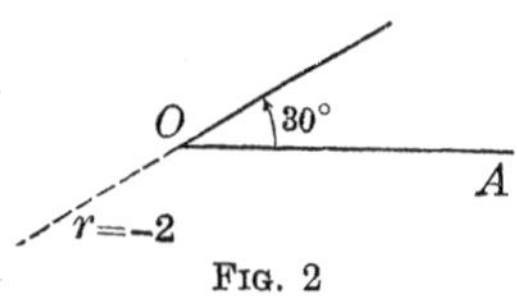

FIG. 2

EXERCISES

For use in these and later exercises the student should procure polar coördinate paper, ruled like a cobweb. Otherwise he should use a scale and protractor.

Plot the following points:

1. $(1, 0°)$.
2. $(0, 1°)$.
3. $(5, 30°)$.
4. $(5, -30°)$.
5. $(2, 200°)$.
6. $(2, -90°)$.
7. $(3, 180°)$.
8. $(4, \frac{1}{2}\pi)$.
9. $(6, \frac{2}{3}\pi)$.

10. What are the coördinates of the vertices of a square whose center is at O, the prime direction being perpendicular to a side, if the length of one side is $2a$?

11. Write down the coördinates of the vertices of an equilateral triangle, the pole being at the center and one vertex lying on OA.

12. The same for a regular heptagon.

13. What loci are represented by the following equations?

(a) $r = 5$; (b) $\cos\theta = 0$; (c) $\theta = 90°$.

2. Circles. Among the simplest curves in polar coördinates are

(a) the circles with center O. The equation of one of them is

(1) $$r = a,$$

where a is the radius.

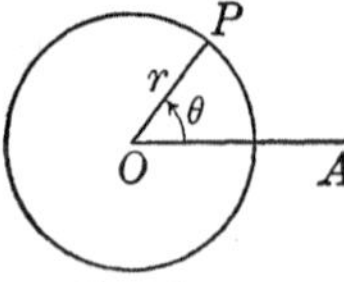

FIG. 3

(*b*) the circles which pass through O. Begin with one whose center lies on OA. If its radius is a, then evidently its equation is

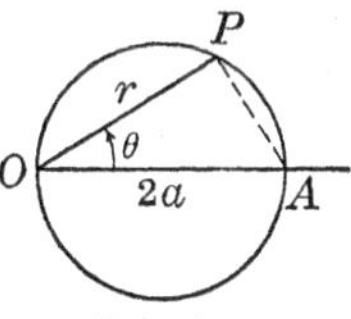

FIG. 4

(2) $$r = 2a\cos\theta.$$

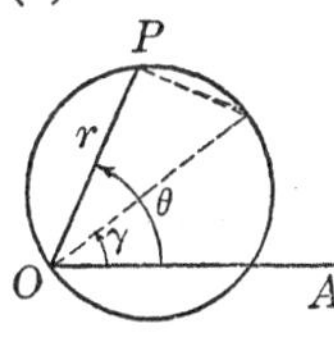

FIG. 5

If the coördinates of the center of an arbitrary circle through O are (a, γ), then the equation is

(3) $$r = 2a\cos(\theta - \gamma).$$

EXERCISES

1. Plot directly each of the following curves (making a convenient numerical choice of a; as, for example, 2 cm.):

(*a*) $r = 2a\sin\theta$;

(*b*) $r = -2a\cos\theta$;

(*c*) $r = -2a\sin\theta$.

2. Obtain each of the equations in Ex. 1 as a special case under (3), by choosing γ properly.

3. Circles are described with their centers at the vertices of the equilateral triangle of Ex. **11,** § 1, each circle passing through the center of the triangle. Find their equations.

4. A circle of radius 2 has its center on OA at a distance 3 from O. Show that its equation is

$$r^2 - 6r\cos\theta + 5 = 0.$$

5. A circle whose radius is 4 has its center at the point $(5, 90°)$. Show that its equation is

$$r^2 - 10r\sin\theta + 9 = 0.$$

6. Show that the equation of any circle is

$$r^2 - 2cr\cos(\theta - \gamma) + c^2 = \rho^2,$$

where ρ denotes the radius and (c, γ) are the coördinates of the center.

What curve is represented by each of the following equations?

7. $r^2 + 8r\cos\theta = 9.$

8. $r^2 + 8r\sin\theta = 9.$

9. $r^2 - 8r\sin\theta = 9.$

10. $r^2 + 2r\cos\theta - 2r\sin\theta = 7.$

11. $r^2 - 2r\cos\theta + 2r\sin\theta = 7.$

12. $r^2 - 6r\cos\theta - 8r\sin\theta = 11.$

3. Straight Lines. Let us consider first a line L which does not pass through O, and assume, to begin with, that L meets the prime direction at right angles at the distance h from O. The equation of L is, evidently,

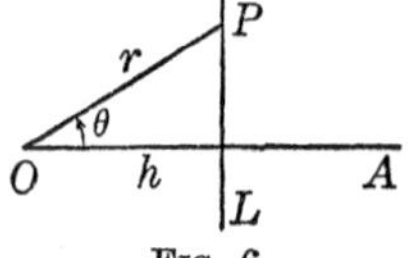

FIG. 6

(1) $$r\cos\theta = h.$$

If L is parallel to the prime direction and at a distance h above it, it is easily shown that its equation is

(2) $$r\sin\theta = h.$$

Let L, now, be any line not going through O. Draw a line through O perpendicular to L, and let B be the point in which it cuts L (Fig. 7). Denote the length of the line-segment OB by h, and the $\measuredangle\, AOB$ by γ. Let $P\colon(r,\ \theta)$ be any point of L.

Then $$\measuredangle\, BOP = \theta - \gamma.$$

Consequently, we have

(3) $$r\cos(\theta - \gamma) = h$$

as the equation of L.

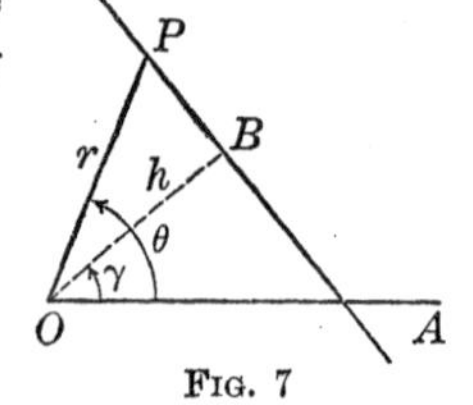

FIG. 7

Rays from O. The equation of a *ray*, or half-line, emanating from O, is

$$\theta = \alpha,$$

where α is a constant angle. Thus $\theta = 0$ is the equation of the prime direction, and $\theta = 90°$ is the equation of a ray drawn from O at right angles to the prime direction. There are two such rays; the equation of the other one is $\theta = -90°$.

To the right-hand side of any of these equations can be added any positive or negative multiple of 360°, without altering the locus.

Lines through O. The equation of the line through O perpendicular to the prime direction is

$$\cot \theta = 0. \tag{4}$$

For then $\theta = 90°$ or $\theta = -90°$

and we have just seen that these are equations of the two rays making up the line.

The equation of any other line through O is

$$\tan \theta = c, \tag{5}$$

where c is a constant. Thus

$$\tan \theta = 1$$

represents a line through O, for the points of which one or the other of the two equations

$$\theta = 45° \qquad \text{or} \qquad \theta = 225°$$

holds.

EXERCISES

1. Establish equation (2).

2. Derive equations (1) and (2) from equation (3).

3. If a line is perpendicular to the prime direction but does not cut it, what is its equation? Let h be its distance from the pole.

4. Find the equation of a line which is parallel to the prime direction and a distance h below it.

5. Find the equation of the line which cuts the prime direction at a distance of 5 units from the pole and makes an angle of $-45°$ with the prime direction.

What does each of the following equations represent? Make a plot in each case.

6. $r \cos \theta = 4$.

7. $r \sin \theta = 4$.

8. $r \cos \theta = -4$.

9. $r \sin \theta = -4$.

10. $r \sin \theta + r \cos \theta = 3$.

11. $r \sin \theta - r \cos \theta = 3$.

12. $4r \sin \theta + 3r \cos \theta = 5$.

13. $5r \sin \theta - 12r \cos \theta = 26$.

14. $\tan \theta = -1$.

15. $2 \cos \theta = 0$.

16. $\theta = 60°$.

17. $\theta = 180°$.

4. Graphs of Equations. If an equation in polar coördinates is given, which cannot be reduced to one of the forms recognized as representing a known curve, it is necessary, in order to determine what curve is defined by the equation, to plot a reasonable number of points whose coördinates satisfy the equation. But considerations of symmetry will often shorten the work.

Example 1. Consider the equation

(1) $$r^2 = 16 \sin \theta.$$

This equation is equivalent to

(2) $$r = 4\sqrt{\sin \theta},$$

where we have taken only the positive square root, since negative values of r have for us no meaning.

When $\theta = 0$, $r = 0$; as θ increases, r increases, and when $\theta = 90°$, $r = 4$. Using a table of sines and a table of square roots, we compute the following coördinates of further points of the curve.

θ	10°	20°	30°	40°	50°	60°	70°	80°
r	1.67	2.34	2.83	3.21	3.50	3.72	3.88	3.97

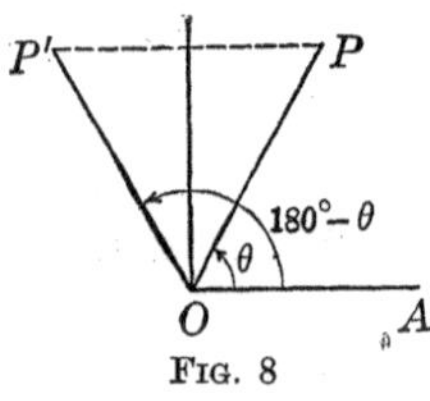

Fig. 8

More computations are unnecessary. For, the curve is symmetric in the ray $\theta = 90°$. To prove this, we note that, if $P: (r, \theta)$ is any point of the curve, then the point $P': (r,\ 180° - \theta)$, which is symmetric to P in the ray $\theta = 90°$, is also a point of the curve, inasmuch as

$$\sin (180° - \theta) = \sin \theta.$$

We now have points on the curve for values of θ from 0° to 180°. These points determine the entire curve, since, if θ is greater than 180° (and less than 360°), $\sin\theta$ is negative and (2) is meaningless.

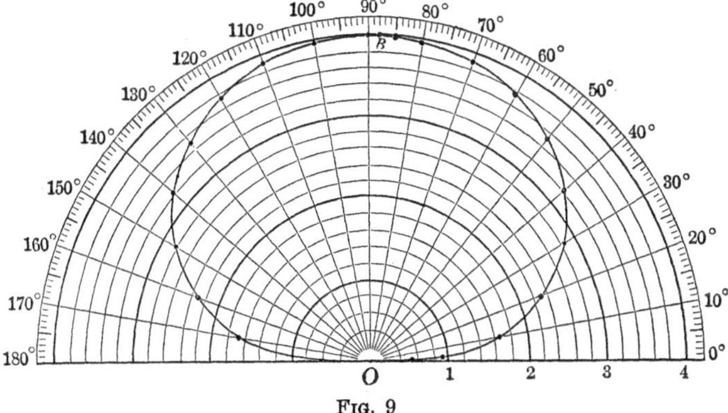

FIG. 9

To make sure that the curve has not sharp corners at O and B, we must compute r for small values of θ and also for values of θ near 90°.

θ	1°	3°	5°	85°	88°
r	.53	.92	1.18	3.99	4.00 –

The corresponding points, when plotted (Fig. 9), show that the curve is smooth at O and B.

If we admit negative r's, we obtain for each point of the present curve a new point symmetric to it in O. We have, then, instead of a single loop, a curve with two loops (Fig. 10) which are symmetric to each other in O, and also in OA.

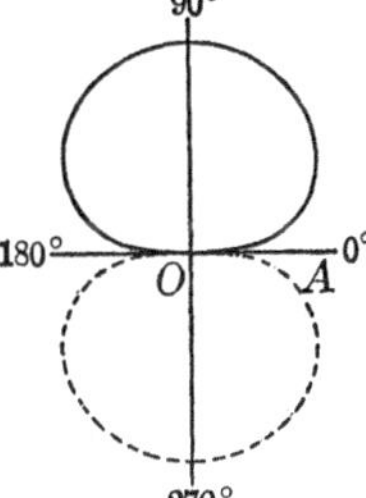

FIG. 10

Example 2. Given the equation

(3) $$r = 10\cos 3\theta.$$

When $\theta = 0$, $r = 10$. But here, as θ increases, r decreases, and when $\theta = 30°$, $r = 0$. The coördinates of intermediate points of the curve are:

θ	5°	10°	15°	20°	25°	27.5°
r	9.7	8.7	7.1	5.0	2.6	1.3

The curve is symmetric in the prime direction. For, if the point $P: (r, \theta)$ is any point of the curve, then the point $P': (r, -\theta)$, which is symmetric to P in OA (Fig. 11), is also a point of the curve, since

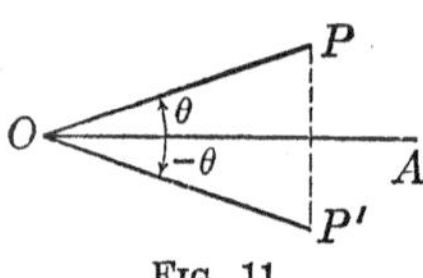

Fig. 11

$$\cos 3(-\theta) = \cos 3\theta.$$

If we plot the points already computed and those symmetric to them, we obtain a piece of the curve (Fig. 12). By plotting, further,—the points for θ equal, say, to 1°, 2°, and 3°,—we would find that the curve is smooth in the point A.

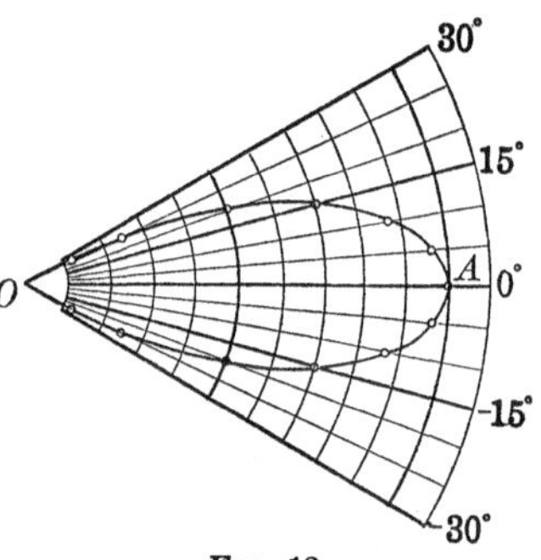

Fig. 12

When θ increases beyond 30°, 3θ is greater than 90° and r becomes negative, so that there are no points on the curve. This situation persists throughout the angle

$$30° < \theta < 90°.$$

In the angle
$$90° < \theta < 150°,$$

however, r is again positive. The piece of the curve which lies in this angle is congruent to the piece OA already plotted and may be obtained by rotating the piece OA through an angle of 120° about the pole. For, if $P: (r, \theta)$ is a point of the curve (Fig. 13), then $P': (r, \theta + 120°)$ is also, since

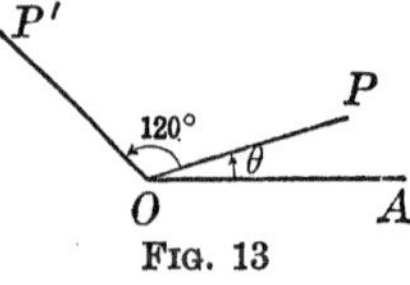

Fig. 13

$$\cos 3(\theta + 120°) = \cos(3\theta + 360°) = \cos 3\theta.$$

Hence every point on the first *lobe* of the curve yields a point on the second lobe by merely rotating the radius vector through 120°.

The second lobe again gives rise to a third lobe congruent to it and advanced by 120°. If this last lobe were again advanced, it would yield the first. Hence the three lobes complete the curve.

If we admit negative r's, the curve is unchanged. The points which we then get, for example, for values of θ between 30° and 90° lie on the third lobe of the curve. Thus, for values of θ from 0° to 360°, each point of the curve is obtained twice, once for $\theta = \theta'$ and once for $\theta = \theta' + 180°$.

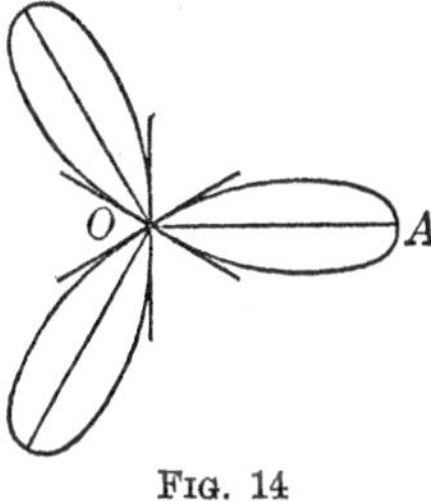

FIG. 14

Tests for Symmetry. Let the student show that the test for symmetry in the ray $\theta = 90°$, given in Example 1, also insures symmetry in the ray $\theta = 270°$, and that the test for symmetry in the prime direction, given in Example 2, also yields symmetry in the ray $\theta = 180°$.

These tests are general, and can be stated as theorems.

THEOREM 1. *A curve is symmetric in the line of the prime direction if, on substituting* $-\theta$ *for* θ *in its equation, the equation is unaltered.*

THEOREM 2. *A curve is symmetric in the line through the pole perpendicular to the prime direction if, on substituting* $180° - \theta$ *for* θ *in its equation, the equation is unaltered.*

EXERCISES

Plot the following curves.

1. The lemniscate (take $a = 5$ cm.),

$$r^2 = a^2 \cos 2\theta.$$

2. The cardioid (take $a = 2\frac{1}{2}$ cm.),

$$r = 2a(1 - \cos\theta).$$

3. The limaçon (a generalization of the cardioid),

$$r = 4 - 3\cos\theta.$$

4. A second type of limaçon,

$$r = 3 - 4\cos\theta.$$

Show that if negative r's are admitted, a piece is added to the curve.

5. $r^2 = 16\cos\theta$. 6. $r = 10\sin 3\theta$. 7. $r^2 = a^2\sin 2\theta$.

8. How are the curves of Exs. 5, 6, and 7 related, respectively, to the curves of Examples 1, 2 of the text and the lemniscate of Ex. 1?

9. $r = \dfrac{3}{1 + \cos\theta}$. 10. $r = \sec^2\dfrac{\theta}{2}$.

11. $r = \dfrac{3}{1 - \frac{1}{2}\cos\theta}$. 12. $r = \dfrac{3}{1 + 2\cos\theta}$.

13. $r = 5\cos 2\theta$.
Show that this curve has two lobes, but that it would have four lobes, if negative r's were admitted.

14. $r = 5\cos 4\theta$. 15. $r = a\cos n\theta$.

16. Show that the curve of Ex. 15 has n lobes; but, if n is even and negative r's are admitted, it has $2n$ lobes.

17. The spiral of Archimedes,

$$r = \theta,$$

taking θ in degrees and $\frac{1}{90}$ cm. as the unit of length.

18. The hyperbolic spiral,

$$r = \frac{1}{\theta},$$

taking θ in radians and 2 cm. as the unit of length.

19. $r = 1 - \theta^2$. 20. $r^2 + \theta^2 = 1$.

5. Conics. The equation of a conic section, when the definition of Ch. VIII, § 7 is used, is simple in polar coördinates.

Let the pole be taken at the focus F, and let the prime direction be chosen perpendicular to the directrix D and away from D; let $KF = m$.

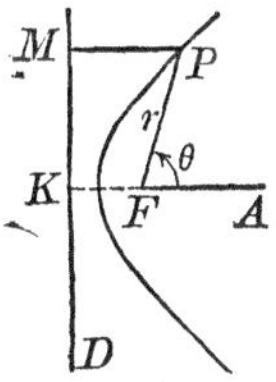

If $P:(r, \theta)$ is a point lying *to the right of D* (Fig. 15) and on the conic, then

$$FP = r, \qquad MP = r\cos\theta + m.$$

Fig. 15

Now, by definition,

$$\frac{FP}{MP} = e,$$

and hence the equation of the locus of P is

$$\frac{r}{r\cos\theta + m} = e,$$

or, if we solve for r,

$$(1) \qquad r = \frac{em}{1 - e\cos\theta}.$$

Ellipses and parabolas lie to the right of D and hence are represented by (1), when $e < 1$ and $e = 1$, respectively.

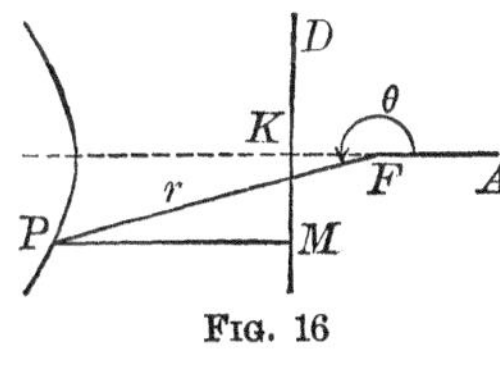

Fig. 16

When $e > 1$, however, (1) represents just the right-hand branch of a hyperbola. The equation of the left-hand branch is

$$(2) \qquad r = -\frac{em}{1 + e\cos\theta}.$$

For, if $P:(r, \theta)$ lies on the left-hand branch (Fig. 16), it is to the left of D and

$$PM = -r\cos\theta - m.$$

We then have, since $FP/PM = e$,

$$\frac{r}{-r\cos\theta - m} = e,$$

which reduces to (2).

If negative r's are admitted, the single formula (1) gives both branches. For, in this case we may take r, for a point

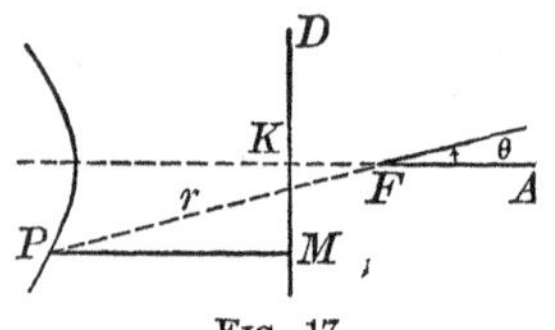

Fig. 17

P on the left-hand branch, as negative (Fig. 17). Then we have

$$FP = -r, \qquad PM = -r\cos\theta - m,$$

and hence
$$\frac{-r}{-m - r\cos\theta} = e,$$

which reduces to (1).

It is seen, then, that the choice we have made, admitting only positive or zero r's, is more discriminating, for we are able to represent a single branch of the curve by a simple formula, — (1) or (2). In analytic geometry we do not usually care to do this, the curve that interests us being the pair of branches. But in applied mathematics it often happens that one branch of a hyperbola plays a rôle and the other has no meaning. Thus when a comet is traveling in a hyperbolic orbit, it is only one branch of the hyperbola which forms the path.*

New Choice of Prime Direction. If the prime direction had not been chosen along KF produced, but at an angle γ with it, as shown in Fig. 18, then evidently $\theta - \gamma$ would take the place of θ in the foregoing formulas, but there would be no other change. The final equations would now read

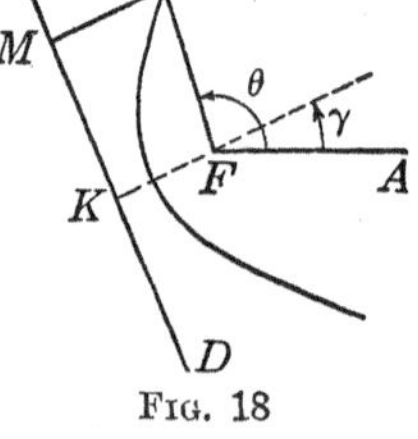

Fig. 18

$$(3) \qquad r = \frac{em}{1 - e\cos(\theta - \gamma)},$$

$$(4) \qquad r = -\frac{em}{1 + e\cos(\theta - \gamma)}.$$

* We note that this is not the only way in which we are able, by simple formulas, to discriminate between the two branches of a hyperbola. Thus the equation

$$-x^2 + y^2 = 1$$

represents a hyperbola on the axis of y. The equation

$$y = \sqrt{1 + x^2}$$

represents one of its branches, and the equation

$$y = -\sqrt{1 + x^2}$$

the other.

Example. What curve is represented by the equation

$$r = \frac{6}{1 + 2 \sin \theta}?$$

The equation can be reduced to the form (3) by choosing γ so that

$$\cos (\theta - \gamma) = - \sin \theta.$$

Obviously, γ must be 270° or, what amounts to the same thing, $-90°$. Moreover, $e = 2$ and $m = 3$.

Thus the equation represents one branch of a hyperbola whose eccentricity is 2. The transverse axis is perpendicular to the prime direction and the branch in question is the one opening downward. The center of the hyperbola is at the point (4, 90°). Its asymptotes make angles of 60° with the transverse axis. The lengths of the semi-axes are $a = 2$ and $b = 2\sqrt{3}$. The vertices are at the points (2, 90°) and (6, 90°). The second focus is at (8, 90°) and the equations of the directrices are $r \sin \theta = 3$ and $r \sin \theta = 5$.

Let the student verify each of these statements, using the formulas of Ch. VIII, § 6, and then draw the curve to the scale of 1 cm. as a unit, marking each of the points mentioned with its coördinates and drawing in the asymptotes and directrices. What is the length of the latus rectum?

EXERCISES

What conic or, in the case of a hyperbola, what branch is represented by each of the following equations? Draw a rough figure showing the position of the curve.

1. $r = \dfrac{3}{1 - \frac{1}{2} \cos \theta}$.

2. $r = \dfrac{2}{1 - \cos \theta}$.

3. $r = \dfrac{12}{1 - 3 \cos \theta}$.

4. $r = -\dfrac{12}{1 + 3 \cos \theta}$.

5. $r = \dfrac{7.2}{1 - .8 \sin \theta}$.

6. $r = \dfrac{4}{1 + \cos \theta}$.

7. $r = \dfrac{24}{1 + 4 \sin \theta}$.

8. $r = \dfrac{-15}{1 - 3 \sin \theta}$.

9. $r = \dfrac{10}{5+3\cos\theta - 4\sin\theta}$. 10. $r = \dfrac{2}{1+\sin\theta+\cos\theta}$.

11. $r = \dfrac{-2}{1+\sin\theta-\cos\theta}$.

12. Draw an accurate figure, to scale, for each of the curves of Exs. 5, 6, and 7, marking the coördinates of *all* the important points and drawing in *all* the important lines.

6. Transformation to and from Cartesian Coördinates. Let P be any point of the plane, whose coördinates, referred to a pair of Cartesian axes, are (x, y). Let the polar coördinates of P be (r, θ), where the origin, O, is taken as the pole, and the positive axis of x as the prime direction. Then it is clear from the figure that

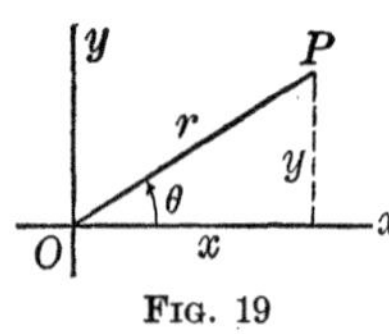

FIG. 19

(1) $$x = r\cos\theta, \qquad y = r\sin\theta.$$

Thus x and y are expressed in terms of r and θ. To express r and θ in terms of x and y, we have, for r:

(2) $$r^2 = x^2 + y^2, \qquad \text{or} \qquad r = \sqrt{x^2+y^2},$$

and, for θ, the pair of equations:

(3) $$\cos\theta = \frac{x}{\sqrt{x^2+y^2}}, \qquad \sin\theta = \frac{y}{\sqrt{x^2+y^2}}$$

For θ we have, also, the equation,

(4) $$\tan\theta = \frac{y}{x}.$$

But not all values of θ satisfying this equation are admissible. Some determine the ray OP, as they should; the others give the opposite ray and are to be excluded. If $\theta = \theta'$ is one admissible value, the others are $\theta = \theta' + 360\,n$, where n is a whole number.

Example 1. What are the polar coördinates of the point $(-5, -5)$?

Here, $r = \sqrt{5^2 + 5^2} = 5\sqrt{2}, \qquad \tan\theta = 1.$

But $\theta = 45°$ is *not* a correct value of θ, for the point lies in the *third* quadrant. The values of θ are, then:

$$\theta = 225° + 360°\, n.$$

It is frequently of importance to obtain, from the equation of a curve in one system of coördinates, its equation in the other system. We illustrate the method of doing this by a number of examples.

Example 2. Find the equation of the equilateral hyperbola

$$x^2 - y^2 = a^2$$

in polar coördinates.

Replacing x and y by their values as given by (1), we have:

$$r^2 \cos^2\theta - r^2 \sin^2\theta = a^2,$$

or

$$r^2 \cos 2\,\theta = a^2.$$

Example 3. Transform the equation of the lemniscate,

$$r^2 = a^2 \cos 2\,\theta,$$

to rectangular coördinates.

We perform the transformation piecemeal, first getting rid of θ. Write the equation as

$$r^2 = a^2(\cos^2\theta - \sin^2\theta),$$

and then replace $\cos\theta$ and $\sin\theta$ by their values, x/r and y/r, from (1); on multiplying both sides of the resulting equation by r^2, we have

$$r^4 = a^2(x^2 - y^2).$$

Finally, we replace r^2 by its value $x^2 + y^2$, and obtain

$$(x^2 + y^2)^2 = a^2(x^2 - y^2).$$

This is an equation of the fourth degree in x and y.

Example 4. Transform the equation of the curve of Example 1, § 4

$$r^2 = a^2 \sin\theta \tag{5}$$

to rectangular coördinates.

Replacing $\sin\theta$ by y/r and multiplying through by r, we have

(6) $$r^3 = a^2 y.$$

This becomes

(7) $$(\sqrt{x^2+y^2})^3 = a^2 y, \qquad \text{or} \qquad \sqrt{(x^2+y^2)^3} = a^2 y.$$

Negative Values of r. If we admit negative values of r, then (2) and (3) become

$$r = \pm\sqrt{x^2+y^2},$$

$$\cos\theta = \frac{x}{\pm\sqrt{x^2+y^2}}, \qquad \sin\theta = \frac{y}{\pm\sqrt{x^2+y^2}},$$

where the plus signs are to be taken if r is positive, the minus signs if r is negative; (1) does not change, as Fig. 20 shows. The admissible solutions for θ of (4) are those determining the ray OP (Fig. 19), if r is positive, or those determining the ray $O\overline{P}$ (Fig. 20), if r is negative.

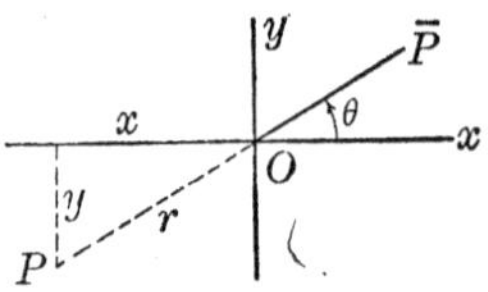

FIG. 20

If in (5) negative values of r are admitted, then (6) becomes, since now $r = \pm\sqrt{x^2+y^2}$,

$$\pm\sqrt{(x^2+y^2)^3} = a^2 y,$$

which may be written as

(8) $$(x^2+y^2)^3 = a^4 y^2.$$

The fact that (5) transforms into (7) when negative r's are excluded and transforms into (8) when negative r's are admitted corresponds to the fact that in the first case the curve (5) consists of a single loop, whereas in the second it is made up of two loops (§ 4).

This situation is not met with in the case of the lemniscate, Example 3, since this curve is the same whether negative r's are excluded or admitted.

EXERCISES

1. Find the Cartesian coördinates of the points:

(*a*) $(2, 60°)$; (*b*) $(5, 120°)$; (*c*) $(10, 225°)$;
(*d*) $(3.281, 110° 32')$; (*e*) $(2.847, 242° 27')$.

Plot the given point each time and check the results by direct measurement.

2. Find the polar coördinates of the points:

(*a*) $(6, 6)$; (*b*) $(-2, -2)$; (*c*) $(2, 3)$;
(*d*) $(-4, 3)$; (*e*) $(7, -8)$; (*f*) $(-12, -5)$.

Transform the following equations to polar coördinates.

3. $x = 3$.
4. $y = -4$.
5. $y = 3x$.
6. $2x - 3y = 8$.
7. $y^2 = 4x$.
8. $xy = a^2$.
9. $x^2 + y^2 - 2x + 4y = 0$.
10. $4x^2 + 3y^2 = 12$.

11. Transform the equation of the cardioid,

$$r = 2a(1 - \cos\theta),$$

to Cartesian coördinates. Of what degree is the resulting equation?

Ans. $(x^2 + y^2 + 2ax)^2 = 4a^2(x^2 + y^2)$, of the fourth degree.

Find the equation in Cartesian coördinates of each of the following curves.

12. $r^2 = a^2 \cos\theta$.
13. $r^2 = a^2 \sin 2\theta$.
14. $r = a \csc\theta$.
15. $r = 4 - 3\cos\theta$.
16. $r = a\cos 3\theta$. *Ans.* $(x^2 + y^2)^2 = ax(x^2 - 3y^2)$.
17. $r = a \sin 3\theta$. *Ans.* $(x^2 + y^2)^2 = ay(3x^2 - y^2)$.
18. $r = a \sin 2\theta$. *Ans.* $\sqrt{(x^2 + y^2)^3} = 2axy$.
19. $r = a \sin 2\theta$, if negative r's are admitted.
Ans. $(x^2 + y^2)^3 = 4a^2x^2y^2$.
20. $r = a\cos 2\theta$. *Ans.* $\sqrt{(x^2 + y^2)^3} = a(x^2 - y^2)$.
21. $r = a\csc^2\dfrac{\theta}{2}$. *Ans.* $y^2 - 4ax - 4a^2 = 0$.

22. $r = \dfrac{em}{1 - e\cos\theta}$, if $e \leq 1$.

Ans. $(1 - e^2)x^2 + y^2 - 2e^2mx - e^2m^2 = 0$.

23. $r = \dfrac{em}{1 - e\cos\theta}$, if $e > 1$. *Ans.* $\sqrt{x^2 + y^2} = e(x + m)$.

24. Same as Ex. 23, if negative r's are admitted.

Ans. That to Ex. 22.

EXERCISES ON CHAPTER X

1. Show that the distance between the two points (r_1, θ_1), (r_2, θ_2) is given by the formula

$$D = \sqrt{r_1^2 + r_2^2 - 2r_1r_2\cos(\theta_1 - \theta_2)}.$$

2. Deduce a formula giving the area of a triangle, one of whose vertices is at the pole.

3. Determine the angle of intersection of the two lines

$$r(\sin\theta + \cos\theta) = 3, \qquad r(4\sin\theta + 3\cos\theta) = 5.$$

Suggestion. Put the equations into the normal form,

$$r\cos(\theta - \gamma) = h,$$

and thus find the value of γ for each line.

4. Show that the line,

$$\frac{x}{a} + \frac{y}{b} = 1,$$

is represented in polar coördinates by the equation

$$r = \frac{ab}{a\sin\theta + b\cos\theta}.$$

5. What lines are represented by the following equations? Plot the line each time.

$$(a)\ r = \frac{2}{\sin\theta + \cos\theta}; \qquad (b)\ r = \frac{2}{2\sin\theta - 3\cos\theta}.$$

Find the equations in polar coördinates of the following conics.

6. $y^2 = 2mx$. *Ans.* $r = 2m\cos\theta\csc^2\theta$.

7. $\frac{x^2}{a^2}+\frac{y^2}{b^2}=1.$ *Ans.* $r^2=\frac{a^2b^2}{a^2\sin^2\theta+b^2\cos^2\theta}.$

8. $\frac{x^2}{a^2}-\frac{y^2}{b^2}=1.$ *Ans.* $r^2=\frac{a^2b^2}{-a^2\sin^2\theta+b^2\cos^2\theta}.$

9. What curves are represented by the following equations?

(*a*) $r^2=\frac{400}{25\sin^2\theta+16\cos^2\theta}$; (*b*) $r^2=\frac{4}{\cos^2\theta-\sin^2\theta}$;

(*c*) $r^2=\frac{6}{2\cos^2\theta+3\sin^2\theta}$; (*d*) $r^2=\frac{1}{\cos^2\theta-2\sin^2\theta}$;

(*e*) $r=\frac{4\cos\theta}{\sin^2\theta}$; (*f*) $r=\frac{4\sin\theta}{\cos^2\theta}$.

10. Transform the equation of the circle,

$$(x-a)^2+(y-b)^2=\rho^2,$$

to polar coördinates; represent the polar coördinates of the center by (c, γ). *Ans.* $r^2-2cr\cos(\theta-\gamma)+c^2=\rho^2$.

Conics

11. A comet moves in a parabolic orbit with the sun as focus. When the comet is 40,000,000 miles from the sun, the line from the sun to it makes an angle of 60° with the axis of the orbit (drawn in the direction in which the curve opens). How near does the comet come to the sun?

12. A comet is observed at two points of its parabolic orbit. The focal radii of these points, neither of which is the vertex of the parabola, make an angle of 90° with one another and have lengths of 10,000,000 and 20,000,000 miles, respectively. Find the equation of the orbit and determine how near the comet comes to the sun.

13. An ellipse which has a focus at the pole and its transverse axis along the prime direction passes through the two points (4, 60°) and (2, 90°). What is its equation? Where is the second focus?

14. A hyperbola has its transverse axis along the prime direction and a focus in the pole. The branch adjacent to this focus goes through the points $(\frac{1}{2}\sqrt{2}, 45°)$, $(\sqrt{2}, 90°)$. Find the equation of this branch.

15. Show that in a parabola a focal radius inclined at an angle of 60° with the direction in which the curve opens is equal in length to the latus rectum.

16. Show that a focal radius of a hyperbola which is parallel to an asymptote is equal in length to a quarter of the latus rectum.

17. Prove that in any conic the sum of the reciprocals of the segments of a focal chord is constant.

18. Prove that the length of the focal chord of any conic is given by the formula

$$\frac{2\,em}{1-e^2\cos^2\theta_0},$$

where θ_0 is the angle which the chord makes with the transverse axis.

19. Show that the sum of the reciprocals of the lengths of two perpendicular focal chords of a conic is constant.

Rotation of the Prime Direction

20. Let the prime direction OA be rotated about O through the angle θ_0 into a new position OA'. Let an arbitrary point have the coördinates (r, θ) with respect to O as pole and OA as prime direction and the coördinates (r', θ') with respect to O as pole and OA' as prime direction. Show that

$$r' = r, \qquad \theta' = \theta - \theta_0.$$

21. Equation (1), § 5, when $e < 1$, represents an ellipse with one focus in O and the other on OA. From it obtain, by rotation of the prime direction, the equation of an ellipse with one focus at O and the other on the ray $\theta = 90°$.

22. Equations (1) and (2), § 5, when $e > 1$, represent a hyperbola with one focus in O and the other on the ray $\theta = 180°$; obtain from them the equations of a hyperbola with one focus at O and the other on the ray $\theta = 90°$.

23. Obtain equations (3) and (4), § 5, from equations (1) and (2), § 5, by a rotation of the prime direction.

24. What does the equation of the line

$$r(\sin\theta + \cos\theta) = 3$$

become, when it is referred to the perpendicular to it from the pole as the new prime direction?

25. By rotating the prime direction through a suitable angle reduce the equation of the circle, $r = 6\cos(\theta - 30°)$, to simpler form.

26. The same for the circle, $r = 4\cos\theta + 3\sin\theta$.

27. The same for each of the conics:

$$(a)\ r = \frac{6}{3 - 3\cos\theta - 4\sin\theta}; \qquad (b)\ r = \frac{1}{\sqrt{2} + \sin\theta - \cos\theta}.$$

28. Prove that the curves $r = a\sin 2\theta$ and $r = a\cos 2\theta$ are the same curves, referred to a common pole and to prime directions making an angle of 45° with one another.

29. Show that the equations $r = a\cos 3\theta$ and $r = a\sin 3\theta$ represent the same curve.

30. Show that the equation of the curve $r = a\cos 3\theta$ remains unchanged if the prime direction is turned through any angle which is an integral multiple of 120°.

31. The same for the curve $r = a\sin 4\theta$, if the angle is 90°.

Pole in an Arbitrary Point

32. Given the point (2, 3) in the Cartesian plane. The polar coördinates of a point P, referred to (2, 3) as pole and to the directed line through (2, 3) in the direction of the positive axis of x as prime direction, are known to be (2, 13°). What are the rectangular coördinates of P?

Ans. $(2 + 2\cos 13°,\ 3 + 2\sin 13°)$.

33. The polar coördinates of a point P, referred to the point (x_0, y_0) as pole and the directed line through (x_0, y_0) in the direction of the positive axis of x as prime direction, are (r, θ). Show that the rectangular coördinates, (x, y), of P are

$$x = x_0 + r\cos\theta, \qquad y = y_0 + r\sin\theta.$$

34. By the *direction* θ is meant the direction which the positive axis of x would assume if it were rotated about one of its points (in the positive sense of rotation) through the angle θ.

From the point (5, 2) one proceeds 2 units distance in the direction 135°, and from the point thus reached one proceeds 3 units distance in the direction 60°. What are the coördinates of the final position?

35. Prove that the equations of Ex. 33 can be considered as the equations of a transformation of coördinates, from (x, y) to (r, θ), which consists first of a change of origin to the point (x_0, y_0) — cf. Ch. XI, § 1 — and then of the introduction of polar coördinates.

36. By shifting the origin to the point (2, 1) and then introducing polar coördinates, identify the locus of the equation

$$y^2 - 2x - 2y + 4 = 0.$$

37. By shifting the origin to a suitable point and then introducing polar coördinates, identify the locus of the equation

$$(x^2 + y^2 - 2x - 2y + 2)^2 = 25(x^2 - y^2 - 2x + 2y).$$

Ans. A lemniscate.

Loci

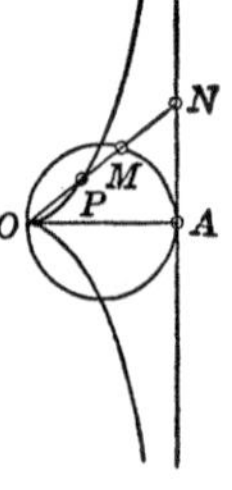

Solve the following problems in loci, using polar coördinates and excluding negative r's. In Exs. 39–41, determine when two equations are necessary to represent the locus.

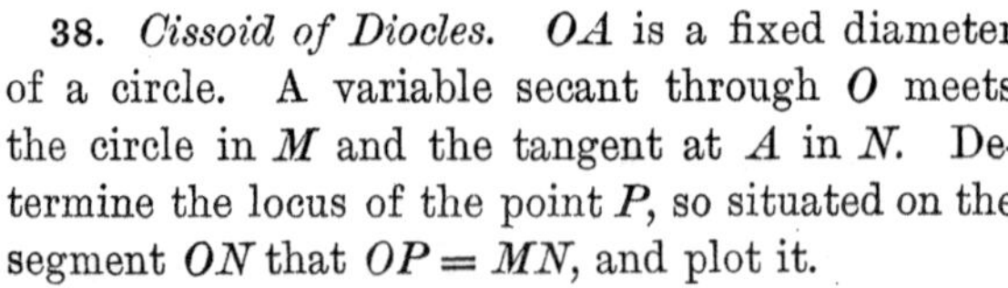

38. *Cissoid of Diocles.* OA is a fixed diameter of a circle. A variable secant through O meets the circle in M and the tangent at A in N. Determine the locus of the point P, so situated on the segment ON that $OP = MN$, and plot it.

39. *Limaçon of Pascal.* A variable secant through a fixed point O of a circle of diameter a meets the circle again in R. The constant length b is laid off in both directions along the

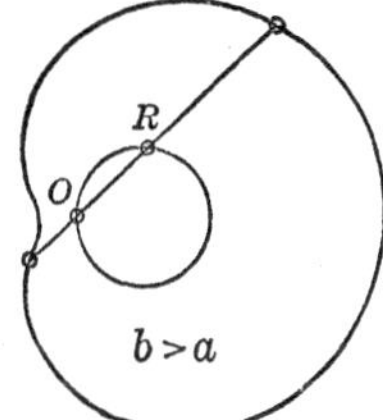

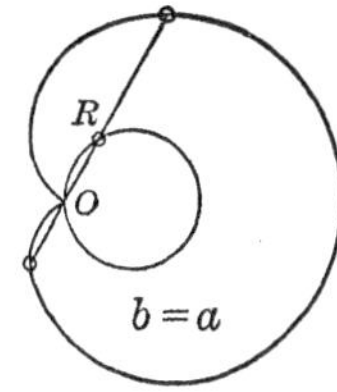

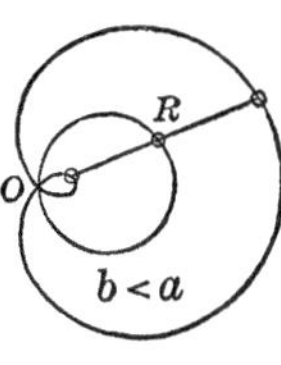

secant from R. Find the locus of the two points thus reached. Show that, if $a = b$, the locus is a cardioid. Plot the locus (a) when $a = 4$, $b = 5$; (b) when $a = 4$, $b = 3$. Cf. § 4, Exs. 2, 3, 4.

40. *Conchoid of Nicomedes.* A variable straight line through a fixed point O meets a fixed straight line, at the distance a from O, in Q. From Q the constant length b is laid off in both directions along QO. Find the locus of the two points thus reached. Plot it for each of the following pairs of values of a and b:

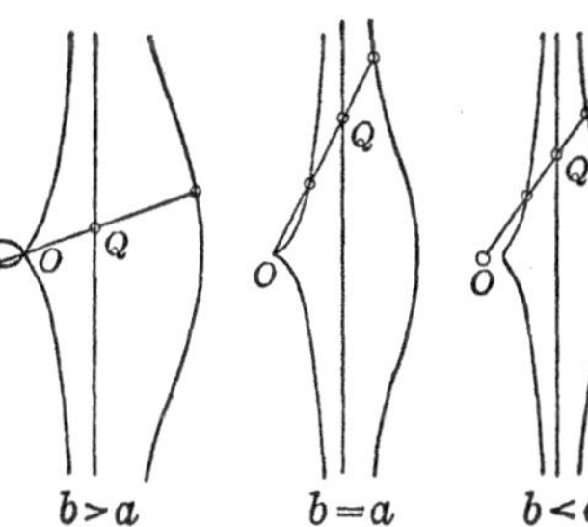

$$a = 4,\ b = 6; \qquad a = b = 4; \qquad a = 4,\ b = 3.$$

41. *Ovals of Cassini.* Given two points, F_1 and F_2, with coördinates (a, 0°), (a, 180°). Determine the locus of a point

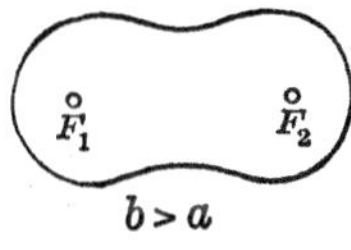

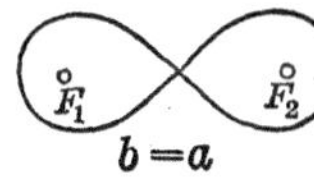

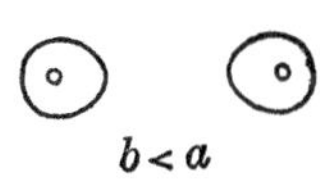

P which moves so that the product of its distances from F_1 and F_2 is constant, and equal to b^2. Show that, if $a^2 = b^2$, the locus is a lemniscate. Plot the locus (a) when $a = 6$, $b = 7$; (b) when $a = 6$, $b = 5$.

CHAPTER XI

TRANSFORMATION OF COÖRDINATES

1. Parallel Axes. It sometimes happens that it is desirable to shift from a given system of Cartesian axes to a new system of axes having the same directions as the old, but with a different origin.

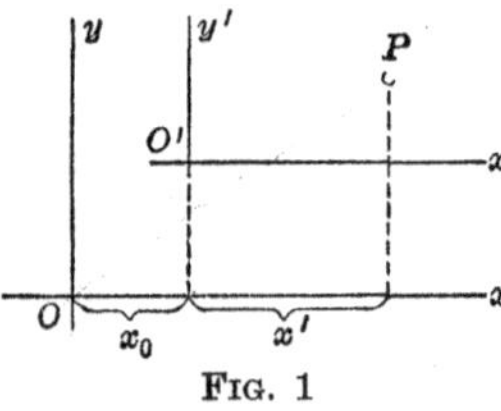

FIG. 1

Let P be any point of the plane; let the coördinates of P, referred to the old axes, be (x, y), and let the coördinates be (x', y') with respect to the new axes. Let the new origin, O', have the coördinates (x_0, y_0) in the old system. Then it is easy to show that

$$(1)\qquad \begin{cases} x = x' + x_0, \\ y = y' + y_0; \end{cases}$$

or

$$(2)\qquad \begin{cases} x' = x - x_0, \\ y' = y - y_0. \end{cases}$$

For, consider the line-segment OP and the broken line, $OO'P$, which has the same extremities. Then

$$\text{Proj. } OP = \text{Proj. } OO' + \text{Proj. } O'P,$$

no matter what direction is chosen, along which the projection is to take place (Introduction, § 3).

If the direction is taken, first, as the positive axis of x and then, again, as the positive axis of y, we obtain, by applying the definition of coördinates (Ch. I, § 1), the equations (1).

Example 1. Find the equation of the curve

$$(3)\qquad y^2 + 2y - 4x + 9 = 0,$$

referred to parallel axes, with the new origin at the point $(2, -1)$.

Here, $x_0 = +2, \quad y_0 = -1,$

and we have

(4) $x = x' + 2, \quad y = y' - 1.$

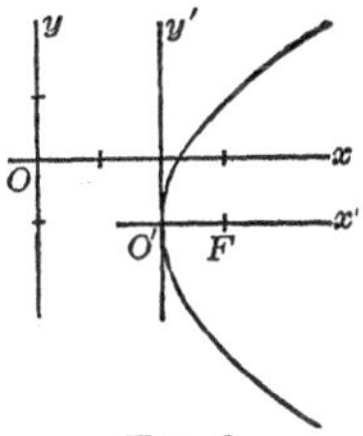

Hence

$$(y' - 1)^2 + 2(y' - 1) - 4(x' + 2) + 9 = 0,$$

FIG. 2

or, on simplification,

$$y'^2 = 4x'.$$

Thus the curve is seen to be a parabola whose vertex is at the new origin. Referred to the old axes, the vertex is at the point $(2, -1)$ and the focus at $(3, -1)$.

Example 2. What curve is represented by the equation

$$9x^2 + 4y^2 + 18x - 16y = 11?$$

We can rewrite the equation in the form:

$$9(x^2 + 2x \qquad) + 4(y^2 - 4y \qquad) = 11.$$

The first parenthesis becomes a perfect square if 1 is added. This means that 9 must be added to each side of the equation.

Again, the second parenthesis becomes a perfect square if 4 is added. This means that 16 must be added to each side of the equation. Hence, finally,

$$9(x^2 + 2x + 1) + 4(y^2 - 4y + 4) = 11 + 9 + 16,$$

or
$$9(x + 1)^2 + 4(y - 2)^2 = 36.$$

If we transform to parallel axes, setting

$$\begin{cases} x' = x + 1, & x_0 = -1, \\ y' = y - 2, & y_0 = 2, \end{cases}$$

the equation becomes

$$9x'^2 + 4y'^2 = 36,$$

or
$$\frac{x'^2}{4} + \frac{y'^2}{9} = 1.$$

This equation represents an ellipse with its center at the new origin, $(x_0, y_0) = (-1, 2)$; its semi-axes are of lengths 2 and 3 and its foci lie on the y'-axis at the points

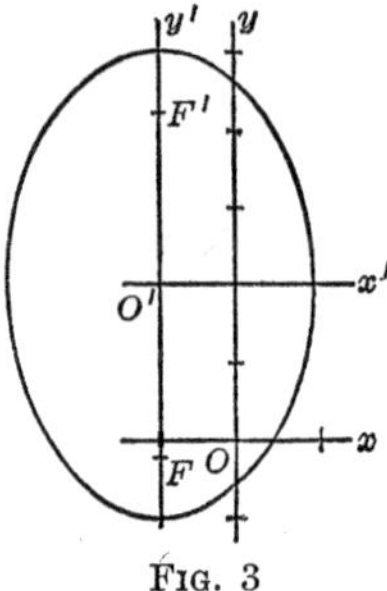

Fig. 3

$$(x', y') = (0, \pm\sqrt{5}),$$

or
$$(x, y) = (-1, 2 \pm \sqrt{5}).$$

If, in Example 1, the position of the new origin (the vertex of the parabola) had not been given, it could have been found by the method employed in Example 2. Equation (3) can be written as

$$y^2 + 2y = 4x - 9.$$

To complete the square of $y^2 + 2y$, add 1 to each side of the equation:

$$y^2 + 2y + 1 = 4x - 9 + 1.$$

Put this into the form:

$$(y + 1)^2 = 4(x - 2).$$

Hence we are led to set

$$x' = x - 2, \qquad y' = y + 1,$$

that is, to transform to parallel axes with the new origin at the point $(2, -1)$. But this is precisely the transformation (4) which was applied in Example 1.

EXERCISES

1. Find the coördinates of the points $(3, 2)$, $(-2, 5)$, $(-4, -1)$, $(0, 0)$, referred to new axes having the same directions as the old, if the new origin is at the point

(a) $(x_0, y_0) = (1, 1)$; (b) $(x_0, y_0) = (5, -3)$.

In each of the following exercises transform the given equation to parallel axes having the same directions, the new origin being at the point specified. Thus identify the curve represented by the equation, and describe carefully its position with respect to the original axes. Draw the curve roughly.

Equation	*New Origin*
2. $y^2 = 2x + 4$,	$(x_0, y_0) = (-2, 0)$.

Ans. A parabola with its vertex at $(-2, 0)$, with its focus at $(-\frac{3}{2}, 0)$, and with $x = -\frac{5}{2}$ as its directrix.

3. $y^2 = 2y + 4x$,	$(x_0, y_0) = (-\frac{1}{4}, 1)$.
4. $3x^2 - 6x - 4y + 11 = 0$,	$(x_0, y_0) = (1, 2)$.
5. $9x^2 + 25y^2 + 18x - 50y - 191 = 0$,	$(x_0, y_0) = (-1, 1)$.

Ans. An ellipse, center at $(-1, 1)$; foci at $(3, 1)$ and $(-5, 1)$; semi-axes of lengths 5 and 3.

6. $x^2 - 4y^2 - 6x - 32y - 59 = 0$,	$(x_0, y_0) = (3, -4)$.

Show that each of the following equations represents a conic section. Draw a rough graph of the conic and find, when they exist, the coördinates of the center and the foci, the equations of the directrices and the asymptotes, and the value of the eccentricity.*

7. $x^2 - 2x - 4y + 5 = 0$.
8. $3x^2 - 6x + 5y + 13 = 0$.
9. $y^2 - 12x + 4y + 28 = 0$.
10. $4y^2 + 3x - 24y + 42 = 0$.
11. $2x^2 + 3y^2 - 4x - 6y - 1 = 0$.
12. $x^2 + 4y^2 + 2x - 24y + 36 = 0$.
13. $4x^2 - 9y^2 - 16x + 18y - 29 = 0$.

Ans. A hyperbola, center at $(2, 1)$; foci at $(2 \pm \sqrt{13}, 1)$; directrices: $x = 2 \pm \frac{9}{13}\sqrt{13}$; asymptotes: $2x + 3y - 7 = 0$, $2x - 3y - 1 = 0$; $e = \frac{1}{3}\sqrt{13}$.

14. $2x^2 - 3y^2 + 20x - 12y - 44 = 0$.
15. $x^2 + 4y^2 + 14x + 45 = 0$.
16. $x^2 - 25y^2 - 50y - 50 = 0$.

2. Rotation of the Axes. Let the new (x', y')-axes have the same origin as the given (x, y)-axes, and let the angle

* Further exercises of this type are Exs. 1–9, of Ch. XII, § 1.

from the positive axis of x *to* the positive axis of x' be denoted by γ (Fig. 4). Let P be any point, whose coördinates, referred to the old and the new axes, are (x, y) and (x', y') respectively, and let M and M' be the projections of P on the axes of x and x' respectively. Then O is joined with P by two broken lines, namely, OMP and $OM'P$. It follows that the projections of these broken lines along any direction are equal:

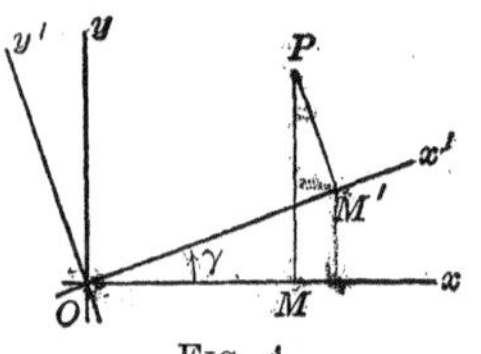

Fig. 4

$$\text{(1)} \qquad \text{Proj. } OM + \text{Proj. } MP = \text{Proj. } OM' + \text{Proj. } M'P.$$

If the direction is taken, first, as the positive axis of x and then, again, as the positive axis of y, we have

$$OM = OM' \cos\gamma - M'P \sin\gamma,$$
$$MP = OM' \sin\gamma + M'P \cos\gamma.$$

But, by the definition of coördinates (Ch. I, § 1),

$$OM = x, \qquad MP = y; \qquad OM' = x', \qquad M'P = y'.$$

The final result is, then, the following:

$$\text{(2)} \qquad \begin{cases} x = x' \cos\gamma - y' \sin\gamma, \\ y = x' \sin\gamma + y' \cos\gamma. \end{cases}$$

To express x' and y' in terms of x and y, these equations can readily be solved for the former variables, regarded as the unknown quantities in the pair of simultaneous equations (2). Or, the formulas can be deduced directly from the figure by taking the projections in equation (1) along the positive axes of x' and y' in turn. The result in either case is

$$\text{(3)} \qquad \begin{cases} x' = \quad x \cos\gamma + y \sin\gamma, \\ y' = - x \sin\gamma + y \cos\gamma. \end{cases}$$

Example 1. Transform the equation of the equilateral hyperbola,

$$x^2 - y^2 = a^2,$$

to new axes with the same origin, the angle *from* the positive axis of x *to* the positive axis of x' being $-45°$.

Here, $\gamma = -45°$, and formulas (2) become

$$x = \frac{1}{\sqrt{2}}(x' + y'), \qquad y = \frac{1}{\sqrt{2}}(-x' + y').$$

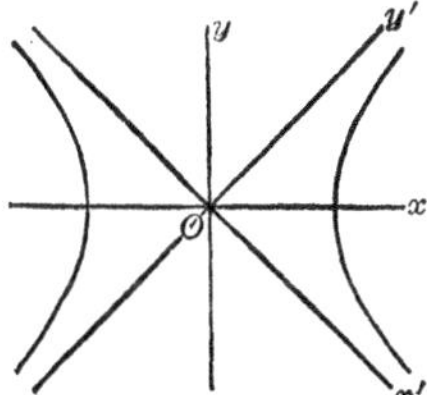

Fig. 5

Hence $$x^2 = \tfrac{1}{2}(x'^2 + 2x'y' + y'^2),$$
$$y^2 = \tfrac{1}{2}(x'^2 - 2x'y' + y'^2).$$

On substituting these values in the given equation we have:

$$2x'y' = a^2.$$

This, then, is the equation of an equilateral hyperbola referred to its asymptotes as the coördinate axes.

If we had rotated the axes through $+45°$ instead of $-45°$, the transformed equation would have read:

$$2x'y' = -a^2.$$

Example 2. Transform the equation

$$\frac{x^2}{A^2} - \frac{y^2}{B^2} = -1 \qquad \text{or} \qquad \frac{y^2}{B^2} - \frac{x^2}{A^2} = 1$$

to new axes, obtained by rotating the given axes about the origin through 90°.

Here, $\gamma = 90°$, and equations (2) become

$$x = -y', \qquad y = x'.$$

Hence $$\frac{x'^2}{B^2} - \frac{y'^2}{A^2} = 1.$$

Thus it appears that the original equation represents a hyperbola with its center at the origin, its transverse axis lying along the axis of y; cf. Ch. VIII, § 8. The length of the major axis is $2B$, that of the minor axis, $2A$. The asymptotes are given by the equations

$$\frac{x'}{B} = -\frac{y'}{A} \qquad \text{and} \qquad \frac{x'}{B} = \frac{y'}{A}.$$

Referred to the original axes, they have the equations

$$\frac{x}{A}=\frac{y}{B} \quad \text{and} \quad \frac{x}{A}=-\frac{y}{B}.$$

EXERCISES

Obtain the equations of transformation in each of the following three cases:

1. When $\gamma = 30°$. *Ans.* $\begin{cases} x = \frac{1}{2}\sqrt{3}x' - \frac{1}{2}y', \\ y = \frac{1}{2}x' + \frac{1}{2}\sqrt{3}y'. \end{cases}$

2. When $\gamma = -120°$. 3. When $\gamma = 135°$.

Draw a figure and deduce from it directly the formulas of transformation in each of the following three cases. Check the results by application of formulas (2) and (3) of the text.

4. $\gamma = 90°$. 5. $\gamma = -90°$. 6. $\gamma = 180°$.

7. Find the coördinates of the points (2, 0), (3, 1), (− 2, 4), (− 5, − 8), referred to new axes obtained by rotating the old through an angle of 45°; of 150°.

8. Show directly by means of a suitable rotation of the axes that the equation $xy = k^2$ represents an equilateral hyperbola referred to its asymptotes as the coördinate axes. Determine the coördinates, referred to these axes, of the vertices and the foci.

9. The same for the equation $xy = -k^2$.

By rotating the axes through an angle of 45° determine the curve represented by each of the equations:

10. $17x^2 - 16xy + 17y^2 = 225.$

11. $3x^2 - 10xy + 3y^2 + 8 = 0.$

12. $x^2 + 4xy + y^2 + 3 = 0.$

13. By rotating the axes through 30° determine the curve represented by the equation

$$\sqrt{3}xy - y^2 = 12.$$

14. By rotating the axes through an angle γ of the first quadrant, whose sine is $\frac{3}{5}$, determine the curve represented by the equation

$$52x^2 - 72xy + 73y^2 = 100.$$

15. Show that the equation of a circle whose center is at the origin is not changed by rotating the axes through any angle. Actually carry through the transformation.

3. The General Case. Let it be required to pass from one system of axes to a new one, in which both the origin and the directions of the axes have been changed. Let the (x, y)-axes, with origin at O, be the given system and the (x', y')-axes, with origin at O', the new system. Let the coördinates of O', referred to the (x, y)-axes, be (x_0, y_0), and let the angle *from* the positive axis of x *to* the positive axis of x' be γ.

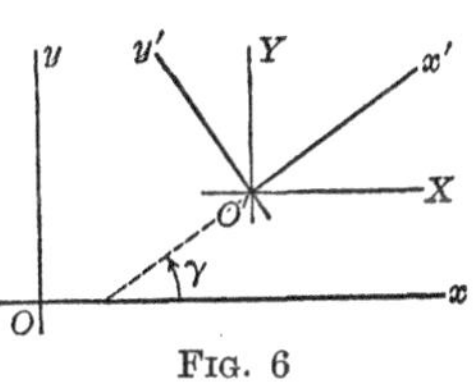

FIG. 6

The transition from one system to the other can be made in two steps:

(*a*) Transform first to a system of parallel axes having the same direction, but with origin at O'. If the new coördinates are denoted by (X, Y), then

$$(1)\qquad \begin{cases} x = X + x_0, \\ y = Y + y_0. \end{cases}$$

(*b*) Now rotate the (X, Y)-axes through the angle γ:

$$(2)\qquad \begin{cases} X = x' \cos\gamma - y' \sin\gamma, \\ Y = x' \sin\gamma + y' \cos\gamma. \end{cases}$$

Combining these results we get, as the final formulas, the following:

$$(3)\qquad \begin{cases} x = x' \cos\gamma - y' \sin\gamma + x_0, \\ y = x' \sin\gamma + y' \cos\gamma + y_0. \end{cases}$$

The formulas for (x', y') in terms of (x, y) are

$$\begin{cases} x' = \quad (x - x_0) \cos\gamma + (y - y_0) \sin\gamma, \\ y' = -(x - x_0) \sin\gamma + (y - y_0) \cos\gamma. \end{cases}$$

These can be written in the form:

$$(4)\qquad \begin{cases} x' = x\cos\gamma + y\sin\gamma - \bar{x}_0, \\ y' = -x\sin\gamma + y\cos\gamma - \bar{y}_0, \end{cases}$$

where

$$(5)\qquad \begin{aligned} \bar{x}_0 &= x_0\cos\gamma + y_0\sin\gamma, \\ \bar{y}_0 &= -x_0\sin\gamma + y_0\cos\gamma. \end{aligned}$$

It is to be noted that, inasmuch as x_0, y_0, and γ are constants, so are $\bar{x}_0$ and $\bar{y}_0$.

Example. Identify the curve represented by the equation

$$(6)\qquad x^2 + 6xy + y^2 - 10x - 14y + 9 = 0$$

by transforming to new axes through the point (2, 1), the angle from the old axis of x to the new being 45°.

Here $x_0 = 2$, $y_0 = 1$, and $\gamma = 45°$. We might substitute these values in formulas (3) and then apply the formulas to the given equation. It is, however, more feasible in general to make the transformation in the two steps (1) and (2).

Formulas (1) are, in this case,

$$x = X + 2, \qquad y = Y + 1.$$

Hence (6) becomes

$$(7)\qquad X^2 + 6XY + Y^2 - 8 = 0.$$

Since $\gamma = 45°$, formulas (2) are:

$$X = \frac{1}{\sqrt{2}}(x' - y'), \qquad Y = \frac{1}{\sqrt{2}}(x' + y').$$

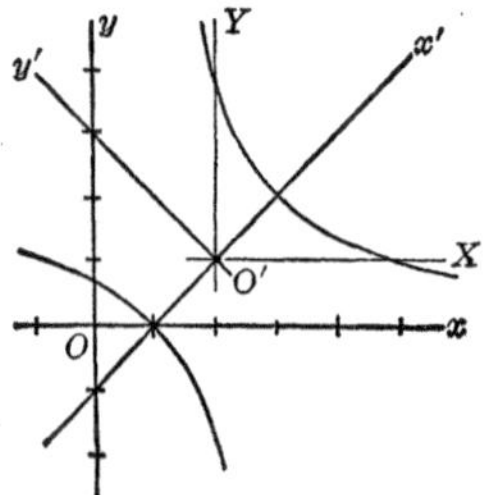

FIG. 7

Then (7) becomes

$$\tfrac{1}{2}(x' - y')^2 + 3(x'^2 - y'^2) + \tfrac{1}{2}(x' + y')^2 - 8 = 0,$$

or, on simplification,

$$2x'^2 - y'^2 = 4.$$

Consequently, equation (6) represents a hyperbola with its center at the point (2, 1) and with its transverse axis inclined at an angle of 45° to the axis of x.

In deducing formulas (3), we first shifted the origin and then rotated the axes. We might equally well have proceeded in the opposite order:

(a) Rotate the (x, y)-axes through the angle γ into the new axes of $\bar{x}$ and $\bar{y}$:

$$(8) \qquad \begin{cases} \bar{x} = \quad x \cos \gamma + y \sin \gamma, \\ \bar{y} = -x \sin \gamma + y \cos \gamma. \end{cases}$$

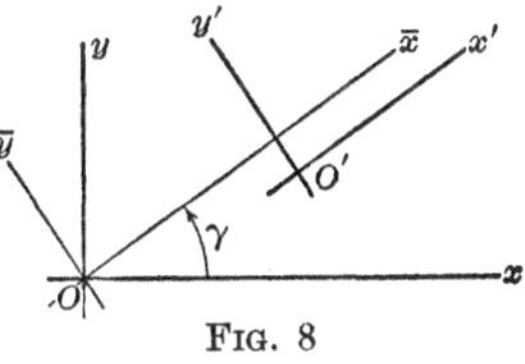

Fig. 8

The coördinates of O', referred to the new axes, are obtained by setting $x = x_0$ and $y = y_0$ in (8); *they are, then, the* $\bar{x}_0$ *and* $\bar{y}_0$ *given by formulas* (5).

(b) Transform from the $(\bar{x}, \bar{y})$-axes to the parallel axes of x' and y', with origin at O'. Since the coördinates of O', referred to the $(\bar{x}, \bar{y})$-axes, are $(\bar{x}_0, \bar{y}_0)$, the equations of this transformation are

$$(9) \qquad \begin{cases} x' = \bar{x} - \bar{x}_0, \\ y' = \bar{y} - \bar{y}_0. \end{cases}$$

Eliminating $\bar{x}$ and $\bar{y}$ from (8) and (9), we obtain, as the final formulas

$$\begin{cases} x' = \quad x \cos \gamma + y \sin \gamma - \bar{x}_0, \\ y' = -x \sin \gamma + y \cos \gamma - \bar{y}_0. \end{cases}$$

But these are precisely the formulas (4) which we had before.

EXERCISES

Obtain the equations of transformation in each of the following cases. First find the formulas for x and y in terms of x' and y', and then solve for x' and y' in terms of x and y.

1. $(x_0, y_0) = (1, 1)$; $\gamma = 45°$.
2. $(x_0, y_0) = (-2, 1)$; $\gamma = 30°$.
3. $(x_0, y_0) = (0, 3)$; $\gamma = -60.°$
4. $(x_0, y_0) = (-5, -3)$; $\gamma = 120°$.

Draw a figure and deduce from it directly the formulas of transformation for each of the following values of γ, the new origin in each case being at an arbitrary point (x_0, y_0). Check the results by the use of formulas (3) and (4) of the text.

5. $\gamma = 90°$. 6. $\gamma = 180°$. 7. $\gamma = -90°$.

8. Find the coördinates of the points (0, 0), (1, 2), (– 3, 4), (– 2, – 5), referred to new axes passing through the point (2, 1), the angle from the old axis of x to the new being 45°.

9. The same, if the new origin is at the point (– 3, 4) and the angle from the old axis of x to the new is – 30°.

Identify the curve represented by each of the following equations by transforming to parallel axes at the point (x_0, y_0) specified, and then rotating the new axes through the given angle γ. Draw a figure in each case.

	Equation	(x_0, y_0)	γ
10.	$5x^2 - 6xy + 5y^2 - 4x - 4y - 4 = 0,$	(1, 1),	45°.
11.	$x^2 - 4xy + y^2 + 10x - 2y + 7 = 0,$	(1, 3),	– 45°.
12.	$x^2 - 10xy + y^2 + 46x + 10y - 47 = 0,$	(2, 5),	135°.

13. Find the curve represented by the equation

$$66x^2 - 24xy + 59y^2 + 108x + 94y + 76 = 0$$

by introducing parallel axes at the point (– 1, – 1) and then rotating these axes through the acute angle whose tangent is $\frac{4}{3}$. Draw a graph.

14. The equation

$$7x^2 - 18xy - 17y^2 - 28x + 36y + 8 = 0$$

represents a conic whose center is at the point (2, 0), and one of whose axes has the slope $-\frac{1}{3}$. Identify the conic, and draw a rough graph of it.

4. Determination of the Transformation from the Equations of the New Axes. Consider the general transformation given by formulas (4) of the preceding paragraph. If, in these formulas, we set $x' = 0$ and $y' = 0$, we obtain

$$\begin{aligned} x \cos\gamma + y \sin\gamma - \bar{x}_0 &= 0, \\ -x \sin\gamma + y \cos\gamma - \bar{y}_0 &= 0. \end{aligned} \tag{1}$$

These are the equations of the new axes, referred to the old system; the first is the equation of the axis of y', the second, that of the axis of x'.

Conversely, if we set the expressions on the left-hand sides of equations (1) equal to x' and y' respectively, we obtain the equations of the transformation.

Problem. Let it be required to find the equations of a transformation which introduces the two perpendicular lines,

$$\begin{aligned} x + 2y + 1 &= 0, \\ 2x - y - 3 &= 0, \end{aligned} \tag{2}$$

as coördinate axes.

The natural procedure, in order to obtain the required equations, would be to set the left-hand sides of equations (2) equal to x' and y'. But, in order to obtain the correct result in this way, we must first *put the left-hand sides of equations* (2) *into the form of those of equations* (1).

These latter are of the form

$$\begin{aligned} ax + by - \overline{x}_0, \\ -bx + ay - \overline{y}_0, \end{aligned} \tag{3}$$

where

$$a^2 + b^2 = 1. \tag{4}$$

The left-hand sides of equations (2) will be of the form (3) if we multiply the second of the equations through by -1:

$$\begin{aligned} x + 2y + 1 &= 0, \\ -2x + y + 3 &= 0. \end{aligned}$$

To bring about the fulfillment of condition (4), we multiply each of these equations through by a constant $\rho \neq 0$:

$$\begin{aligned} \rho x + 2\rho y + \rho &= 0, \\ -2\rho x + \rho y + 3\rho &= 0. \end{aligned} \tag{2a}$$

Thereby we have not changed the lines which the original equations (2) represent.

The value of ρ is to be determined so that condition (4) is satisfied by the left-hand sides of equations ($2\,a$), that is, so that

$$(\rho)^2+(2\,\rho)^2=1 \qquad \text{or} \qquad 5\,\rho^2=1.$$

Hence $\rho=\pm 1/\sqrt{5}$.

We choose $\rho=1/\sqrt{5}$. If this value is substituted for ρ in equations ($2\,a$), these equations will be precisely of the form (1). Hence the equations of a transformation introducing the lines (2) as the axes are

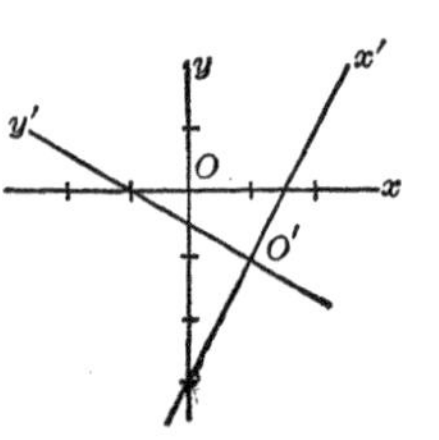

FIG. 9

$$(5) \qquad \begin{cases} x'=\dfrac{x+2\,y+1}{\sqrt{5}}, \\ y'=\dfrac{-2\,x+y+3}{\sqrt{5}}. \end{cases}$$

The first of the lines (2) is the axis of y'; the second, the axis of x'.

The new origin is at the point $(1, -1)$; for, this is the point of intersection of the lines (2). The old origin $(x, y)=(0, 0)$ has, according to (5), the coördinates $\left(\dfrac{1}{\sqrt{5}}, \dfrac{3}{\sqrt{5}}\right)$, referred to the new axes, and must lie, then, in the *first* quadrant formed by these axes. Consequently, the new axes must be directed as shown in the figure.

The slope of the axis of x', the second of the lines (2), is 2 and so its slope angle is $63°\,26'$, or $243°\,26'$. It is clear from the figure that it is the first of these angles which is the angle γ.

We obtain a second transformation, for which the lines (2) are the new axes, by taking the value $-1/\sqrt{5}$ for ρ. For this transformation the directions of both axes are opposite to those for (5), and γ has the value $243°\,26'$.

For both transformations the first of the lines (2) is the axis of y'; the second, the axis of x'. By reversing the rôles of the lines, two more transformations can be obtained. Thus, there are in all four transformations introducing a given pair of mutually perpendicular lines as coördinate axes.

EXERCISES

In each of the following exercises find the equations of a transformation which introduces the given perpendicular lines as new axes. Find the position of the new origin and the value of γ; draw an accurate figure, and indicate the directions of the new axes.

	Axis of y'	*Axis of x'*
1.	$3x+4y-11=0$;	$4x-3y+2=0$.
2.	$x+y-3=0$;	$x-y-1=0$.
3.	$2x-3y=0$.	$3x+2y+5=0$.
4.	$5x-2y=0$;	$2x+5y=0$.
5.	$x-2=0$;	$y+3=0$.
6.	$y-8=0$;	$x-5=0$.

7. If the lines of Ex. 1 are introduced as axes, what does the equation

$$(4x-3y+2)^2=3x+4y-11$$

become? What curve does it represent? Draw a rough graph.

By a suitable transformation of axes determine the nature and position of each of the following curves. In each case draw a figure showing accurately the new axes, properly directed; then sketch the curve.

8. $(2x-3y)^2+6x+4y+10=0$.
9. $(x+y)^2-2x+2y=0$.
10. $3(5x-2y)^2+(2x+5y)^2=1$.
11. $(x+y-1)^3-5(x-y-1)=0$.

12. Find the equations of all four transformations which introduce the lines

$$5x-12y+7=0, \qquad 12x+5y-17=0$$

as axes of coördinates. Draw the four corresponding figures, and find the four values of γ.

13. Obtain the equations required in Ex. 1 by finding the coördinates of the new origin and a value for γ and then applying formulas (3) of § 3.

5. Reversal of One Axis. There is one more case which deserves mention. Suppose that the sense of one axis is reversed, while the other axis remains unchanged. Let the axis which is reversed be the axis of x (Fig. 10). Then, evidently,

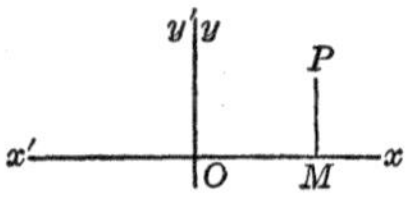

Fig. 10

$$(1) \qquad \begin{cases} x = -x', \\ y = y'; \end{cases} \qquad \begin{cases} x' = -x, \\ y' = y. \end{cases}$$

If the sense of the axis of y had been reversed, the axis of x remaining unchanged, we should have had:

$$(2) \qquad \begin{cases} x = x', \\ y = -y'; \end{cases} \qquad \begin{cases} x' = x, \\ y' = -y. \end{cases}$$

Consider, for example, the equation of a parabola in the normal form,

$$y^2 = 2\,mx.$$

If the sense of the axis of x is reversed, the equation becomes

$$y'^2 = -2\,mx'.$$

We could use this result to interpret the equation,

$$y^2 = -2\,mx,$$

if we knew the parabola only in its normal form. Taking the axis of x' opposite to the axis of x, and starting with the known parabola

$$y'^2 = 2\,mx',$$

we see that the transformed equation,

$$y^2 = -2\,mx,$$

represents a parabola on the negative axis of x, its vertex being at the origin.

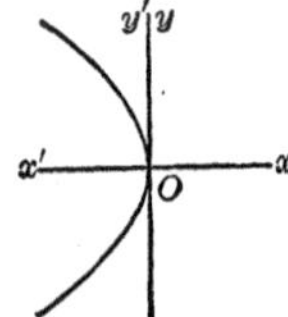

Fig. 11

EXERCISES

1. Assuming that the equation of the parabola in the form

$$x^2 = 2\,my$$

is known, interpret the equation

$$x^2 = -\,2\,my$$

by the method of the text.

2. Plot the so-called *semi-cubical parabola*

$$y^2 = x^3.$$

From the graph determine the curve defined by the equation

$$y^2 = -\,x^3.$$

EXERCISES ON CHAPTER XI

Change of Origin

In each of the following exercises prove, by making a suitable transformation to parallel axes, that the given equation represents two straight lines. Find the equations of the lines, referred to the original axes.

1. $x^2 - y^2 - 4\,x + 6\,y - 5 = 0.$
2. $4\,x^2 - 9\,y^2 + 8\,x + 18\,y - 5 = 0.$
3. $4\,x^2 - 16\,y^2 - 12\,x + 9 = 0.$

What does each of the following equations represent?

4. $x^2 + 2\,y^2 - 10\,x + 12\,y + 43 = 0.$

Ans. The point $(5, -3)$.

5. $3\,x^2 + 4\,y^2 - 6\,x + 16\,y + 21 = 0.$

6. By completing the cube for the terms in x in the equation

$$y = x^3 + 3\,x^2 + 3\,x - 2,$$

and by making the transformation to parallel axes which is suggested by the result, determine the curve defined by the equation. Draw the curve roughly.

By the method of Ex. 6, identify and plot the locus of each of the following equations.

7. $6y = x^3 + 6x^2 + 12x + 14$.

8. $x = 2y^3 - 6y^2 + 6y + 6$.

9. $x^3 - 3x^2 - 6x + xy^2 - y^2 + 8 = 0$.

10. Determine the position of the point (x_0, y_0) such that, if the straight lines

$$3x - 4y - 2 = 0, \qquad x + 2y = 4$$

are referred to parallel axes at (x_0, y_0), their equations will contain no constant terms. What will these equations be?

11. Show that, if the two intersecting straight lines

$$A_1x + B_1y + C_1 = 0, \qquad A_2x + B_2y + C_2 = 0$$

are referred to parallel axes at their point of intersection, the equations of the lines become

$$A_1x' + B_1y' = 0, \qquad A_2x' + B_2y' = 0.$$

12. Determine the position of the point (x_0, y_0) such that, if the curve

$$xy - 2x - y - 2 = 0$$

is referred to parallel axes at (x_0, y_0), its equation will contain no linear terms in x and y. Identify and plot the curve.

13. Identify and plot roughly the locus of the equation

$$6xy + 7x - 5y + 3 = 0.$$

14. Determine the point (x_0, y_0) such that, if the curve

$$3x^2 - 7xy - 6y^2 - 19x + 2y + 20 = 0$$

is referred to parallel axes at (x_0, y_0), its equation will contain no linear terms in x and y. Show that the equation will also contain no constant term and hence that it will represent two straight lines. Find the equations of these lines with respect to the original axes.

Rotation of Axes

15. Given the two perpendicular lines through the origin of slopes $\frac{1}{2}$ and -2. Find the equations of a transformation introducing these lines as axes.

16. The equation $2x^2 + 3xy - 2y^2 = 0$ represents two perpendicular lines through the origin. Show that it may be transformed into the equation $x'y' = 0$ by a suitable rotation of axes.

Identify and plot roughly the curve defined by each of the following equations. Cf. Ex. 16.

17. $2x^2 + 3xy - 2y^2 = 6$. **18.** $12x^2 - 7xy - 12y^2 = 25$.

19. Determine the equations of a rotation of axes whereby the axis of y comes into coincidence with the line $4x + 3y = 0$.

20. Identify and plot roughly the curve defined by the equation $(4x + 3y)^2 = 125x$. Cf. Ex. 19.

The same for each of the following equations.

21. $(x + y)^2 = 4\sqrt{2}\,y$.

22. $4x^2 + 4xy + y^2 = 5\sqrt{5}\,x$.

23. $16x^2 + 24xy + 9y^2 = 60x - 80y$.

24. The line $2x - y = 0$ is an axis of the conic

$$6x^2 - 4xy + 3y^2 = 6.$$

By a suitable rotation of axes determine the nature and position of the conic.

Show that, by a suitable rotation of axes, each of the following equations becomes linear in y and hence capable of solution for y, without radicals.

25. $x^2 - y^2 + 2x - 3 = 0$.

26. $4x^2 - 4xy + y^2 + 3x - y = 2$.

27. $2x^2 + 3xy - 2y^2 + 6x - 2y = 8$.

General Transformation of Axes

28. Prove that the straight lines

$$3x - 4y - 2 = 0, \qquad x + 2y = 4,$$

when referred to suitable axes, will have equations the first of which is $x' = 0$, while the second contains no constant term.

29. Find the equations of the circle

$$x^2 + y^2 + 6x - 8y + 6 = 0$$

and the line

$$5x + 12y - 13 = 0,$$

when they are referred to axes through the center of the circle parallel and perpendicular, respectively, to the line.

30. Find the equations of the two circles

$$x^2 + y^2 + 4x - 6y - 4 = 0,$$
$$x^2 + y^2 - 6x + 4y - 4 = 0,$$

when they are referred to the point midway between their points of intersection as origin and the line joining the points of intersection as axis of x'.

31. What will the equations of the circles of Ex. 30 become, if they are referred to the mid-point between their centers as origin and the line of the centers as axis of y'?

32. A transformation consists of a change of origin to the point (x_0, y_0), of a rotation of the new axes through the angle γ, and of a reversal of the sense of the axis of x thus obtained. Show that the equations of the transformations are

$$\begin{cases} x = -x' \cos\gamma - y' \sin\gamma + x_0, \\ y = -x' \sin\gamma + y' \cos\gamma + y_0. \end{cases}$$

CHAPTER XII

THE GENERAL EQUATION OF THE SECOND DEGREE

1. Change of Origin of Coördinates. The aim of this chapter is twofold: To determine what curves are represented by equations of the second degree in x and y; and to develop methods by means of which the curve represented by any particular equation may be easily identified and its size and position accurately described. The methods used consist primarily in transformations of coördinates. We begin, then, by investigating what can be accomplished by a change of origin, *i.e.* a transformation to parallel axes.

Example 1. Let it be required to identify and to describe accurately the curve represented by the equation

$$(1) \qquad 5x^2 - 4y^2 - 20x - 24y + 4 = 0.$$

Completing the square of the terms in x and then of the terms in y, according to the method of Ch. XI, § 1, we obtain:

$$5(x-2)^2 - 4(y+3)^2 = -20.$$

On setting

$$x' = x - 2, \qquad y' = y + 3,$$

that is, on changing the origin of coördinates to the point $(2, -3)$, this equation becomes

$$5x'^2 - 4y'^2 = -20,$$

or

$$\frac{x'^2}{4} - \frac{y'^2}{5} = -1.$$

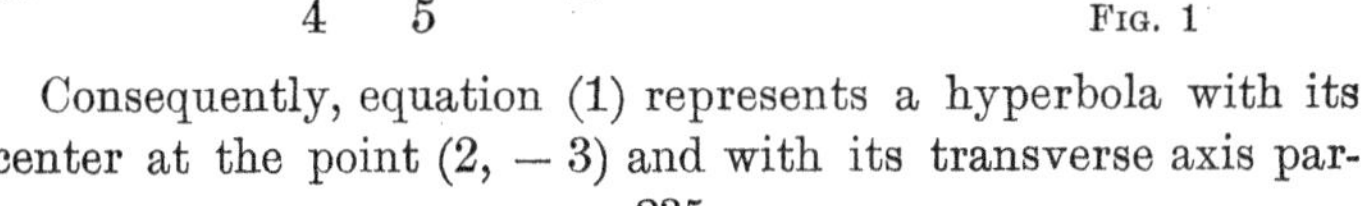

Fig. 1

Consequently, equation (1) represents a hyperbola with its center at the point $(2, -3)$ and with its transverse axis parallel to the axis of y. The coördinates of the foci, referred to the new axes, are $(0, \pm 3)$; consequently, when referred to the original axes, they are $(2, 0)$ and $(2, -6)$. The equations of the asymptotes, with respect to the new axes, are

$$\sqrt{5}\,x' - 2\,y' = 0, \qquad \sqrt{5}\,x' + 2\,y' = 0;$$

hence they are, with respect to the original axes,

$$\sqrt{5}\,x - 2\,y - 2\sqrt{5} - 6 = 0, \qquad \sqrt{5}\,x + 2\,y - 2\sqrt{5} + 6 = 0.$$

The semi-axis major is $\sqrt{5}$, and the semi-axis minor, 2; the eccentricity has the value $\frac{3}{5}\sqrt{5}$.

Example 2. Consider the equation

$$3\,xy - 6\,x + 3\,y - 10 = 0. \tag{2}$$

We rewrite this equation, first, in the form

$$3(xy - 2\,x + y \qquad) = 10,$$

and then as

$$3[x(y-2) + (y \qquad)] = 10.$$

If -2 is added to the y in the second parenthesis and, in equalization, $3 \cdot 1 \cdot (-2)$ or -6 is added to the right-hand side, this equation becomes

$$3\,(x+1)(y-2) = 4.$$

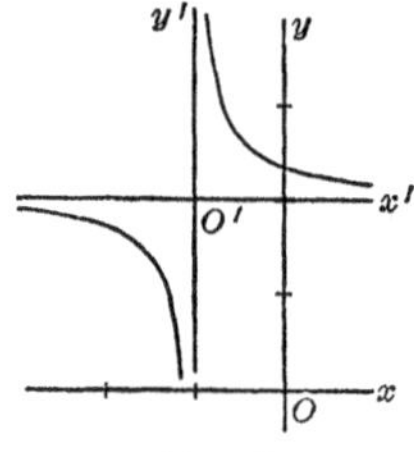

FIG. 2

We now change the origin to the point $(-1, 2)$ by setting

$$x' = x + 1, \qquad y' = y - 2.$$

The equation thus becomes

$$3\,x'y' = 4.$$

Accordingly, (2) represents a rectangular hyperbola with the lines $x + 1 = 0$ and $y - 2 = 0$ as asymptotes.

Example 3. The equation,

$$x^2 + 2\,x - 2\,y - 1 = 0, \tag{3}$$

can, according to the method of Ch. XI, § 1, be put into the form

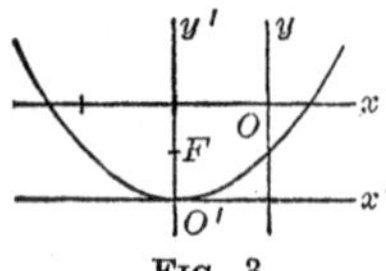

FIG. 3

$$(x+1)^2 - 2\,(y+1) = 0,$$

and hence represents a parabola with vertex at the point $(-1, -1)$ and with axis parallel to the axis of y.

Example 4. Consider, now, an equation in which all three quadratic terms are present:

$$(4) \qquad 6x^2 - xy - 2y^2 + 4x + 9y - 10 = 0.$$

In this case, completing the squares of the terms $6x^2 + 4x$ and of the terms $-2y^2 + 9y$ does not help. Let us make an arbitrary change of origin, setting

$$(5) \qquad x = x' + x_0, \qquad y = y' + y_0,$$

and aim to determine the new origin, (x_0, y_0), so that in the resulting equation the *linear* terms in x' and y' do not appear.

Setting in (4) the values of x and y as given by (5) and collecting terms, we have

$$(6) \quad 6x'^2 - x'y' - 2y'^2 + (12x_0 - y_0 + 4)x' + (-x_0 - 4y_0 + 9)y' + F' = 0,$$

where

$$(7) \qquad F' = 6x_0^2 - x_0y_0 - 2y_0^2 + 4x_0 + 9y_0 - 10.$$

If the terms in x' and y' are to drop from this equation, x_0 and y_0 must be so chosen that

$$(8) \qquad \begin{aligned} 12x_0 - y_0 + 4 &= 0, \\ -x_0 - 4y_0 + 9 &= 0. \end{aligned}$$

Solving (8) simultaneously for x_0, y_0, we have

$$x_0 = -\tfrac{1}{7}, \qquad y_0 = \tfrac{16}{7}.$$

The value of F', for these values of x_0 and y_0, is 0. Consequently, we have shown that equation (4), when referred to a new origin at $(-\frac{1}{7}, 2\frac{2}{7})$, becomes

$$6x'^2 - x'y' - 2y'^2 = 0.$$

The left-hand side of this equation can be factored:

$$(9) \qquad (3x' - 2y')(2x' + y') = 0.$$

Thus (4) represents two straight lines through the point $(-\frac{1}{7}, 2\frac{2}{7})$ with slopes $\frac{3}{2}$ and -2.

If in (9) we set

$$x' = x + \tfrac{1}{7}, \qquad y' = y - \tfrac{16}{7},$$

i.e. if we transform back to the original axes, we obtain, finally,

$$(3x - 2y + 5)(2x + y - 2) = 0.$$

This equation is seen to be precisely equation (4), with the left-hand side factored into two linear factors. The two straight lines represented by (4) have, then, the equations

$$3x - 2y + 5 = 0, \qquad 2x + y - 2 = 0.$$

It should be noted that the constant term (7) of equation (6) is the value of the left-hand side of (4) for $x = x_0$, $y = y_0$. In this example, this constant term took on the value zero when x_0, y_0 were chosen so that the coefficients of the linear terms in x' and y' vanished. This does not, however, occur in general, as we shall see in the next paragraph.

EXERCISES

In each of the following exercises identify and plot roughly the curve represented by the given equation. If the curve is a conic section, find, when they exist, the coördinates of the center and the foci, the equations of the directrices and the asymptotes, and the value of the eccentricity.

1. $4x^2 + 9y^2 - 16x - 18y - 11 = 0.$
2. $18x^2 + 12y^2 - 12x + 12y - 19 = 0.$
3. $4x^2 + 3y^2 + 16x - 6y + 31 = 0.$ *Ans.* No locus.
4. $25x^2 - 4y^2 + 50x - 8y - 79 = 0.$
5. $4x^2 - 9y^2 + 20x + 12y + 33 = 0.$
6. $7x^2 - 5y^2 + 2x - 4y - 1 = 0.$
7. $y^2 - 8x + 6y + 49 = 0.$
8. $3x^2 - 6x - 5y + 3 = 0.$
9. $2y^2 + 4x + 3y - 8 = 0.$
10. $xy + 2x - 3y - 11 = 0.$
11. $5xy - 5x + y + 1 = 0.$

12. $3xy + x - 18y - 6 = 0.$

13. $2x^2 + 5xy - 3y^2 + 3x + 16y - 5 = 0.$

14. $x^2 + 4xy + 3y^2 - 2x - 2y = 0.$

15. $3x^2 - xy + 5y^2 - 6x + y + 3 = 0.$ *Ans.* The point (1, 0).

16. Prove that every equation of the form

$$y = Ax^2 + Bx + C, \qquad A \neq 0,$$

or of the form

$$x = Ay^2 + By + C, \qquad A \neq 0,$$

represents a parabola with its axis parallel to an axis of coördinates.

17. Show that every equation of the form

$$bxy + dx + ey + f = 0, \qquad b \neq 0,$$

represents either a rectangular hyperbola with its asymptotes parallel to the axes, or two perpendicular straight lines parallel to the axes. Prove that the latter case occurs if and only if $bf = de$.

18. Given the equation

$$ax^2 + cy^2 + dx + ey + f = 0,$$

where neither a nor c is 0: $ac \neq 0$.

(*a*) If $ac > 0$, prove that the equation represents an ellipse, or a point, or that it has no locus.

(*b*) If $ac < 0$, show that the equation represents a hyperbola, or a pair of intersecting straight lines.

2. Rotation of Axes. *Example* 1. Let it be required to identify the curve defined by the equation

$$(1) \qquad 5x^2 - 6xy + 5y^2 - 8 = 0.$$

We transform (1) by a rotation of the (x, y)-axes through an arbitrary angle γ into the (x', y')-axes. For x and y in (1) we set, then, according to Ch. XI, § 2,

$$(2) \qquad \begin{aligned} x &= x' \cos\gamma - y' \sin\gamma, \\ y &= x' \sin\gamma + y' \cos\gamma, \end{aligned}$$

and obtain, after collecting terms and simplifying,

$$(5-6\cos\gamma\sin\gamma)x'^2-6x'y'(\cos^2\gamma-\sin^2\gamma)+(5+6\sin\gamma\cos\gamma)y'^2-8=0,$$

or, on replacing the trigonometric functions of γ by functions of 2γ,

(3) $(5-3\sin 2\gamma)x'^2-6x'y'\cos 2\gamma+(5+3\sin 2\gamma)y'^2-8=0.$

We now choose γ so that the coefficient of $x'y'$ will become 0:

$$\cos 2\gamma = 0.$$

Values of 2γ satisfying this equation are 90°, 270°, 450°, 630°; the corresponding values of γ are 45°, 135°, 225°, 315°. We choose, arbitrarily, the smallest of these values, namely, $\gamma=45°$. Equation (3) thus becomes

$$2x'^2+8y'^2-8=0,$$

or

(4) $$\frac{x'^2}{4}+\frac{y'^2}{1}=1.$$

Consequently, equation (1) represents an ellipse with its center at the origin and with the transverse axis inclined at an angle of 45° to the axis of x.

Example 2. Consider the equation

(5) $$5x^2-6xy+5y^2-4x-4y-4=0.$$

We proceed, as in § 1, Example 4, transforming (5) to arbitrary parallel axes, $\bar{x}$, $\bar{y}$, and then choosing the new origin, (x_0, y_0), so that in the equation resulting from (5) the linear terms in $\bar{x}$ and $\bar{y}$ drop out. We find that the new origin must be at the point (1, 1), and that the resulting equation then becomes

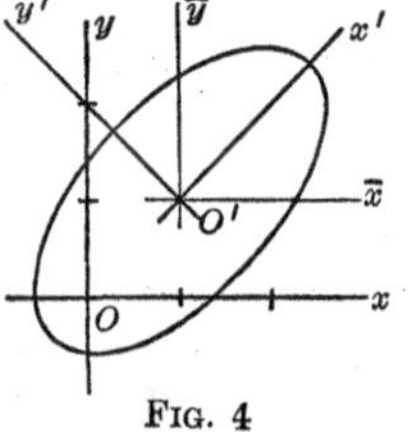

FIG. 4

(6) $$5\bar{x}^2-6\bar{x}\bar{y}+5\bar{y}^2-8=0,$$

where the constant term, -8, is found as the value of the left-hand side of (1) for $x=1$, $y=1$; cf. end of § 1.

Now (6) is the same equation in $\bar{x}$, $\bar{y}$ as (1) is in x, y.

Hence, it follows that (5) represents an ellipse with its center at the point (1, 1) and with the transverse axis inclined at an angle of 45° to the axis of x.

The procedure, then, for any equation similar in form to (5) consists first in transforming to parallel axes so that the linear terms in x and y drop out, and then in rotating the new axes so that the quadratic term in x, y drops out. We shall show later that this procedure is always valid except in one case.

EXERCISES

Identify the curves represented by the following equations. Draw a graph in each case, showing the original and the new axes and the curve.

1. $5x^2 - 6xy + 5y^2 - 32 = 0.$
2. $5x^2 + 26xy + 5y^2 - 72 = 0.$
3. $7x^2 + 2xy + 7y^2 + 2 = 0.$
4. $5x^2 + 2\sqrt{3}xy + 7y^2 - 16 = 0.$
5. $2x^2 + 4\sqrt{3}xy - 2y^2 - 16 = 0.$
6. $3x^2 - 2xy + 3y^2 - 4x - 4y = 0.$
7. $x^2 + 6xy + y^2 - 10x - 14y + 14 = 0.$
8. $4x^2 + 16xy + 4y^2 - 4x - 8y + 13 = 0.$
9. Show that, if $b \neq \pm 2a$, the equation

$$ax^2 + bxy + ay^2 + f = 0, \qquad bf \neq 0,$$

represents an ellipse, or a hyperbola, with its center at the origin and with its axes bisecting the angles between the coördinate axes.

3. Continuation. General Case. We propose to develop and simplify the method of § 2, Example 1, for the removal of the term in xy. Take the equation

$$(1) \qquad Ax^2 + Bxy + Cy^2 + F' = 0, \qquad B \neq 0,$$

and rotate the axes through the arbitrary angle γ, by means of formulas (2), § 2. The resulting equation can be written as

$$ax'^2 + bx'y' + cy'^2 + F' = 0, \tag{2}$$

where

$$\begin{aligned} a &= A\cos^2\gamma + B\sin\gamma\cos\gamma + C\sin^2\gamma,\\ b &= -(A-C)\sin 2\gamma + B\cos 2\gamma, \\ c &= A\sin^2\gamma - B\sin\gamma\cos\gamma + C\cos^2\gamma. \end{aligned} \tag{3}$$

Since γ is to be chosen so that $b = 0$,

$$-(A - C)\sin 2\gamma + B\cos 2\gamma = 0, \tag{4}$$

$$\cot 2\gamma = \frac{A - C}{B}. \tag{5}$$

Of the values of 2γ which satisfy this equation, we choose arbitrarily that one which lies between 0° and 180°. Then γ is a positive *acute* angle.

If the axes are rotated through this angle γ, (2) becomes

$$ax'^2 + cy'^2 + F' = 0. \tag{6}$$

The values of a and c are still to be determined. There is a simpler way of doing this than substituting the value found for γ in the formulas for a and c, as given by the first and last equations of (3). First, add these two equations; the result is

$$a + c = A + C. \tag{7}$$

Thus we have one very simple equation for the two unknown quantities a and c.

Next, subtract the second of the two equations from the first:

$$a - c = (A - C)\cos 2\gamma + B\sin 2\gamma. \tag{8}$$

Square both sides of (8) and both sides of the second equation of (3):

$$b = -(A - C)\sin 2\gamma + B\cos 2\gamma,$$

and add the equations thus obtained; the final result is

$$(a - c)^2 + b^2 = (A - C)^2 + B^2. \tag{9}$$

But γ was chosen so that $b = 0$. Consequently, (9) becomes

$$(a - c)^2 = (A - C)^2 + B^2,$$

or

$$(10) \qquad a - c = \pm \sqrt{(A - C)^2 + B^2}.$$

Thus we have a second simple equation for the two unknowns a and c.

From equations (7) and (10) the values of a and c are easily found in terms of the known coefficients, A, B, and C, of (1). There are, however, two values of each, due to the double sign before the radical in (10). Which values should we take?

If in (8) we substitute for $A - C$ its value as given by (4), we obtain

$$a - c = B\frac{\cos^2 2\gamma}{\sin 2\gamma} + B \sin 2\gamma,$$

or

$$a - c = \frac{B}{\sin 2\gamma}.$$

But 2γ lies, by choice, between $0°$ and $180°$ and, consequently, $\sin 2\gamma$ is positive. It follows that $a - c$ *must have the same sign as* B.

Accordingly, if we rewrite (10) as

$$a - c = \pm B\sqrt{\left(\frac{A - C}{B}\right)^2 + 1},$$

the plus sign must be chosen. Hence, always,

$$(11) \qquad a - c = B\sqrt{\left(\frac{A - C}{B}\right)^2 + 1}.$$

From equations (7) and (11) unique values for a and c can now be found.

Example. Consider the equation

$$(12) \qquad 7x^2 - 8xy + y^2 + 14x - 8y - 2 = 0.$$

By shifting the origin properly, to the point $(-1, 0)$, (12) becomes

$$(13) \qquad 7x'^2 - 8x'y' + y'^2 - 9 = 0.$$

Next, rotate the new axes through the positive acute angle given by formula (5), which in this case is

$$\cot 2\gamma = \frac{7-1}{-8} = -\frac{3}{4},$$

so that γ has the value 63° 26′. Thus (13) becomes

$$(14) \qquad ax''^2 + cy''^2 - 9 = 0.$$

The values of a and c are determined from equations (7) and (11), which are, here,

$$a + c = 8,$$

$$a - c = -8\sqrt{\left(\frac{6}{-8}\right)^2 + 1} = -10.$$

Then the values for a and c are: $a = -1$, $c = 9$. Consequently, (14) becomes

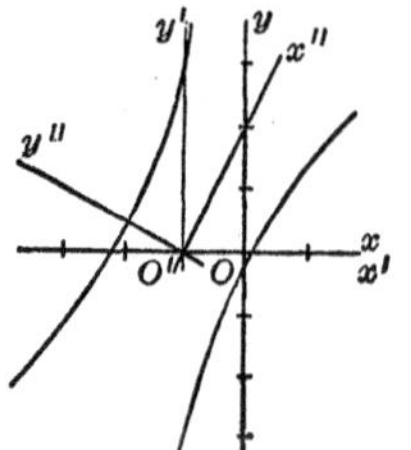

FIG. 5

$$-x''^2 + 9y''^2 - 9 = 0,$$

or

$$(15) \qquad \frac{x''^2}{9} - \frac{y''^2}{1} = -1.$$

Equation (15) represents a hyperbola with its transverse axis along the axis of y''. Hence (12) represents a hyperbola with its center at $(-1, 0)$ and with its transverse axis inclined at an angle of 63° 26′ + 90° = 153° 26′ with the axis of x.

The Expression $B^2 - 4AC$. If from (9) we subtract the square of (7), we obtain

$$(16) \qquad b^2 - 4ac = B^2 - 4AC,$$

or, since we chose γ so that $b = 0$,

$$(17) \qquad -4ac = B^2 - 4AC.$$

The Case $B^2 - 4AC > 0$. If $B^2 - 4AC$ is positive, ac is, by (17), negative; hence a and c have opposite signs. Thus, if $F' \neq 0$, (6) represents a hyperbola. If $F' = 0$, (6) becomes

$$(18) \qquad ax'^2 + cy'^2 = 0.$$

Since a and c have opposite signs, the left-hand side of (18) can be written as the difference of two squares and then factored. Therefore, (18) represents two straight lines (Ch. IX, § 4) which intersect at the origin.

These results for (6) are true for the original equation (1). They hold not only if $B \neq 0$, — the case which we have been treating, — but also if $B = 0$. For, if $B = 0$, (1) is itself in the form (6), and hence may be considered directly. We have, then, the following theorem.

THEOREM 1. *When* $B^2 - 4AC > 0$, *the equation*

$$Ax^2 + Bxy + Cy^2 + F' = 0$$

represents a hyperbola, if $F' \neq 0$; *if* $F' = 0$, *it represents two intersecting straight lines.*

The Case $B^2 - 4AC < 0$. In this case, according to (17), ac is positive, and a and c have the same signs. Then if $F' \neq 0$, (6) represents an ellipse, or, in the case that a, c, and F' are all of the same sign, has no locus. If $F' = 0$, (6) reduces to (18). But now the left-hand side of (18) can be written as the sum of two squares, since a and c have the same sign. Hence it is satisfied only by $x = 0$, $y = 0$. It represents, then, a single point, or, as we may say, a *null ellipse*.*

Not merely a and c have the same signs in this case, but also A and C. For, if A and C have not the same signs, the product $AC < 0$ or $= 0$; consequently, $B^2 - 4AC > 0$, — a contradiction. It follows, further, from (7), that A and C have the same signs as a and c.

We can now characterize more fully the two cases which arise when $F' \neq 0$. We have seen that equation (6) has no locus, if F' is of the same sign as a and c, or, as we can now say, if F' is of the same sign as A and C, *i.e.* if AF' (or CF') > 0. On the other hand, (6) represents an ellipse, if F' is opposite in sign to A and C, *i.e.* if AF' (or CF') < 0.

We summarize our results in the form of a theorem.

* Cf. null circle, Ch. IV, § 2.

THEOREM 2. *When* $B^2 - 4AC < 0$, *the equation*

$$Ax^2 + Bxy + Cy^2 + F' = 0,$$

if $F' \neq 0$, *represents an ellipse or has no locus, according as* AF' *(or* CF'*) is negative or positive; if* $F' = 0$, *the equation represents a single point.*

The Case $B^2 - 4AC = 0$. If $B^2 - 4AC = 0$, there is no need of rotating the axes. Consider, for example, the equation

$$(19) \qquad 9x^2 - 6xy + y^2 - 4 = 0,$$

for which $B^2 - 4AC = 36 - 4 \cdot 9 = 0$. This equation can be written in the form

$$(3x - y)^2 - 4 = 0,$$

or

$$(3x - y - 2)(3x - y + 2) = 0,$$

and hence represents two parallel lines of slope 3.

EXERCISES

Identify the curves represented by the following equations. Draw a graph in each case, showing the original and the new axes and the curve.

1. $2x^2 + 4xy + 5y^2 + 4x + 16y + 2 = 0$.
2. $3x^2 + 12xy + 8y^2 + 6x + 16y + 38 = 0$.
3. $73x^2 + 72xy + 52y^2 + 74x - 32y - 47 = 0$.
4. $2x^2 + 3xy - 2y^2 - 16x - 12y + 22 = 0$.
5. $x^2 - 5xy + 13y^2 - 3x + 21y = 0$.
6. $15xy - 8y^2 + 450y - 450 = 0$.
7. $20x^2 - 16xy + 8y^2 + 52x - 40y + 5 = 0$.
8. $8x^2 + 8xy - 7y^2 + 36y + 36 = 0$.
9. $7x^2 - 3xy + 3y^2 + 5x + 15y + 35 = 0$. *Ans.* No locus.
10. $12x^2 - 20xy - 36y^2 - 22x - 26y - 9 = 0$.
11. $3x^2 + 2xy + 2y^2 + 10x = 0$.
12. $x^2 + 3xy - y^2 + 2x - 10y = 0$.

4. The General Equation, $B^2 - 4AC \neq 0$. We consider here the general equation of the second degree:

$$(1) \qquad Ax^2 + Bxy + Cy^2 + Dx + Ey + F = 0,$$

assuming that $B^2 - 4AC \neq 0$. From the results of the preceding paragraph, we should expect that, *in general*, (1) represents an ellipse or has no locus, if $B^2 - 4AC < 0$, and represents a hyperbola, if $B^2 - 4AC > 0$. Accordingly, we shall call (1) an equation of *elliptic type* or of *hyperbolic type*, according as $B^2 - 4AC$ is *negative* or *positive*.*

To remove the terms in x and y from (1), we set

$$x = x' + x_0, \qquad y = y' + y_0$$

in (1), obtaining

$$(2) \qquad Ax'^2 + Bx'y' + Cy'^2 + (2Ax_0 + By_0 + D)x' + (Bx_0 + 2Cy_0 + E)y' + F' = 0,$$

where

$$(3) \qquad F' = Ax_0^2 + Bx_0y_0 + Cy_0^2 + Dx_0 + Ey_0 + F$$

is the value of the left-hand side of (1), formed for $x = x_0$, $y = y_0$.

Setting the coefficients of x' and y' in (2) equal to zero:

$$(4) \qquad \begin{aligned} 2Ax_0 + By_0 + D &= 0, \\ Bx_0 + 2Cy_0 + E &= 0, \end{aligned}$$

and solving these equations simultaneously for x_0 and y_0, we have

$$(5) \qquad x_0 = \frac{2CD - BE}{B^2 - 4AC}, \qquad y_0 = \frac{2AE - BD}{B^2 - 4AC}.$$

Since it has been assumed that the denominator, $B^2 - 4AC$, of these fractions is not 0, it is always possible to solve equations (4), and the solution (5) is unique.

If the new origin (x_0, y_0) is taken at the point (5), equation (2) becomes

$$(6) \qquad Ax'^2 + Bx'y' + Cy'^2 + F' = 0.$$

* If $B^2 - 4AC = 0$, we shall say that (1) is of *parabolic type*. This case will be treated in the next paragraph.

Equation (6) is exactly the equation treated in § 3. Therefore the theorems of § 3 are valid for it and, consequently, for the original equation (1).

The value of F' as given by (3) can be put in a more convenient form. Multiply the first of the equations (4) by x_0, the second by y_0, and add:

$$2Ax_0^2 + 2Bx_0y_0 + 2Cy_0^2 + Dx_0 + Ey_0 = 0.$$

Multiply this equation by $-\frac{1}{2}$ and add it to (3):

$$F' = \tfrac{1}{2}Dx_0 + \tfrac{1}{2}Ey_0 + F.$$

Finally, substitute the values of x_0 and y_0 as given by (5). The result is

$$F' = -\frac{4ACF - B^2F - AE^2 - CD^2 + BDE}{B^2 - 4AC}.$$

The numerator of the fraction is known as the *discriminant* of equation (1) and is denoted by Δ:

(7) $$\Delta = 4ACF - B^2F - AE^2 - CD^2 + BDE.$$

In terms of Δ, F' has the value

(8) $$F' = -\frac{\Delta}{B^2 - 4AC}.$$

It is clear that if $F' = 0$, then $\Delta = 0$, and conversely. In stating the theorems of § 3 for equation (1) above, we can, therefore, replace $F' \neq 0$ and $F' = 0$ by $\Delta \neq 0$ and $\Delta = 0$ respectively. Furthermore, in case $B^2 - 4AC$ is negative and F' and Δ are not 0, Δ has the same sign as F'. In this case, then, AF' (or CF') is positive or negative, according as $A\Delta$ (or $C\Delta$) is positive or negative.

We now restate, for equation (1), the theorems of § 3.

THEOREM 3. *An equation* (1) *of hyperbolic type:*

$$B^2 - 4ac > 0,$$

represents a hyperbola, if $\Delta \neq 0$. *If* $\Delta = 0$, *it represents two intersecting straight lines.*

THEOREM 4. *An equation* (1) *of elliptic type:*

$$B^2 - 4AC < 0,$$

if $\Delta \neq 0$, *represents an ellipse or has no locus, according as* $A\Delta$ (*or* $C\Delta$) *is negative or positive; if* $\Delta = 0$, *the equation represents a single point.*

If an equation of the form (1) is given, and $B^2 - 4AC \neq 0$, the type of curve which the equation represents can be determined by finding the sign of $B^2 - 4AC$ and by ascertaining whether or not $\Delta = 0$. Further investigation is necessary only in case $B^2 - 4AC < 0$ and $\Delta \neq 0$; the sign of $A\Delta$ (or $C\Delta$) must then be determined.

For example, the equation

$$x^2 - 3xy + 2y^2 + x - 5y + 3 = 0$$

represents a hyperbola, inasmuch as $B^2 - 4AC = 9 - 4 \cdot 2 = 1 > 0$, and $\Delta = -15 \neq 0$.

To find the position and size of an ellipse or a hyperbola defined by an equation of the form (1), it is necessary to carry through in detail the work of changing the origin and rotating the axes. If, however, $\Delta = 0$, it is sufficient merely to make the proper change of origin. The equation then takes on the form (6), where $F' = 0$. In the elliptic case, it represents a single point, the new origin. In the hyperbolic case, it can be factored into two linear equations, which determine the two lines typical of this case.

EXERCISES

Determine the nature of the curve defined by each of the following equations. In case the equation represents two straight lines or a single point, find the equations of the lines, or the coördinates of the point, referred to the (x, y)-axes.

1. $4x^2 - 5xy + y^2 + 11x - 8y = 0$.

2. $3x^2 - 4xy + 2y^2 - 2x = 0$.

3. $3x^2 + 2xy + y^2 - 8x - 4y + 6 = 0$.

Ans. The point (1, 1).

4. $2x^2 + 3xy - 2y^2 - 11x - 2y + 12 = 0.$
Ans. The lines $2x - y - 3 = 0,\quad x + 2y - 4 = 0.$

5. $x^2 + xy + y^2 + 3y + 4 = 0.$ *Ans.* No locus.

6. $3x^2 - xy - 2y^2 - 5x - 2y - 56 = 0.$

7. $2x^2 - xy + y^2 - 7y + 10 = 0.$

8. $4x^2 - 3xy + 9y^2 + 17x - 12y + 19 = 0.$

9. $10x^2 - 9xy - 9y^2 + 14x + 21y - 12 = 0.$

10. $4x^2 - 2xy + y^2 - 4x + y + 5 = 0.$

11. $2x^2 - 3xy + y^2 - 6x + 5y + 4 = 0.$

12. Prove that the general equation is of hyperbolic type, if $AC < 0$, *i.e.* if A and C are of opposite signs.

13. The same, if $B \neq 0$ and $AC = 0$.

5. The General Equation, $B^2 - 4AC = 0$. *First Method.* If $B^2 - 4AC$ has the value 0, the equation

$$(1)\qquad Ax^2 + Bxy + Cy^2 + Dx + Ey + F = 0$$

is said to be of *parabolic type.* The method used in the case $B^2 - 4AC \neq 0$, which begins with shifting the origin so that the linear terms in x and y drop out, is inapplicable here, since equations (4) of § 4, for the determination of the new origin, have in general no solution if $B^2 - 4AC = 0$.*

Let us begin, not with a change of origin, but with a rotation of axes, assuming that $B \neq 0$. Applying to (1) the transformation (2) of § 2, we obtain

$$(2)\qquad ax'^2 + bx'y' + cy'^2 + dx' + ey' + F = 0,$$

where a, b, c are as given by formulas (3) of § 3, and

$$(3)\qquad \begin{aligned} d &= D\cos\gamma + E\sin\gamma, \\ e &= -D\sin\gamma + E\cos\gamma. \end{aligned}$$

* They have no solution if the lines

$$2Ax + By + D = 0, \qquad Bx + 2Cy + E = 0$$

are parallel; infinitely many solutions, if these lines are identical.

Since formulas (3) of § 3 are valid, so are the equations which were deduced from them; in particular,

(4) $$a + c = A + C,$$

(5) $$b^2 - 4ac = B^2 - 4AC.$$

Here, $B^2 - 4AC = 0$, and hence $b^2 - 4ac = 0$. It follows, then, that we can make $b = 0$ by choosing γ so that $b = 0$, or by choosing γ so that either $a = 0$ or $c = 0$. The second of these two methods is in the end the simpler. We will follow it and, in particular, choose to make $a = 0$.

If $a = 0$, we have, by the first of the formulas (3) of § 3,

$$A\cos^2\gamma + B\sin\gamma\cos\gamma + C\sin^2\gamma = 0.$$

Divide by $\sin^2\gamma$, substitute for C its value $\dfrac{B^2}{4A}$,* and clear of fractions; the result is

$$4A^2\cot^2\gamma + 4AB\cot\gamma + B^2 = 0,$$

or

$$(2A\cot\gamma + B)^2 = 0.$$

Hence

(6) $$\cot\gamma = -\frac{B}{2A}.$$

We choose that value of γ satisfying (6) which lies between 0° and 180°.

If the axes are rotated through this angle γ, then $a = 0$, $b = 0$ and, from (4), $c = A + C$. Thus (2) becomes

(7) $$(A + C)y'^2 + dx' + ey' + F = 0,$$

where the values of d and e are to be computed from (3).

Equation (7) can now be treated by the method of § 1, Example 3.

Example. The equation

(8) $$3x^2 + 12xy + 12y^2 + 10x + 10y - 3 = 0$$

is of parabolic type, since $B^2 - 4AC = 144 - 4 \cdot 3 \cdot 12 = 0$.

* $A \neq 0$, for otherwise $B = 0$; and we have assumed $B \neq 0$.

Here, (6) becomes

$$\cot\gamma = -\tfrac{12}{6} = -2,$$

whence

$$\gamma = 153^\circ\,26', \qquad \sin\gamma = \tfrac{1}{5}\sqrt{5}, \qquad \cos\gamma = -\tfrac{2}{5}\sqrt{5}.$$

Rotate the axes through this angle γ and compute the values which $A + C$, d, and e have in this case. There results, as the equation into which (8) transforms,

$$15\,y'^2 - 2\sqrt{5}\,x' - 6\sqrt{5}\,y' - 3 = 0.$$

This equation can be rewritten as

$$15\left(y'^2 - \tfrac{2}{5}\sqrt{5}\,y' + \tfrac{1}{5}\right) = 2\sqrt{5}\,x' + 3 + 3,$$

or

$$15\left(y' - \tfrac{1}{5}\sqrt{5}\right)^2 = 2\sqrt{5}\left(x' + \tfrac{3}{5}\sqrt{5}\right),$$

or, finally, as

$$15\,y''^2 = 2\sqrt{5}\,x'', \tag{9}$$

where

$$(x', y') = \left(-\tfrac{3}{5}\sqrt{5}, \tfrac{1}{5}\sqrt{5}\right) \tag{10}$$

has been introduced as new origin of coördinates.

It follows from (9) that equation (8) represents a parabola with its vertex at the new origin (10) and with its axis inclined at an angle of 153° 26′ to the axis of x.

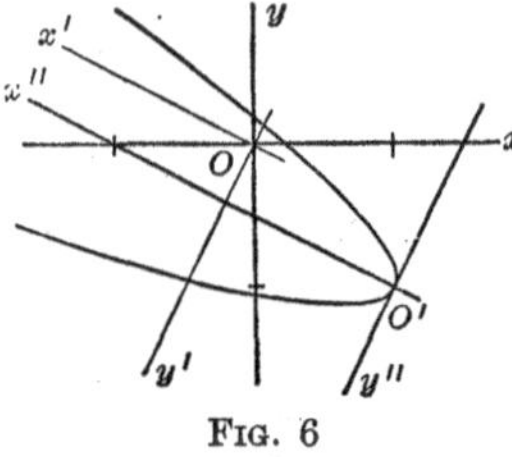

FIG. 6

To find the coördinates of the vertex (10) with respect to the original axes, we substitute in formulas (2) of § 2, first, the values which $\sin\gamma$ and $\cos\gamma$ have in this case:

$$x = -\frac{2\,x' + y'}{\sqrt{5}}, \qquad y = \frac{x' - 2\,y'}{\sqrt{5}},$$

and then the values for x', y' given by (10). We obtain, as the desired coördinates, $(x, y) = (1, -1)$.

We return now to the general case. If $d \neq 0$, equation (7) represents a parabola; cf. Ex. 16, § 1. If $d = 0$, (7) can be written in the form

$$y'^2 + 2\,ky' = l, \tag{11}$$

where we have divided through by $A + C$* and introduced simpler notations for the resulting constants. Equation (11) becomes immediately

(12) $$(y' + k)^2 = k^2 + l;$$

consequently, it represents two parallel lines, a single line, or has no locus, according as $k^2 + l$ is positive, zero, or negative.

To obtain the condition for the exceptional case, $d = 0$, in terms of the coefficients of (1), we note that, since $B^2 - 4AC = 0$, $AC > 0$ and A and C are of the same sign. We assume that A and C are positive; if they were negative, equation (1) could be multiplied through by -1. Since

(13) $$B = \pm 2\sqrt{AC},$$

(6) can be written as

$$\cot \gamma = \mp \sqrt{\frac{C}{A}},$$

whence it can be shown that

$$\sin \gamma = \frac{\sqrt{A}}{\sqrt{A + C}}, \qquad \cos \gamma = \mp \frac{\sqrt{C}}{\sqrt{A + C}}.$$

Hence, from (3),

$$d = \frac{E\sqrt{A} \mp D\sqrt{C}}{\sqrt{A + C}}, \qquad e = \frac{-D\sqrt{A} \mp E\sqrt{C}}{\sqrt{A + C}}.$$

If $d = 0$,

(14) $$E\sqrt{A} \mp D\sqrt{C} = 0.$$

Squaring and replacing $\mp 2\sqrt{AC}$ by $-B$, we have

(15) $$AE^2 + CD^2 - BDE = 0.$$

Now Δ may be written as

$$\Delta \equiv F(4AC - B^2) - (AE^2 + CD^2 - BDE).$$

Since we are treating the case $B^2 - 4AC = 0$, it follows from (15) that $\Delta = 0$; conversely, if $\Delta = 0$, then (15), and hence (14), holds and $d = 0$. Thus the condition for the exceptional case is $\Delta = 0$.

* $A + C \neq 0$, since otherwise (7), and therefore (1), would not be of the second degree.

We collect our results in the form of a theorem.

THEOREM 5. *An equation* (1) *of parabolic type:*

$$B^2 - 4AC = 0,$$

represents a parabola, if $\Delta \neq 0$. *If* $\Delta = 0$, *it represents two parallel lines, a single line, or has no locus.*

We have proved the theorem on the assumption that $B \neq 0$. If $B = 0$, then either $A = 0$ or $C = 0$, and the content of the theorem is easily verified.

If $\Delta = 0$, the given equation may be treated directly, without change of axes. An equation of this type is

$$8x^2 + 24xy + 18y^2 - 14x - 21y + 3 = 0.$$

It can be written in the form

$$2(2x + 3y)^2 - 7(2x + 3y) + 3 = 0;$$

the left-hand side can then be factored:

$$[2(2x + 3y) - 1][(2x + 3y) - 3] = 0.$$

The equation, therefore, represents the two parallel lines

$$4x + 6y - 1 = 0, \qquad 2x + 3y - 3 = 0.$$

Second Method. We notice that equation (8) can be written in the form

$$(16) \qquad 3(x + 2y)^2 + 10x + 10y - 3 = 0.$$

There are two linear expressions in (16), namely, that in the parenthesis and that consisting of the remaining terms in the equation. If the lines represented by these expressions, set equal to zero, were perpendicular, (16) could be simplified by introducing these lines as coördinate axes. As the equation stands, these lines are not perpendicular. We can, however, rewrite it in a form in which they will be.

Add an arbitrary constant k to the expression in the parenthesis in (16) and, in equalization, subtract $6k(x + 2y) + 3k^2$ from the remaining terms:

$$(17) \quad 3(x + 2y + k)^2 + (10 - 6k)x + (10 - 12k)y - 3 - 3k^2 = 0.$$

Determine k so that the two lines defined by the linear expressions in (17) are perpendicular, that is, so that

$$-\frac{10-6k}{10-12k}=2.$$

Thus $\qquad -10+6k=20-24k \quad \text{and} \quad k=1.$

For $k=1$, (17) becomes

(18) $$3(x+2y+1)^2+2(2x-y-3)=0,$$

and the two lines in question are perpendicular. The equations of a transformation introducing these lines as axes were found in § 4 of Ch. XI, and are given by formulas (5) of that paragraph.

Referred to the new axes, (18) becomes

$$3(\sqrt{5}\,x')^2+2(-\sqrt{5}\,y')=0,$$

or $$15x'^2-2\sqrt{5}\,y'=0.$$

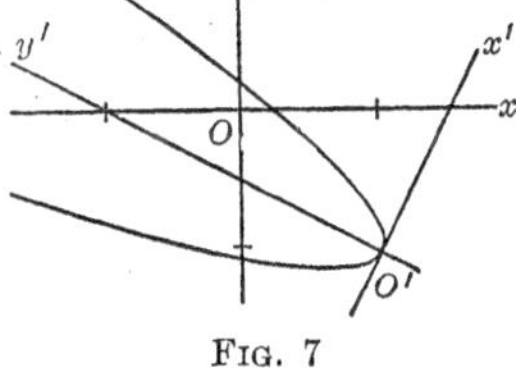

FIG. 7

Hence we have shown again that (8) represents a parabola with its vertex at the point $(1, -1)$ and with its axis inclined at an angle of $153°26'$ to the axis of x.*

The first of the two methods described is more direct and more in keeping with previous methods. Its application to a particular equation, however, is handicapped by the early

* To treat the general equation (1) by this method, assume that A and C are positive. Then, since $B=\pm 2\sqrt{AC}$, (1) can be written as

$$(\sqrt{A}x \pm \sqrt{C}y)^2+Dx+Ey+F=0.$$

From this point the discussion proceeds as in the example in the text.

It can be shown that the exceptional case arises when and only when

$$\sqrt{A}x \pm \sqrt{C}y=0, \qquad Dx+Ey+F=0$$

are parallel; *i.e.* when and only when

$$E\sqrt{A} \mp D\sqrt{C}=0.$$

But this is precisely the equation (14) obtained by the first method. From it follows that $\Delta=0$ is the condition for the exceptional case. Thus we have, in sketch, the proof of Theorem 5 by the second method.

introduction of radicals. The second method avoids this disadvantage, and is the more elegant, though perhaps theoretically the more difficult, of the two.

EXERCISES

Identify and plot roughly the curve defined by each of the following equations. If a change of axes is necessary, show the new axes on the graph.

1. $16x^2 - 24xy + 9y^2 - 38x - 34y + 71 = 0.$
2. $9x^2 - 24xy + 16y^2 + 3x - 4y - 6 = 0.$
3. $25x^2 + 120xy + 144y^2 + 86x - 233y + 270 = 0.$
4. $5x^2 - 20xy + 20y^2 + 2x + y + 3 = 0.$
5. $25x^2 + 30xy + 9y^2 + 10x + 6y + 1 = 0.$
6. $x^2 - 2xy + y^2 + 3x - y - 4 = 0.$
7. $x^2 - 4xy + 4y^2 + 3x - 6y - 10 = 0.$
8. $27x^2 - 36xy + 12y^2 - 40x + 18y + 32 = 0.$
9. $2x^2 + 12xy + 18y^2 + x + 13y + 9 = 0.$
10. $4x^2 + 12xy + 9y^2 + 2x + 3y + 2 = 0.$ *Ans.* No locus.

6. Summary. Invariants. The content of Theorems 3, 4, 5 we summarize in the following table.

	$B^2 - 4AC < 0$	$B^2 - 4AC = 0$	$B^2 - 4AC > 0$
$\Delta \neq 0$	Ellipse, if $A\Delta < 0$ No locus, if $A\Delta > 0$	Parabola	Hyperbola
$\Delta = 0$	Point (Null Ellipse)	Two parallel lines, a single line, or no locus	Two intersecting lines

The ellipse, parabola, and hyperbola are plane sections of a right circular cone (Ch. VIII, § 10). Now, the section of the cone by a plane through the vertex is a point, a single line, or two intersecting lines. If the vertex of the cone is carried off in the direction of the axis indefinitely, the cone approaches a cylinder as its limit, and the plane approaches a position parallel to the rulings of the cylinder. But the section of a cylinder by a plane parallel to the rulings is two parallel lines, a single line, or nothing. These sections of the cone or cylinder are called *degenerate*; those first mentioned, *non-degenerate.*

From the above table we can now draw a general conclusion.

THEOREM 6. *An equation of the second degree, if it has a locus, represents a conic section, which is non-degenerate if* $\Delta \neq 0$, *and degenerate if* $\Delta = 0$.

Invariants. We have seen that the value of the quantity $A + C$ is unchanged by a rotation of axes [§ 3, (7) and § 5, (4)]. This is true also of the value of the quantity $B^2 - 4AC$ [§ 3, (16) and § 5, (5)]. We say that $A + C$ and $B^2 - 4AC$ are *invariant* under a rotation of axes. They are also invariant under a change of origin, since we saw, in § 4, that the quadratic terms in the general equation are not affected by a change of origin.

Consequently, $A + C$ and $B^2 - 4AC$ are invariant under any change of axes. For, any change of axes consists of a change of origin, combined with a rotation of axes.

It can be shown that the discriminant Δ is also invariant under any change of axes.

The importance which these quantities, Δ, $B^2 - 4AC$, and $A + C$, have assumed in the course of the treatment is closely related to the fact that they are *invariants* with respect to any change of axes. For, it is clear that a quantity whose value varies with the choice of axes can have no particular significance in a theory which deals primarily with properties of the curve which are independent of the choice of axes, whereas it is to be expected that an invariant quantity would play an important rôle.

EXERCISES ON CHAPTER XII

In each of the following exercises, determine the nature of the curve represented by the given equation, and then find its position. Draw a figure, showing the curve, the original axes, and any new axes used.

1. $11x^2 + 6xy + 3y^2 - 12x - 12y - 12 = 0.$
2. $7x^2 - 8xy + y^2 + 14x - 8y + 16 = 0.$
3. $8x^2 + 8xy + 2y^2 - 6x - 3y - 5 = 0.$
4. $4x^2 + 8xy + 4y^2 + 13x + 3y + 4 = 0.$
5. $9x^2 - 8xy + 24y^2 - 32x - 16y + 138 = 0.$
6. $x^2 + xy - 2y^2 - 11x - y + 28 = 0.$
7. $9x^2 + 24xy + 16y^2 + 8x - 6y + 3 = 0.$
8. $3x^2 + 4xy + 10x + 12y + 7 = 0.$
9. $3x^2 + 4xy + 2y^2 + 8x + 4y + 6 = 0.$
10. $9x^2 - 15xy + y^2 + 63x = 0.$
11. $10x^2 - 12xy + 5y^2 - 84x + 56y - 14 = 0.$
12. $25x^2 - 20xy + 4y^2 + 20x - 10y + 5 = 0.$
13. $6x^2 + 12xy + y^2 - 36x - 6y = 0.$
14. $32x^2 + 48xy + 18y^2 - 57x - 24y + 6 = 0.$
15. $4x^2 - 7xy - 2y^2 + 22x + y + 10 = 0.$
16. $7x^2 - 18xy - 17y^2 - 28x + 36y + 8 = 0.$
17. $9x^2 - 12xy + 4y^2 + 4x - 59y + 38 = 0.$
18. $7x^2 - 5xy + y^2 - 42x + 15y + 63 = 0.$
19. $14x^2 + 24xy + 21y^2 + 52x + 66y + 14 = 0.$
20. $9x^2 + 12xy + 4y^2 - 4x - 7y - 4 = 0.$
21. $20x^2 + 23xy + 6y^2 + 11x + 10y - 4 = 0.$
22. $25x^2 - 7xy + y^2 - 107x + 16y + 13 = 0.$
23. $49x^2 - 28xy + 4y^2 - 42x + 12y + 9 = 0.$
24. $4x^2 + 6xy + 5y^2 + 2x + 7y + 3 = 0.$
25. $x^2 - 2xy - 6x + 4y + 4 = 0.$
26. $2x^2 - xy + y^2 - 7y + 6 = 0.$

27. Show that an equation of the second degree represents an equilateral hyperbola or two perpendicular lines, if and only if $A + C = 0$.

28. If the equation (1), § 3, represents a hyperbola, show that the asymptotes are defined by the equation

$$Ax^2 + Bxy + Cy^2 = 0.$$

29. If the general equation of the second degree represents a hyperbola, prove that the asymptotes have the directions of the lines defined by the equation of Ex. 28.

30. Prove that every equation of the form

$$AB(x^2 - y^2) - (A^2 - B^2)xy = C,$$

where $C \neq 0$ and not both A and B are 0, represents a rectangular hyperbola with the lines

$$Ax + By = 0, \qquad Bx - Ay = 0$$

as asymptotes.

31. Show that the equation of every rectangular hyperbola can be written in the form

$$AB(x^2 - y^2) - (A^2 - B^2)xy + Dx + Ey + F = 0.$$

32. Find the equation of each of the rectangular hyperbolas

(*a*) $\qquad 12x^2 - 7xy - 12y^2 - 17x + 31y - 13 = 0,$

(*b*) $\qquad 6x^2 + 5xy - 6y^2 - 39x + 26y - 13 = 0,$

referred to the asymptotes as axes.

33. Show that for just one value of λ the equation

$$\lambda x^2 + 4xy + y^2 - 4x - 2y - 3 = 0$$

represents two straight lines. Find the equations of the lines.

34. If the general equation of the second degree represents an ellipse or hyperbola, what is the condition that the center be at the origin?

35. If the general equation represents a parabola, show that the vertex is at the origin if and only if

$$AD^2 + CE^2 + BDE = 0 \qquad \text{and} \qquad F = 0.$$

Suggestion. Write the equation in the form (7), § 5.

36. Prove that, if the equation (1), § 3, represents an ellipse or hyperbola, the axes are defined by the equation

$$B(x^2 - y^2) - 2(A - C)xy = 0.$$

37. Show that, in case $B^2 - 4AC = 0$ and $\Delta = 0$, the general equation represents two parallel lines, a single line, or has no locus, according as the expression

$$D^2 + E^2 - 4(A + C)F$$

is positive, zero, or negative.

Suggestion. Consider equation (7), § 5, where $d = 0$, and use (15), § 5, in simplifying the result.

38. Prove that the expression in Ex. 37 can be replaced by $D^2 - 4AF$, if $A \neq 0$; and by $E^2 - 4CF$, if $C \neq 0$.

Definition. Two conics are said to be similar and similarly placed, if their eccentricities are equal and their corresponding axes are parallel.

39. Prove that the conics,

$$11x^2 + 6xy + 3y^2 - 12x - 12y - 12 = 0,$$
$$11x^2 + 6xy + 3y^2 - 34x - 18y + 29 = 0,$$

are similar and similarly placed.

40. Show that, if the coefficients of the quadratic terms in two equations which represent non-degenerate conics are respectively equal or proportional, the conics are of the same type. Prove further that, if they are ellipses or parabolas, they are similar and similarly placed, and that, if they are hyperbolas, they are similar and similarly placed or each is similar and similarly placed to the conjugate of the other.

CHAPTER XIII

A SECOND CHAPTER ON LOCI. AUXILIARY VARIABLES. INEQUALITIES

1. Extension of the Method for the Determination of Loci. If we look back over the locus problems which we have thus far solved, we find that there is invariably but a *single* essential condition governing the motion of the point tracing the locus. For example, the sum or the difference of two distances or of their squares is required to be constant; or the slope of one line is given as proportional to that of another. This single condition it is comparatively simple to express analytically and thus to determine the locus.

In a great many problems, however, the motion is governed by not just one, but by two or more essential conditions, interdependent on one another. An example of such a problem is the following: A triangle has a fixed base AB and its vertex V moves on an indefinite straight line L, parallel to the base. Find the locus of the point of intersection P of the altitudes.

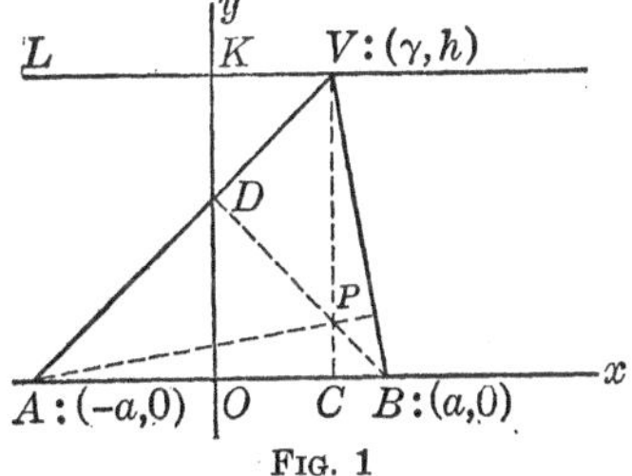

FIG. 1

Here the motion of P is governed by *two* essential conditions; first, by the motion of V, and secondly, by the fact that P is the point of intersection of the altitudes. The two conditions are interrelated, since the position of V determines the position of the altitudes and hence of their point of intersection. Consequently, they are in substance equivalent to a *single* condition, which, when expressed analytically, would give the equation of the locus.

Our problem, then, is to reduce the *two* conditions to a *single* condition. This it is, in general, very difficult to do *geometrically*. *Analytically*, however, the task is simpler. For, conditions expressed in analytical form, in terms of equations, are usually more easily combined than when they are in geometrical form.

Accordingly, we proceed to express analytically the two conditions governing the motion of P. Take the mid-point of the base of the triangle as the origin and the axis of x along the base. Let the length of the base be $2a$ and the distance of L above the base, h. The coördinates of A and B are $(-a, 0)$ and $(a, 0)$. Denote those of P by (X, Y).

The first of the two conditions is that V move along L. But then the distance, KV, of V from the axis of y varies. Accordingly, we can express the motion of V along L by taking the abscissa of V as a *variable*. Denote this variable by γ. The coördinates of V are, then, (γ, h).

We now have coördinates for the three vertices of the triangle. Hence we can find the coördinates (X, Y) of the point of intersection of the altitudes. Thus we shall have expressed the condition that P be this point.

The coördinates (X, Y) of P will be obtained in terms of the constants, a and h, and the variable γ. If we eliminate γ from the two equations which give the values of X and Y, the resulting equation will contain only a and h, and X and Y, and will be the equation of the locus of P.

The variable γ is known as an *auxiliary variable*, or *parameter*. It helps in expressing analytically the conditions governing the generation of the locus. The method involving its use, which we have just described, is general in scope, and may be applied with advantage to any locus problem containing multiple conditions.

2. One Auxiliary Variable. In illustrating the method by examples, let us first complete the problem of the previous paragraph.

Consider P as the point of intersection of VC and BD. The equation of VC is

(1) $$x = \gamma.$$

The slope of AV is

$$\frac{h}{\gamma + a};$$

hence the equation of the perpendicular, BD, to AV is

(2) $$y - 0 = -\frac{\gamma + a}{h}(x - a).$$

Solving equations (1) and (2) simultaneously, we obtain the coördinates of P,

(3) $$X = \gamma, \qquad Y = \frac{a^2 - \gamma^2}{h},$$

in terms of the constants a and h and the auxiliary variable γ.

By eliminating γ from equations (3), we obtain

(4) $$X^2 = -hY + a^2$$

as the equation of the locus.

The locus of P is, then, a parabola, with its axis along the perpendicular bisector of the base of the triangle; it goes through the extremities of the base and opens away from the line L. Every point of it is included in the locus.*

* In the locus problems considered hitherto, particularly in Ch. V, care was taken to emphasize that two things are necessary: (a) to determine the curve, or curves, on which points of the locus lie; (b) to show, conversely, that every point lying on the curve, or curves, obtained is a point of the locus. In the problems of the present chapter, — for example, in the one above, — part (b) of the proof is usually omitted. It consists, as a rule, in retracing the steps of part (a) and so presents, in general, no difficulty. And it is more important, now, that the student gain facility in deducing the equation of the curve, or the equations of the curves, which turn out, in the great majority of cases, to be precisely the locus.

Remark. It was not necessary to find the actual coördinates (3) of P. The fact that P is the point of intersection of the altitudes might have been expressed by writing down the conditions that (X, Y) satisfy equations (1) and (2), namely,

$$X = \gamma, \qquad hY = -(\gamma + a)(X - a).$$

If we eliminate γ from these equations, we obtain equation (4) of the locus.

Example 2. A straight line L passes through a fixed point P_0; find the locus of the mid-point P of the portion of L intercepted by two given perpendicular lines, neither of which goes through P_0.

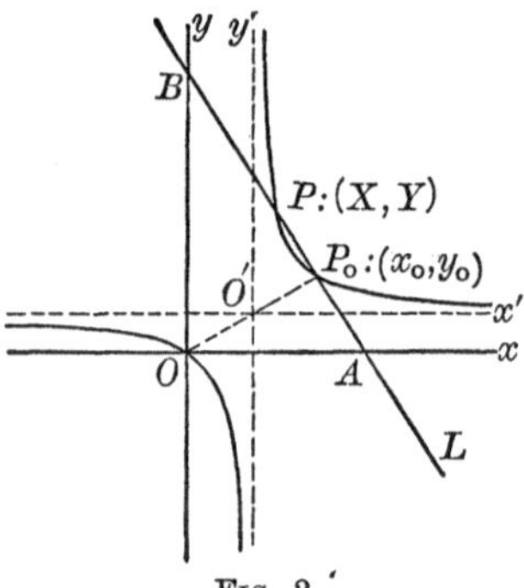

Fig. 2

Take the two given lines as axes, and let the coördinates of P_0, referred to them, be (x_0, y_0). The conditions governing the motion of P are, first, the rotation of L about P_0, and secondly, the fact that P is the mid-point of the segment AB.

We express the rotation of L by taking its slope, λ, as auxiliary variable. The equation of L is, then,

$$y - y_0 = \lambda(x - x_0).$$

The coördinates of the points of intersection of L with the axes are:

$$A: \left(x_0 - \frac{y_0}{\lambda}, 0\right); \qquad B: (0, y_0 - \lambda x_0).$$

Hence the coördinates of P, the mid-point of AB, are

$$X = \frac{1}{2}\left(x_0 - \frac{y_0}{\lambda}\right), \qquad Y = \tfrac{1}{2}(y_0 - \lambda x_0).$$

To eliminate λ from these two equations, we might solve the first for λ and substitute its value in the second. But we notice an easier method; rewriting the equations in the form

$$2X - x_0 = -\frac{y_0}{\lambda},$$

$$2Y - y_0 = -\lambda x_0,$$

and multiplying together the left-hand sides and then the right-hand sides, we obtain the equation

(5) $$(2X - x_0)(2Y - y_0) = x_0 y_0,$$

devoid of λ.

The equation of the locus, in this form, or better, in the form:

$$4\left(X - \frac{x_0}{2}\right)\left(Y - \frac{y_0}{2}\right) = x_0 y_0,$$

suggests that we change to parallel axes, with the new origin at $\left(\frac{x_0}{2}, \frac{y_0}{2}\right)$:

$$x' = X - \frac{x_0}{2}, \qquad y' = Y - \frac{y_0}{2}.$$

The locus, referred to the new axes, has the equation,

$$4x'y' = x_0 y_0,$$

and is, therefore, a rectangular hyperbola. It follows from (5) that the hyperbola goes through O and P_0.

To describe the locus independently of the coördinate system: Let O be the point of intersection of the given lines; the locus is a rectangular hyperbola through O and P_0, with its center at the mid-point of OP_0 and with its asymptotes parallel to the given lines.

EXERCISES

1. Given a line L parallel to the axis of x. Through the origin draw a variable line meeting L in Q, and on this variable line mark the point P whose ordinate equals the abscissa of Q. What is the locus of P?

Ans. The parabola $y^2 = hx$, where h is the algebraic distance from the axis of x to L.

2. A line-segment AB of fixed length moves so that its extremities lie always on two perpendicular lines. Find the locus of the point dividing AB in the ratio $2:1$, using as auxiliary variable the angle which the moving line makes with one of the two perpendicular lines.

Ans. An ellipse, center in the point of intersection of the given lines, axes along them, with length and breadth in the ratio $2:1$.

3. Determine the locus described in Ex. 2, when the given ratio is $m_1 : m_2$.

4. The line L is the perpendicular bisector of the fixed horizontal line-segment AB. The points R and S are taken on L, with R always below S, so that the distance RS is one half the distance AB. Find the locus of the point of intersection of AR and BS, taking the axes of x and y along AB and L. *Ans.* The hyperbola, $2xy + x^2 = a^2$, through A and B.

5. A variable line is drawn through a fixed point P_1 meeting a fixed line L in P_2. Points P are taken on this line so that the product of the distances P_1P and P_1P_2 is constant. Find the locus of these points.

Ans. Two circles, tangent at P_1 to the line through P_1 parallel to L.

6. Find the locus of points from which the tangents drawn to a parabola are perpendicular.

Suggestion. Use the equation of the tangent with given slope, Ch. IX, § 6, eq. (10).

7. The same for the ellipse.

8. The same for the hyperbola.

9. Determine the locus of the mid-points of all the chords drawn from the vertex of a parabola.

3. Coördinates of a Point Tracing a Curve, as Auxiliary Variables. In the problems of the previous paragraph one of the conditions governing the motion of the point tracing the locus was the *auxiliary motion* of a line or of a second point. In each case this auxiliary motion could be expressed analytically by the introduction of *one* auxiliary variable.

Suppose, now, that the auxiliary motion consists of the tracing of a given curve, not a straight line, by a point R. Let the curve be, for example, the circle

$$(1) \qquad x^2 + y^2 = a^2.$$

The motion of R on the circle might be represented analytically by the introduction of a single auxiliary variable, *e.g.* one of the coördinates of R; but it is in general simpler, analytically, to represent the motion by *two* auxiliary variables, namely, by both the coördinates (x', y') of R. These will be connected by the equation,

$$(2) \qquad x'^2 + y'^2 = a^2,$$

which states that R is on the circle.

The reason for this choice of auxiliary variables lies partly in the fact that we thereby avoid radicals;* partly in the principle of *algebraic symmetry*. By this term we mean to signalize the fact that equation (1) bears equally on x and y, and so it is well to carry the solution through in such a manner that it, too, will bear equally on the two coördinates of each of the principal points involved.†

Example 1. Let AA' be a fixed diameter of a given circle and let RR' be a variable chord perpendicular to AA'. What is the locus of the point of intersection, P, of AR and $A'R'$?

Choose the center of the circle as the origin and the axis of x along AA'. Then (1) is the equation of the circle.

* If we had taken x' as a single auxiliary variable, the coördinates of R would be $(x', \pm\sqrt{a^2 - x'^2})$.

† It is possible to represent the motion of R by a single auxiliary variable and at the same time to avoid radicals and preserve symmetry, by choosing as the auxiliary variable the angle θ which the radius drawn to R makes with a fixed direction, *e.g.* the axis of x; the coördinates of R are then: $x' = a\cos\theta$, $y' = a\sin\theta$. We prefer, however, to use as auxiliary variables the coördinates of R connected by equation (2).

Take, as the auxiliary motion, the tracing of the circle by the point R and, as auxiliary variables, the coördinates (x', y') of R. These are connected by the equation (2). The coördinates of R' are, evidently, $(x', -y')$.

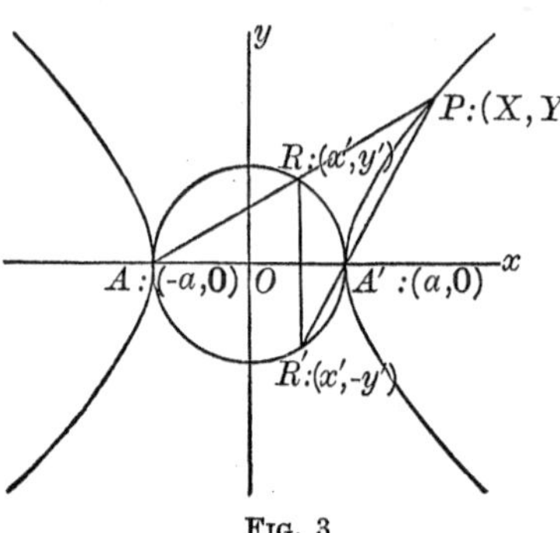

FIG. 3

The equations of AR and $A'R'$ are

$$y - 0 = \frac{y'}{x' + a}(x + a),$$

$$y - 0 = \frac{-y'}{x' - a}(x - a).$$

Since P is the point of intersection of AR and $A'R'$, its coördinates (X, Y) satisfy both these equations:

$$Y = \frac{y'}{x' + a}(X + a). \tag{3}$$

$$Y = \frac{-y'}{x' - a}(X - a). \tag{4}$$

We have, then, three equations, (2), (3), and (4), involving, besides the constant a, the coördinates (X, Y) of the moving point and the auxiliary variables x', y'. To obtain an equation in X, Y alone, we must eliminate x', y'. We shall do this by solving two of these equations, preferably (3) and (4), simultaneously for x' and y', and substituting the values obtained for them in the third equation, (2).

To this end we rewrite equations (3) and (4) as follows:

$$Yx' - (X + a)y' = -aY, \tag{3a}$$

$$Yx' + (X - a)y' = aY. \tag{4a}$$

Hence $$x' = \frac{a^2}{X}, \qquad y' = a\frac{Y}{X}.$$

Substituting these values in (2) and reducing, we obtain

$$X^2 - Y^2 = a^2$$

as the equation of the locus.

The locus is thus seen to be a rectangular hyperbola with the given diameter of the circle as major axis. It is evident from the figure, however, that if R is restricted to the upper half of the circle, and R' to the lower half, only the upper half of one branch and the lower half of the other belong to the locus. It is only when R and R' are each permitted to trace both halves of the circle that the locus consists of the entire hyperbola.

Remark. The points A and A' do not belong to the locus. For, the only possible way in which P can take on the position of the point A', for example, is for R and R' to coincide in A'; but then there is no chord RR' and also no line $A'R'$, so that no point P on the locus is determined.

Let us return now to equations (3a) and (4a). In solving them for x', we actually obtain

$$YXx' = a^2Y.$$

But $Y \neq 0$, since P cannot lie on the axis of x, in either A or A'; hence we were justified in dividing by Y, and the result, $x' = a^2/X$, is correct.

In subsequent problems we shall lay no stress on exceptional points such as A and A'. Their importance for the student at this stage is relatively small.

Example 2. A point R traces a parabola. Find the locus of the point of intersection, P, of the line through the focus and R with the line through the vertex perpendicular to the tangent at R.

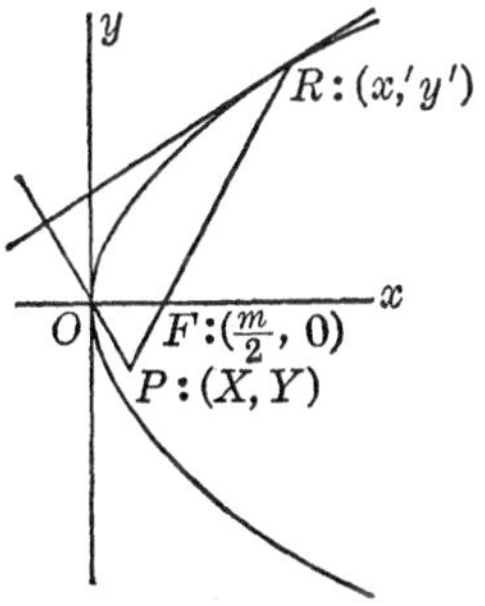

The parabola, referred to the coördinate axes shown in the figure, has the equation

$$y^2 = 2mx.$$

FIG. 4

The motion of R can be expressed by taking, as auxiliary variables, its coördinates (x', y'), connected by the relation

$$y'^2 = 2mx', \tag{5}$$

which states that R, in moving, stays always on the parabola.

The slope of the tangent at R is m/y', Ch. IX, § 2, eq. (5); consequently, the line through O perpendicular to the tangent is

$$y = -\frac{y'}{m}x$$

As the equation of FR we have

$$y = \frac{y}{x' - \frac{m}{2}}\left(x - \frac{m}{2}\right).$$

The equations expressing the fact that $P:(X, Y)$ is the point of intersection of these two lines are, therefore,

$$(6) \qquad Y = -\frac{y'}{m}X,$$

$$(7) \qquad \left(x' - \frac{m}{2}\right)Y = y'\left(X - \frac{m}{2}\right).$$

From equations (5), (6), and (7) we have to eliminate x' and y'. Solving (6) and (7) for x' and y', we have:

$$y' = -m\frac{Y}{X}, \qquad x' = \frac{m(m - X)}{2X}.$$

Substituting these values for y' and x' in (5) and reducing the result, we obtain

$$X^2 + Y^2 - mX = 0$$

as the equation of the locus.

The locus is therefore a circle, passing through the vertex of the parabola and having its center at the focus. The vertex, O, is not a point of the locus.*

Elimination of x', y'. In each of the above examples we eliminated the auxiliary variables x', y' by solving the last two of a set of three equations for x', y' and substituting the values thus obtained for x', y' in the first equation, — the

* It is an exceptional point, similar in type to the exceptional points, A and A', of Example 1. For, when R is at O, FR and OP coincide and consequently determine no point on the locus.

equation stating that the point (x', y') lies on the given curve. This method is valuable because of its general applicability. The student should, however, be on the alert for short cuts in the elimination. For example, he might have noticed by close inspection that, in Example 1, x', y' can be eliminated easily from equations (2), (3), (4) by multiplying equations (3) and (4) together:

$$Y^2 = -\frac{y'^2}{x'^2 - a^2}(X^2 - a),$$

and by noting, from equation (2), that the quantity

$$\frac{-y'^2}{x'^2 - a^2}$$

has unity as its value.

EXERCISES

1. Let AA' be the major axis of an ellipse and RR' be a variable chord perpendicular to AA'. Find the locus of the point of intersection of AR and $A'R'$.

2. Given a fixed diameter of a circle and a variable chord parallel to it. Find the locus of the point of intersection of the line through the mid-point of the chord and one extremity of the diameter with the radius drawn to the corresponding extremity of the chord. What is the locus if the radius is drawn to either extremity of the chord?

Ans. Part of a parabola; the parabola.

3. Find the locus of the point of intersection of the line drawn through a given focus of an ellipse perpendicular to a variable tangent with the line joining the center to the point of tangency.

Ans. The directrix corresponding to the given focus.

4. Let R be a point tracing an ellipse. Find the locus of the point of intersection of the line drawn through the center perpendicular to the tangent at R with the line drawn through R parallel to the conjugate axis.

Ans. An ellipse, similar to and having the same axes as the given ellipse, but with foci on the opposite axis.

5. The preceding problem for a hyperbola.

6. Find the locus of the point of intersection of the line drawn through a variable point R of a parabola parallel to the axis and the line through the vertex perpendicular to the tangent at R.

7. A variable tangent to an ellipse meets the transverse axis in the point T. Determine the locus of the point of intersection of the line drawn through T parallel to the conjugate axis and the line joining the point of contact of the tangent to a vertex.

8. The preceding problem for a hyperbola.

9. Let RR' be an arbitrary chord of an ellipse parallel to the conjugate axis; let the normal at R meet the line joining the center to R' in the point S. Find the locus of the mid-point of RS.

10. The preceding problem for a hyperbola.

4. Other Problems Involving Two or More Auxiliary Variables. There are problems in which it is convenient to use two auxiliary variables other than those of the type which we considered in the preceding paragraph.

Example. The points A and B are fixed and the line L is perpendicular to AB at its mid-point, O; R and S are two points on L, both on the same side of AB and moving so that the product of their distances from O is constant, and equal to b^2. Find the locus of the point of intersection, P, of AR and BS.

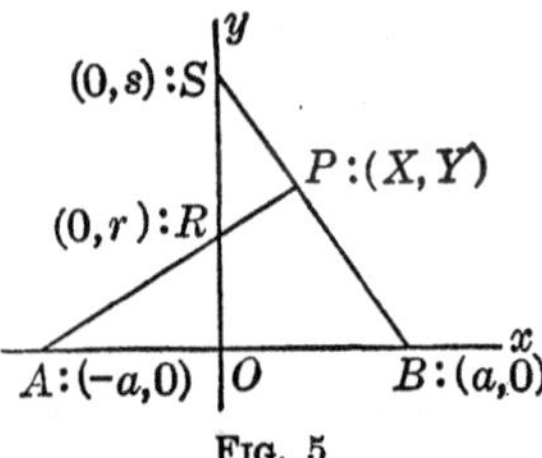

Fig. 5

Take the axes as shown in the figure and let $AB = 2a$. The motions of R and S can be represented by taking their ordinates, which we denote by r and s, as auxiliary variables. The condition that R and S are on the same side of AB and relatively so situated that $OR \cdot OS = b^2$ is then given by the equation,

$$(1) \qquad rs = b^2.$$

The equations of AR and BS are

$$\frac{x}{-a} + \frac{y}{r} = 1, \qquad \frac{x}{a} + \frac{y}{s} = 1.$$

Since $P:(X, Y)$ is the point of intersection of these lines, we have

$$(2) \qquad 1 + \frac{X}{a} = \frac{Y}{r},$$

$$(3) \qquad 1 - \frac{X}{a} = \frac{Y}{s}.$$

To eliminate the auxiliary variables r and s from equations (1), (2), (3) is now our problem. We notice that in the product of equations (2) and (3):

$$1 - \frac{X^2}{a^2} = \frac{Y^2}{rs},$$

r and s enter only in the form rs, and that the value of rs is given by (1) as b^2. We have, therefore, as the equation of the locus

$$\frac{X^2}{a^2} + \frac{Y^2}{b^2} = 1.$$

The locus of P is, therefore, an ellipse, with its axes along AB and L, and passing through the points A and B. These points are not, however, points of the locus.

EXERCISES

1. What is the locus of P in the problem in the text, if R and S are always on opposite sides of AB?

2. The points P_1 and P_2 are fixed, and the lines L_1 and L_2 are perpendicular to P_1P_2 in P_1 and P_2 respectively; Q_1 and Q_2 are two points on L_1 and L_2 respectively, both on the same side of P_1P_2 and moving so that the product of their distances from P_1 and P_2 respectively is constant. Find the locus of the point of intersection, P, of P_1Q_2 and P_2Q_1.

3. What is the locus of P in the preceding example, if Q_1 and Q_2 are always on opposite sides of P_1P_2?

4. Do Ex. 2, § 2, using the intercepts of the moving line AB on the two given perpendicular lines as auxiliary variables.

5. The same for Ex. 3, § 2.

6. The points R and S move, one on each of two fixed perpendicular lines, so that the segment RS subtends always a right angle at a fixed point, not at the intersection of the two lines. Find the locus of the mid-point of RS.

Ans. Perpendicular bisector of the line-segment joining the fixed point with the intersection of the fixed lines.

7. Two right angles, having their vertices in fixed points A and B, rotate about these points, so that the point of intersection of two of their sides traces a line parallel to AB. What is the locus of the point of intersection of the other two sides?

5. Use of the Formula for the Sum of the Roots of a Quadratic Equation. The sum of the roots of a quadratic equation (Ch. IX, § 5),

$$Ax^2 + Bx + C = 0, \qquad A \neq 0,$$

is the negative of the ratio of the coefficients of the terms in x and x^2; that is,

$$(1) \qquad x_1 + x_2 = -\frac{B}{A},$$

where x_1 and x_2 are the roots.

As a simple example of the way in which this fact may be used to advantage, let us find the coördinates of the point P midway between the points of intersection, P_1 and P_2, of a line and a conic. Take, for example, the line

$$(2) \qquad 2x - y = 1,$$

and the ellipse

$$(3) \qquad 3x^2 + 4y^2 = 3.$$

The coördinates (x_1, y_1) and (x_2, y_2), of P_1 and P_2, are the simultaneous solutions of equations (2) and (3). Substituting in (3) the value of y from (2) and collecting terms, we have the quadratic equation,

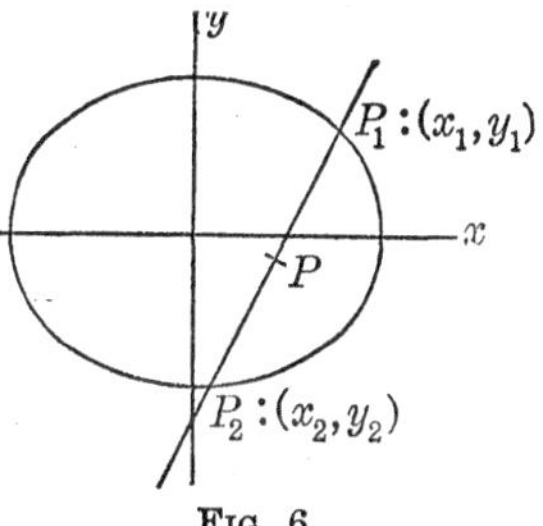

FIG. 6

$$19x^2 - 16x + 1 = 0,$$

for the determination of x_1 and x_2.

We are interested, not in the actual values of x_1 and x_2, but in half their sum; for this is the abscissa of the mid-point, P, of P_1P_2. By (1) the sum is $\frac{16}{19}$. Then the abscissa of P is $\frac{8}{19}$; and, since P lies on the line (2), its ordinate is

$$y = 2(\tfrac{8}{19}) - 1 = -\tfrac{3}{19}.$$

Consider now the following locus problem: A variable tangent to the parabola,

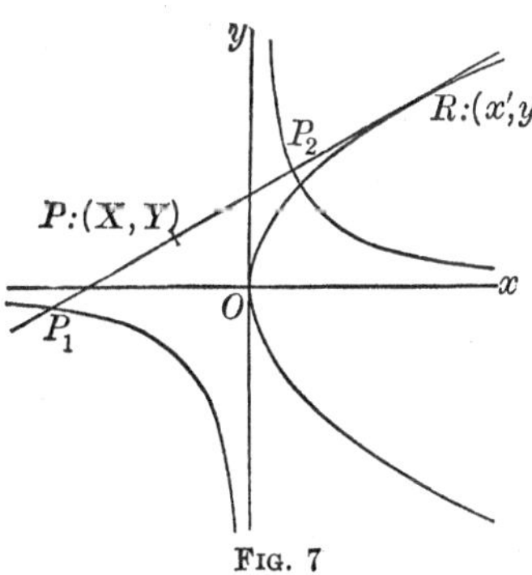

FIG. 7

$$y^2 = 2mx,$$

meets the hyperbola,

$$xy = c^2,$$

in the points P_1 and P_2. What is the locus of the mid-point, P, of P_1P_2?

As auxiliary variables we take the coördinates (x', y') of the point R tracing the parabola; they satisfy the equation of the parabola:

(4) $$y'^2 = 2mx'.$$

The tangent at R has the equation

(5) $$y'y = m(x + x').$$

To find the coördinates of P_1 and P_2, we solve (5) simultaneously with the equation of the hyperbola. Eliminating y, we have:

$$mx^2 + mx'x - c^2y' = 0.$$

Half the sum of the roots of this equation is X; hence, by (1),

$$(6) \qquad X = -\frac{x'}{2}.$$

Since Y is the ordinate of the point on the line (5) whose abscissa is given by (6), we have

$$(7) \qquad y'Y = m\left(-\frac{x'}{2} + x'\right) = \frac{mx'}{2}.$$

We now have three equations, (4), (6), and (7), from which to eliminate the auxiliary variables x' and y'. We solve (6) and (7) for x' and y', obtaining

$$x' = -2X, \qquad y' = -\frac{mX}{Y}.$$

Substituting these values of x' and y' in (4) and simplifying the result, we have

$$4Y^2 = -mX.$$

Consequently, the locus of P is a parabola with vertex at the origin and opening out along the negative axis of x as axis. The origin is not a point of the locus.

EXERCISES

1. A variable tangent to the circle

$$x^2 + y^2 = a^2$$

meets the hyperbola

$$2xy = a^2$$

in the points P_1 and P_2. Find the equation of the locus of the mid-point of P_1P_2. Plot the locus.

Ans. $\frac{1}{x^2} + \frac{1}{y^2} = \frac{4}{a^2}$, a curve which does not, despite its appearance, consist of two conjugate rectangular hyperbolas.

2. Two equal parabolas have the same axis and vertex, but open in opposite directions. Find the locus of the mid-points of the chords of one which, when produced, are tangent to the other.

3. Find the locus of the mid-points of the focal chords — chords through the focus — of a parabola.

Ans. A parabola with its vertex in the focus of the given parabola, with the same axis, but half the size.

4. Determine the locus of the mid-points of one set of focal chords of an ellipse.

5. The same for a hyperbola.

6. A variable tangent to a parabola meets the tangents at the extremities of the latus rectum in the points P_1 and P_2. Find the locus of the mid-point of P_1P_2.

7. The asymptotes of a hyperbola intercept the segment P_1P_2 on a variable tangent. What is the locus of the mid-point of P_1P_2? *Ans.* The hyperbola itself.

6. Loci of Inequalities. Though we are concerned primarily in mathematics with equalities, it is not infrequent that inequalities become important. Accordingly, it is not out of place to consider here the loci of some inequalities.

Example 1. The equation $x-1=0$ represents all the points of the line parallel to and one unit to the right of the axis of y, and no other points. Consequently, the inequality, $x-1 \neq 0$, represents all the points of the plane not on this line. In particular,

$$x-1>0$$

represents all the points to the right of it and

$$x-1<0$$

represents all the points to the left of it.

Example 2. What is the locus of points whose coördinates satisfy the inequality

$$5x+12y+6>0? \tag{1}$$

The equation obtained by replacing the sign $>$ by the sign of equality represents the line L shown in the figure. From

Example 1 we should expect that the locus of (1) would consist of all the points on one side of L. This is, in fact, the case: If the quantity,

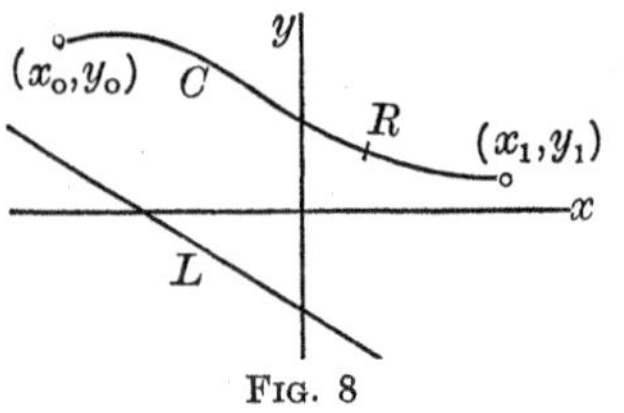

FIG. 8

$$F \equiv 5x + 12y + 6,$$

is positive for a certain point (x_0, y_0), then it is positive for all points on the same side of L as (x_0, y_0).

We prove this by showing that the opposite assumption leads to a contradiction. Suppose that F becomes negative for some point (x_1, y_1) on the same side of L as (x_0, y_0). Join (x_0, y_0) to (x_1, y_1) by any curve C not cutting L, and let a point (x, y) trace this curve. For (x_0, y_0), F is positive; but when (x, y) has reached (x_1, y_1), F has become negative. Consequently, for some intermediate point R on C, F has the value zero, inasmuch as its value changes continuously as (x, y) moves along C. Hence R must lie on L, — a contradiction, since we took C as a curve never cutting L.

To ascertain on which side of L the points represented by (1) lie, we have but to find the value of F for one point not on L. In this case the simplest point to take is the origin. But, when $x = 0$ and $y = 0$, F is positive. Therefore the locus of (1) consists of all points on the same side of L as the origin.

Example 3. What is the locus of the inequality

(2) $$y^2 > 2x?$$

The equation, obtained by replacing the sign $>$ by the equality sign, represents a parabola. By the reasoning of Example 2, then, the inequality represents all the points within, or all the points without, the parabola. The latter is clearly the case, since (2) is not satisfied by the coördinates of the point (1, 0), — a point which is within the parabola.

EXERCISES

Find the loci of the following inequalities. Draw a figure in each case and shade the area of points represented by the inequality.

1. $x + 2 > 0$.
2. $2y + 3 < 0$.
3. $x + y + 1 > 0$.
4. $3x - 4y - 2 > 0$.
5. $2x - 3y < 0$.
6. $x^2 + y^2 < 1$.
7. $y^2 + 7x > 0$.
8. $3x^2 + 4y^2 > 8$.
9. $x^2 - y^2 < 1$.
10. $3x^2 - 2y^2 < -6$.

7. Locus of Two or More Simultaneous Inequalities.

Example 1. Find the locus of points whose coördinates satisfy simultaneously the two inequalities

(1) $$5x + 12y + 6 > 0,$$

(2) $$3x - 4y - 2 > 0.$$

Denote the left-hand sides of (1) and (2) by F_1 and F_2, respectively. By Example 2 of § 6, the points whose coördinates satisfy (1) are all the points which are on the same side of the line $L_1 : F_1 = 0$ as the origin; similarly, the points whose coördinates satisfy (2) are all the points which are on the opposite side of the line $L_2 : F_2 = 0$ from the origin. The points whose coördinates satisfy (1) and (2) are the points common to these two sets, namely, those of region I of the figure.

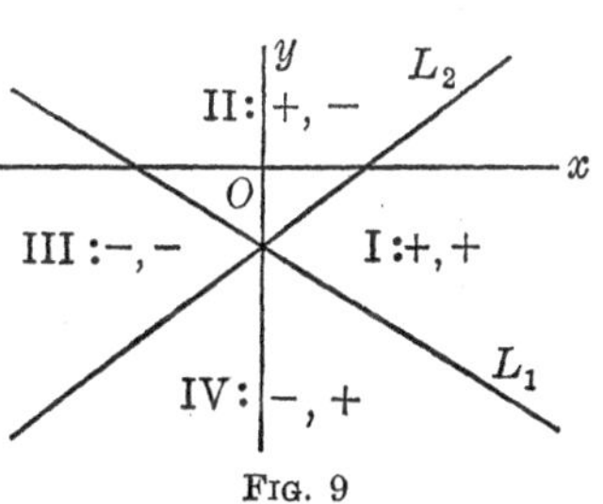

FIG. 9

Lying between the lines L_1 and L_2 there are four regions, I, II, III, IV. It is clear from the foregoing that the pairs of simultaneous inequalities representing these regions are:

$$\text{I}: \begin{cases} F_1 > 0, \\ F_2 > 0; \end{cases} \quad \text{II}: \begin{cases} F_1 > 0, \\ F_2 < 0; \end{cases} \quad \text{III}: \begin{cases} F_1 < 0, \\ F_2 < 0; \end{cases} \quad \text{IV}: \begin{cases} F_1 < 0, \\ F_2 > 0. \end{cases}$$

Example 2. Find the points satisfying simultaneously the inequalities:

$$y^2 - 2x < 0, \qquad x + y - 1 < 0.$$

The equations obtained by replacing the signs $<$ by signs of equality represent a parabola and a line intersecting it. The locus of the first inequality is the interior of the parabola; that of the second is the half-plane bounded by the line and containing the origin. Common to these two regions is the finite region contained between the parabola and the line; this, then, is the locus of the two inequalities taken simultaneously.

EXERCISES

Find the locus of points whose coördinates satisfy simultaneously the following sets of inequalities. Draw a figure in each case, and shade the region represented.

1. $\begin{cases} 4x - 3 < 0, \\ 3x + 2y - 6 < 0. \end{cases}$

2. $\begin{cases} 5x - 12y + 26 > 0, \\ 3x + 4y - 10 > 0. \end{cases}$

3. $\begin{cases} 2x - y + 3 < 0, \\ 4x - 2y + 9 > 0. \end{cases}$

4. $\begin{cases} x^2 + y^2 < 4, \\ x - 3 > 0. \end{cases}$ *Ans.* No locus.

5. $\begin{cases} 3x^2 + 4y^2 - 12 > 0, \\ 2x - 3y + 12 > 0. \end{cases}$

6. $\begin{cases} x^2 - 7y < 0, \\ x^2 - y^2 + 1 > 0. \end{cases}$

7. $\begin{cases} 2x - y - 3 < 0, \\ x + 3y - 5 < 0, \\ 5x + y + 3 > 0. \end{cases}$

8. $\begin{cases} x^2 + y^2 - 1 > 0, \\ y > 0, \\ 2x - y > 0. \end{cases}$

Each of the following pairs of curves divide the plane (minus the points on the curves) into a number of regions. Find the pairs of simultaneous inequalities representing these regions.

9. $5y + 8 = 0, \qquad 3x + 8y - 2 = 0.$

10. $5x - 12y + 26 = 0, \qquad 3x + 4y - 10 = 0.$

11. $2x^2 + y^2 = 8, \qquad 4x - 3y - 2 = 0.$

12. $y^2 = 2mx, \qquad x = 0.$

13. $x^2 - y^2 = 4, \qquad 2x - y - 2 = 0.$

14. $x^2 + 4y = 0, \qquad 2x - 3y - 6 = 0.$

15. $y^2 + 8x = 0, \qquad x^2 + y^2 = 9.$

8. Bisectors of the Angles between Two Lines. The two lines

$$L_1: \qquad 5x + 12y + 6 = 0,$$
$$L_2: \qquad 3x - 4y - 2 = 0,$$

are given. It is required to find the equations of the lines bisecting the angles between them.

We solve this problem by finding the locus of the point $P:(X, Y)$ moving so that its distance D_1 from L_1 equals its distance D_2 from L_2:

$$D_1 = D_2.$$

According to Ch. II, § 8, D_1 and D_2 are

$$D_1 = \pm \frac{5X + 12Y + 6}{13}, \qquad D_2 = \pm \frac{3X - 4Y - 2}{5},$$

where, in each case, that sign is to be chosen which will make the distance positive.

The lines L_1 and L_2 are those of Example 1, § 7. It follows, from the results there given in connection with Fig. 9, that the signs which must be taken to make D_1 and D_2 both positive are:

If P is in I, $+$ for D_1, $+$ for D_2;
if P is in II, $+$ for D_1, $-$ for D_2;
if P is in III, $-$ for D_1, $-$ for D_2;
if P is in IV, $-$ for D_1, $+$ for D_2.

For, if P lies, for example, in the region I, then the numerators in the expressions for D_1 and D_2 are both positive and the $+$ sign must be taken in each case to make D_1 and D_2 positive.

If, now, P is in I or III and $D_1 = D_2$, we have

$$5(5X + 12Y + 6) = 13(3X - 4Y - 2),$$

or, on reducing,

$$(1) \qquad X - 8Y - 4 = 0.$$

Thus (1) is that bisector of the angles between L_1 and L_2 which lies in the regions I and III.

If P is in II or IV and $D_1 = D_2$, we have

$$5(5X + 12Y + 6) = -13(3X - 4Y - 2)$$

or

$$16X + 2Y + 1 = 0. \tag{2}$$

This is the bisector which lies in the regions II and IV.

Simplification. We now give in condensed form the method of finding the bisectors. By equating D_1 and D_2, we have

$$\pm\frac{5X + 12Y + 6}{13} = \pm\frac{3X - 4Y - 2}{5}.$$

If we take both signs positive or both negative and reduce the result, we get (1). If we take the plus sign on the right and the minus sign on the left or vice versa, and then simplify, we get (2). The equations (1) and (2) represent the bisectors; which equation represents a chosen bisector is easily determined by making a plot.

EXERCISES

1. Find the equations of the bisectors of the angles between the following pairs of lines, and draw a figure which shall indicate each bisector.

(a) $\begin{cases} 5x - 12y + 26 = 0, \\ 3x + 4y - 10 = 0; \end{cases}$ (b) $\begin{cases} 4x + 3y - 2 = 0, \\ x - y - 4 = 0. \end{cases}$

2. Find the equation of that bisector of the angle between the two lines,

$$4x - 3y + 3 = 0 \quad \text{and} \quad 3x - 4y - 6 = 0,$$

which passes through the region between the two lines which contains the origin.

3. Find the equations of the circles tangent to the lines of Ex. 1, Part (a), and having their centers on the line $y = 8$.

4. Find the equations of the circles tangent to the lines of Ex. 2 and passing through the point (1, 0).

Given the triangle ABC with the sides

$$\begin{aligned} AB: &\quad 3x + 4y - 3 = 0, \\ BC: &\quad 3x - 4y - 3 = 0, \\ CA: &\quad 12x - 5y + 15 = 0. \end{aligned}$$

5. Prove that the bisectors of the interior angles of the triangle meet in a point. Find its coördinates. *Ans.* $(-\frac{4}{11}, 0)$.

6. Find the equation of the circle inscribed in the triangle.
Ans. $121(x^2 + y^2) + 88x - 65 = 0$.

7. Show that the bisector of the interior angle at the vertex A and the bisectors of the exterior angles at the vertices B and C meet in a point. Find its coördinates. *Ans.* $(1, -5)$.

8. Find the equation of the circle tangent to BC, and to AB and AC produced. *Ans.* $x^2 + y^2 - 2x + 10y + 10 = 0$.

9. How many circles are there tangent to three lines? Draw a figure showing these circles.

10. Given the triangle with vertices A, B, and C in the three points $(1, 0)$, $(-2, 4)$, and $(-5, -8)$. Prove analytically that the bisector of the interior angle at A divides the side BC into segments proportional to AB and AC.

See also Exs. 26–30 at the end of the chapter.

EXERCISES ON CHAPTER XIII

1. Let AA' be a fixed diameter of a circle and R a point tracing the circle. Find the locus of the point of intersection of $A'R$ and the line through A perpendicular to the tangent at R.

2. Find the locus of the point of intersection of the normals to an ellipse and to the auxiliary circle at corresponding points. Take the eccentric angle (Ch. VII, § 10) as the auxiliary variable.

3. The circle $x^2 + y^2 = a^2$ cuts the axis of y in $A:(0, a)$. A point S traces the tangent at A and the second tangent from S touches the circle in R. Find the locus of the point of intersection of the altitudes of the triangle ARS.

Suggestion. Take the abscissa of S and the coördinates of R as auxiliary variables, and use the fact that OS is perpendicular to AR.

4. Let AA' be a fixed diameter of a circle and R a point tracing the circle. Find the locus of the point of intersection of AR and the line joining A' to the point of intersection of the tangents at A and R.

5. The normal to a hyperbola at a variable point R meets the transverse axis in N. Determine the locus of the mid-point of RN.

6. Find the locus of the point of intersection of the line drawn through one focus of an ellipse perpendicular to a variable tangent and the line drawn through the point of tangency parallel to the transverse axis.

Ans. An ellipse, center in the focus chosen, with axes having the same directions as those of the given ellipse.

7. Find the locus of the point of intersection of the line drawn through one vertex of a hyperbola perpendicular to a variable tangent and the line drawn through the point of tangency parallel to the transverse axis.

8. Find the locus of the point of intersection of the line drawn through the focus of a parabola perpendicular to a variable tangent and the line joining the vertex with the point of tangency.

Ans. An ellipse, whose minor axis is the line-segment joining the vertex of the parabola to the focus.

9. Two lines, passing through the points A and B respectively, are originally in coincidence along AB. They are made to rotate in the same direction about A and B respectively, the first twice as fast as the second. What is the locus of their point of intersection?

10. Find the locus of the center of a circle which touches one of two perpendicular lines and intercepts a segment of constant length on the other.

11. Find the locus of the point of intersection of the line drawn through the vertex of the parabola $y^2 = 2\,mx$ perpendicular to a variable tangent and the line drawn through the point of tangency perpendicular to the axis.

Ans. The semi-cubical parabola, $my^2 = 2\,x^3$.

12. A vertex O of a quadrilateral and the directions of the sides through O are fixed. The two angles adjacent to O are right angles and the diagonal joining their vertices has a fixed direction. Find the locus of the fourth vertex.

Ans. Straight line through O, perpendicular to the line through O which makes an angle with the fixed direction equal to the sum of the two angles which the sides through O make with the fixed direction.

13. A parallelogram has sides of constant length a and b and has one vertex fixed at a point O. It opens and closes so that the two sides through O are always equally inclined to a fixed line through O. Taking the angle which these sides make with the fixed line as auxiliary variable, find the locus of the vertex opposite to O.

14. Each of two straight lines moves always parallel to itself so that the product of the distances of the lines from a fixed point O is constant. Find the locus of their point of intersection, taking the axes so that O is the origin and the directions of the two lines are equally inclined to the axis of x.

Ans. Two conjugate hyperbolas, center at O, with asymptotes parallel to the fixed directions.

15. Find the locus of the center of a circle which passes through a fixed point on one of two perpendicular lines and intercepts a segment of constant length on the other.

16. Find the locus of points from which it is possible to draw two perpendicular normals to a parabola.

17. Find the locus of the point of intersection of the tangents to an ellipse at points subtending a right angle at the center.

18. The preceding problem for the hyperbola.

19. Determine the locus of the mid-point of a variable chord of an ellipse drawn from a vertex.

20. Find the locus of the mid-point of a variable chord of a parabola which subtends a right angle at the vertex.

21. A point R traces an ellipse, of which A and A' are the vertices. Find the locus of the point of intersection of the lines drawn through A and A' perpendicular to AR and $A'R$ respectively.

Ans. An ellipse, similar to and with the same center as the given ellipse, but with opposite transverse and conjugate axes.

22. The asymptotes of a hyperbola intercept the segment AB on a variable tangent. What is the locus of the point dividing AB in a given ratio, $m_1 : m_2$?

Ans. A similar hyperbola, with the same transverse and conjugate axes.

Exercises 23–25. Determine the equations of the desired loci by use of rectangular coördinates. To identify the locus from its equation introduce polar coördinates.

23. Find the locus of the point of intersection of a variable tangent to a rectangular hyperbola with the line through the center perpendicular to the tangent. *Ans.* A lemniscate.

24. Find the locus of the point of intersection of a variable tangent to the circle $x^2 + y^2 + 2ax = 0$ and the perpendicular to this tangent from the origin. *Ans.* A cardioid.

25. What is the locus of the mid-points of the chords of the circle $x^2 + y^2 = a^2$ which, when produced, are tangent to the hyperbola $2xy = c^2$? *Ans.* A lemniscate.

26. Show that the line

$$x \cos 30° + y \sin 30° = 5$$

is 5 units distant from the origin and that the perpendicular from the origin to it makes with the positive axis of x an angle of 30°. Prove that the distance of the point (x_0, y_0) from the line is

$$-(x_0 \cos 30° + y_0 \sin 30° - 5),$$

if (x_0, y_0) is on the same side of the line as the origin, and is

$$x_0 \cos 30^\circ + y_0 \sin 30^\circ - 5,$$

if (x_0, y_0) is on the opposite side of the line from the origin.

27. State and prove for the line

$$(1) \qquad x \cos \phi + y \sin \phi = p, \qquad p \geq 0$$

the results corresponding to those given in the preceding exercise for the particular line for which $\phi = 30^\circ$, $p = 5$. Prove that the equation of every line can be written in the form (1).

28. Two lines, with their equations in the form (1), are given. Let $\alpha = 0$, $\beta = 0$ be the abridged notation (Ch. IX, § 3) for these equations. Prove that the bisectors of the angles between the two lines are given by the equations $\alpha - \beta = 0$ and $\alpha + \beta = 0$. Show that, if neither line goes through the origin, the bisector $\alpha - \beta = 0$ passes through that opening between the lines in which the origin lies.

29. The equations of the sides of a triangle, given in the form (1), are $\alpha = 0$, $\beta = 0$, and $\gamma = 0$. Assuming that the origin lies within the triangle, find the equations of the bisectors of the interior angles and prove that they meet in a point.

30. Prove that the bisectors of two exterior angles of the triangle of the preceding exercise and the bisector of the interior angle at the third vertex meet in a point.

CHAPTER XIV

DIAMETERS. POLES AND POLARS

1. Diameters of an Ellipse. By the axes of an ellipse we may mean either the transverse and conjugate axes, indefinite straight lines, or the major and minor axes, the segments of these lines intercepted by the ellipse; cf. the dual definition, Ch. VII, § 1.

By a *diameter* of an ellipse we may mean, also, one of two things, either *an indefinite straight line through the center of the ellipse*, or *the segment of this line intercepted by the ellipse*; and we agree to adopt this dual definition. The length of the segment is called the *length* of the diameter; its end points, the *extremities* of the diameter.

Problem. What is the locus of the mid-points of a set of parallel chords of an ellipse?

In the special case of a circle, the locus is a diameter, considered as a line-segment. This is true, also, for the general ellipse. For, if the chords are parallel to an axis of the ellipse, the theorem is geometrically obvious; if they are oblique to the axes, as is generally the case, we resort to an analytical proof.

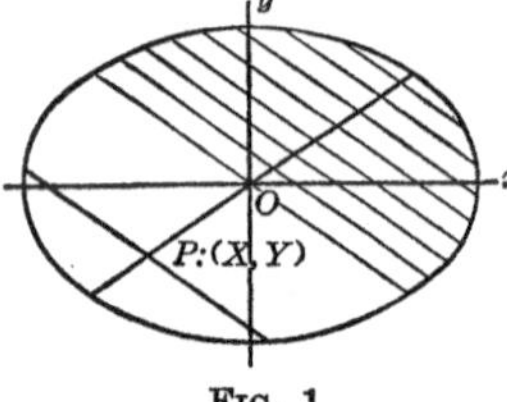

FIG. 1

Let the ellipse be

$$\frac{x^2}{a^2}+\frac{y^2}{b^2}=1, \tag{1}$$

and let $\lambda\ (\neq 0)$ be the slope of the chords. Consider a variable chord of slope λ moving always parallel to itself. Its motion we express analytically by taking β, its intercept on the axis of y, as auxiliary variable. The equation of the chord is, then,

(2) $$y = \lambda x + \beta,$$

where λ is constant and β is variable.

The work now proceeds according to the method of Ch. XIII, § 5. If in (1) we set for y its value as given by (2), we obtain the equation

$$b^2x^2 + a^2(\lambda x + \beta)^2 = a^2b^2,$$

or $$(a^2\lambda^2 + b^2)x^2 + 2\,a^2\lambda\beta x + a^2(\beta^2 - b^2) = 0,$$

whose roots are the abscissæ of the two points of intersection of the line (2) with the ellipse. Half the sum of these roots is X, the abscissa of the mid-point, P, of the chord. Hence, by the formula, Ch. XIII, § 5, (1), for the sum of the roots of a quadratic equation,

(3) $$X = -\frac{a^2\lambda\beta}{a^2\lambda^2 + b^2}.$$

Since, moreover, $P: (X, Y)$ lies on the chord (2), we have

(4) $$Y = \lambda X + \beta.$$

It remains to eliminate β from (3) and (4). Substituting its value, as given by (4), into (3) and simplifying the resulting equation, we obtain

(5) $$b^2X + a^2\lambda Y = 0.$$

This is the equation of a line through the center of the ellipse, that is, a diameter. It is clear geometrically, however, that it is not the indefinite line which is the locus, but merely the portion of it lying within the ellipse. We have thus obtained the following result.

Theorem 1. *The locus of the mid-points of a set of parallel chords of the ellipse* (1) *is a diameter, considered as a line-segment (exclusive of the end points). If the slope of the chords is* $\lambda (\neq 0)$, *the slope* λ' *of the diameter is*

(6) $$\lambda' = -\frac{b^2}{a^2\lambda}.$$

EXERCISES

1. Find the locus of the mid-points of the chords of the ellipse

$$3x^2 + 4y^2 = 12,$$

which are inclined at an angle of 135° to the axis of x. First draw an accurate figure, showing the chords and the locus; then solve the problem analytically, using the method, but not the formulas, of the text.

2. Prove the converse of Theorem 1, namely, that every diameter of the ellipse (1) bisects some set of parallel chords. Show that, if $\lambda'(\neq 0)$ is the slope of the diameter, then the chords which it bisects are of slope λ, where

$$\lambda = -\frac{b^2}{a^2\lambda'}.$$

3. Prove analytically that the tangent to an ellipse at an extremity of a diameter is parallel to the chords which the diameter bisects.

Suggestion. Let (x_1, y_1) be the coördinates of the extremity of the diameter and find, by using (6), the slope λ of the chords in terms of x_1 and y_1.

2. Conjugate Diameters of an Ellipse. Two mutually perpendicular diameters of a circle have the property that each bisects the chords parallel to the other. The axes of an ellipse have this same property. Are there other pairs of diameters of the ellipse which have it? This question is answered in the affirmative by the following theorem.

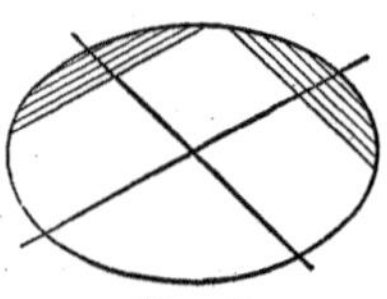

Fig. 2

Theorem 2. *If one diameter bisects the chords parallel to a second, the second diameter bisects the chords parallel to the first.*

The two diameters stand in a *reciprocal* relationship; each bisects the chords parallel to the other. We call them a *pair of conjugate diameters*, and say that each is *conjugate* to the other.

We now prove Theorem 2. Let the diameter D' bisect the chords parallel to the diameter D; to prove that D bisects the chords parallel to D'.

Denote the slopes of D and D' by λ and λ'. By hypothesis, the diameter of slope λ' bisects the chords of slope λ; consequently, by Th. 1, § 1,

$$\lambda' = -\frac{b^2}{a^2\lambda}.$$

But then

$$\lambda = -\frac{b^2}{a^2\lambda'}.$$

This equation says that the diameter of slope λ bisects the chords of slope λ'; that is, D bisects the chords parallel to D', q. e. d.

Incidentally, we have also proved the following theorem.

THEOREM 3. *Two diameters D and D' of the ellipse*

$$\frac{x^2}{a^2} + \frac{y^2}{b^2} = 1 \tag{1}$$

are conjugate, if and only if they are the axes or have slopes λ and λ' related by the equation

$$\lambda\lambda' = -\frac{b^2}{a^2}. \tag{2}$$

The symmetry of (2) in λ and λ' corresponds to the symmetry in the geometrical relationship of D and D'.

To each diameter D there corresponds a conjugate diameter — the diameter parallel to the chords which D bisects. There are, then, infinitely many pairs of conjugate diameters. Since, by (2), the product of the slopes of any pair, other than the axes, is negative, the two diameters of the pair pass through *different* quadrants.

The axes are the only mutually perpendicular pair, unless the ellipse becomes a circle. For, if any other pair were perpendicular, the product, $\lambda\lambda'$, of their slopes would be -1, and this is impossible, according to (2), unless $b^2 = a^2$; but then the ellipse becomes a circle.

We now find the conjugate diameters which are equally inclined to the axes. It is evident, geometrically, that these will also be the conjugate diameters of equal lengths. If they exist, their slopes must be equal except for sign: $\lambda' = -\lambda$. Hence, by (2),

$$\lambda^2 = \frac{b^2}{a^2} \qquad \text{and} \qquad \lambda = \pm \frac{b}{a}$$

We see, then, that *there is a single pair of conjugate diameters which are equally inclined to the axes, or have equal lengths.* They are the diagonals of the rectangle circumscribed about the ellipse (Fig. 3). We denote them by D_1 and D_1'.

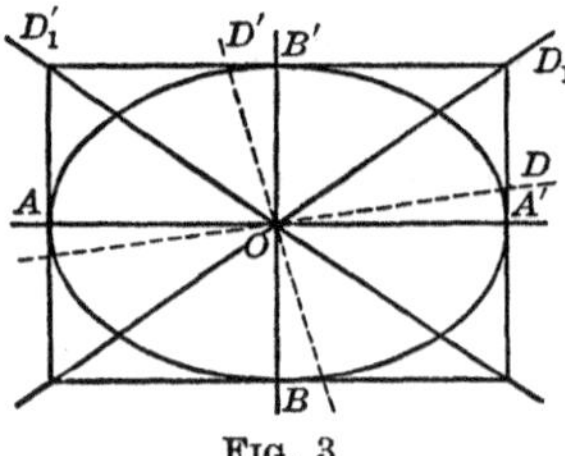

FIG. 3

Now let a diameter D, starting from coincidence with the transverse axis AA', rotate about O into coincidence with D_1; then the conjugate diameter, D', starts from coincidence with the conjugate axis BB', and rotates into coincidence with D_1'. But D' rotates more quickly than D,* so that the angle from D to D', at first 90°, becomes obtuse and steadily increases. When D continues to rotate from D_1 to BB', then D' rotates from D_1' to AA'; but now D' rotates less quickly than D, so that the angle from D to D' decreases and becomes again 90° in the final position.

EXERCISES

1. Draw accurately an ellipse whose axes are 10 cm. and 7 cm. Construct the axes and the pair of conjugate diameters equally inclined to the axes. Then draw the diameters in-

* Since the slope of D_1 is $\frac{b}{a} < 1$, $\measuredangle A'OD_1 < 45°$ and $\measuredangle B'OD_1' > 45°$; hence D has a smaller angle through which to rotate than D'. Consequently, it is to be expected that D' will rotate more quickly than D. A proof of the fact may easily be given later, when the student studies the Calculus.

clined at angles of 10°, 20°, 30°, 40°, 50°, 60°, 70°, and 80° to the transverse axis. For each of these diameters compute the slope, and the angle of inclination, of the conjugate diameter and construct it. Find the angles between the successive diameters of this new set, and also the angles between the successive pairs of conjugate diameters. Mark clearly the pairs, and study the results and the figure in light of the text.

2. Prove that, if one of a pair of conjugate diameters of an ellipse has the slope e or $-e$, where e is the eccentricity, the other joins two extremities of the latera recta.

3. If the equal conjugate diameters of an ellipse form with one another an angle of 60°, what is the eccentricity of the ellipse?

4. The axes of an ellipse are the axes of coördinates and the slopes of two conjugate diameters are $\frac{2}{3}$ and $-\frac{3}{8}$. What is the eccentricity?

5. The same, if the slopes of two conjugate diameters are $-\frac{5}{3}$ and $\frac{3}{4}$.

6. Prove that the line joining a focus to the point of intersection of the corresponding directrix and a diameter is perpendicular to the conjugate diameter.

3. Diameters of a Hyperbola. A diameter of a hyperbola is defined in the same way as a diameter of an ellipse, § 1. Certain diameters of a hyperbola, however, do not meet the curve. Special definitions of the length and extremities of such a diameter must, then, be adopted. These we shall consider later.

The locus of the mid-points of a set of parallel chords of slope $\lambda\ (\neq 0)$ of the hyperbola

$$\frac{x^2}{a^2} - \frac{y^2}{b^2} = 1 \tag{1}$$

can be found by the method of § 1. It is, however, unnecessary to repeat the work there given. For, this work becomes valid immediately for the hyperbola (1) if, in it, we replace b^2 by $-b^2$. It follows, then, that the locus now required is the diameter

$$(2) \qquad -b^2x + a^2\lambda y = 0 \qquad \text{or} \qquad b^2x - a^2\lambda y = 0.$$

The locus consists of *all* the points of this diameter only if the given chords connect points of opposite branches of the

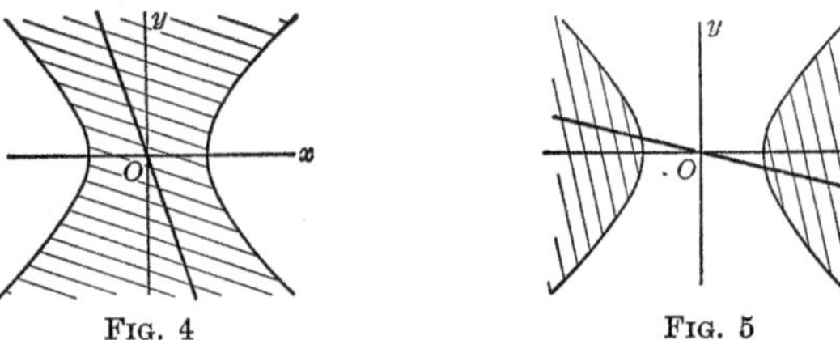

FIG. 4 FIG. 5

hyperbola (Fig. 4). If the chords connect points on the same branch (Fig. 5), the locus is merely the points of the diameter which lie within the curve. The result can be stated as follows.

THEOREM 4. *The locus of the mid-points of a set of parallel chords of the hyperbola* (1) *is a diameter, or so much of a diameter as lies within the curve. If the slope of the chords is* λ ($\neq 0$), *the slope* λ' *of the diameter is*

$$(3) \qquad \lambda' = \frac{b^2}{a^2\lambda}.$$

There are chords of an ellipse with any given direction. This is not true, however, for a hyperbola. For, there are no chords of a hyperbola parallel to an asymptote, since a line parallel to an asymptote meets the curve in but one point. Consequently, the slope, λ, of the chords of Theorem 4 cannot have either of the values, $\pm b/a$.

EXERCISES

1. A set of parallel chords of the rectangular hyperbola

$$x^2 - y^2 = 6$$

are inclined at an angle of 30° to the positive axis of x. What is the inclination of the diameter which bisects them?

First draw an accurate figure, showing the chords and the diameter; then solve the problem analytically, without reference to the formulas of the text.

2. If a set of parallel chords has a slope nearly equal to that of an asymptote S, then the diameter D bisecting the chords has a slope nearly equal to that of S, and when the chords approach a limiting position of parallelism to S, then D approaches S as its limit. Draw a figure showing the reasonableness of this theorem and then prove the theorem analytically by use of (3).

3. Prove the converse of Theorem 4, namely: Every diameter of a hyperbola, not an asymptote, bisects some set of parallel chords. Cf. § 1, Ex. 2.

4. Show that the mid-point of a chord of a hyperbola is also the mid-point of the chord of the conjugate hyperbola which lies on the same line. Hence show that the mid-points of the chords of a given slope lie on one and the same diameter, whether the chords are chords of the given hyperbola or of its conjugate.

5. Prove Ex. 3, § 1, for a hyperbola.

4. Conjugate Diameters of a Hyperbola. Let the diameter, D', of slope λ', bisect the chords of the hyperbola

$$\frac{x^2}{a^2} - \frac{y^2}{b^2} = 1 \tag{1}$$

which are parallel to the diameter D, of slope $\lambda (\neq 0)$. Then, by Th. 4, § 3,

$$\lambda' = \frac{b^2}{a^2\lambda} \qquad \text{or} \qquad \lambda\lambda' = \frac{b^2}{a^2}.$$

Since these equations are symmetric in λ and λ', it follows that the diameter D bisects the chords parallel to D'.

Thus Theorem 2, § 2, is established for the hyperbola, and the two diameters D and D' are, in the sense of that theorem, *conjugate* diameters; each bisects the chords parallel to the other.

We have also proved, incidentally, the following theorem.

THEOREM 5. *Two diameters, D and D', of the hyperbola* (1) *are conjugate if and only if they are the axes or have slopes, λ and λ', related by the equation*

$$\lambda\lambda' = \frac{b^2}{a^2}. \tag{2}$$

There are infinitely many pairs of conjugate diameters, as in the case of the ellipse. But here the two diameters of a pair, not the axes, pass through the *same* quadrants, since the product, $\lambda\lambda'$, of their slopes is positive.

The value, b^2/a^2, of this product is the square of the slope of an asymptote. The slope of an asymptote, therefore, is a mean proportional between the slopes of any two conjugate diameters, not the axes. Consequently, two such conjugate diameters, D and D', are always separated by the asymptote S which lies in the same quadrants with them, and the nearer D lies to S on the one side, the nearer D' will lie to S on the other side. If D approaches S as a limiting position, then so will D'. Thus, an asymptote is often spoken of as a self-conjugate diameter; actually, however, it has no conjugate, since, as we have seen, there are no chords parallel to it.

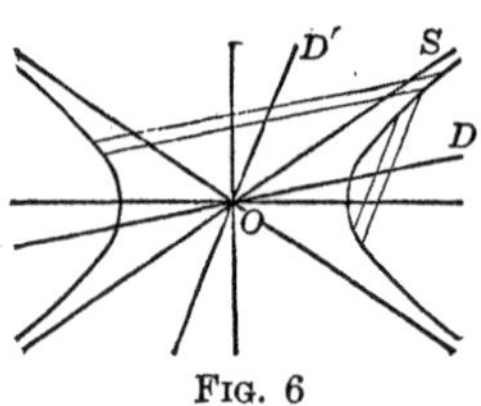

FIG. 6

It is now clear that if a diameter, D, starting from coincidence with the transverse axis, rotates in one direction about O into coincidence with an asymptote, then the conjugate diameter, D', starting from the conjugate axis, will rotate in the *opposite* direction about O into coincidence with the same asymptote.

Conjugate Hyperbolas. Consider now the hyperbola,

$$\frac{x^2}{a^2} - \frac{y^2}{b^2} = -1, \tag{3}$$

conjugate to the hyperbola (1). Since the two hyperbolas have the same center, they have the same diameters, considered as indefinite straight lines. Moreover, *they have the same pairs of conjugate diameters.* For, the chords of (3) of a given slope and the chords of (1) of the same slope are bisected by one and the same diameter, Ex. 4, § 3.

We see now a suitable definition for the extremities of a diameter which does not meet the given hyperbola. They shall be the points in which the diameter meets the conjugate hyperbola (Fig. 7), and the distance between these points shall be the length of the diameter.*

Conjugate Diameters of a Rectangular Hyperbola. A special ellipse, all of whose conjugate diameters are mutually perpendicular, is the circle. There is no special hyperbola with this property, since two conjugate diameters of a hyperbola, other than the axes, always pass through the same quadrants. For this reason, too, there are no conjugate diameters equally inclined to the axes.

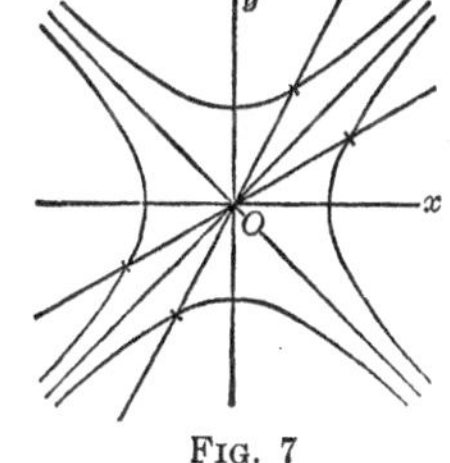

FIG. 7

There may, however, be conjugate diameters, each of which has the same inclination to one axis as the other has to the other axis. The product, $\lambda\lambda'$, of the slopes of two such diameters is 1; hence, by (2), such diameters exist only if

$$1 = \frac{b^2}{a^2} \qquad \text{or} \qquad a^2 = b^2,$$

that is, only if the hyperbola is rectangular. In this case $\lambda\lambda' \equiv 1$, and every pair of conjugate diameters are in the required relation. Consequently, the two diameters are equally inclined to the asymptotes, inasmuch as the asymptotes are now the bisectors of the angles between the axes. They are also equal in length, as considerations of symmetry immediately show (Fig. 7). We have thus proved the following theorem.

* An asymptote, considered as a diameter, we shall not think of as having length or extremities.

THEOREM 6. *Two conjugate diameters of a rectangular hyperbola are always equally inclined to the asymptotes and always equal in length.*

It can be shown that the rectangular hyperbola is the only one with either of these properties. Cf. Ex. 6, below, and § 6, Ex. 5. Thus the rectangular hyperbola plays a rôle among the hyperbolas which is somewhat similar to that played by the circle among the ellipses.

EXERCISES

1. Draw accurately the hyperbola for which $2a = 10$ cm. and $2b = 7$ cm. Construct the axes, AA' and BB', the asymptote S passing through the first quadrant, and the diameters D_1, D_2, D_3 inclined at angles of 10°, 20°, 30° to the transverse axis. Compute the slopes and angles of inclination of the conjugate diameters, D_1', D_2', D_3', and draw these diameters. Find the angles between the successive diameters, BB', D_1', D_2', D_3', S, and compare them with the corresponding angles between the diameters, AA', D_1, D_2, D_3, S. Study the results and the figure in light of the text.

2. Prove Ex. 2, § 2, for the hyperbola.

3. Prove that the asymptotes of the hyperbola

$$\frac{x^2}{a^2} - \frac{y^2}{b^2} = 1$$

are conjugate diameters of the ellipse

$$\frac{x^2}{a^2} + \frac{y^2}{b^2} = 1.$$

4. The axes of a hyperbola are the axes of coördinates, and the slopes of two conjugate diameters are 2 and $\frac{3}{2}$. What is the eccentricity of the hyperbola? Two answers.

5. Prove Ex. 6, § 2, for the hyperbola.

6. Show that two conjugate diameters of a hyperbola are never equally inclined to the asymptotes unless the hyperbola is rectangular.

5. Diameters of a Parabola. When one focus and the corresponding directrix of a *central conic* — an ellipse or a hyperbola — are held fast and the center is allowed to recede indefinitely along the transverse axis, the limit of the conic is a parabola and the limit of the diameters of the conic is a set of lines parallel to the axis of the parabola. Accordingly, by a *diameter* of a parabola we shall mean *any line in the direction of the axis of the parabola.*

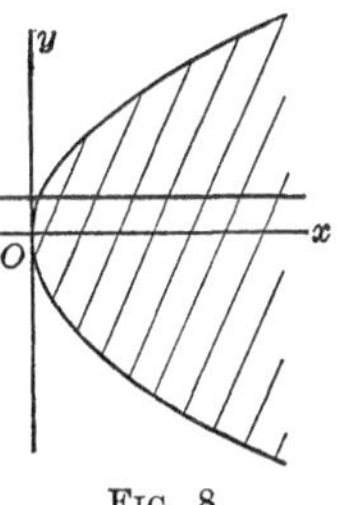

Fig. 8

If this definition is really in accord with that of a diameter of a central conic, we should find that the mid-points of a set of parallel chords of a parabola lie on a line in the direction of the axis. This is the case. If the chords are perpendicular to the axis, their mid-points evidently lie on the axis; if the slope of the chords is $\lambda(\neq 0)$, and the equation of the parabola is

(1) $$y^2 = 2\,mx,$$

the mid-points of the chords lie on the line

(2) $$y = \frac{m}{\lambda},$$

as may easily be shown.

EXERCISES

1. Establish the result embodied in formula (2).

2. What is the equation of the diameter of the parabola

$$y^2 + 6\,x = 0,$$

which bisects the chords of slope $\frac{1}{2}$?

3. Prove Ex. 3, § 1, for the parabola.

4. There are no conjugate diameters for a parabola. Why?

6. Extremities and Lengths of Conjugate Diameters. *Ellipse.* Let the coördinates of one extremity of a diameter D (not an axis) of the ellipse

(1) $$\frac{x^2}{a^2} + \frac{y^2}{b^2} = 1$$

be (x_1, y_1). The slope of D is then $\lambda = y_1/x_1$. From (2), § 2, the slope λ' of the conjugate diameter D' is

$$\lambda' = -\frac{b^2}{a^2\lambda} = -\frac{b^2x_1}{a^2y_1}.$$

Consequently, the equation of D' can be written in the form

$$\frac{x_1x}{a^2} + \frac{y_1y}{b^2} = 0. \tag{2}$$

It follows that D' is parallel to the tangent to the ellipse at (x_1, y_1). In other words, *the tangents at the extremities of a diameter are parallel to the conjugate diameter.*

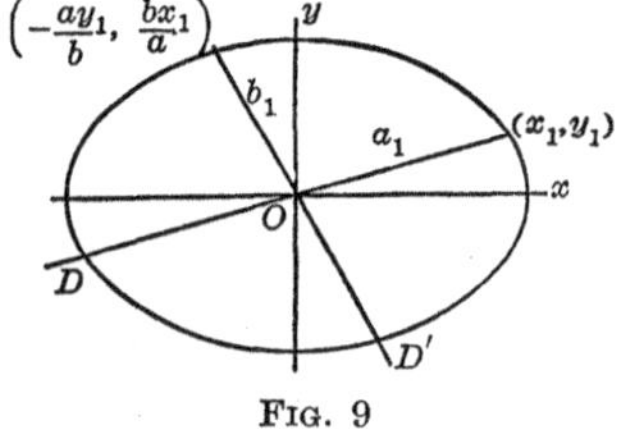

FIG. 9

The coördinates of the extremities of D' may be found by solving equations (1) and (2) simultaneously. The solutions are found to be

$$\left(-\frac{ay_1}{b}, \frac{bx_1}{a}\right), \qquad \left(\frac{ay_1}{b}, -\frac{bx_1}{a}\right).$$

We summarize the foregoing results in a theorem.

THEOREM 7. *If (x_1, y_1) is one extremity of a diameter D of the ellipse* (1), *then* (2) *is the equation of the conjugate diameter D', and one extremity (x_1', y_1') of D' is*

$$x_1' = -\frac{ay_1}{b}, \qquad y_1' = \frac{bx_1}{a}. \tag{3}$$

Suppose, now, that we denote the length of D by $2a_1$ and that of D' by $2b_1$. Then it can be shown, by application of (3) and the equation which states that (x_1, y_1) is on the ellipse, that

$$a_1^2 = b^2 + e^2x_1^2 \qquad \text{and} \qquad b_1^2 = a^2 - e^2x_1^2. \tag{4}$$

We have, then,

$$a_1^2 + b_1^2 = a^2 + b^2 \quad \text{or} \quad (2a_1)^2 + (2b_1)^2 = (2a)^2 + (2b)^2.$$

This result we express as a theorem.

THEOREM 8. *The sum of the squares of the lengths of any two conjugate diameters of an ellipse is constant, and equals the sum of the squares of the axes.*

Hyperbola. Of two conjugate diameters, D and D', of the conjugate hyperbolas

$$(5) \qquad (a) \quad \frac{x^2}{a^2}-\frac{y^2}{b^2}=1, \qquad\qquad (b) \quad \frac{x^2}{a^2}-\frac{y^2}{b^2}=-1,$$

one meets the one hyperbola; the other, the other hyperbola. Suppose that D meets $(5\,a)$ and D' meets $(5\,b)$, and that the coördinates of an extremity of D on $(5\,a)$ are (x_1, y_1), while those of an extremity of D' on $(5\,b)$ are (x_1', y_1').

Then the equations of D' and D are, respectively,

$$(6) \qquad (a) \quad \frac{x_1x}{a^2}-\frac{y_1y}{b^2}=0, \qquad\qquad (b) \quad \frac{x_1'x}{a^2}-\frac{y_1'y}{b^2}=0,$$

as is evident from analogy to the corresponding equation (2) in the case of the ellipse. From (6) it follows that the tangents, at the extremities of a diameter, to the hyperbola on which these extremities lie are parallel to the conjugate diameter.

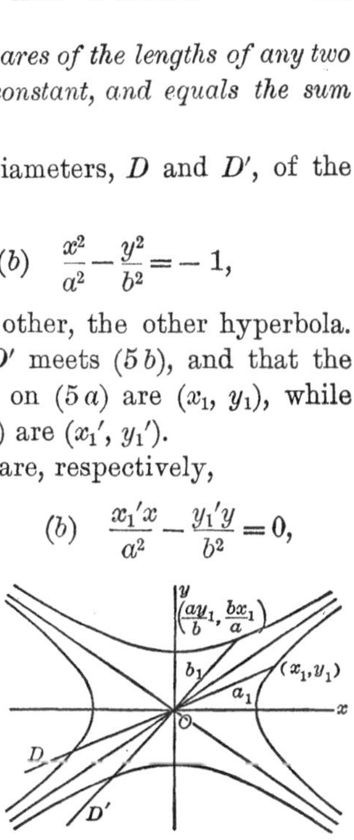

FIG. 10

The coördinates of the extremities of D' can be found by solving equations $(6\,a)$ and $(5\,b)$ simultaneously. The solutions are

$$\left(\frac{ay_1}{b}, \frac{bx_1}{a}\right) \qquad \text{and} \qquad \left(-\frac{ay_1}{b}, -\frac{bx_1}{a}\right).$$

One of these extremities is (x_1', y_1'); let us say, the first one. Then the values of x_1' and y_1' in terms of x_1 and y_1, and vice versa, are

$$(7) \quad (a) \quad x_1'=\frac{ay_1}{b},\ y_1'=\frac{bx_1}{a}; \qquad (b) \quad x_1=\frac{ay_1'}{b},\ y_1=\frac{bx_1'}{a},$$

where the equations (b) are obtained by solving equations (a) for x_1 and y_1.

The student should note the symmetry of formulas (6) and (7) in (x_1, y_1) and (x_1', y_1'). The results embodied in these formulas we state as a theorem.

THEOREM 9. *If* (x_1, y_1) *is one extremity of a diameter* D *meeting the hyperbola* (5 a), *and* (x_1', y_1') *is a properly chosen one of the extremities of the conjugate diameter* D', *then* (6) *are the equations of* D *and* D', *and* (7) *give the relations between the two extremities.*

Let $2a_1$ be the length of D and $2b_1$ that of D'. It can be shown that

$$(8) \qquad \begin{cases} a_1^2 = e^2x_1^2 - b^2 = e^2x_1'^2 + a^2, \\ b_1^2 = e^2x_1^2 - a^2 = e^2x_1'^2 + b^2. \end{cases}$$

Hence $$a_1^2 - b_1^2 = a^2 - b^2$$

and we have the following theorem.

THEOREM 10. *The difference of the squares of the lengths of any two conjugate diameters of a hyperbola is constant, and equals the difference of the squares of the axes.*

EXERCISES

Establish the following formulas.

1. Formulas (3).
2. Formulas (4).
3. Formulas (7).
4. Formulas (8).

5. Show that two conjugate diameters of a hyperbola are never equal unless the hyperbola is rectangular and that in this case they are always equal.

6. Prove that the product of the focal radii to any point of an ellipse equals the square of half the diameter conjugate to the diameter through the point.

7. State and prove the corresponding theorem for the hyperbola.

7. Physical Meaning of Conjugate Diameters. *Ellipse.* Consider a flat bar of iron, on the end $ABCD$ of which a circle is drawn. Let $\overline{D}$ and $\overline{D}'$ be any two mutually perpendicular, and therefore conjugate, diameters of the circle. Imagine that the bar is subjected to heavy pressure. Then the lengths of all lines parallel to AB and CD will be shortened in the same ratio, and the lengths of lines parallel to AD and BC will all be lengthened in the same ratio; the two ratios will not, however, be equal.* The circle will thereby be carried over into an ellipse, and the diameters $\overline{D}$ and $\overline{D'}$ will become conjugate diameters of this ellipse. A proof of these facts will be given shortly.

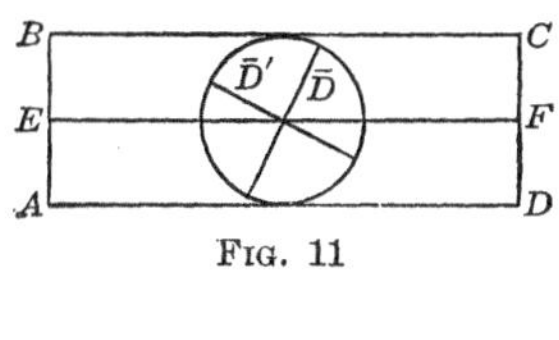

FIG. 11

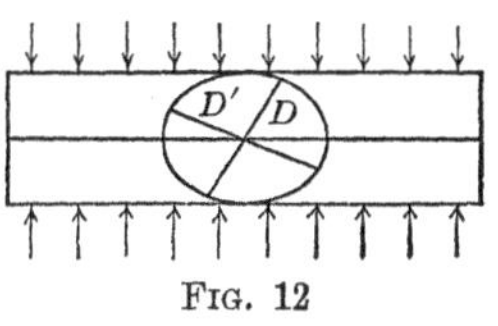

FIG. 12

The student can perform a suggestive experiment by taking an ordinary four-sided eraser, drawing a circle and the diameters $\overline{D}$, $\overline{D'}$ on one of the broader faces of it, and then pinching the eraser in a vise. The circle will go over into an oval that looks like an ellipse, and $\overline{D}$ and $\overline{D'}$ will remain sensibly straight lines.

If the vise is set too hard, the bulging will be considerable. But imagine the ends of the eraser cut off square and the eraser then fitted snugly into a tube or chamber of rectangular cross-section, with the broader faces and the ends in contact with the walls of the chamber. Let the chamber be closed at one end by a rigid, plane diaphragm, against which the eraser is to be pressed.

If, now, a plunger, which just fits the chamber, is introduced and pressed down, the deformation will be much like that described in the opening paragraph; the circle will become a true ellipse, and $\overline{D}$, $\overline{D'}$, remaining straight lines, will become

* Near the ends AB and CD of the cross-section these statements will be only approximately true, since there will be a slight bulging; and, indeed, there will also be a slight bulging of the ends themselves. But near the middle of the cross-section the deformation will be, to a high degree of approximation, as described.

conjugate diameters of the ellipse. But there will be one essential difference, in that in the first case lines parallel to AD are lengthened, whereas in this second case they remain unchanged; cf. Figs. 11 and 12, which have been drawn for the second case, rather than for the first. The first case can, however, be reduced geometrically to the second if, after the deformation has been made, the new figure is reduced in scale, so that lines parallel to AD again assume their original lengths.

We shall confine ourselves to the second case. The deformation of the plane of the circle may, in this case, be called a *compression in one direction* or *a simple compression.* All line-segments in the direction of compression are shortened in the same ratio, *the ratio of compression.* All line-segments in the perpendicular direction remain the same in length; they are all moved parallel to themselves, with the exception of one which remains fixed. In the case described this one rests against the diaphragm, either along AD or BC. If, however, the diaphragm is replaced by a second plunger, the fixed line might be AD or BC or any parallel line such as EF, depending on the manner in which the pressures on the two plungers are applied. This line, perpendicular to the direction of compression and having all its points fixed under the compression, we shall call the *central line.*

In studying the effects of a compression let us take the central line as the axis of x and the ratio of compression as l; l is a positive constant < 1. We prove first that the compression carries a straight line $\bar{L}$ into a straight line L. This is obvious if $\bar{L}$ is parallel to either axis. If $\bar{L}$ is any other line, the similar triangles in Fig. 13 show that it goes over into a line L, and that *if $\bar{L}$ is of slope λ, L is of slope $l\lambda$.*

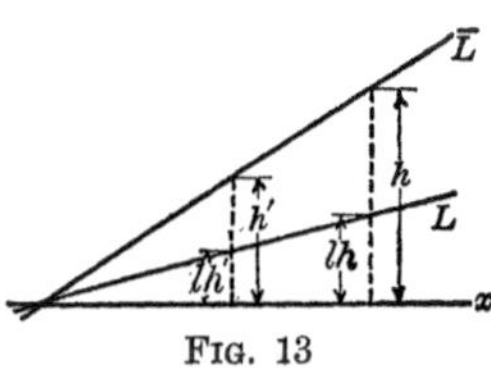

FIG. 13

Next, consider an arbitrary circle, with center $\bar{O}$ (Fig. 14). The diameter $\overline{A'A}$ of the circle which is parallel to the central line, the axis of x, goes over into a parallel line-segment, $A'A$, of equal length, and the mid-point, $\overline{O}$, of $\overline{A}'\overline{A}$ goes over into the mid-point, O', of $A'A$.

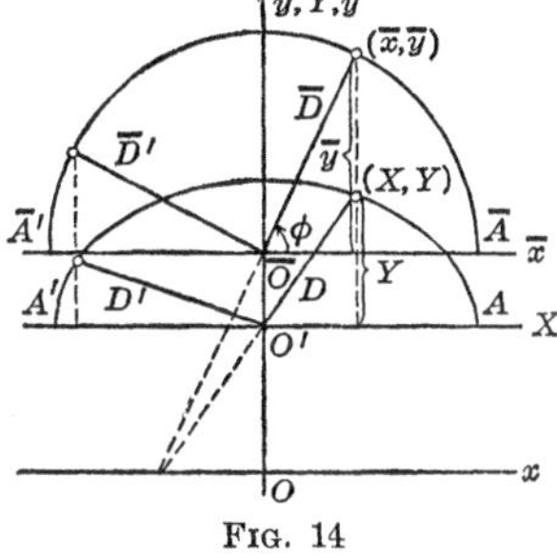

FIG. 14

Let the circle be referred to axes $(\overline{x}, \overline{y})$ with the origin at $\overline{O}$, the axis of $\overline{x}$ lying along $\overline{A}'\overline{A}$. Its equation will be

(1) $$\overline{x}^2 + \overline{y}^2 = a^2.$$

Let the curve into which the circle is deformed be referred to axes (X, Y) with the origin at O', the axis of X lying along $A'A$. Then any point $(\overline{x}, \overline{y})$ on the circle goes over into a point (X, Y) such that

$$X = \overline{x}, \qquad Y = l\overline{y}, \qquad \text{or} \qquad \overline{x} = X, \qquad \overline{y} = \frac{Y}{l}.$$

It follows, then, that the circle (1) is transformed into the curve

$$X^2 + \frac{Y^2}{l^2} = a^2,$$

or

(2) $$\frac{X^2}{a^2} + \frac{Y^2}{b^2} = 1, \qquad b = la.$$

Thus the circle is seen to be carried into an ellipse.

It remains to prove that the lines D and D', into which two conjugate diameters $\overline{D}$ and $\overline{D}'$ of the circle are carried by the compression, are conjugate diameters of the ellipse. If the angle ϕ is as shown in Fig. 14, the slopes of $\overline{D}$ and $\overline{D}'$ are $\tan\phi$ and $\tan(\phi + 90°) = -\cot\phi$. Hence the slopes of D and D' are

$$\lambda = l \tan\phi \qquad \text{and} \qquad \lambda' = -l\cot\phi.$$

Then $$\lambda\lambda' = -l^2 \tan\phi \cot\phi = -l^2,$$

or, since, by (2), $l = b/a$,

$$\lambda\lambda' = -\frac{b^2}{a^2}.$$

Consequently, D and D', according to Th. 3, § 2, are conjugate diameters of the ellipse, q. e. d.

If the center $\overline{O}$ of the circle lies on the central line (Fig. 15), the circle is the auxiliary circle of the ellipse (Ch. VII, 10), and the angle ϕ is the eccentric angle for the extremity P of the diameter D. The eccentric angle for the extremity P' of the conjugate diameter D' is, clearly, $\phi + 90°$, or $\phi + n\,90°$, where n is an odd number. Consequently, we have proved the following theorem.

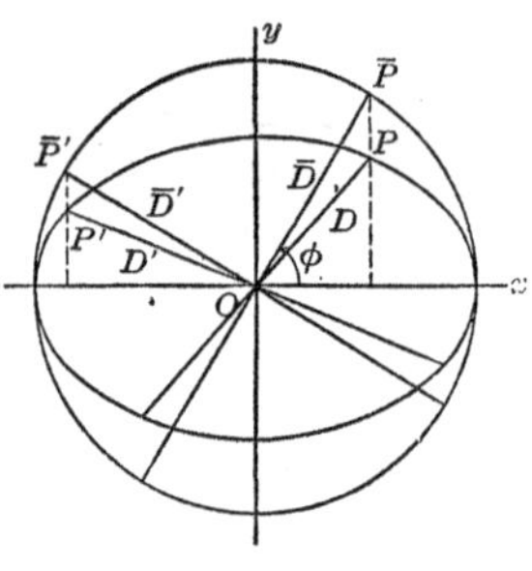

FIG. 15

THEOREM 11. *The eccentric angles for two points of an ellipse which are extremities of two conjugate diameters differ by* 90°, *or by an odd multiple of* 90°.

The theorem is essentially the geometrical equivalent of the physical property of conjugate diameters which we have been discussing. It furnishes a method of constructing rapidly as many pairs of conjugate diameters of an ellipse as may be desired.

The parametric representation of the ellipse can be used to great advantage throughout the study of conjugate diameters. The extremity P of the diameter D (Fig. 15) has, by Ch. VII, § 10, the coördinates

$$(3) \qquad x = a\cos\phi, \qquad y = b\sin\phi.$$

Then the extremity P' of the conjugate diameter D' has, by Th. 11, the coördinates

$$(4) \qquad x' = -a\sin\phi, \qquad y' = b\cos\phi.$$

Hence we obtain, for the squares of the half-lengths of D and D':

$$a_1^2 = a^2\cos^2\phi + b^2\sin^2\phi, \qquad b_1^2 = a^2\sin^2\phi + b^2\cos^2\phi.$$

Therefore, $$a_1^2 + b_1^2 = a^2 + b^2,$$

and we have a simple proof of Th. 8, § 6.

Hyperbola. Consider, now, a set of four steel girders A, B, C, and D in the form of a square, and a cross girder E through the center of the square parallel to A and C. Suppose that tie-rods are spanned into the frame along the diagonals of the square and along pairs of lines making equal angles with the diagonals. These pairs of tie-rods, then, lie along conjugate diameters of a rectangular hyperbola, of which the diagonals of the square are the asymptotes (Th. 6, § 4).

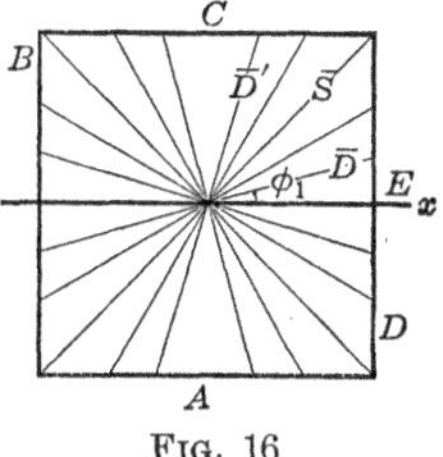

Fig. 16

Suppose that the girder E is firmly set in masonry, so that it is immovable, and suppose that equal tensions are exerted on the girders A and C as shown. Then the square is elongated into a rectangle, except for a slight bulging; the diagonal rods come to lie along the diagonals of the rectangle and the other pairs of tie-rods take on the positions of pairs of lines which are conjugate diameters in a hyperbola having the diagonals of the rectangle as asymptotes, as we shall presently show.

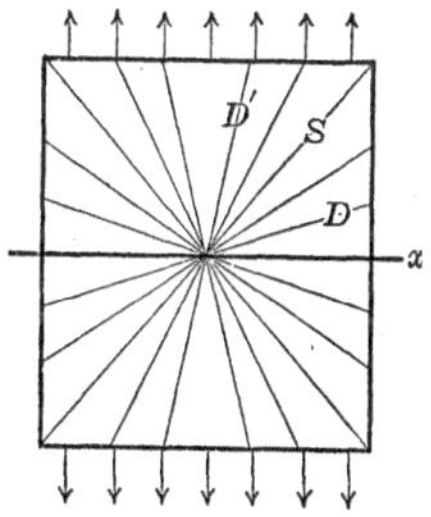

Fig. 17

In this case we speak of an *elongation in one direction* or *a simple elongation.* The line of the girder E is the *central line* of the elongation, and the ratio $l (> 1)$, in which all distances perpendicular to E are stretched, is the *ratio of elongation.*

Let us take the central line as axis of x. Since the slope of the diagonal $\bar{S}$ of the square is 1, the slope of the diagonal S of the rectangle is l. If the angle ϕ_1 is as shown, Fig. 16, the slopes of the two lines $\bar{D}$ and $\bar{D}'$ making equal angles with $\bar{S}$ are $\tan \phi_1$ and $\tan(90 - \phi_1) = \cot \phi_1$. Hence the slopes, λ and λ', of the lines D and D', into which $\bar{D}$ and $\bar{D}'$ are carried, are

$$\lambda = l \tan \phi_1 \qquad \text{and} \qquad \lambda' = l \cot \phi_1.$$

But then $$\lambda\lambda' = l^2,$$

and, according to Th. 5, § 4, the lines D and D' are conjugate diameters of a hyperbola, of which the diagonals of the rectangle are the asymptotes. The ratio b/a of the axes of the hyperbola equals the ratio of elongation, l:

$$\frac{b}{a} = l.$$

Finally, let us show that the elongation carries the rectangular hyperbola

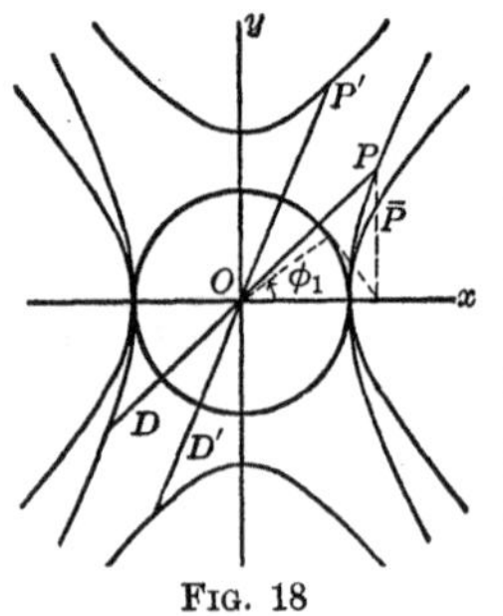

Fig. 18

(5) $$\frac{x^2}{a^2} - \frac{y^2}{a^2} = 1$$

into the hyperbola

(6) $$\frac{x^2}{a^2} - \frac{y^2}{b^2} = 1.$$

The two hyperbolas have the same auxiliary circle, and the same eccentric angle, ϕ_1, for points, $\bar{P}: (x_1, \bar{y}_1)$ and $P: (x_1, y_1)$, with the same abscissa. Hence, according to the method of parametric representation of a hyperbola (Ch. VIII, § 9), the coördinates of $\bar{P}$ and P are

(5 a) $$x_1 = a \sec \phi_1, \qquad \bar{y}_1 = a \tan \phi_1;$$

(6 a) $$x_1 = a \sec \phi_1, \qquad y_1 = b \tan \phi_1.$$

Therefore $$y_1 = \frac{b}{a}\bar{y}_1,$$

or, since $b/a = l$, $$y_1 = l\bar{y}_1.$$

Hence the elongation does carry the hyperbola (5) into the hyperbola (6).

Let P, with coördinates (6 a), be an extremity of the diameter D. Then the coördinates of an extremity P' of the conjugate diameter D' are, by (7), § 6,

(7) $$x_1' = \frac{ay_1}{b} = a \tan \phi_1, \qquad y_1' = \frac{bx_1}{a} = b \sec \phi_1.$$

Here again, then, the use of the eccentric angle gives symmetry to the results.

One-Dimensional Strains. In the case of the ellipse we might equally well have subjected the given circle to an elongation, and in the case of the hyperbola we might have compressed the given equilateral hyperbola, instead of elongating it. Compression and elongation in one direction are but two types of a single kind of deformation, known as a *one-dimensional strain.* If *the coefficient l of the strain* is greater than unity, the strain is an elongation; on the other hand, if $l < 1$, the strain is a compression.

EXERCISES

1. Repeat Ex. 1 of § 2, drawing the auxiliary circle and constructing the diameters conjugate to the given diameters by application of Theorem 11.

2. Draw in pencil the asymptotes and a number of pairs of conjugate diameters, including the axes, of a rectangular hyperbola. Construct in ink the lines into which the given lines are carried by the compression of ratio $\frac{2}{3}$ which has an axis of the hyperbola as central line. What does the resulting figure represent?

3. Prove Th. 10, § 6, by means of formulas ($6a$) and (7) of the present paragraph.

8. Harmonic Division. Let P_1P_2 be a line-segment, and let Q_1 be one of its points. Then Q_1 divides P_1P_2 internally in a certain ratio, μ (Ch. I, § 6):

$$\frac{P_1Q_1}{Q_1P_2} = \mu.$$

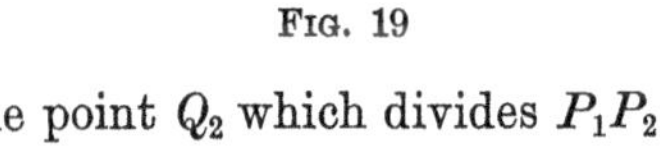

Q_2 P_1 Q_1 M P_2

FIG. 19

On P_1P_2 produced construct the point Q_2 which divides P_1P_2 externally in the same ratio:

$$\frac{Q_2P_1}{Q_2P_2} = \mu.$$

The points Q_1 and Q_2 are said to divide the segment P_1P_2 *harmonically*; they divide P_1P_2 internally and externally in the same ratio:

$$(1) \qquad \frac{P_1Q_1}{Q_1P_2} = \frac{Q_2P_1}{Q_2P_2}.$$

Let us start with μ as given and trace the changes in Q_1 and Q_2 as μ varies. If $\mu = 0$, Q_1 and Q_2 coincide in P_1. As μ increases from 0 to 1, Q_1 moves to the right from P_1 to the mid-point M of P_1P_2, and Q_2 moves to the left and recedes indefinitely. If $\mu = 1$, Q_1 is at M; but Q_2 has disappeared. *Thus there is no point which, with the mid-point of a segment, divides the segment harmonically.* As μ increases from 1 without limit, Q_1 proceeds from M toward P_2 as its limit, and Q_2 appears again from the extreme right, continually moving in and approaching P_2 as its limit.

The proportion (1) may be written in the form:

$$(2) \qquad \frac{P_1Q_1}{Q_2P_1} = \frac{Q_1P_2}{Q_2P_2};$$

this new proportion says that P_1 and P_2 divide the segment Q_1Q_2 harmonically. Thus we have the following theorem.

THEOREM 1. *If the points Q_1 and Q_2 divide the line-segment P_1P_2 harmonically, then, reciprocally, the points P_1 and P_2 divide the line-segment Q_1Q_2 harmonically.*

In other words, the relationship between the two pairs of points is symmetric.

Suppose that P_1 and P_2 have the coördinates (x_1, y_1) and (x_2, y_2). Then the coördinates (x_1', y_1') and (x_2', y_2') of Q_1 and Q_2 are given by formulas (1) and (2) of § 6, Ch. I. If, in each of these formulas, we divide the numerator and denominator by m_2 and then set $m_1/m_2 = \mu$, we obtain, as the desired coördinates:

$$Q_1: \qquad x_1' = \frac{x_1 + \mu x_2}{1 + \mu}, \qquad y_1' = \frac{y_1 + \mu y_2}{1 + \mu};$$

$$Q_2: \qquad x_2' = \frac{x_1 - \mu x_2}{1 - \mu}, \qquad y_2' = \frac{y_1 - \mu y_2}{1 - \mu}.$$

EXERCISES

1. Four points P_1, P_2, Q_1, Q_2 on the axis of x have, respectively, the abscissas 3, 8, 5, -7. Show that Q_1, Q_2 divide P_1P_2 harmonically, and find the common ratio μ of internal and external division. Find, also, the value of the ratio, μ', for the division by P_1 and P_2 of the segment Q_1Q_2.

2. Find the point on the axis of x which, with the point $(-1, 0)$, divides harmonically the segment of the axis joining the points $(-8, 0)$, $(3, 0)$.

3. Exercise 1, for the four points P_1, P_2, Q_1, Q_2 with the respective coördinates $(2, 3)$, $(-1, 9)$, $(1, 5)$, $(5, -3)$.

4. Find the point which, with the point $(2, 1)$, divides harmonically the line-segment joining the points $(5, -2)$, $(1, 2)$.

9. Polar of a Point. Consider the following locus problem. The ellipse

$$(1)\qquad \frac{x^2}{a^2}+\frac{y^2}{b^2}=1$$

and the point $P_1:(x_1, y_1)$ are given. A line L is drawn through P_1 meeting the ellipse in Q_1 and Q_2, and on L the point $P:(X, Y)$ is marked which, with P_1, divides Q_1Q_2 harmonically. What is the locus of P, as L revolves about P_1?

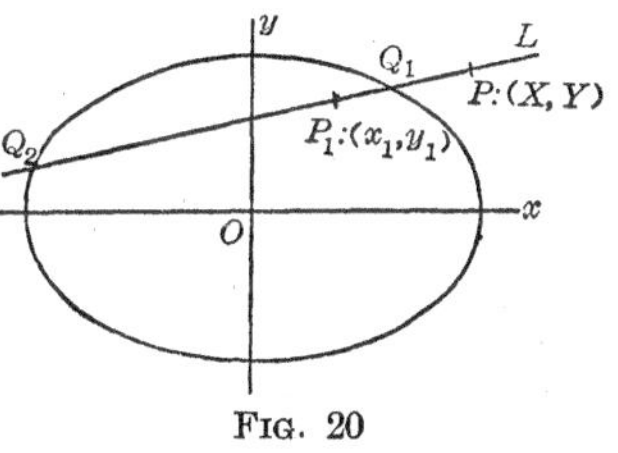

FIG. 20

Since P_1, P divide Q_1Q_2 harmonically, Q_1, Q_2 divide P_1P harmonically. Hence the coördinates (x_1', y_1'), (x_2', y_2') of Q_1, Q_2 are:

$$Q_1:\qquad x_1'=\frac{x_1+\mu X}{1+\mu},\qquad y_1'=\frac{y_1+\mu Y}{1+\mu},$$

$$Q_2:\qquad x_2'=\frac{x_1-\mu X}{1-\mu},\qquad y_2'=\frac{y_1-\mu Y}{1-\mu}.$$

As L rotates, the ratio μ varies; it is, then, an auxiliary variable expressing analytically the rotation of L.

The coördinates of Q_1 and Q_2 satisfy (1). Substituting them in turn in (1) and clearing each of the resulting equations of fractions, we have

$$b^2(x_1 + \mu X)^2 + a^2(y_1 + \mu Y)^2 = a^2b^2(1 + \mu)^2,$$
$$b^2(x_1 - \mu X)^2 + a^2(y_1 - \mu Y)^2 = a^2b^2(1 - \mu)^2.$$

To eliminate μ, we subtract the second equation from the first, thus getting

$$4\,\mu b^2 x_1 X + 4\,\mu a^2 y_1 Y = 4\,\mu a^2 b^2,$$

or, finally,
$$\frac{x_1 X}{a^2} + \frac{y_1 Y}{b^2} = 1.$$

The locus of P is, therefore, a straight line, or a portion of a straight line.* This line is known as *the polar of the point* P_1 *with respect to the ellipse.* Hence we may say:

The polar of the point (x_1, y_1) *with respect to the ellipse* (1) *has the equation*

$$\text{(2)} \qquad \frac{x_1 x}{a^2} + \frac{y_1 y}{b^2} = 1.$$

This equation is identical *in form* with the equation of the tangent to the ellipse at the point (x_1, y_1), Ch. IX, § 2, (12). But in the present problem (x_1, y_1) is, in general, not on the curve, and then (2) represents a line which is not a tangent.

If, in particular, $P_1 : (x_1, y_1)$ is on the ellipse, then (2) does represent the tangent at P_1. Accordingly, we should like to say: *The polar of a point on the ellipse is the tangent at the point.* Now there is trouble, geometrically, when P_1 is on the ellipse. For then Q_1 or Q_2 coincides with P_1, and P coincides with them, so that, actually, no polar is defined. Suppose, however, that $\bar{P}_1$ is a point near to P_1, but not on the curve. Then it can be shown (Exs. 1, 2) that the limiting position of the polar of $\bar{P}_1$, when $\bar{P}_1$ approaches P_1 as its limit, is the tangent at P_1. Hence the above statement is substantiated,

* If P_1 is inside the conic, the locus is the entire line, but if P_1 is outside the conic, the locus consists of only those points of the line which are inside the conic.

not as a conclusion, but as a proper definition of the polar of a point on the ellipse.

If P_1 is at the origin, it is always the mid-point of Q_1Q_2, and so there is never a point P which, with P_1, divides Q_1Q_2 harmonically. Consequently, *the origin has no polar.*

The foregoing discussion is valid for the hyperbola

$$\frac{x^2}{a^2}-\frac{y^2}{b^2}=1, \tag{3}$$

if in the equations we replace b^2 by $-b^2$. Thus, *the polar of the point* (x_1, y_1) *with respect to the hyperbola* (3) *has the equation*

$$\frac{x_1x}{a^2}-\frac{y_1y}{b^2}=1. \tag{4}$$

The polar of a point on the hyperbola is defined as the tangent at the point. The center of the hyperbola has no polar.

We can now state the following theorem.

THEOREM 2. *Given a central conic, C. Every point in the plane, except the center of C, has a polar with respect to C.*

Let the student show that *the polar of the point* (x_1, y_1) *with respect to the parabola*

$$y^2=2mx \tag{5}$$

has the equation

$$y_1y=m(x+x_1). \tag{6}$$

If we define the polar of a point on the parabola as the tangent at the point, equation (6) shows that there are no exceptions in this case. Accordingly, we have the theorem:

THEOREM 3. *Every point in the plane has a polar with respect to a parabola.*

From the definition of a polar it is evident that the polar of a point internal to a conic does not cut the conic, and that the polar of a point external to a conic does cut the conic. In the intermediate case, when the point lies on the conic, the polar is a tangent.

EXERCISES

1. Show that the polar of a point P_1 external to a conic is the line L_1 drawn through the points of contact of the tangents to the conic from P_1.

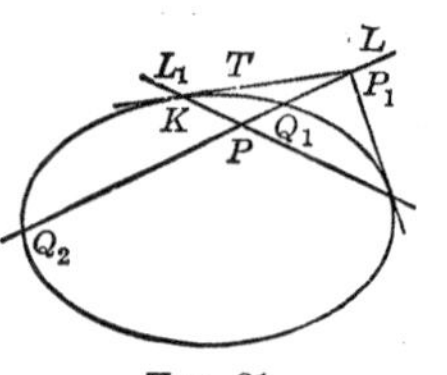

FIG. 21

Suggestion. Prove that P (Fig. 21) approaches K as its limit, when the line L, rotating about P_1, approaches T.

2. Prove that, if the point P_1 of Ex. 1 approaches a point on the conic as its limit, then its polar, L_1, will approach the tangent at this point.

3. Establish formula (6).

In each of the following exercises, find the equation of the polar of the given point with respect to the given conic and draw a figure, showing the conic, point, and polar.

	Conic	*Point*
4.	$x^2 + y^2 = 9$,	(0, 2).
5.	$3x^2 + 5y^2 = 15$,	(5, 6).
6.	$x^2 - y^2 = 16$,	(2, 1).
7.	$2y^2 - 5x = 0$,	$(-3, 4)$.

8. Prove that in any conic the polar of a focus is the corresponding directrix.

9. Prove that the polar of a point P_1 with respect to a circle, center at O, is perpendicular to the line OP_1.

10. Show, further, that the product of the distances of O from P_1 and the polar of P_1 is the square of the radius of the circle.

11. On the basis of the results of Exs. 9, 10, discuss the variation in position of the polar of a point P with respect to a circle, (*a*) when P moves on a straight line through the center of the circle; (*b*) when P traces a circle, concentric with the given circle.

Find the equation of the polar of the point (x_1, y_1) with respect to each of the following conics.

12. The hyperbola: $xy = k$.

13. The circle: $(x - \alpha)^2 + (y - \beta)^2 = \rho^2$.

14. The conic: $(1 - e^2)x^2 + y^2 - 2mx + m^2 = 0$.

10. Pole of a Line. If, with respect to a given conic, the line L is the polar of the point P, the point P is known as *the pole of the line L.*

Given a conic and a line L; to find the pole, P, of L with respect to the conic.

Let the conic be the ellipse,

(1) $$2x^2 + 3y^2 = 6,$$

and L, the line,

(2) $$4x - 3y - 2 = 0.$$

If we denote the coördinates of P by (x_1, y_1), the polar of P with respect to (1) is

(3) $$2x_1x + 3y_1y - 6 = 0.$$

But the polar of P was given as the line (2). Equations (2) and (3), then, represent the same line. Consequently, by Ch. II, § 10, Th. 5,

$$\frac{2x_1}{4} = \frac{3y_1}{-3} = \frac{-6}{-2}.$$

Then $$x_1 = 6, \qquad y_1 = -3$$

and so the point $(6, -3)$ is the pole of the line (2) with respect to the ellipse (1).

We now raise the question: Has every line a pole with respect to a given conic? Let us answer this question first for the central conics. Equations (2) and (4) of § 9, which represent the polars of a given point with respect to the central conics (1) and (3) of § 9, are never satisfied by $x = 0$, $y = 0$, no matter where the given point lies. Consequently, the polar of a point with respect to a central conic never passes through the center of the conic. In other words, *a diameter of a central conic has no pole.*

We proceed to show that every other line has a pole, giving the proof in the case of the hyperbola

$$\frac{x^2}{a^2}-\frac{y^2}{b^2}=1. \tag{4}$$

Any line not a diameter of (4), that is, not passing through the origin, can be represented by an equation of the form *

$$Ax+By=1. \tag{5}$$

If (x_1, y_1) is the pole of this line, the line also has the equation

$$\frac{x_1x}{a^2}-\frac{y_1y}{b^2}=1.$$

Hence $$\frac{x_1}{a^2}:A=-\frac{y_1}{b^2}:B=1:1,$$

and $$x_1=a^2A, \qquad y_1=-b^2B.$$

We see, then, that the line (5) has always a definite pole, namely, the point $(a^2A, -b^2B)$, q. e. d.

In the case of the ellipse the proof is similar.

As regards the parabola,

$$y^2=2mx,$$

the equation of the polar of (x_1, y_1):

$$y_1y=mx+mx_1,$$

has one term which can never drop out, no matter where (x_1, y_1) lies, — namely, the term mx. Thus the polar can never be parallel to the axis of x, or coincide with it. In other words, *a diameter of a parabola has no pole.* It can be shown, however, that every other line has a pole. The proof is left to the student; cf. Ex. 1.

The foregoing results we now summarize in the form of a theorem.

* A and B are not both zero.

THEOREM 4. *Given a conic C. Every line of the plane, which is not a diameter of C, has a pole with respect to C.*

By comparing this theorem with Theorems 2, 3, § 9, we see that the lines which have no poles with respect to a conic go through the point which has no polar, provided these lines intersect.

EXERCISES

1. Give the proof of Theorem 4 for the parabola.

In each of the following exercises find the pole of the given line with respect to the given conic.

	Conic	*Line*
2.	$x^2 + y^2 = 8,$	$2x - 3y - 2 = 0.$
3.	$5x^2 - 6y^2 - 30 = 0,$	$4x + 2y - 7 = 0.$
4.	$3y^2 - 8x = 0,$	$2x - 3 = 0.$
5.	$7x^2 + 2y^2 = 14,$	$6x + 5y - 8 = 0.$

6. Prove that the pole of any line through the focus of a conic is a point on the corresponding directrix.

7. Given the circle $x^2 + y^2 = a^2$. Prove that the pole, with respect to this circle, of a line moving so that it is always tangent to a concentric circle traces a second concentric circle. Cf. Exs. 9–11, § 9.

11. Properties of Poles and Polars. The poles and polars * discussed in this paragraph are all taken with reference to an arbitrarily given conic. For the sake of brevity mention of the conic is, in general, suppressed.

THEOREM 5a. *If a point P_1 lies on the polar of a second point P_2, then, conversely, P_2 lies on the polar of P_1.*

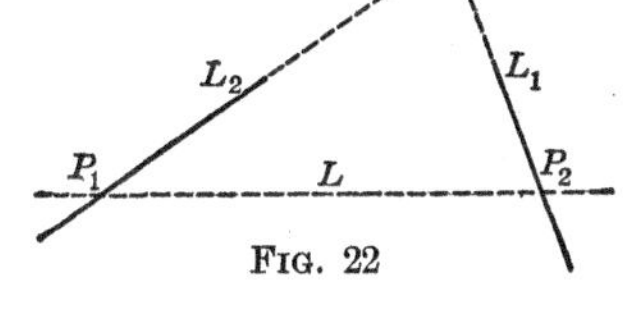

FIG. 22

Let the polars of the points P_1 and P_2 be L_1 and L_2. Then the theorem says that, if P_1 lies on L_2, P_2 lies on L_1. But this

* Only those points which have polars and those lines which have poles are considered.

is the same as saying that, if L_2 goes through P_1, L_1 goes through P_2, or vice versa. This second form of the statement we enunciate as a theorem.

THEOREM 5*b*. *If a line L_1 goes through the pole of a second line L_2, then, conversely, L_2 goes through the pole of L_1.*

Since the two theorems are equivalent in content, and differ only in point of view, a proof of one also proves the other. We choose to prove Theorem 5*a*, and to give the proof in the case that the given conic is the hyperbola

$$\frac{x^2}{a^2}-\frac{y^2}{b^2}=1.$$

The proofs in the other two cases are similar.

Let P_1 and P_2 have the coördinates (x_1, y_1) and (x_2, y_2). Then L_1 and L_2 have the equations

$$\frac{x_1x}{a^2}-\frac{y_1y}{b^2}=1 \quad \text{and} \quad \frac{x_2x}{a^2}-\frac{y_2y}{b^2}=1.$$

The condition that P_1 lies on L_2 is

$$\frac{x_2x_1}{a^2}-\frac{y_2y_1}{b^2}=1,$$

and the condition that P_2 lies on L_1 is

$$\frac{x_1x_2}{a^2}-\frac{y_1y_2}{b^2}=1.$$

But these two conditions are the same. Hence, if P_1 lies on L_2, then P_2 lies on L_1, and conversely, q. e. d.

Suppose, now, that we join the points P_1 and P_2 of Fig. 22 by the line L. Since P_1 lies on L, it follows, by Th. 5*a*, that the pole, P, of L lies on L_1. Similarly, since P_2 lies on L, P lies also on L_2. Hence, P is the point of intersection of L_1 and L_2. Thus we have the theorem:

THEOREM 6*a*. *The pole of the line joining two points, P_1 and P_2, is the point of intersection of the polars of P_1 and P_2.*

Starting again, we bring the lines L_1 and L_2 of Fig. 22 to intersection in P. By Th. 5*b*, since each of the lines L_1 and

L_2 goes through P, it follows that the polar, L, of P goes through each of the poles, P_1 and P_2, of L_1 and L_2. Consequently, L is the line joining P_1 and P_2 and we have proved Theorem 6*b*:

THEOREM 6*b*. *The polar of the point of intersection of two lines, L_1 and L_2, is the line joining the poles of L_1 and L_2.*

By application of either Ths. 5*a*, 5*b* or Ths. 6*a*, 6*b*, the student can easily prove the following theorems.

THEOREM 7*a*. *If a number of points all lie on a line, L, their polars all go through a point, namely, the pole of L.*

THEOREM 7*b*. *If a number of lines all go through a point, P, their poles all lie on a line, namely, the polar of P.*

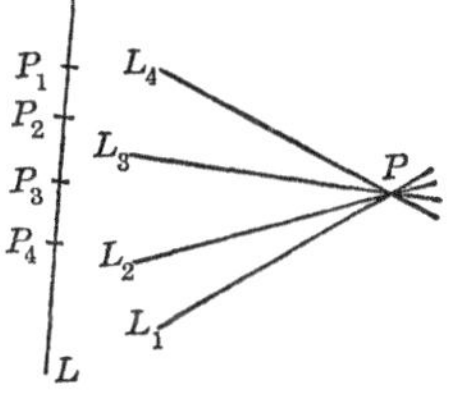

FIG. 23

Finally, take a line L which cuts the given conic in two points, P_1 and P_2. Since L is the line joining P_1 and P_2, the pole, P, of L is the point of intersection of the polars of P_1 and P_2 (Th. 6*a*), that is, of the tangents to the conic at P_1 and P_2. Thus we have proved the theorem:

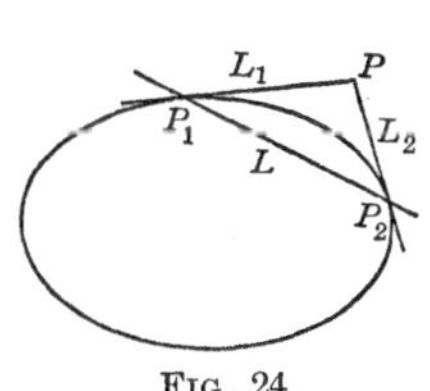

FIG. 24

THEOREM 8*a*. *The pole of a line intersecting the given conic in two points, P_1 and P_2, is the point of intersection of the tangents to the conic at P_1 and P_2.*

Let the student prove the mate of this theorem, namely:

THEOREM 8*b*. *The polar of a point external to the given conic is the line joining the points of contact of the tangents to the conic from the point.*

Theorem 8*a* furnishes a means of constructing the pole of a line which meets the given conic; Theorem 8*b*, a means of constructing the polar of a point external to the conic.

To construct the pole, P, of a given line, L, which does not meet the conic (Fig. 25), choose any two points, P_1 and P_2, on

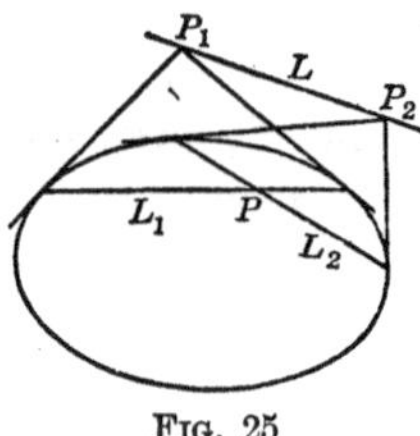

FIG. 25

L, and construct their polars, L_1 and L_2. Since L is the line joining P_1 and P_2, its pole, P, is the point of intersection of L_1 and L_2.

The student should now establish the analogous construction for the polar of a given point which is internal to the conic.*

EXERCISES

1. Prove Theorems $7a$, $7b$.

2. Prove Theorem $8b$.

3. Show how to construct the polar of a given point which is internal to the conic. Prove the validity of the construction.

4. On the basis of Theorem $8b$ develop in detail a method for finding the equations of the tangents to a conic from an external point.

By means of this method find the equations of the tangents required in each of the following exercises of Ch. IX, § 7.

5. Exercise 5.

6. Exercise 6.

7. Exercise 9.

8. Exercise 12.

12. Relative Positions of Pole and Polar. *Central Conics.* The following theorem is instructive concerning the relative positions of pole and polar with regard to a central conic.

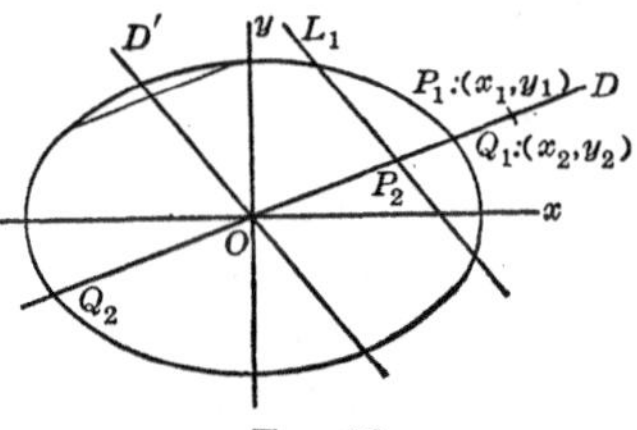

FIG. 26

THEOREM 9. *Let the point P_1 and the line L_1 be pole and polar in a central conic, center at O; let D be the diameter through*

* These methods are not very serviceable if *accurate* constructions are desired, since they involve the construction, not only of the tangent at a given point of the conic, but also of the tangents from an external point; cf. § 13. They are, however, useful in rough work.

P_1, D' the diameter conjugate to D, and P_2 the point of intersection of L_1 with D. Then L_1 is parallel to D' and the half-length, d, of D is a mean proportional between OP_1 and OP_2:

$$(1) \qquad OP_1 \cdot OP_2 = d^2.$$

We will prove the second part of the theorem first. By the definition of the polar of a point, P_1P_2 is divided harmonically by the points, Q_1 and Q_2, in which D meets the conic. Consequently, by (1), § 8,

$$Q_1P_1 \cdot Q_2P_2 = Q_2P_1 \cdot P_2Q_1.$$

Expressing each of the four distances in terms of OP_1, OP_2, and $OQ_1 = OQ_2 = d$, we have

$$(OP_1 - d)(OP_2 + d) = (OP_1 + d)(d - OP_2).$$

On multiplying out and reducing, we obtain equation (1), q. e. d.

We will give the proof of the first part of the theorem in the case that the conic is the ellipse

$$\frac{x^2}{a^2} + \frac{y^2}{b^2} = 1.$$

Let the coördinates of P_1 be (x_1, y_1) and those of Q_1, (x_2, y_2). Then the equations of L_1 and D' are, respectively, by § 9, (2), and § 6, (2),

$$(2) \qquad \frac{x_1x}{a^2} + \frac{y_1y}{b^2} = 1 \qquad \text{and} \qquad \frac{x_2x}{a^2} + \frac{y_1y}{b^2} = 0.$$

Since P_1 and Q_1 are on a line with O, their coördinates are proportional:

$$x_1 : y_1 = x_2 : y_2,$$

and hence so are the left-hand sides of equations (2). Consequently, L_1 and D' are parallel, q. e. d.

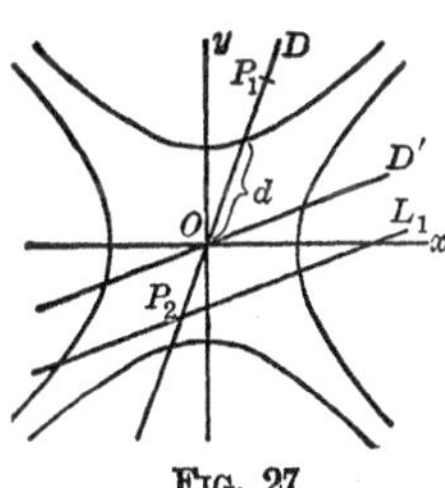

FIG. 27

The proof of the second part of the theorem assumes that D meets the conic. This is not true, however, if the conic is a hyperbola and P_1 lies in an opening between the asymptotes not containing a branch of the hyperbola. In this case, too, the theorem is valid, but the points P_1 and P_2, instead of being on the same side of O, are on opposite sides.*

COROLLARY. *The points P_1 and P_2 are on the same or opposite sides of O, according as D meets or does not meet the conic.*

Given, now, a point, P, and its polar, L, with respect to a central conic. If P traces a diameter D, then L moves always parallel to the conjugate diameter.

In the case that D intersects the conic, and P is an intersection, L is the tangent at P. If P then moves in along D toward the center as its limit, L ceases to meet the conic, and recedes indefinitely. On the other hand, if P moves out along D, receding indefinitely, L moves in toward the center, and approaches the diameter conjugate to D as its limit.

The case in which the conic is a hyperbola and D intersects the conjugate hyperbola remains. If P is at one of the intersections, L is the tangent to the conjugate hyperbola at the other; cf. Ex. 6. If P then moves in toward the center, L moves away from it, and so forth, as before.

Parabola. Corresponding to Theorem 9, we have the following theorem, the proof of which is left to the student.

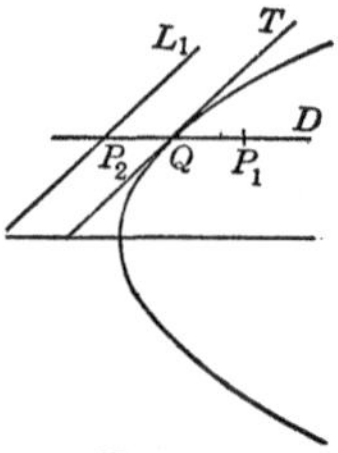

FIG. 28

THEOREM 10. *Let P_1 and L_1 be pole and polar in a parabola, and let the diameter through P_1 meet L_1 in P_2 and the parabola in Q. Then L_1 is parallel to the tangent at Q, and Q is the mid-point of P_1P_2.*

Consequently, if a point P traces a diameter

* Let the student give an *analytical* proof of these facts and hence of the corollary; cf. Exs. 1, 2. There is no *geometrical* proof, analogous to that of the text. The rôles of Q_1 and Q_2 in that proof cannot be played here by the points in which D meets the *conjugate* hyperbola; these points do not divide P_1P_2 harmonically: L_1 is the polar of P_1 with respect to the given hyperbola, and not with respect to its conjugate.

D of a parabola, its polar L moves always parallel to the tangent at the point in which D meets the curve. If P moves along D in either direction, receding indefinitely, then L moves in the opposite direction, and recedes indefinitely.

Theorems 9 and 10 furnish new methods for the construction of the polar of a given point or the pole of a given line. These we shall consider in the next paragraph.

EXERCISES

1. Give an analytical proof of the second part of Theorem 9, in the case that D meets the conic.

2. The same if the conic is a hyperbola and D does not meet it.

3. Theorem 9 is no longer valid if the conic is a hyperbola and P_1 lies on an asymptote. Prove that, in this case, L_1 is parallel to the asymptote, and that the product of the distances OP_1 and OP_2 is constant, where P_2 is the point in which L_1 intersects the other asymptote.

4. Prove Theorem 10.

5. A pair of conjugate hyperbolas and a point P are given. Show that the polars of P with respect to the two hyperbolas are parallel to, and equally distant from, the diameter conjugate to the diameter through P.

By applying Th. 9, the Corollary, and Th. 10, prove the following theorems.

6. Let C be a hyperbola, C' the conjugate hyperbola, and D a diameter meeting C'. Then the polar of an extremity of D with respect to C is the tangent to C' at the other extremity of D.

7. The polar of a point with respect to a central conic (not a circle) is perpendicular to the line joining the point to the center, if and only if the point is on an axis of the conic.

8. If a line is normal to a parabola at one extremity of the latus rectum, its pole lies on the diameter passing through the other extremity.

13. Construction Problems. In problems of construction relating to conics, diameters play an important rôle. To construct a diameter of a conic, one has but to draw two parallel chords and join their mid-points.

Center, Axes, Foci. Consider first a central conic, drawn on paper. Construct two diameters; their point of intersection will be the center, O, of the conic. With O as center, describe a circle cutting the conic in four points; the lines through O parallel to the sides of the rectangle determined by the four points will be the axes.

If the conic is an ellipse, the lengths a and b are now known, and $c = \sqrt{a^2 - b^2}$ can be constructed by means of a right triangle; cf. Ch. VII, Fig. 2. Thus the foci will be located.

If the conic is a hyperbola, only a is known. But then b can be found by reversing the construction of Ch. VIII, § 9. Hence the asymptotes and foci may be accurately constructed.

Let a parabola be given. Construct a diameter and two chords perpendicular to it; the line joining the mid-points of these chords is the axis of the parabola. The construction of the focus we postpone until we have given that of a tangent.

Tangents. To construct the tangent to a central conic at a point P, construct the center O, and then draw OP and a chord parallel to OP. Let K be the mid-point of this chord. Then the line through P parallel to OK is the tangent at P. Why?

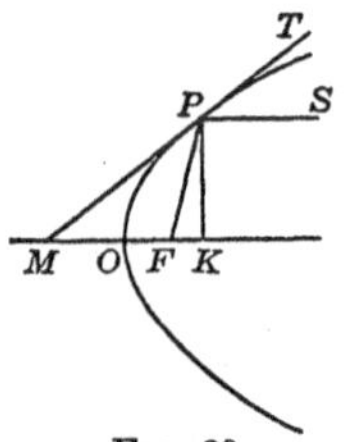

If the conic is a parabola, construct the axis. Let K be the foot of the perpendicular dropped from P on the axis, and make OM equal to OK (Fig. 29). Then, by Ch. VI, § 3, Ex. 8, MP is the tangent at P. The focus can now be constructed by use of the focal property, namely, by constructing the focal radius PF as the line making $\measuredangle\, MPF = \measuredangle\, TPS$.

Fig. 29

Of course, if the focus, or foci, of a conic are given, the tangent at a point can be constructed by means of the focal property of the conic.

To construct the tangents to a conic from an external point is a more difficult problem. We shall give presently a solution involving poles and polars and refer the student elsewhere for one based on more elementary, though less elegant, principles.*

Poles and Polars. Given an elementary construction for the tangents from an external point, we can carry through accurately the constructions of § 11 for the polar of a given point and the pole of a given line.

We are more interested, however, in the constructions of poles and polars, based on the theorems of § 12. We will describe, for example, the construction by this method of the polar of a given point P_1 with respect to a central conic. Draw the diameter D through P_1 (Fig. 26) and, by drawing a chord parallel to D and bisecting it, construct the diameter D' conjugate to D. On a separate sheet construct the third proportional to the length OP_1 and the half length, d, of D.† Lay off the resulting length from O on D in the proper direction, according to the Corollary of Theorem 9, and through the point thus reached draw the line parallel to D'. This line is the polar of P_1.

The construction is the same whether P_1 lies inside, on, or outside the conic.

Tangents from an External Point. Let the point be P and construct its polar L by the method just described. The lines joining P to the points of intersection of L with the conic are, by § 11, Th. 8*b*, the required tangents.

* Cf., for example, Wentworth's Plane and Solid Geometry, Ed. of 1900, pp. 416, 435, 455.

† If D does not meet the conic (as may happen in the case of a hyperbola) and the conjugate hyperbola is not given, the length d (Fig. 27) is unknown, so that a separate construction for it is necessary. Here d equals the b_1 of the formula, $a_1^2 - b_1^2 = a^2 - b^2$, of § 6, Th. 10. The lengths a, b, and a_1 are known or can be constructed by methods already given; the length $k = \sqrt{a^2 - b^2}$ is found by using a right triangle and, finally, that of $b_1 = \sqrt{a_1^2 - k^2}$, in the same way.

EXERCISES

1. Using a templet, draw an ellipse. Carry through in detail the constructions for (*a*) the center, (*b*) the axes, (*c*) the foci. Devise a method for constructing the directrices.

2. The same problem for the hyperbola. Construct also the asymptotes.

3. Construct the axis, a tangent, the focus, and the directrix of a parabola.

4. Construct the tangent to a hyperbola at a given point by use of the focal property. Use a templet to draw the hyperbola and consider that the foci are given.

5. The same for an ellipse.

6. Perform in detail the construction, based on Theorem 9, § 12, of the pole of a given line with respect to a central conic.

7. Carry through carefully the construction, based on Theorem 10, § 12, of the polar of a given point with respect to a parabola.

8. The same for the pole of a given line.

EXERCISES ON CHAPTER XIV

Diameters

1. Prove that two similar ellipses with the same center and the same transverse axis have the same pairs of conjugate diameters.

2. A line meets a hyperbola in the points P_1 and P_2 and meets the asymptotes in the points Q_1 and Q_2. Prove that the segments P_1P_2 and Q_1Q_2 have the same mid-points.

3. Using the result of Ex. 2, show that any two hyperbolas with the same asymptotes have the same pairs of conjugate diameters.

4. Prove that the line-segment joining two extremities of conjugate diameters of a hyperbola is parallel to one asymptote and is bisected by the other.

5. The chords of an ellipse from a vertex to the extremities of the minor axis are parallel to a pair of conjugate diameters. Prove this theorem.

6. Two chords connecting a point of a central conic with the ends of a diameter are called *supplemental* chords. Show that chords of this nature are always parallel to a pair of conjugate diameters.

7. Show that, if a parallelogram has its vertices on a central conic, its center is at the center of the conic; hence prove, by Ex. 6, that the sides of the parallelogram are parallel to a pair of conjugate diameters.

8. Prove that the angle which a diameter of an ellipse, not an axis, subtends at a vertex is the supplement of the angle which the conjugate diameter subtends at an extremity of the minor axis.

9. A parallelogram is circumscribed about an ellipse by drawing the tangents at the ends of a pair of conjugate diameters. Prove that the area of this parallelogram is the same, no matter what pair of conjugate diameters is chosen.

Suggestion. Compute the area of the triangle with one diameter as base and an extremity of the other as vertex.

10. State and prove the corresponding theorem for the hyperbola.

11. Show that in the case of the hyperbola the parallelogram of the two preceding exercises always has its vertices on the asymptotes.

12. Prove that the segment of a tangent to a hyperbola cut off by the asymptotes is equal in length to the diameter parallel to it.

13. Prove that the tangents to a central conic at the extremities of a chord meet on the diameter bisecting the chord.

14. Show that a line through a focus of a central conic perpendicular to a diameter meets the conjugate diameter on a directrix.

15. Prove that, if P and P' are extremities of a pair of conjugate diameters of a central conic, the normals at P and P' and the line through the center perpendicular to PP' meet in a point.

Poles and Polars

16. Find the polar of a focus of a central conic with respect to the auxiliary circle.

17. Prove, for a central conic, that the line-segment joining any point to the intersection of the polar of the point with a directrix subtends a right angle at the corresponding focus.

18. The same for a parabola.

19. Show, for a central conic, that any chord through a focus is perpendicular to the line joining the focus to the pole of the chord.

20. The same for a parabola.

21. Two rectangular hyperbolas are so situated that the axes of one are the asymptotes of the other. Prove that the polars of a point with respect to the two hyperbolas are always perpendicular.

22. The perpendicular from a point P on the polar of P with respect to a central conic meets the transverse axis in A and the conjugate axis in B. Show that $PA : PB = b^2 : a^2$.

23. The segment of the axis of a parabola intercepted by the polars of two points is equal to the projection on the axis of the line-segment joining the two points.

Locus Problems

24. A line is drawn through the focus of a central conic perpendicular to a variable diameter. Find the locus of the point in which it intersects the conjugate diameter.

25. A point moves so that its polar with respect to an ellipse forms a triangle of constant area with the axes of the ellipse. What is its locus?

Ans. A pair of conjugate rectangular hyperbolas with the axes of the ellipse as asymptotes.

26. Find the locus of the poles, with respect to a central conic, of the tangents to a circle whose center is the center of the conic.

27. Find the locus of the poles, with respect to the circle $x^2 + y^2 = a^2$, of the tangents to the parabola $y^2 = 2\,mx$.

28. Find the locus of the poles, with respect to the parabola $y^2 = 2\,mx$, of the tangents to the parabola $y^2 = -\,2\,mx$.

29. Find the locus of the mid-point of a chord of an ellipse, if the pole of the chord traces the auxiliary circle.

30. The same for a hyperbola.

CHAPTER XV

TRANSFORMATIONS OF THE PLANE. STRAIN

1. Translations. *Definition.* By a *translation* of a plane region S is meant a displacement of S whereby each point of S is carried in a given (fixed) direction by one and the same given distance. Thus, when a window is raised, a pane of glass in the window experiences a translation.

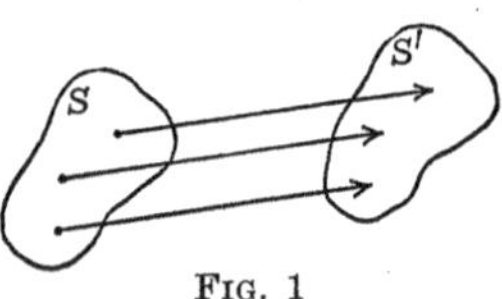

FIG. 1

It is not important what particular region S is considered. Indeed, it is usually desirable to consider the whole unbounded plane as S. The essential thing is the above law which connects the initial position of an arbitrary point of S with its final position.

Analytic Representation. Let $P:(x, y)$ be an arbitrary point of the plane, and let $P':(x', y')$ be the point into which P is carried by the translation. Let a and b be respectively the projections of the directed line-segment PP' on the axes of x and y. Then

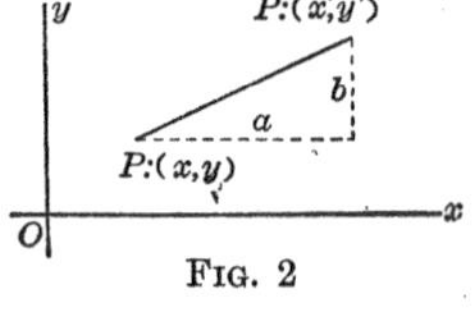

FIG. 2

$$(1) \qquad \begin{cases} x' = x + a, \\ y' = y + b. \end{cases}$$

These formulas are the same as those which represent a transformation of coördinates, the new axes being parallel to the old and having the same respective directions. But the interpretation of the formulas is wholly different. There, the point P remained unchanged. It had new coördinates assigned to it by referring it to a new set of axes. Here, the axes do not change. It is the point P that changes. The point P is picked up and set down in a new place, namely, at P'.

Example 1. Represent analytically the translation whereby the plane is carried in the direction of the positive x-axis a distance of 2 units:

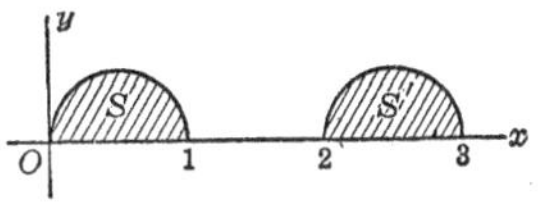

Solution: $\quad x' = x + 2, \qquad y' = y.$

FIG. 3

Example 2. Let the curve C:

$$(2) \qquad y = x^3 - x + \tfrac{3}{2}$$

be carried in the direction of the negative axis of y a distance of $\frac{3}{2}$ units. What will be the equation of the new curve, C''?

The formulas representing the translation are:

$$x' = x, \qquad y' = y - \tfrac{3}{2}.$$

Hence $\quad x = x', \qquad y = y' + \frac{3}{2},$

and equation (2) goes over into

$$y' + \tfrac{3}{2} = x'^3 - x' + \tfrac{3}{2}$$

or

$$y' = x'^3 - x'.$$

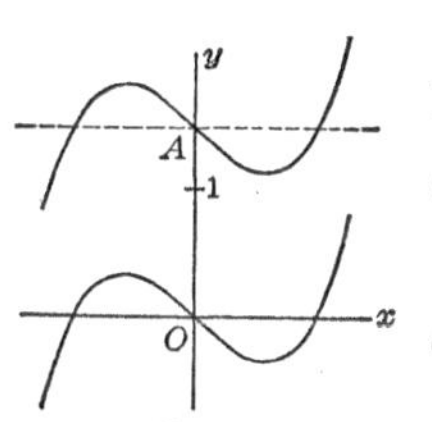

FIG. 4

The new curve, C', is evidently symmetric in the origin. But the shape of C' is the same as the shape of C. Hence C is symmetric in the point $A: (0, \frac{3}{2})$, which corresponds to the origin.

Example 3. A freight train is running northwest at the rate of 30 miles an hour. If (x, y) are, at noon, the coördinates of an arbitrary point of the floor of one of the platform cars, referred to axes directed east and north respectively, determine the coördinates (x', y') of the same point t hours later.

Here, the components of PP', after one hour has elapsed, are clearly:

$$a = 30 \cos 135° = -15\sqrt{2}, \qquad b = 30 \sin 135° = 15\sqrt{2}.$$

After t hours they are

$$a = -15\sqrt{2}\,t, \qquad b = 15\sqrt{2}\,t.$$

Hence

$$x' = x - 15\sqrt{2}\,t, \qquad y' = y + 15\sqrt{2}\,t.$$

EXERCISES

1. Express analytically the translation which carries the origin into the point $(2, -1)$, and hence show that the point $(-1, 2)$ is carried into $(1, 1)$. Draw a figure showing what happens to the unit circle, $x^2 + y^2 = 1$.

2. Apply the translation of Ex. 1 to the curve

$$y = 2x^2 + 8x + 9.$$

3. Determine a translation which will carry the curve

$$y = 4x^2 - 8x + 3$$

into a parabola whose equation is in a normal form.

4. An aëroplane is flying at the rate of 120 miles an hour on a straight, horizontal course having a direction 30° south of east. If (x, y) are, at a given instant, its coördinates, referred to axes directed east and north respectively, determine its coördinates (x', y') after t minutes have elapsed.

Prove analytically (*i.e.* by means of the representation (1) of the text) the following theorems.

5. A translation carries a straight line, in general, into a parallel straight line. What are the exceptions?

6. A translation carries a circle into a circle of the same radius.

7. A translation carries two mutually perpendicular right lines into two mutually perpendicular right lines.

2. Rotations. Let the plane be rotated about the origin through an angle θ. What will be the coördinates, (x', y'), of the point P' into which a given point P, with the coördinates (x, y), is carried?

The solution can be read off at sight from the figure. We have:

$$x = OM = OM_1; \qquad y = MP = M_1P';$$
$$x' = OM', \qquad y' = M'P'.$$

Now, $\qquad \text{Proj } OP' = \text{Proj } OM_1 + \text{Proj } M_1P',$

and if we take the projections first along the axis of x, and then along the axis of y, we obtain immediately the desired relations:

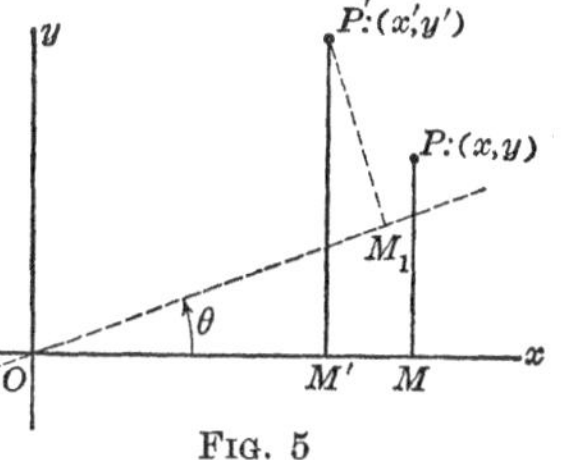

FIG. 5

(1) $$\begin{cases} x' = x \cos\theta - y \sin\theta, \\ y' = x \sin\theta + y \cos\theta. \end{cases}$$

It is easy to solve these equations algebraically for x and y; or the formulas for x and y, in terms of x' and y', can be written down directly by projecting the broken line $OM'P'$ along OM_1 and perpendicularly to OM_1:

(2) $$\begin{cases} x = x' \cos\theta + y' \sin\theta, \\ y = -x' \sin\theta + y' \cos\theta. \end{cases}$$

Example. It is clear geometrically that a circle with its center at the origin must be carried over into itself by any of the above rotations. Let us see what the analytic effect on its equation is if such a rotation is performed.

The equation of the given circle is

$$x^2 + y^2 = \rho^2.$$

Replacing x and y by their values from (2), we have:

$$(x' \cos\theta + y' \sin\theta)^2 + (-x' \sin\theta + y' \cos\theta)^2 = \rho^2,$$

or

$$\left.\begin{matrix}\cos^2\theta \\ +\sin^2\theta\end{matrix}\right| x'^2 + \left.\begin{matrix}2\sin\theta\cos\theta \\ -2\sin\theta\cos\theta\end{matrix}\right| x'y' + \left.\begin{matrix}\sin^2\theta \\ +\cos^2\theta\end{matrix}\right| y'^2 = \rho^2.$$

Hence $\qquad x'^2 + y'^2 = \rho^2,$

and we get the same circle, as we should.

EXERCISES

1. Write down directly from a figure the formulas which represent a rotation of 90° about the origin, and verify the result by substituting $\theta = 90°$ in (1).

2. Show that, if the curve $xy = 2a^2$ is rotated about the origin through an angle $\theta = -45°$, its equation goes over into the usual form of the equation of an equilateral hyperbola.

3. Rotate the parabola $y^2 = 2mx$ through $-90°$ about the origin.

4. Prove analytically that, if an arbitrary straight line be rotated about the origin through 90°, the new line will be perpendicular to the old one.

5. Prove that, if an arbitrary line be rotated about the origin through the angle θ, the angle from this line to the new line will be θ.

3. Transformations of Similitude. Let the plane be *stretched*, like an elastic membrane, uniformly in all directions away from the origin. This tranformation is evidently represented analytically by the equations:

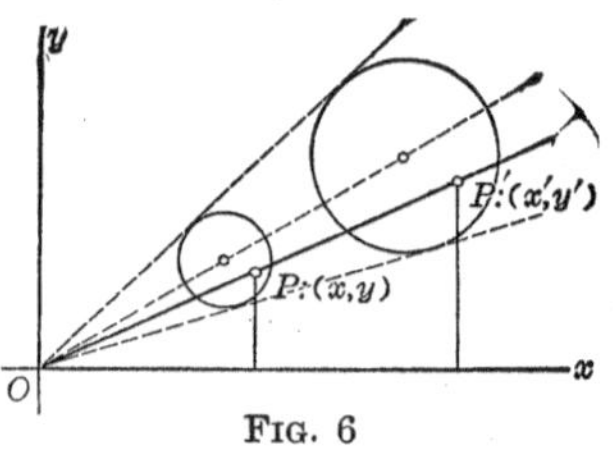

FIG. 6

$$(1) \qquad \begin{cases} x' = kx, \\ y' = ky, \end{cases}$$

where k is a constant greater than unity. If k is positive, but less than unity, the transformation represents a *shrinking* toward the origin. The stretchings and shrinkings defined by (1) are known as *transformations of similitude.*

These transformations, like the translations and the rotations, preserve the *shapes* of all figures; but, unlike those transformations, they alter the *sizes* of figures.

Example. The equilateral hyperbola

$$x^2 - y^2 = a^2$$

is carried by (1), if k is taken equal to $\frac{1}{a}$:

$$x' = \frac{x}{a}, \quad y' = \frac{y}{a}, \qquad \text{or} \qquad x = ax', \quad y = ay',$$

into the curve

$$a^2x'^2 - a^2y'^2 = a^2,$$

or

$$x'^2 - y'^2 = 1.$$

Thus all equilateral hyperbolas are seen to be similar to one another, since each can be transformed by (1) into the particular equilateral hyperbola

$$x^2 - y^2 = 1.$$

Inverse of a Transformation. The transformation,

$$(2) \qquad \begin{cases} x = \dfrac{x'}{k}, \\ y = \dfrac{y'}{k}, \end{cases}$$

obtained by solving the formulas (1) for x, y, is called the *inverse* of the transformation (1). In general, if a given transformation carries (x, y) into (x', y'), the transformation carrying (x', y') into (x, y) is known as the *inverse* of the given transformation. Thus, the rotation (2), § 2, is the inverse of the rotation (1), § 2.

It is clear that the effect of the inverse transformation, if performed after the given one, is to nullify the given one. Thus (1), § 2, rotates all figures through the angle θ, and then (2), § 2, rotates them through the angle $-\theta$, *i.e.* back into their original positions.

EXERCISES

1. Show that the parabola $y^2 = 2mx$, $0 < m$, can be transformed by (1) into the parabola $y^2 = x$. What value must be taken for k?

2. Show that the effect of performing transformation (1) and then transformation (2) is to leave the plane unchanged.

4. Reflections in the Axes. Let the plane be reflected in the axis of x. In other words, let it be rotated through 180° about the axis of x. Let $P:(x, y)$ be an arbitrary point, and let $P':(x', y')$ be the point into which P is carried. Then, obviously,

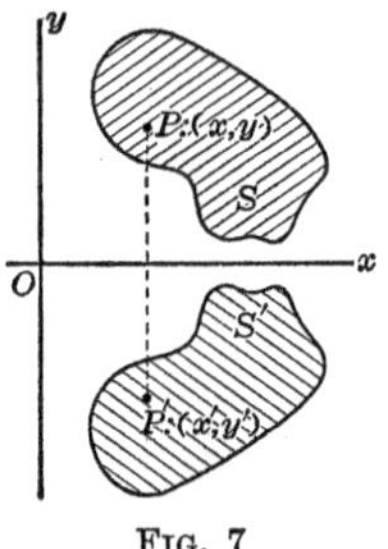

FIG. 7

$$(1) \qquad \begin{cases} x' = x, \\ y' = -y. \end{cases}$$

Similarly, a reflection in the axis of y is represented by the formulas:

$$(2) \qquad \begin{cases} x' = -x, \\ y' = y. \end{cases}$$

The condition that a curve be symmetric in one of the axes (cf. Ch. V, § 2) is obtained at once from these transformations. Thus the curve C will be symmetric in the axis of x if the curve C', into which C is carried by (1), is the same curve as C; and the test for this is, that the equation of C be essentially unchanged when the transformation (1) is performed on it.

For example, if C is the curve

$$y^4 + x^2 = 2y^2 + x^3,$$

its equation is unchanged by (1), and hence C is symmetric in the axis of x. But it is changed by (2), and C is, therefore, not symmetric in the axis of y.

Isogonal Transformations. A transformation is said to be *equiangular* or *isogonal* if the angle which any two intersecting curves, C_1 and C_2, make with each other is the same as the angle which the transformed curves, C'_1 and C'_2, make with each other.

All of the transformations considered thus far are evidently isogonal. We turn now to a transformation which is not.

EXERCISE

Show that the equations of the inverse of the reflection (1) [or (2)] are precisely of the same form as the equations of the reflection. A transformation for which this is true is said to be *involutory.*

5. Simple Elongations and Compressions. Let the plane be stretched directly away from the axis of x, so that each point is carried, along a parallel to the axis of y, to twice its original distance from the axis of x (Fig. 8). Evidently, the analytic condition is that

$$x' = x, \qquad y' = 2y.$$

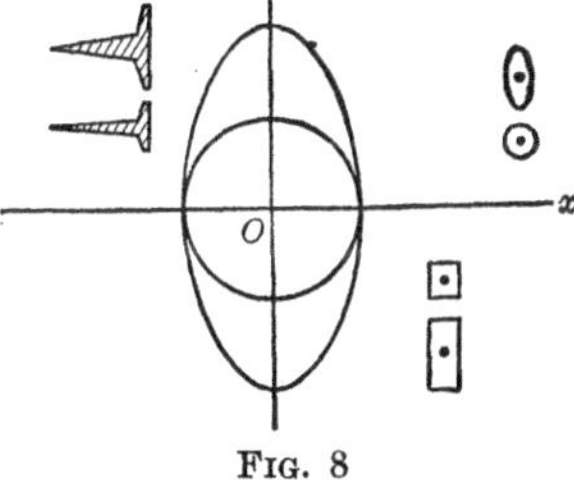

FIG. 8

More generally, if a point $P:(x, y)$ is to be carried to l times its original distance from the axis of x, where l may have any positive constant value, not unity, the transformation will be given by the formulas:

$$(1) \qquad \begin{cases} x' = x, \\ y' = l y. \end{cases}$$

When l is greater than unity, these formulas represent an elongation; when l is less than unity, they represent a compression.

If the elongation is away from the axis of y or the compression is toward it, then

$$(2) \qquad \begin{cases} x' = k x, \\ y' = y, \end{cases}$$

where k is greater than unity in the first case and less than unity, but positive, in the second.

These transformations were discussed geometrically in Ch. XIV, § 7. There we called them *one-dimensional,* or *simple, elongations* and *compressions*; or, jointly, *one-dimensional strains.*

Example 1. Let the circle

i) $$x^2 + y^2 = 1$$

be subjected to the transformation (1). Then it goes over into

$$x'^2 + \frac{y'^2}{l^2} = 1.$$

Thus the circle i) is carried into the ellipse

ii) $$\frac{x^2}{1} + \frac{y^2}{b^2} = 1, \qquad b = l,$$

whose axis lying along the axis of x is identical with the corresponding diameter of the circle, but whose axis lying along the axis of y is the corresponding diameter of the circle stretched in the ratio $l:1$.

Example 2. Let the ellipse ii) be subjected to the transformation (2). Then

$$\frac{x'^2}{k^2} + \frac{y'^2}{b^2} = 1.$$

Thus the ellipse ii) is carried into the ellipse

iii) $$\frac{x^2}{a^2} + \frac{y^2}{b^2} = 1, \qquad a = k.$$

From these examples we see that the particular circle i) can be carried by means of two one-dimensional strains into an arbitrary ellipse iii) whose axes lie along the axes of coördinates.

Exercise. Show that the circle i) can be carried into the ellipse iii) by a single one-dimensional strain and a transformation of similitude.

Product of Two Transformations. The combined effect of the two transformations of Examples 1 and 2 can be represented analytically as follows. First, we have

(*a*) $$x' = x. \qquad y' = ly.$$

Next, the point (x', y') is carried into (x'', y'') by the transformation

(*b*) $$x'' = kx', \qquad y'' = y'.$$

Eliminating the intermediate stage (x', y'), we get:

(*c*) $$x'' = kx, \qquad y'' = ly.$$

A transformation, as (*c*), which arises as the result of two successive transformations, as (*a*) and (*b*), is called the *product* of these transformations. Similarly, (*a*) and (*b*) are spoken of as the *factors* of (*c*); or (*c*) is said to be *factored* into (*a*) and (*b*).

Let the student verify the fact that, if the circle i) is subjected to the transformation (*c*), it is carried over into the same ellipse iii) into which i) was carried by the successive applications of the transformations (*a*) and (*b*).

Properties of the Transformation. One of the most important properties of one-dimensional strains is that, like the transformations previously studied, *they carry straight lines over into straight lines.*

This was proved geometrically on p. 304. The transformation considered there is given analytically by (1). It was proved also that, if $\overline{L}$ is a line of slope λ, the slope of the line into which $\overline{L}$ is carried by (1) is*

(3) $$\lambda' = l\lambda.$$

From the theorem contained in formula (3), it is seen that *a one-dimensional strain carries parallel lines into parallel lines.*

Consider an arbitrary curve, C. Its slope at any one of its points, P, is

$$\lambda = \lim_{Q \doteq P} \frac{MQ}{PM}.$$

Perform the transformation (1) on C. Then PM remains unchanged in length; but MQ goes over into

$$M'Q' = l\,MQ.$$

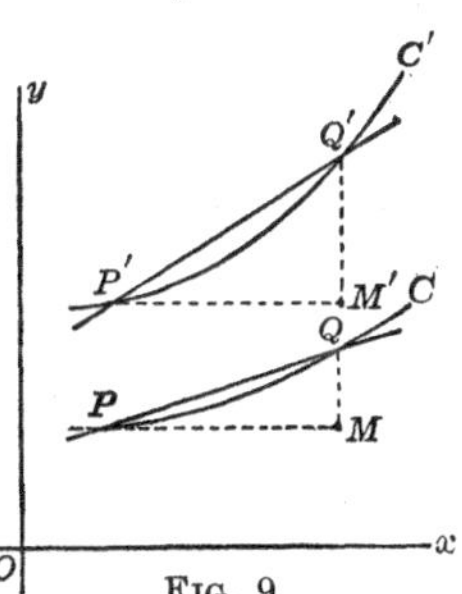

FIG. 9

Hence the slope, λ', of C' is

$$\lambda' = \lim_{Q' \doteq P'} \frac{M'Q'}{P'M'} = \lim_{Q \doteq P} \frac{l\,MQ}{PM} = l \lim_{Q \doteq P} \frac{MQ}{PM},$$

or $$\lambda' = l\lambda.$$

We have thus extended the validity of formulas (3).

* The proofs on p. 304 were given for compressions, but they are valid, also, for elongations.

It follows from this extension that, if two curves, C_1 and C_2, are tangent to each other, the transformed curves, C'_1 and C'_2, will also be tangent. For, if C_1 and C_2 are tangent, they have the same slope λ at the point of tangency. Hence, at the corresponding point, C'_1 and C'_2 will each have the slope $\lambda' = l\lambda$ and consequently will be tangent to each other.

Angles are not in general preserved by a one-dimensional strain. It is true that right angles whose sides are parallel respectively to the coördinate axes go over into right angles satisfying the same condition. But consider, for example, the angle between a line $\bar{L}$ and the axis of x (p. 304, Fig. 13). $\bar{L}$ is carried into L by the transformation (1), and the axis of x remains fixed. It is clear, then, that the new angle is not equal to the original one.

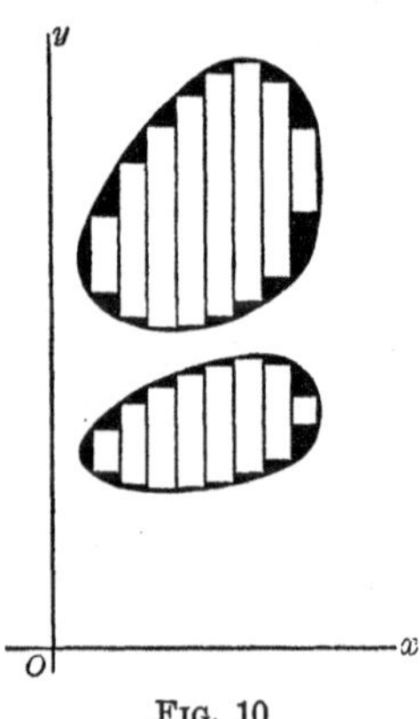

Fig. 10

The areas of figures, also, are changed, and changed in precisely the ratio of l (or k) : 1. This is obviously true for rectangles whose sides are parallel to the coördinate axes. The area, A, of any other figure is the limit approached by the sum, B, of the areas of rectangles inscribed as shown in the drawing:

$$A = \lim B.$$

By the transformation (1), A is carried into A' and B into B'; evidently

$$A' = \lim B'.$$

Since the area of each rectangle represented in the sum B' is l times the area of the rectangle from which it originated,

$$B' = lB.$$

Hence
$$A' = \lim lB = l \lim B,$$
or
$$A' = lA, \qquad \text{q. e. d.}$$

Example. The area of the circle i) is π. It follows, then, that the area of the ellipse ii) is πb. Applying the method again to this ellipse, we obtain as the area of the ellipse iii) the value πab. We have thus obtained the following result.

The area of the ellipse

$$\frac{x^2}{a^2}+\frac{y^2}{b^2}=1$$

is

$$A=\pi ab.$$

EXERCISES

1. Show that the circle $x^2+y^2=a^2$ can be carried into the ellipse

$$\frac{x^2}{a^2}+\frac{y^2}{b^2}=1$$

by a one-dimensional strain; cf. p. 306.

2. Prove that the rectangular hyperbola $x^2-y^2=a^2$ can be carried into the hyperbola

$$\frac{x^2}{a^2}-\frac{y^2}{b^2}=1$$

by a one-dimensional strain.

3. In Examples 1, 2 of p. 338 can the order of the transformations be reversed? Prove your answer.

4. A one-dimensional strain changes, in general, the shapes of curves. Is this true in all cases? For example, in the case of a parabola?

5. Prove analytically that the transformation (1) carries a straight line into a straight line.

6. Show analytically that, if a line L is carried by (1) into a line L', the slopes of L and L' are connected by formula (3).

7. Find the equations of the transformation which is the product of the two transformations:

$$x'=x-1,\ y'=y+2, \qquad x''=x',\ y''=-y'.$$

8. The same for the rotation about the origin through 45°, followed by the translation which carries the origin into the point $(3, -1)$.

9. The same for the transformation of similitude which doubles all the lengths, followed by a reflection in the axis of y.

10. Prove analytically, for the following transformations, that the product of a transformation and its inverse is the *identical transformation*, $x' = x$, $y' = y$.

(*a*) translations; (*b*) rotations;
(*c*) reflections; (*d*) one-dimensional strains.

11. Factor the transformation

$$x' = 4x, \qquad y' = 2y,$$

(*a*) into two one-dimensional strains;

(*b*) into a one-dimensional strain and a transformation of similitude.

12. Prove that the rotation about the origin through 180° — also called the *reflection in the origin* — is the product of the reflections in the axes.

13. Factor the transformation

$$x' = 4x, \qquad y' = -2y$$

into two one-dimensional strains and the reflection in the axis of x. How else can it be factored?

Express each of the following transformations as the product of two or more simple transformations.

14. $x' = -3x, \qquad y' = -2y.$

15. $x' = -x + 2, \qquad y' = y - 3.$

16. $x' = 5x + 2, \qquad y' = -3x - 1.$

17. $x' = \dfrac{x - y}{\sqrt{2}} - 6, \qquad y' = \dfrac{x + y}{\sqrt{2}} + 3.$

6. The General Affine Transformation. By the title is meant the transformation

$$(1) \qquad \begin{cases} x' = ax + by + c, \\ y' = a'x + b'y + c', \end{cases}$$

where $\qquad \Delta \equiv ab' - a'b \neq 0.$

All of the foregoing transformations come under this type, but there are transformations comprised under (1), for example,

$$(2) \qquad x' = 2x - 3y + 1, \qquad y' = -x + 4y - 2,$$

which are not of any of the above forms. We shall prove the following theorem.

THEOREM 1. *The transformation of the plane defined by means of equations* (1) *can be generated by a succession of the transformations studied in* §§ 1–5. *In other words, it can be* FACTORED *into transformations of the type of those of* §§ 1–5.

Proof. If c and c' are not both 0, let the (x, y)-plane be subjected to a translation:

$$(3) \qquad \begin{cases} x_1 = x + \xi, \\ y_1 = y + \eta, \end{cases} \qquad \begin{cases} x = x_1 - \xi, \\ y = y_1 - \eta, \end{cases}$$

where ξ, η are arbitrary and shall be determined presently. Thus equations (1) are replaced by the following:

$$(4) \qquad \begin{cases} x' = ax_1 + by_1 - (a\xi + b\eta - c), \\ y' = a'x_1 + b'y_1 - (a'\xi + b'\eta - c'). \end{cases}$$

We now determine ξ and η so that both parentheses will disappear. This is done by solving the simultaneous linear equations:

$$a\xi + b\eta - c = 0,$$
$$a'\xi + b'\eta - c' = 0.$$

The solution is always possible and unique, since, by hypothesis, $\Delta \equiv ab' - a'b \neq 0$.

We thus have a simpler pair of equations to study, namely,

$$(5) \qquad \begin{cases} x' = ax_1 + by_1, \\ y' = a'x_1 + b'y_1. \end{cases}$$

It will be sufficient, then, if we can prove our theorem for the case that $c = 0$ and $c' = 0$, *i.e.* for the pair of equations

$$(6) \qquad \begin{cases} x' = ax + by, \\ y' = a'x + b'y, \end{cases}$$

for we have just seen that we can pass from equations (1) to equations (5) by a *translation*, and this is one of the transformations admitted by the theorem.

Consider an arbitrary circle with its center at the origin:

$$(7) \qquad x^2 + y^2 = \rho^2.$$

Let us see into what curve it is carried by (6).

To do this, solve equations (6) for x and y. The result is the inverse of (6), namely:

$$(8) \qquad \begin{cases} x = A\,x' + B\,y', \\ y = A'x' + B'y', \end{cases}$$

where $A = \dfrac{b'}{\Delta}, \quad B = -\dfrac{b}{\Delta}, \quad A' = -\dfrac{a'}{\Delta}, \quad B' = \dfrac{a}{\Delta},$

and $AB' - A'B = \dfrac{ab' - a'b}{\Delta^2} = \dfrac{1}{\Delta} \neq 0.$

Next, substitute these values for x and y in (7):

$$(9) \quad (A^2 + A'^2)\,x'^2 + 2(AB + A'B')\,x'y' + (B^2 + B'^2)\,y'^2 = \rho^2.$$

The locus of this equation is an ellipse with its center at the origin. For, first, the equation has a locus, since all the points (x', y') into which the points (x, y) of the circle (7) are carried by (6) lie on (9). Secondly, the locus does not extend to infinity in any direction.*

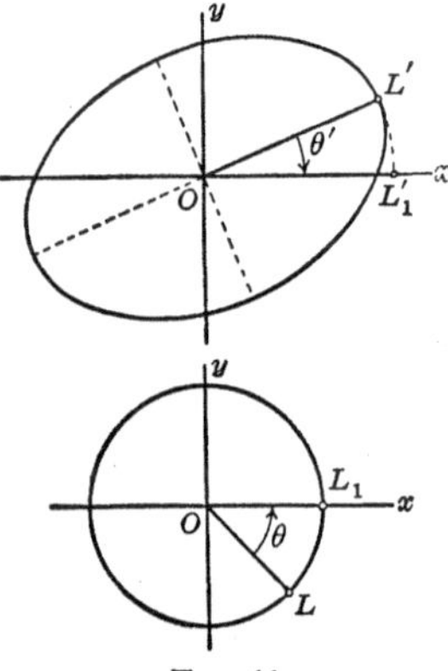

FIG. 11

In general, this ellipse will not be a circle. Let L' be the extremity of an axis. Since the transformation (8) carries any straight line through the origin into a straight line through the origin, OL' will correspond to a certain radius OL of the circle (7).

We now rotate the (x, y)-plane

* Or, the expression $B^2 - 4AC$ (Ch. XII, § 3), formed for (9), is negative:

$$4(AB + A'B')^2 - 4(A^2 + A'^2)(B^2 + B'^2) = -4(AB' - A'B)^2 < 0.$$

about the origin through an angle θ such that OL comes to lie along the positive axis of x:

(10) $$\begin{cases} x_1 = x\cos\theta - y\sin\theta, \\ y_1 = x\sin\theta + y\cos\theta. \end{cases}$$

Furthermore, we rotate the (x', y')-plane about the origin through such an angle θ' that OL' comes to lie along the positive axis of x:

(11) $$\begin{cases} x_1' = x'\cos\theta' - y'\sin\theta', \\ y_1' = x'\sin\theta' + y'\cos\theta'. \end{cases}$$

What is the final result? Obviously the following. An arbitrary point (x_1, y_1) of the plane is carried by the inverse of (10) into a point (x, y); this point is carried by (6) into a point (x', y'); and finally the point (x', y') thus obtained is carried by (11) into (x_1', y_1'). To write out these transformations explicitly would be a long piece of work; but it is not necessary to do so. For, first of all, each is linear and leaves the origin unchanged. Hence the final transformation, carrying the point (x_1, y_1) directly into the point (x_1', y_1'), is also linear,* and it leaves the origin unchanged. It is, then, of the form:

(12) $$\begin{cases} x_1' = \alpha x_1 + \beta y_1, \\ y_1' = \gamma x_1 + \delta y_1. \end{cases}$$

Consider next what we know about this transformation.

i) It carries the positive axis of x_1 over into itself. Hence, when $y_1 = 0$, y_1' must also vanish, no matter what value x_1 may have; thus

$$0 = \gamma x_1,$$

and, consequently, $$\gamma = 0.$$

ii) The axes of the ellipse which corresponds to the circle,

(13) $$x_1^2 + y_1^2 = \rho^2,$$

are the coördinate axes.

* Cf. Ex. 1 at the end of the chapter.

To find the equation of the ellipse, solve (12) for x_1, y_1, remembering that $\gamma = 0$:

$$(14)\qquad \begin{cases} x_1 = \mathfrak{A}x_1' + \mathfrak{B}y_1', \\ y_1 = \qquad\quad \mathfrak{D}y_1', \end{cases}$$

$$\mathfrak{D} = \frac{1}{\delta}, \qquad \mathfrak{A} = \frac{1}{\alpha}, \qquad \mathfrak{B} = -\frac{\beta}{\alpha\delta}.$$

Thus (13) is seen to go into the ellipse

$$(15)\qquad \mathfrak{A}^2 x_1'^2 + 2\,\mathfrak{A}\mathfrak{B}x_1'y_1' + (\mathfrak{B}^2 + \mathfrak{D}^2)y_1'^2 = \rho^2.$$

Since the axes of the ellipse (15) lie along the coördinate axes, the term in $x_1'y_1'$ must disappear, and so we must have

$$\mathfrak{A}\mathfrak{B} = 0.$$

Now, $\mathfrak{A}$ cannot be 0, for $\mathfrak{A} = 1/\alpha$. Hence $\mathfrak{B}$ must vanish:

$$\mathfrak{B} = 0.$$

It follows, then, that

$$\beta = 0,$$

and thus (12) reduces to the transformation:

$$(16)\qquad \begin{cases} x_1' = \alpha x_1, \\ y_1' = \delta y_1. \end{cases}$$

Moreover, since the positive axis of x_1 goes over into the positive axis of x_1, we see that

$$0 < \alpha.$$

But δ may be negative. In this case, a reflection in the axis of x_1 will change the sign of δ, and hence the case is reduced to the one in which δ is positive:

$$0 < \delta.$$

Finally, the transformation (16) is the product of two one-dimensional strains:

$$\begin{cases} x' = \alpha x, \\ y' = y, \end{cases} \qquad \begin{cases} x' = x, \\ y' = \delta y, \end{cases}$$

one along the axis of x and one along the axis of y.

Let us now recapitulate. The point (x, y) is carried into the point (x_1, y_1) by the rotation (10); (x_1, y_1) is carried into (x_1', y_1')

by (16), which is a product of two one-dimensional strains and, perhaps, the reflection in the axis of x; finally, (x_1', y_1') is carried into (x', y') by the inverse of (11), another rotation. The transformation (6) is, then, the product of these transformations in the order enumerated. The proof of Theorem I is thus complete.

Properties which all the component transformations of (1) have in common are also properties of (1). Consequently, *the general affine transformation carries straight lines into straight lines, parallel lines into parallel lines, and tangent curves into tangent curves.* It does not, in general, preserve angles or areas.

Isogonal Transformations. Of the component transformations of (6), the rotations (10) and (11) always preserve angles. This is true of the transformation (16) if and only if it is a transformation of similarity, with or without a reflection in the axis of x, *i.e.* if and only if $\delta = \pm\, \alpha$. Hence the most general *isogonal* transformation of the form (6) is the product of the rotation (10), the transformation

$$x_1' = \rho\, x_1, \qquad y_1' = \pm\, \rho\, y_1, \qquad \rho = \alpha > 0,$$

and the inverse of the rotation (11). This product is easily found to be

$$\begin{aligned} x' &= \quad \rho[x \cos(\theta \mp \theta') - y \sin(\theta \mp \theta')], \\ y' &= \pm \rho[x \sin(\theta \mp \theta') + y \cos(\theta \mp \theta')]. \end{aligned}$$

But the angle $\theta - \theta'$ (or $\theta + \theta'$) is no more general than a single angle, which we may denote by ϕ. Thus the result can be written as

$$x' = \rho(x \cos\phi - y \sin\phi), \qquad y' = \pm\, \rho(x \sin\phi + y \cos\phi).$$

Replacing x, y by x_1, y_1, so that this transformation of the form (6) reverts to the form (5), and then applying the translation (3), we obtain as the most general transformation of the form (1) which is isogonal:

$$(17) \qquad \begin{cases} x' = \quad \rho(x \cos\phi - y \sin\phi) + c, \\ y' = \pm\, \rho(x \sin\phi + y \cos\phi) + c'. \end{cases}$$

Here

(18) $a=\rho\cos\phi,\quad b=-\rho\sin\phi,\quad a'=\pm\rho\sin\phi,\quad b'=\pm\rho\cos\phi,$
and hence,

(19) $$b'=\pm a,\qquad a'=\mp b.$$

Conversely, every transformation (1), for which (19) is true for one set of signs, can be written in one of the forms (17) and hence is isogonal. For, if a, b, a', b' are given, satisfying (19) for one set of signs, values of ρ, $\cos\phi$, and $\sin\phi$ can be found, so that equations (18) hold for the same set of signs. These values are, namely,

$$\rho=\sqrt{a^2+b^2},\qquad \cos\phi=\frac{a}{\sqrt{a^2+b^2}},\qquad \sin\phi=\frac{-b}{\sqrt{a^2+b^2}}.$$

The following theorem summaries our results.

THEOREM 2. *The transformation* (1) *is isogonal when and only when either*

$$b'=a \text{ and } a'=-b \quad\text{or}\quad b'=-a \text{ and } a'=b.$$

If it is isogonal, it can be written in one of the forms (17).

Homogeneous and Non-Homogeneous Transformations. A polynomial in x and y is *homogeneous*, if its terms are all of the same degree in x and y.* Thus, the left-hand sides of formulas (6) are homogeneous polynomials of the first degree. Accordingly, a transformation of the form (6) is called a *homogeneous affine transformation*; and, in distinction, a transformation of the form (1), where c and c' are not both zero, a *non-homogeneous* affine transformation.

Since (1) can be reduced to (6) by means of a translation, we have the theorem.

THEOREM 3. *A non-homogeneous affine transformation is the product of a translation and the corresponding homogeneous transformation.*

* Only those terms with non-vanishing coefficients need be considered, since a term whose coefficient vanishes has the value of 0, and 0 is not defined as having a degree. For example, if, in the polynomial $x^2+2xy+ax$, a has the value 0, the polynomial is homogeneous of the second degree.

EXERCISES

1. Prove analytically that the affine transformation (1) carries a straight line into a straight line.

2. Show that, if L is a line of slope λ, the line into which L is carried by (1) has the slope

$$\lambda' = \frac{a' + b'\lambda}{a + b\lambda}.$$

3. Using the result of Ex. 2, prove that (1) carries parallel lines into parallel lines.

7. Factorization of Particular Transformations. We proceed to illustrate the theory of the preceding section by carrying it through, step by step, for a particular case.

Let the given transformation be

$$(1) \qquad x' = x + 3y, \qquad y' = -3x - y.$$

The first step is to find the ellipse into which the circle,

$$(2) \qquad x^2 + y^2 = \rho^2,$$

is carried by (1). Solving equations (1) for x and y, and substituting the values obtained, namely,

$$x = -\tfrac{1}{8}x' - \tfrac{3}{8}y', \qquad y = \tfrac{3}{8}x' + \tfrac{1}{8}y',$$

in (2), we have, finally,

$$(3) \qquad 5x'^2 + 6x'y' + 5y'^2 = 32\rho^2.$$

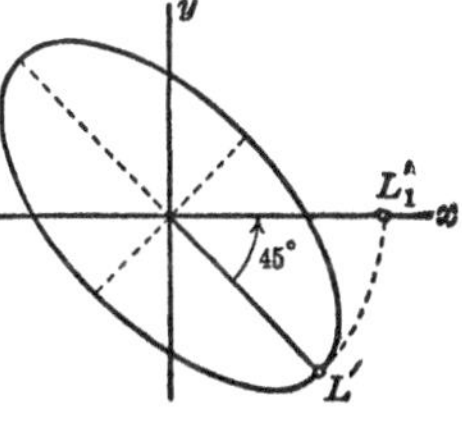

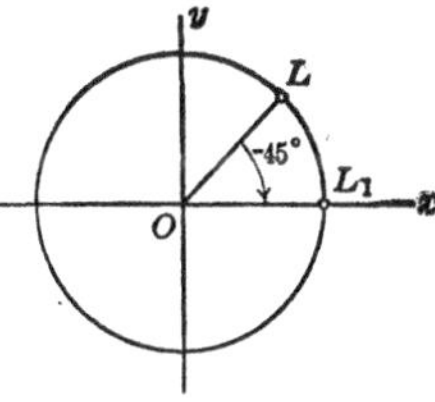

Fig. 12

This ellipse is as shown. One axis, OL', lies along the line $x' + y' = 0$. By adding equations (1), we have

$$x' + y' = -2x + 2y,$$

and hence the radius OL of the circle which is carried into OL' is along the line $x - y = 0$. Furthermore, if L' lies in the fourth quadrant, as in the figure, then L lies in the first — not in the third.

Consequently, a rotation of the (x, y)-plane about O through $-45°$:

(4) $$x_1 = \frac{x+y}{\sqrt{2}}, \qquad y_1 = \frac{-x+y}{\sqrt{2}},$$

brings OL to lie along the positive axis of x. And a rotation of the (x', y')-plane about O through $+45°$:

(5) $$x_1' = \frac{x'-y'}{\sqrt{2}}, \qquad y_1' = \frac{x'+y'}{\sqrt{2}},$$

does the same for OL'.

The next step is to find the transformation carrying (x_1, y_1) directly into (x_1', y_1'). This we do by eliminating x, y and x', y' from (1), (4), (5). Thus,

$$\begin{cases} x_1' = \dfrac{1}{\sqrt{2}}(x+3y+3x+y) = 4\dfrac{x+y}{\sqrt{2}} = 4x_1, \\ y_1' = \dfrac{1}{\sqrt{2}}(x+3y-3x-y) = 2\dfrac{-x+y}{\sqrt{2}} = 2y_1, \end{cases}$$

or

(6) $$x_1' = 4x_1, \qquad y_1' = 2y_1.$$

Finally, we solve (5) for x' and y':

(7) $$x' = \frac{x_1'+y_1'}{\sqrt{2}}, \qquad y' = \frac{-x_1'+y_1'}{\sqrt{2}}.$$

The transformation (1) is now seen to be the product of the transformations (4), (6), and (7); (4) carries (x, y) into (x_1, y_1) by a rotation about O through $-45°$; (6) carries (x_1, y_1) into (x_1', y_1') by two one-dimensional strains; (7) carries (x_1', y_1') into (x', y') by another rotation about O through $-45°$.

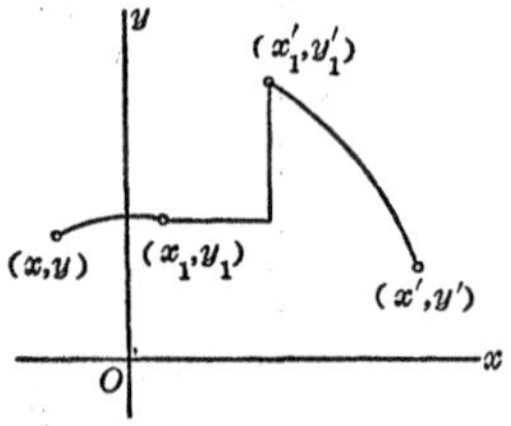

FIG. 13

Simplifications in Technique. Instead of seeking the ellipse in the (x', y')-plane into which the circle (2) is carried, we might equally well ask for the ellipse in the (x, y)-plane which is carried into the circle

(8) $$x'^2 + y'^2 = \rho^2.$$

The rôles of the two planes are merely reversed. Adopting this procedure obviates the necessity of solving (1) for the values of x, y to be substituted in (2). For now (2) is replaced by (8) and x' and y' are given by (1). Thus one step in the process is eliminated. The others remain unchanged.

Another simplification arises in factoring a transformation of the form

(9) $$\begin{cases} x' = x + 3y + 4, \\ y' = -3x - y - 2. \end{cases}$$

Instead of proceeding as in § 6, we set

$$\begin{cases} x' = \bar{x} + 4, \\ y' = \bar{y} - 2, \end{cases} \quad \text{and} \quad \begin{cases} \bar{x} = x + 3y, \\ \bar{y} = -3x - y. \end{cases}$$

Thus (9) is the product of the transformation (1) and the translation which carries the origin into the point $(4, -2)$.

EXERCISES

Factor the following transformations, using the simplified method.

1. $x' = x + 3y$, $y' = -3x - y$.
2. $x' = 3x - 2y$, $y' = -2x + 3y$.
3. $x' = 5x + 11y$, $y' = 10x + 2y$.
4. $x' = 6x + 18y - 2$, $y' = 17x + y + 3$.
5. $x' = 111x + 4y + 1$, $y' = 52x + 78y - 7$.

8. Simple Shears. In a rectangle with its center at the origin and with its sides parallel to the coördinate axes, draw the lines parallel to the axis of x. Twist this rectangle as shown in the figure, leaving the line along the axis of x fixed and sliding each parallel line along itself into a new position. Thinking of the lines as representing the edges of a pack of cards or of a block of paper is an aid in visualizing the motion.

If the line one unit above the x-axis slides to the right through the distance k, then the line y units above the x-axis will evidently slide to the right through the distance ky, while a line which is, say, 2 units below the x-axis will slide a distance $2k$ to the left. In other words, the algebraic distance through which each line slides is equal to k times the algebraic distance of the line from the x-axis.

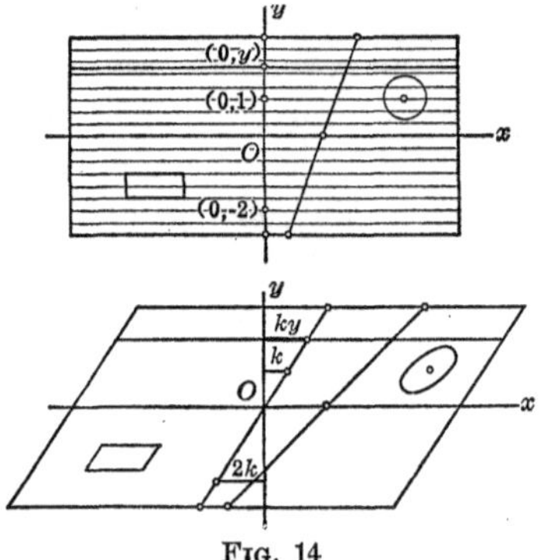

FIG. 14

This is true, also, of the motion of each point of the rectangle, since the lines slide as units. If, then, the whole plane is twisted according to this law, an arbitrary point (x, y) will be carried along a parallel to the axis of x through the algebraic distance ky. Hence

$$(1) \qquad \begin{cases} x' = x + ky, \\ y' = y, \end{cases}$$

are the equations of the transformation.

Thus far we have assumed that k is positive. It may equally well be negative. Then points above the axis of x are shifted to the left, and points below it to the right.

If the sliding were along parallels to the axis of y, the transformation would be

$$(2) \qquad \begin{cases} x' = x, \\ y' = lx + y, \end{cases}$$

where l is any constant, not zero.

These transformations are known as *simple shears*, and the motions which they generate are called *shearing motions*.

Example 1. Subject the curve

$$(3) \qquad y = x^3 + 2x$$

to the shear

$$x' = x, \qquad y' = -2x + y.$$

Here

(4) $$x = x', \qquad y = 2x' + y',$$

and (3) becomes

$$2x' + y' = x'^3 + 2x',$$

or

(5) $$y' = x'^3.$$

Conversely, the curve (5) is carried by the shear (4) into the curve (3). The shear (4) adds to the ordinate of a point (x', y') the amount $2x'$ equal to the corresponding ordinate of the line $y' = 2x'$. Consequently, the ordinates of the curve (3) can be obtained by adding to the ordinates of the line $y' = 2x'$ the corresponding ordinates of the curve (5), whose graph is known. Thus the curve (3) can be easily plotted. It is tangent to the line $y = 2x$ at the origin.

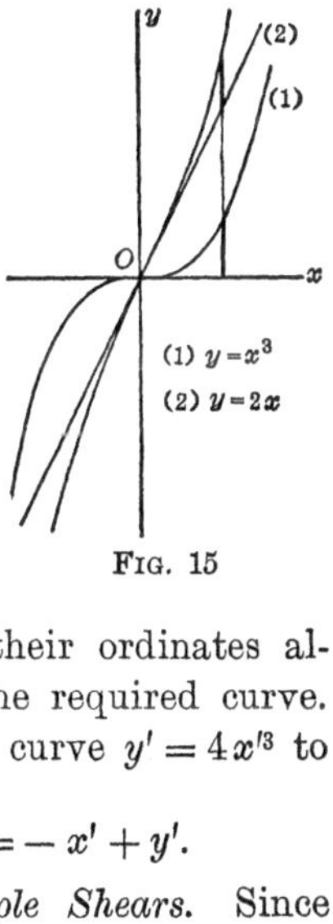

(1) $y = x^3$
(2) $y = 2x$

Fig. 15

Example 2. Construct the curve

$$y = 4x^3 - x.$$

This is done by plotting the line $y = -x$ and the curve $y = 4x^3$, and then adding their ordinates algebraically for a new ordinate — that of the required curve. The process is equivalent to subjecting the curve $y' = 4x'^3$ to the shear

$$x = x', \qquad y = -x' + y'.$$

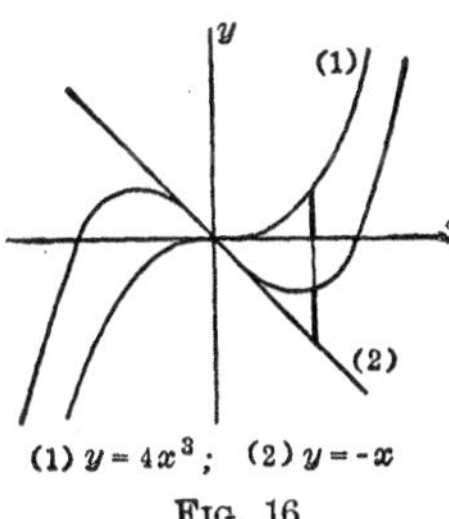

(1) $y = 4x^3$; (2) $y = -x$

Fig. 16

Properties of Simple Shears. Since the transformations (1) and (2) are special affine transformations (cf. § 6), it follows that simple shears carry straight lines into straight lines, parallel lines into parallel lines, and tangent curves into tangent curves. They do not in general preserve angles.

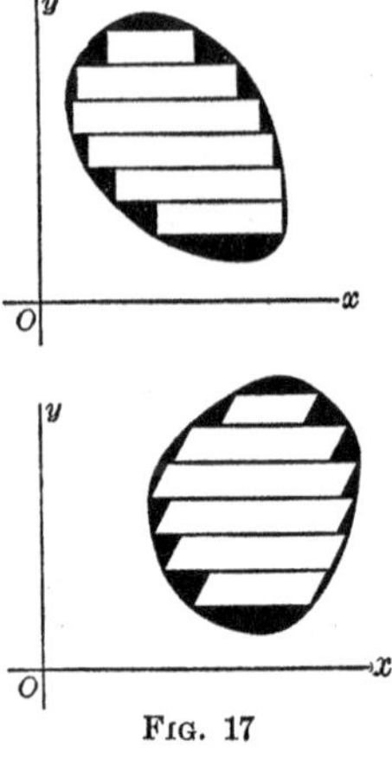

FIG. 17

Simple shears do, however, *preserve areas.* For, first, this is true for any rectangle whose base is parallel to the direction of shearing, since such a rectangle is carried into a parallelogram with the same lengths of base and altitude; cf. Fig. 14. Secondly, the area A of any other figure can be considered as the limit of the sum of the areas of rectangles of the type just described, which are inscribed in the figure as shown. But this sum is equal always to the sum of the areas of the corresponding parallelograms, whose limit is the area A' of the transformed figure. Consequently, $A = A'$, q. e. d.

EXERCISES

Construct the following curves.

1. $y = 2x^3 + \frac{3}{2}x.$
2. $y = 2x^3 - \frac{3}{2}x.$
3. $y = -x^3 + x.$
4. $x = y^3 + 3y.$
5. $3x = y^3 - 6y.$
6. $2x = -4y^3 - 3y.$

The same for the following curves, making use of a translation as well as a shear.

7. $y = x^3 + x - 2.$
8. $2y = 2x^3 - x + 3.$
9. $x = 2y^2 - 3y + 1.$
10. $3x = -6y^2 - 2y - 7.$

11. Construct the curve

$$y = 2x^3 - 6x^2 + 7x - 1,$$

beginning by putting the equation into the form:

$$y - b = 2(x - a)^3 + k(x - a).$$

The same for the curves:

12. $y = x^3 + 6x^2 + 10x + 7.$

13. $2x = 2y^3 - 6y^2 + 9y - 3.$

Factor the two following shears by the method of §§ 6, 7.

14. $x' = x + \frac{2}{3}\sqrt{3}\,y, \qquad y' = y.$

15. $x' = x, \qquad y' = -\frac{2}{3}\sqrt{3}\,x + y.$

9. Second Method of Factorization. Homogeneous Strains.

THEOREM. *The homogeneous affine transformation*

$$(1) \qquad \begin{cases} x' = a\,x + b\,y, \\ y' = a'x + b'y, \end{cases} \qquad \Delta \equiv ab' - a'b \neq 0,$$

can be factored into one-dimensional strains and simple shears, with the addition, in certain cases, of a reflection in one or both axes.

Case 1: *a and b' not both* 0. In proving the theorem we begin with the case in which a and b' are not both zero, and assume first that $a \neq 0$. A simple shear which suggests itself as a probable component of (1) is

$$(2) \qquad \begin{cases} x_1 = x + \dfrac{b}{a}y, \\ y_1 = y. \end{cases}$$

Eliminating x, y from (1) and (2), we obtain

$$(3) \qquad \begin{cases} x' = ax_1, \\ y' = a'x_1 + \dfrac{\Delta}{a}y_1. \end{cases}$$

This transformation suggests as a factor the second shear:

$$(4) \qquad \begin{cases} x' = x_2, \\ y' = kx_2 + y_2, \end{cases}$$

where the value of k is to be determined. Elimination of x', y' from (3) and (4) gives

$$(5) \qquad \begin{cases} x_2 = ax_1, \\ y_2 = (a' - ka)x_1 + \dfrac{\Delta}{a}y_1. \end{cases}$$

An obvious choice for k is that which makes $a' - ka = 0$; then $k = a'/a$.

We have now factored (1) into the transformations (2), (5), (4), where $k = a'/a$, namely into:

$$(6)\qquad \begin{cases} x' = x_2, \\ y = \dfrac{a'}{a} x_2 + y_2, \end{cases} \qquad \begin{cases} x_2 = a x_1, \\ y_2 = \dfrac{\Delta}{a} y_1, \end{cases} \qquad \begin{cases} x_1 = x + \dfrac{b}{a} y, \\ y_1 = y. \end{cases}$$

The first and last of these transformations are simple shears. The second can be factored into two one-dimensional strains, or, in case a or Δ/a or both are negative, into these and a reflection in one or the other or both axes. Thus the theorem is proved in this case.

The proof is similar in the case that $b' \neq 0$. The factors of (1) are

$$(7)\qquad \begin{cases} x' = x_2 + \dfrac{b}{b'} y_2, \\ y' = y_2, \end{cases} \qquad \begin{cases} x_2 = \dfrac{\Delta}{b'} x_1, \\ y_2 = b' y_1, \end{cases} \qquad \begin{cases} x_1 = x, \\ y_1 = \dfrac{a'}{b'} x + y. \end{cases}$$

Case 2: $a = b' = 0$. Here (1) becomes

$$(8)\qquad \begin{cases} x' = by, \\ y' = a'x. \end{cases} \qquad \Delta \equiv -a'b \neq 0.$$

This transformation can be factored into a rotation about the origin through 90°:

$$(9)\qquad \bar{x} = -y, \qquad \bar{y} = x,$$

and the transformation

$$x' = -b\bar{x}, \qquad y' = a\bar{y}.$$

It can be shown that the rotation (9) is the product of three simple shears, namely,

$$\begin{cases} \bar{x} = x_2, \\ \bar{y} = x_2 + y_2, \end{cases} \qquad \begin{cases} x_2 = x_1 - y_1, \\ y_2 = y_1, \end{cases} \qquad \begin{cases} x_1 = x, \\ y_1 = x + y, \end{cases}$$

and this completes the proof of the theorem.

Homogeneous Strains. The extension to space of the transformations (1) is given by the formulas

$$\begin{cases} x' = a\ x + b\ y + c\ z, \\ y' = a'\ x + b'\ y + c'\ z, \\ z' = a''x + b''y + c''z. \end{cases}$$

The three-dimensional case admits a treatment similar to the foregoing, and the results are like those obtained above.

Transformations of the form (1) or (9) are known in physics as *homogeneous strains.* They are of particular importance in the theory of elasticity. For, it can be shown that, if an elastic body, such as a solid piece of rubber or of steel, is slightly deformed from its normal shape, the displacement of its points can be represented to a high degree of approximation by a transformation of the form (1) or (9).

It is a fact, which we shall not attempt to prove, that a transformation (1) representing in the above sense a slight deformation cannot have a reflection in an axis as one of its component transformations.

One-dimensional strains — simple elongations and compressions — and simple shears are often given the single name, *simple strains.* Adopting this terminology, we can say: *Every homogeneous strain representing in the above sense a slight deformation can be generated by a succession of simple strains without reflections.*

EXERCISES

Factor the following homogeneous strains by the method of this paragraph.

1. The strain of Ex. 1, § 7. 2. The strain of Ex. 3, § 7.

3. $x' = 6y,\ y' = -2x + y.$ 4. $x' = 3x - 5y,\ y' = 4x + 3y.$

The following homogeneous strains represent slight displacements. Factor them and note that a reflection in an axis never appears as a component transformation.

5. $x' = 1.01x + .02y, \qquad y' = .03x + .98y.$

6. $x' = .9x - .1y, \qquad y' = .2x + 1.1y.$

7. $x' = (1 + \alpha)x + \beta y, \qquad y' = \gamma x + (1 + \delta)y,$

where $\alpha, \beta, \gamma, \delta$ are small quantities, such as .01 or $-.08$.

Factor the following transformations.

8. The transformation of Ex. 4, § 7.

9. $x' = -2x + y + 3, \qquad y' = 3x - 2y + 1.$

EXERCISES ON CHAPTER XV

1. Prove that the product of any two affine transformations is an affine transformation.

Definition. The general affine transformation (1), § 6, is called *non-singular*, if

$$\Delta \equiv ab' - a'b \neq 0;$$

if $\Delta = 0$, it is called *singular*. The expression Δ is known as the *determinant* (cf. Ch. XVI) of the transformation.

2. The inverse of any non-singular affine transformation is non-singular. This was proved incidentally on p. 344 in the case of a homogeneous transformation. Prove it in the general case.

3. Show that the product of two non-singular affine transformations is non-singular.

Suggestion. Prove that the determinant of the product transformation is the product of the determinants of the given transformations.

4. The transformation,

$$x' = 2x - y, \qquad y' = 4x - 2y,$$

is singular. Verify this and show that the transformation carries all the points of the plane into points of the line $2x' - y' = 0$. Has the transformation any inverse?

5. The product of the general rotation about the origin, followed by the general translation, is a transformation known as the general *rigid motion*. Find its equations.

Ans. $x' = x\cos\theta - y\sin\theta + a$; $y' = x\sin\theta + y\cos\theta + b.$

6. The product of the transformation of similitude of § 3 and the general rigid motion of the preceding exercise is known as *the general transformation of similitude*. Find its

equations, and show that it is identical with the general isogonal transformation for which $b' = a$, $a' = -b$ (§ 6, Th. 2).

7. A non-singular affine transformation carries the four collinear points P_1, P_2, Q_1, Q_2, into the four collinear points P'_1, P'_2, Q'_1, Q'_2. Prove that, if Q_1, Q_2 divide P_1, P_2 harmonically, Q'_1, Q'_2 will divide $P'_1P'_2$ harmonically.

Suggestion. Prove the theorem first for the transformations considered in §§ 1–5.

8. Find the equations of a rotation of the plane about the point (x_0, y_0) through the angle θ.

9. The plane is stretched uniformly in all directions away from the point (x_0, y_0). Find the equations representing the transformation.

10. Deduce the equations of the reflection in the line

$$Ax + By + C = 0.$$

11. Deduce the formulas representing a one-dimensional strain away from the line of Ex. 10.

12. Find the equations of the simple shear which leaves each point of the line of Ex. 10 fixed.

13. Let the simple shear (1), § 8, be factored into the three transformations of § 6, namely (10), (16), and the inverse of (11). Prove that $\sin 2\theta' = \sin 2\theta$, but that the only allowable solutions of this equation are $\theta' = 90° - \theta$ and $\theta' = 270° - \theta$, together with those equivalent to them. Show that, if the first of these solutions is chosen, $\alpha = \tan\theta$, $\delta = \cot\theta$, whereas, if the second is taken, $\alpha = -\tan\theta$, $\delta = -\cot\theta$. Prove that, in either case, $2\cot 2\theta = k$.

Suggestion. Form the product of the three transformations and demand that it be identical with the transformation (1), § 8.

CHAPTER XVI

DETERMINANTS AND THEIR APPLICATIONS

I. Determinants

1. Simultaneous Linear Equations. The solution of the simultaneous equations,

$$(1) \qquad \begin{aligned} a_1x + b_1y &= k_1, \\ a_2x + b_2y &= k_2, \end{aligned}$$

is

$$(2) \qquad x = \frac{k_1b_2 - k_2b_1}{a_1b_2 - a_2b_1}, \qquad y = \frac{a_1k_2 - a_2k_1}{a_1b_2 - a_2b_1},$$

provided $a_1b_2 - a_2b_1 \neq 0$.*

If we have three simultaneous linear equations in three unknowns,

$$(3) \qquad \begin{aligned} a_1x + b_1y + c_1z &= k_1, \\ a_2x + b_2y + c_2z &= k_2, \\ a_3x + b_3y + c_3z &= k_3, \end{aligned}$$

and first eliminate z, obtaining two equations in x and y, and then from these equations eliminate y, we find, as the value of x,

$$(4) \qquad x = \frac{k_1b_2c_3 + k_2b_3c_1 + k_3b_1c_2 - k_3b_2c_1 - k_2b_1c_3 - k_1b_3c_2}{a_1b_2c_3 + a_2b_3c_1 + a_3b_1c_2 - a_3b_2c_1 - a_2b_1c_3 - a_1b_3c_2}.$$

Similarly, we can find the values of y and z. These will also be in the form of quotients, with the same denominator as in (4),

* If $a_1b_2 - a_2b_1 = 0$ (but a_1 and b_1, and a_2 and b_2, are not both zero), the two straight lines represented by equations (1) are either parallel or coincident (Ch. 2, § 10, Ths. 3, 5); in the former case the equations have *no* solution, in the latter, *infinitely many* solutions. Both cases are exceptional to the general case, $a_1b_2 - a_2b_1 \neq 0$, in which the solution (2) is *unique*.

and the solution is valid subject to the condition that this denominator is not zero.

2. Two- and Three-Rowed Determinants. The expressions in the numerators and denominators of the quotients in (2) and (4) are of so great importance that they are given a name. They are called *determinants*, — those in (2), *determinants of the second order*, and those in (4), *determinants of the third order*. A determinant, then, is a *polynomial* of the above type.

The determinant of the second order,

$$a_1b_2 - a_2b_1,$$

can easily be remembered by means of the diagram

(1)

in which the lines and the signs show how the terms of the determinant are to be obtained.

The diagram

(2)

fulfills the same purpose for the determinant of the third order,

$$a_1b_2c_3 + a_2b_3c_1 + a_3b_1c_2 - a_3b_2c_1 - a_2b_1c_3 - a_1b_3c_2.$$

The four quantities a_1, a_2, b_1, b_2, arranged in a square as in (1), form what is known as a *square array of the second order.* Similarly, the system of nine quantities, which forms the basis of the diagram (2), is known as a *square array of the third order.* The square array is not itself the determinant. It is merely a convenient arrangement of the given four, or nine, quantities, from which the value of the determinant can be written down. However, it is common practice to use, as a symbol or notation for the determinant, the square array inclosed between vertical bars, and to write, accordingly:

$$(3)\qquad \begin{vmatrix} a_1 & b_1 \\ a_2 & b_2 \end{vmatrix} = a_1b_2 - a_2b_1,$$

$$(4)\qquad \begin{vmatrix} a_1 & b_1 & c_1 \\ a_2 & b_2 & c_2 \\ a_3 & b_3 & c_3 \end{vmatrix} = a_1b_2c_3 + a_2b_3c_1 + a_3b_1c_2 - a_3b_2c_1 - a_2b_1c_3 - a_1b_3c_2.$$

These symbols for the determinants are sometimes abbreviated still further. Instead of the first, we often find $|a_1\ b_2|$ or merely $|a\ b|$, and for the second, $|a_1\ b_2\ c_3|$ or $|a\ b\ c|$.*

The solution of the equations (1), § 1, we can now write in the form

$$x = \frac{\begin{vmatrix} k_1 & b_1 \\ k_2 & b_2 \end{vmatrix}}{\begin{vmatrix} a_1 & b_1 \\ a_2 & b_2 \end{vmatrix}}, \qquad y = \frac{\begin{vmatrix} a_1 & k_1 \\ a_2 & k_2 \end{vmatrix}}{\begin{vmatrix} a_1 & b_1 \\ a_2 & b_2 \end{vmatrix}},$$

or more compactly,

$$(5)\qquad x = \frac{|k\ b|}{|a\ b|}, \qquad y = \frac{|a\ k|}{|a\ b|}, \qquad |a\ b| \neq 0.$$

The solution of the equations (3), § 1, becomes

$$(6)\quad x = \frac{|k\ b\ c|}{|a\ b\ c|}, \quad y = \frac{|a\ k\ c|}{|a\ b\ c|}, \quad z = \frac{|a\ b\ k|}{|a\ b\ c|}, \qquad |a\ b\ c| \neq 0.$$

The value of x is as given by (4), § 1. The determinant $|a\ b\ c|$ in the denominator is evidently the determinant of the coefficients of x, y, z in the given equations, and the determinant $|k\ b\ c|$ in the numerator is obtained from $|a\ b\ c|$ by replacing, respectively, a_1, a_2, a_3 — the coefficients of x — by k_1, k_2, k_3 — the constant terms. Similarly, the numerator $|a\ k\ c|$ of the value of y is obtained from $|a\ b\ c|$ by replacing

* The vertical bars must not be confused with the absolute value signs. They have nothing to do with these. The context will always show which meaning is intended.

the b's — the coefficients of y — by the k's; and likewise for z.*

The four, or nine, quantities from which the determinant is formed are known as the *elements* of the determinant. The rows and columns in which they are arranged are called the *rows* and *columns* of the determinant. The diagonal containing the elements a_1, b_2 (c_3) is the *principal diagonal*; the other, the *secondary diagonal*. The determinants are often called *two-* and *three-rowed determinants*, instead of determinants of the second and third orders.

EXERCISES

Evaluate the following determinants.

1. $\begin{vmatrix} 2 & 3 \\ 3 & 5 \end{vmatrix}$. **2.** $\begin{vmatrix} 1 & 3 \\ -2 & 7 \end{vmatrix}$. **3.** $\begin{vmatrix} 0 & -4 \\ 2 & 7 \end{vmatrix}$. **4.** $\begin{vmatrix} 2A & B \\ B & 2C \end{vmatrix}$.

5. $\begin{vmatrix} 3 & 5 & 2 \\ 2 & 1 & 3 \\ 4 & 3 & 7 \end{vmatrix}$. **6.** $\begin{vmatrix} 2 & 4 & 3 \\ -3 & 1 & 2 \\ 1 & 5 & -6 \end{vmatrix}$. **7.** $\begin{vmatrix} 2 & -1 & 3 \\ -1 & 5 & 0 \\ 0 & 3 & 2 \end{vmatrix}$.

Solve the following simultaneous equations by means of determinants. Check your answers.

Remark. The constant terms in (1) and (3), § 1, are on the *right-hand* sides of the equations. The formulas (5) and (6), for the solution of the equations, are subject, then, to the arrangement of the equations in this form.

8. $2x - y = 3,$
$3x + 4y = 5.$

9. $5x + 3y - 2 = 0,$
$4x + 2y + 3 = 0.$

10. $3x + y + 2z = 2,$
$2x + 4y + z = 5,$
$4x + 5y + 3z = 6.$

11. $2x - 3y + z = -4,$
$4x + 2y - 3z = 11,$
$3x - y + 2z = 5.$

12. $3x + 2y - 5z - 1 = 0,$
$5x - 8y + z - 9 = 0,$
$3x - 2y - 7 = 0.$

13. $4x + 5y + 2z - 7 = 0,$
$2x - 4y + z + 9 = 0,$
$5x - 2y - 3z + 5 = 0.$

* Formulas (6) are proved later, in § 8.

3. Determinants of the Fourth and Higher Orders. Given sixteen quantities arranged in a square array:

$$(1) \qquad \begin{matrix} a_1 & b_1 & c_1 & d_1 \\ a_2 & b_2 & c_2 & d_2 \\ a_3 & b_3 & c_3 & d_3 \\ a_4 & b_4 & c_4 & d_4 \end{matrix}$$

What shall we mean by the determinant symbolized or denoted by

$$\begin{vmatrix} a_1 & b_1 & c_1 & d_1 \\ a_2 & b_2 & c_2 & d_2 \\ a_3 & b_3 & c_3 & d_3 \\ a_4 & b_4 & c_4 & d_4 \end{vmatrix},$$

or, more simply, by $|a_1\ b_2\ c_3\ d_4|$ or $|a\ b\ c\ d|$?

If we were to proceed as before, we should write down four simultaneous linear equations in four unknowns, with the elements of (1) as the coefficients of the unknowns and k_1, k_2, k_3, k_4 as the constant terms, and then solve the equations. The value of each unknown would be a quotient, and all four quotients would have the same denominator, which we should then define as the determinant $|a\ b\ c\ d|$. As a matter of fact, this denominator and each of the numerators contains 24 terms. The prospect of solving the equations is, then, forbidding.

Why not form the products suggested by a diagram based on (1), similar to the diagram for the three-rowed determinant, prefix the proper signs, and call the result the determinant? Unfortunately this method yields but 8 terms, whereas according to our prediction the determinant, properly defined, contains 24.

We adopt here a new method of attack. Let us inspect more closely the relationship between the square arrays of orders two and three and the corresponding determinants. Consider a specimen term of the determinant (4), § 2. It contains just one a, just one b, and just one c; furthermore, each of the subscripts 1, 2, 3 appears just once. In other words, *the term is the product of three elements, one from each row and one from each column of the square array.* Moreover, every product of this type is present as some term in the determinant, as can be shown by writing down all such products and comparing them with the terms of the determinant.

By analogy, then, to form the determinant $|a\ b\ c\ d|$, we should write down all the products of elements of (1), each of which contains just one factor from each row and just one factor from each column of (1), that is, all the products of the form $a_i b_j c_k d_l$, where i, j, k, l are the numbers 1, 2, 3, 4 in all possible orders. There are 24 such products. For, we can choose the first factor, say from the column of a's, in four ways — from any one of the four rows; and then the second factor, say from the column of b's, in three ways — from any one of the three remaining rows; and the third factor, in two ways; the fourth is then uniquely determined. The number of possible products is, therefore, $4 \cdot 3 \cdot 2 \cdot 1 = 4! = 24$.

It remains to determine the signs to be given to the 24 products. Toward this end, let us write down the subscripts of the terms of (4), § 2, in the order in which they occur, *when the letters a, b, c are in their natural order.* For the terms with plus signs we have

$$1\ 2\ 3, \qquad 2\ 3\ 1, \qquad 3\ 1\ 2,$$

and for the terms with minus signs,

$$3\ 2\ 1, \qquad 2\ 1\ 3, \qquad 1\ 3\ 2.$$

The first set 1 2 3 is normal. In the second set, 2 3 1, 2 and 3 each precede 1, and we say that there are two *inversions* from the normal order. In 3 1 2, 3 precedes 1 and 2, — again two inversions. In the three sets for the negative terms the number of inversions is respectively three, one, and one.

It appears, then, that *the number of inversions in the set of subscripts for a term with a plus sign is even* (*or zero*), *whereas for a term with a minus sign, this number is always odd.*

Proceeding according to this rule, we should give to each of the 24 products, $a_i b_j c_k d_l$, formed from (1) a plus sign or a minus sign, according as $i\ j\ k\ l$ presents an even or an odd number of inversions from the normal order 1 2 3 4. Thus the product $b_2d_1a_3c_4$ would be taken as plus, since, when the factors are arranged in the order of the letters, viz. — $a_3b_2c_4d_1$, the number of inversions in the subscripts 3 2 4 1 is even, namely 4. The product $a_4b_2c_3d_1$ would be taken as minus, since the number of inversions in 4 2 3 1 is odd, namely 5.

We can now give a complete definition of the determinant of the fourth order.

Definition. *Form all the products of elements of* (1) *which contain just one factor from each row and one factor from each column of* (1)*; to each product* $a_ib_jc_kd_l$ *prefix a plus sign or a minus sign, according as the number of inversions of* $i\ j\ k\ l$ *from the normal order* 1 2 3 4 *is even or odd. The sum of the products, thus signed, is the determinant.*

Determinants of the fifth, sixth, and higher orders are similarly defined. Let the student think through the definition for a five-rowed determinant, and let him show, also, that in the case of two-and three-rowed determinants the definition yields precisely the expressions which were defined as these determinants in § 2.

The signed products which make up a determinant are known as the *terms* of the determinant. Thus, $+a_3b_2c_4d_1$ and $-a_4b_2c_3d_1$ are terms of $|a\ b\ c\ d|$.

EXERCISES

1. What is the number of inversions of each of the following orders, from the normal order?

(*a*) 3 1 4 2; (*c*) 2 5 3 1 4; (*e*) 3 1 6 4 5 2;
(*b*) 2 4 3 1; (*d*) 4 3 5 2 1; (*f*) 6 5 4 3 2 1.

2. Write out all the terms of $|a\ b\ c\ d|$. To how many products have you prefixed plus signs? To how many, minus signs?

3. How many terms has a determinant of the fifth order? Prove your answer.

4. The same for a determinant of the nth order.

5. Show that the sign to be prefixed to the product of the elements of the principal diagonal is always the plus sign, no matter what the order of the determinant.

4. Evaluation of a Determinant by Minors. Fix the attention on a particular element of a determinant Δ. Cross out the row and column in which this element stands. There will remain a determinant of order one less than that of Δ. This determinant is known as the *minor* of the element chosen.

For example, the minor of a_2 in the determinant

$$(1) \qquad \begin{vmatrix} a_1 & b_1 & c_1 \\ a_2 & b_2 & c_2 \\ a_3 & b_3 & c_3 \end{vmatrix}$$

is the determinant

$$A_2 = \begin{vmatrix} b_1 & c_1 \\ b_3 & c_3 \end{vmatrix}.$$

Consider the product a_2A_2. The terms in this product are $a_2b_1c_3$ and $-a_2b_3c_1$, and by (4), § 2, these are terms of (1) except for sign; moreover, they are, except for sign, all the terms of (1) which contain a_2.

Again, the terms of b_2B_2, where B_2 is the minor of b_2 in (1), are $a_1b_2c_3$ and $-a_3b_2c_1$. These are precisely terms of (1) and, in fact, all the terms of (1) which contain b_2.

In general, let m be an element of a determinant Δ and let M be its minor. Then the terms of the product mM are terms of Δ, except perhaps for sign; furthermore, they are, except perhaps for sign, all the terms of Δ which contain m.

For, if we take the factor m from a term of Δ which contains m, the product which remains contains just one element from each row and column other than the row and column in which m stands, and is, therefore, a product occurring in the determinant M. And, if we take the factor m from *all* the terms of Δ containing m, the products which remain are *all* the products of the type described and hence are *all* the products occurring in M, q. e. d.

It will be shown later (§ 7) that the terms of mM as they stand, or the terms of mM with *all* their signs changed, are *precisely* terms of Δ, according as the sum of the number of the row and the number of the column in which m stands is even or odd. Assuming this, we can now state the theorem:

THEOREM 1. *If m is the element in the i-th row and j-th column of Δ, and M is its minor, $+mM$ or $-mM$, according as $i+j$ is even or odd, consists of all the terms of Δ which contain m.*

Thus, in the case of the element a_2 of (1), $i=2, j=1$, and $i+j=3$; accordingly, $-a_2A_2$ gives all the terms of (1) containing a_2. For b_2, $i=2$, $j=2$, and $i+j=4$, and so $+b_2B_2$ consists of all the terms of (1) containing b_2.* Similarly, if C_2 is the minor of c_2, $-c_2C_2$ consists of all the terms of (1) containing c_2.

The sum

$$-a_2A_2+b_2B_2-c_2C_2 \tag{2}$$

is precisely the value of the determinant (1). For, it consists of all the terms of (1) containing a_2 or b_2 or c_2, *i.e.* containing an element of the second row, and every term of (1) contains such an element. The student should also verify the statement by comparing the terms of (2), when expanded, with those of (1).

In (2) we have the sum of the products of the elements of the second row by their minors, each product having the proper sign according to Theorem 1. We say that (2) is the *evaluation* or *expansion of the determinant* (1) *by the minors of the second row.*

Similarly, the sum,

$$c_1C_1-c_2C_2+c_3C_3, \tag{3}$$

of the products of the elements of the third column of (1) by their minors, where the signs have been determined by Theorem

* Compare these results with those obtained directly at the beginning of the section.

1, is precisely the determinant (1). We speak of (3) as the evaluation of (1) by the minors of the third column.

The reasoning here is perfectly general, applying to a determinant of any order and to any row or column of the determinant. The result we summarize as follows:

EVALUATION OF A DETERMINANT BY THE MINORS OF A ROW OR A COLUMN. *Single out a row or a column of a determinant. Multiply each element of it by the minor of the element and prefix to the product the proper sign, as determined by Theorem* 1. *The sum of the signed products is the determinant.*

We now have a feasible means of finding the values of determinants of the fourth and higher orders. For example, the determinant

$$(4)\qquad \begin{vmatrix} 2 & 5 & -2 & 8 \\ 4 & -9 & 3 & -7 \\ -3 & 6 & -4 & 4 \\ -1 & 4 & -3 & 5 \end{vmatrix},$$

evaluated by the minors of the first row, is equal to

$$2\begin{vmatrix} -9 & 3 & -7 \\ 6 & -4 & 4 \\ 4 & -3 & 5 \end{vmatrix} - 5\begin{vmatrix} 4 & 3 & -7 \\ -3 & -4 & 4 \\ -1 & -3 & 5 \end{vmatrix} + (-2)\begin{vmatrix} 4 & -9 & -7 \\ -3 & 6 & 4 \\ -1 & 4 & 5 \end{vmatrix} - 8\begin{vmatrix} 4 & -9 & 3 \\ -3 & 6 & -4 \\ -1 & 4 & -3 \end{vmatrix}.$$

When the value of each of the three-rowed determinants is computed, by the above method or by that of the diagram of § 2, this becomes

$$2(44) - 5(-34) - 2(-1) - 8(19),$$

which yields finally, as the value of the given determinant, 108.

EXERCISES

1. Given the determinant $|a\ b\ c|$ and the three quantities k_1, k_2, k_3. Prove that

$$k_1A_1 - k_2A_2 + k_3A_3 = |k\ b\ c|.$$

2. (Generalization of Ex. 1.) Given a determinant Δ of the nth order and the n quantities $k_1, k_2, \cdots, k_n$. Single out a column (or row) of Δ, form the minors of its elements and prefix to each the sign prescribed by Th. 1. Multiply each signed minor by the corresponding k and take the sum of these products. Prove that this sum is equal to the determinant obtained from Δ by replacing the column (or row) in question by $k_1, k_2, \cdots, k_n$.

By the method of this section, evaluate the following determinants

3. That of Ex. 5, § 2.

4. That of Ex. 7, § 2.

5. $$\begin{vmatrix} 2 & 3 & 1 & 5 \\ 5 & 2 & -2 & 1 \\ 3 & 4 & 6 & 2 \\ -1 & 5 & 2 & 3 \end{vmatrix}.$$

6. $$\begin{vmatrix} 3 & -2 & 5 & 1 \\ 4 & -3 & 7 & -2 \\ -6 & 2 & -3 & 0 \\ 5 & 3 & -2 & 2 \end{vmatrix}.$$

7. $$\begin{vmatrix} 1 & 0 & -2 & 0 & 0 \\ 2 & -1 & 4 & 3 & 2 \\ -5 & 2 & 0 & 2 & -1 \\ 0 & 3 & -1 & 4 & 5 \\ 2 & 0 & 3 & 1 & 4 \end{vmatrix}.$$

5. Simplified Evaluation by Minors. Given the determinant,

$$\begin{vmatrix} 3 & 2 & 0 & 1 \\ 2 & -1 & 3 & 4 \\ 4 & 5 & 0 & 2 \\ 2 & 6 & 0 & 3 \end{vmatrix}.$$

Three of the four elements in the third column are zero. Accordingly, if we expand the determinant by the minors of the third column, three of the four resulting products have zero factors and drop out, so that there is left, merely,

$$-3\begin{vmatrix} 3 & 2 & 1 \\ 4 & 5 & 2 \\ 2 & 6 & 3 \end{vmatrix}.$$

Hence $-3(7) = -21$ is the value of the determinant.

It is clear from this example that a determinant which has the property that all but one of the elements in some row or in some column are zero is very simply evaluated. Consequently, if a determinant which has not this property can be transformed into an equal determinant which has the property, a simple method is at hand for the evaluation of all determinants.

The transformation in question is always possible. It is based on the following theorem.

THEOREM 2. *If the elements of a row (or column) of a determinant are each multiplied by the same quantity and are then added to the corresponding elements of a second row (or column), the value of the determinant is unchanged.*

Let us first try to appreciate the value of the theorem, postponing the proof until later. Consider the determinant (4) of § 4, namely,

$$(1)\qquad \begin{vmatrix} 2 & 5 & -2 & 8 \\ 4 & -9 & 3 & -7 \\ -3 & 6 & -4 & 4 \\ -1 & 4 & 3 & 5 \end{vmatrix}.$$

By application of the theorem we proceed to transform this determinant into an equal determinant with the first three elements of the first column all zero.

Rewrite (1), putting in, to begin with, only the last row:

$$(2)\qquad \begin{vmatrix} & & & \\ & & & \\ & & & \\ -1 & 4 & -3 & 5 \end{vmatrix}.$$

Multiply the elements of the last row of (1) by 2 and add the numbers obtained to the elements of the first row of (1); the result is 0, 13, -8, 18 as a new first row, to be put into (2). Similarly, multiply the last row by 4 and add to the second row, thus getting 0, 7, -9, 13 as a new second row, to be written in (2). Finally, the last row multiplied by -3 and added to the third row gives 0, – 6, 5, – 11 as a new third row. Thus (2) has become

$$(3) \qquad \begin{vmatrix} 0 & 13 & -8 & 18 \\ 0 & 7 & -9 & 13 \\ 0 & -6 & 5 & -11 \\ -1 & 4 & -3 & 5 \end{vmatrix},$$

a determinant whose value, by Th. 2, is equal to that of (1).

Expansion of (3) by the minors of the first column gives

$$-(-1)\begin{vmatrix} 13 & -8 & 18 \\ 7 & -9 & 13 \\ -6 & 5 & -11 \end{vmatrix}.$$

The evaluation of this three-rowed determinant by means of the schematic diagram of § 2 involves the multiplication of large numbers. This may be avoided as follows. Apply Theorem 2 so as to introduce 1 or – 1 as an element; for instance, by multiplying the last row through by 2 and adding to the first row:

$$\begin{vmatrix} 1 & 2 & -4 \\ 7 & -9 & 13 \\ -6 & 5 & -11 \end{vmatrix}.$$

Now rewrite the determinant, putting in just the first column. Multiply the first column by – 2 and add to the second column, for a new second column; and multiply the first column by 4 and add to the third column, for a new third column. The result is the equal determinant

$$\begin{vmatrix} 1 & 0 & 0 \\ 7 & -23 & 41 \\ -6 & 17 & -35 \end{vmatrix},$$

which, on expansion by the minors of the first row, has the value

$$\begin{vmatrix} -23 & 41 \\ 17 & -35 \end{vmatrix},$$

or $$(-23)(-35)-17\cdot 41 = 805 - 697 = 108.*$$

Thus the determinant (1) has the value 108.

* Here, too, the long multiplications could be replaced by simpler ones, through the application of the above method of reduction.

EXERCISES

Evaluate, by the above method, the following determinants.

1. $\begin{vmatrix} 8 & 9 & 7 \\ 7 & 6 & 5 \\ 6 & 7 & 4 \end{vmatrix}$.
2. $\begin{vmatrix} 10 & 12 & 15 \\ 21 & 26 & 30 \\ 30 & 33 & 37 \end{vmatrix}$.
3. $\begin{vmatrix} 5 & 7 & -3 \\ 8 & 9 & 5 \\ -6 & 0 & 1 \end{vmatrix}$.
4. That of Ex. 5, § 4.
5. That of Ex. 6, § 4.
6. That of Ex. 7, § 4.

6. Fundamental Properties of Determinants. The following theorems are fundamental in the transformation and evaluation of determinants. They lead up to a simple proof of Theorem 2.

THEOREM 3. *If all the elements of a row (or column) are multiplied by the same quantity, the value of the determinant is multiplied by this quantity.*

For, each term of a determinant Δ contains as a factor just one element from the row (or column) in question, and consequently, when the elements of this row (or column) are all multiplied by the same quantity, m, the terms of Δ will all be multiplied by m, and the resulting determinant will have the value $m\Delta$.

The theorem is often of use in evaluating a determinant. For example, if Δ is

$$\begin{vmatrix} 35 & 14 \\ 20 & 42 \end{vmatrix},$$

then

$$\Delta = 7\begin{vmatrix} 5 & 2 \\ 20 & 42 \end{vmatrix} = 7 \cdot 5\begin{vmatrix} 1 & 2 \\ 4 & 42 \end{vmatrix} = 7 \cdot 5 \cdot 2\begin{vmatrix} 1 & 1 \\ 4 & 21 \end{vmatrix} = 70 \cdot 17 = 1190.$$

THEOREM 4. *If two rows (or columns) of a determinant are identical, element for element, the determinant has the value zero.*

Let us assume that it is two rows which are identical. The proof is similar, if it is two columns.

For a determinant of the second order, the theorem is obvious:

$$\begin{vmatrix} a & b \\ a & b \end{vmatrix} = ab - ab = 0.$$

Consider, next, a determinant of the third order, with two rows identical. Expand the determinant by the minors of the third, or odd, row. Each of these minors has its two rows identical and is, therefore, zero, since the theorem has been proved for two-rowed determinants. Consequently, the given determinant is zero.

Similarly, having proved the theorem for three-rowed determinants, we can prove it for a four-rowed determinant. For, we have but to expand the four-rowed determinant by the minors of a row which is not one of the two identical rows. This expansion will have the value zero, since each of the minors in question is a three-rowed determinant with two identical rows.

The process perpetuates itself. Hence the theorem is true for a determinant of any order.

The method of proof used here is known as *mathematical induction*. The fact that the theorem is true for a two-rowed determinant leads up to its truth for a three-rowed determinant, etc.

Corollary. *If the elements of two rows (or columns) of a determinant are proportional, the determinant has the value zero.*

For, each element of one of the two rows (or columns) in question is by hypothesis a multiple, m, of the corresponding element of the other. Thus m can be taken out from the first of the two rows (or columns) as a factor (Th. 3). The two rows (or columns) are then identical, and Theorem 4 can be applied.

Theorem 5. *If each element of a row (or column) is the sum of two quantities, the determinant can be written as the sum of two determinants.*

Denote the determinant by Δ and the elements of the column (or row) in question by $m_1 + m_1'$, $m_2 + m_2'$, $\cdots$. Denote by $\bar{\Delta}$ the determinant obtained by replacing all the m''s in Δ by zeros, and by $\bar{\Delta}'$ the determinant obtained by replacing all the

m's in Δ by zeros. We shall prove that

$$\Delta = \bar{\Delta} + \bar{\Delta}'.$$

For example,

$$\begin{vmatrix} m_1 + m_1' & b_1 & c_1 \\ m_2 + m_2' & b_2 & c_2 \\ m_3 + m_3' & b_3 & c_3 \end{vmatrix} = \begin{vmatrix} m_1 & b_1 & c_1 \\ m_2 & b_2 & c_2 \\ m_3 & b_3 & c_3 \end{vmatrix} + \begin{vmatrix} m_1' & b_1 & c_1 \\ m_2' & b_2 & c_2 \\ m_3' & b_3 & c_3 \end{vmatrix}.$$

Proof. Every term of Δ, since it contains just one element from the column (or row) under discussion, is the sum of two quantities, one containing an m and the other an m'. All the quantities containing m's form the determinant $\bar{\Delta}$, and all those containing m''s, the determinant $\bar{\Delta}'$. Hence, $\Delta = \bar{\Delta} + \bar{\Delta}'$.

Or, expand Δ by the elements of the column (or row) in question, denoting the minors of these elements by $M_1, M_2, \cdots$. The result is

$$(1) \quad \begin{aligned} \Delta &= \pm[(m_1 + m_1')M_1 - (m_2 + m_2')M_2 + \cdots] \\ &= \pm[m_1M_1 - m_2M_2 + \cdots] \pm [m_1'M_1 - m_2'M_2 + \cdots]. \end{aligned}$$

The values of $\bar{\Delta}$ and $\bar{\Delta}'$ can be obtained from (1) by replacing, first, the m''s, and then the m's, by zeros:

$$\bar{\Delta} = \pm[m_1M_1 - m_2M_2 + \cdots], \qquad \bar{\Delta}' = \pm[m_1'M_1 - m_2'M_2 + \cdots].$$

Hence $$\Delta = \bar{\Delta} + \bar{\Delta}'.$$

The proof of Theorem 2, § 5, is now simple. The determinant Δ', which is obtained from the given determinant Δ by adding to the elements $m_1, m_2, \cdots$ of, let us say, a column the corresponding elements $p_1, p_2, \cdots$ of a second column, each multiplied by a quantity k, contains the column $m_1 + kp_1$, $m_2 + kp_2, \cdots$. Hence Δ' equals the sum of two determinants, the first of which is Δ. The second has the two columns $kp_1, kp_2, \cdots$ and $p_1, p_2, \cdots$ and is therefore zero (Th. 4, Cor.). Consequently, $\Delta' = \Delta$.

For example, if Δ is the three-rowed determinant (4), § 2, and to the elements of the second column are added those of the first, each multiplied by k, we have

$$\Delta' = \begin{vmatrix} a_1 & b_1 + ka_1 & c_1 \\ a_2 & b_2 + ka_2 & c_2 \\ a_3 & b_3 + ka_3 & c_3 \end{vmatrix} = \begin{vmatrix} a_1 & b_1 & c_1 \\ a_2 & b_2 & c_2 \\ a_3 & b_3 & c_3 \end{vmatrix} + \begin{vmatrix} a_1 & ka_1 & c_1 \\ a_2 & ka_2 & c_2 \\ a_3 & ka_3 & c_3 \end{vmatrix} = \Delta + 0 = \Delta.$$

EXERCISES

1. Prove the theorem: *If all the elements of a row (or column) are zero, the determinant has the value zero.*

2. Given the determinant $|a\ b\ c|$. Using Ex. 1, § 4, show that

$$b_1A_1 - b_2A_2 + b_3A_3 = 0.$$

3. (Generalization of Ex. 2.) *If to the minors of a column (or row) of a determinant are prefixed the signs prescribed by Theorem* 1, *and if each signed minor is then multiplied by the corresponding element of a different column (or row), the sum of the resulting products has the value zero.* Prove this theorem. Cf. Ex. 2, § 4.

Evaluate the following determinants, making as much use as possible of Ths. 2–5.

4. $\begin{vmatrix} 8 & 6 & 6 \\ 2 & 9 & 2 \\ 6 & 6 & 1 \end{vmatrix}$.   5. $\begin{vmatrix} 2 & 3 & -2 \\ -4 & 8 & 24 \\ 16 & 20 & 16 \end{vmatrix}$.   6. $\begin{vmatrix} 4 & 15 & -6 \\ -6 & 12 & 9 \\ 2 & 38 & -3 \end{vmatrix}$.

7. Interchanges of Rows and of Columns. Given the first n integers in natural order:

(1) $$1\ 2\ 3 \cdots l\ \ l+1 \cdots n;$$

in this order (1) interchange two *successive* integers, l and $l+1$:

(2) $$1\ 2\ 3 \cdots l+1\ \ l \cdots n.$$

Consider, now, the n integers in an arbitrary order,

(3) $$p\ q\ r \cdots\cdots\cdots t,$$

and compare the number of inversions of this order from the order (2) with the number of its inversions from the order (1).

Each pair of integers is in the same order in (2) as it was in (1), except the pair l, $l+1$. Consequently, if a pair of integers in (3), not the pair l, $l+1$, presents an inversion from the order (1), it also presents an inversion from the order (2), and vice versa. But the pair l, $l+1$ in (3) presents an inversion from *one* of the orders, (1) and (2), and *not* from the other. Hence, we conclude:

LEMMA 1. *The total number of inversions from the order* (2), *which* (3) *presents, differs by one from the total number of inversions from the order* (1), *which it presents.*

For example, 2 3 1 4 5 has two inversions from the natural order 1 2 3 4 5 and three from the order 1 2 4 3 5.

In the general determinant of the nth order,

$$\Delta = \begin{vmatrix} a_1 & b_1 & c_1 & \cdot & \cdot & \cdot & \cdot & k_1 \\ a_2 & b_2 & c_2 & \cdot & \cdot & \cdot & \cdot & k_2 \\ a_3 & b_3 & c_3 & \cdot & \cdot & \cdot & \cdot & k_3 \\ \cdot & \cdot & \cdot & \cdot & \cdot & \cdot & \cdot & \cdot \\ \cdot & \cdot & \cdot & \cdot & \cdot & \cdot & \cdot & \cdot \\ \cdot & \cdot & \cdot & \cdot & \cdot & \cdot & \cdot & \cdot \\ a_n & b_n & c_n & \cdot & \cdot & \cdot & \cdot & k_n \end{vmatrix},$$

the normal order for the subscripts is the order of the rows, namely the order (1). Accordingly, if in Δ two adjacent rows, the lth and $(l+1)$st, are interchanged, the normal order for the subscripts in the new determinant, Δ', is the order (2).

The terms in Δ and in Δ' are the same, except perhaps for sign. To determine the sign of a term $a_p b_q c_r \cdots k_t$, as a term of Δ, the number of inversions which the subscripts

$$(3) \qquad p\ q\ r \cdot\cdot\cdot\cdot\cdot\cdot\cdot\cdot\cdot\cdot\ t$$

present from the order (1) is counted; to determine its sign, as a term of Δ', the number of inversions of (3) from the order (2) is counted. We have just shown that the two results differ always by unity. Consequently, the term in question has one sign in Δ and the opposite sign in Δ'. Therefore, $\Delta' = -\Delta$. We have thus proved the theorem:

THEOREM 6. *If two adjacent rows of Δ are interchanged, the sign of Δ is changed.*

Suppose, now, that we carry a row over m rows. This can be effected by m interchanges of adjacent rows; for example, if the row is to be carried downward, by interchanging it with the row just below it, then with the row just below its new position, etc. Since each interchange of adjacent rows changes the sign of Δ, the final determinant will be equal to Δ or $-\Delta$, according as m is even or odd. This result we state in the form of a theorem:

THEOREM 7. *If a row of Δ is carried over m rows, the result is Δ or $-\Delta$, according as m is even or odd.*

Finally, interchange any two rows. If there are m rows between the two, the interchange can be effected by carrying one of the rows over these m and then by carrying the second one over this one and the m, *i.e.* over $m+1$ rows. Thereby the determinant experiences $m+m+1=2m+1$ changes of sign, *i.e.* an odd number. Thus we have the result:

THEOREM 8. *If any two rows of Δ are interchanged, the sign of Δ is changed.*

New Rules for Determining the Sign of a Term. We first state the following lemma:

LEMMA 2. *Take the first n integers in an arbitrary order:*

$$p\ q\ r \cdots i\ j \cdots t, \tag{4}$$

and in this order interchange two ADJACENT *integers, i and j:*

$$p\ q\ r \cdots j\ i \cdots t. \tag{5}$$

Then the number of inversions of (5) *from the natural order differs from the number of inversions of* (4) *from the natural order by one.*

The proof of this lemma is exactly like the proof of Lemma 1.

An arbitrary term of Δ, without its sign, can be written in the form

(6) $$v_p w_q x_r \cdot \cdot \cdot \cdot \cdot \cdot \cdot z_t,$$

where

(7) $$v \ w \ x \cdot \cdot \cdot \cdot \cdot \cdot \cdot \cdot z$$

are the letters $a\ b\ c \cdot \cdot \cdot \cdot \cdot k$ in some order and

(8) $$p \ q \ r \cdot \cdot \cdot \cdot \cdot \cdot \cdot \cdot \cdot t$$

are the subscripts $1\ 2\ 3 \cdot \cdot \cdot \cdot \cdot n$ in some order. Let N be the number of inversions of the letters (7) from the natural order and let M be the number of inversions of the subscripts (8) from the natural order. $N + M$ is the total number of inversions in letters and subscripts.

If we interchange two *adjacent* factors in (6), the effect is to interchange two *adjacent* letters in (7) and two *adjacent* subscripts in (8). Hence, by Lemma 2, N is changed by one * and M by one; consequently, the sum $N + M$ is changed by 2 or left unchanged. But any reordering of the factors in (6) can be effected by a number of interchanges of adjacent factors. It follows, then, that any reordering of the factors of (6) changes $N + M$ by an even number or leaves it unchanged.

That is, *the evenness or oddness of the total number of inversions in letters and subscripts in a term of Δ is independent of the order of the factors in the term.*

We may, therefore, arrange the factors with the letters in the natural order and count the inversions in the subscripts, as in the definition, § 3, or we may arrange the factors with the subscripts in the natural order and count the inversions in the letters, or we may leave the factors unarranged and count the inversions in both letters and subscripts. The result will always be even or always be odd, no matter which of the three methods is used, and consequently the sign to be given to the term will always turn out to be the same.

* Lemma 2 is stated in terms of integers; it holds equally well for letters.

In the above methods of determining the sign of a term the letters and subscripts (or the columns and rows) enter symmetrically. The columns and rows also play the same rôles in the choice of the factors which constitute the term. In other words, the formation of a determinant from its square array bears equally on the rows and columns of the array. We have, then, the following theorem.

THEOREM 9. *If the rows and columns of* Δ *are interchanged,* Δ *is unchanged.*

Consequently, Theorems 6, 7, and 8, which have been proved for rows, are true also for columns.

Completion of the Proof of Theorem 1. If m is the element in the ith row and jth column of Δ, we have to show that $+mM$ or $-mM$ gives terms of Δ, according as $i+j$ is even or odd; or more briefly, that $(-1)^{i+j}mM$ always gives terms of Δ.

If $i=1, j=1$, *i.e.* if m is the element in the upper left-hand corner of Δ, the natural orders of letters and subscripts in M are

$$b\ c \cdot \cdot \cdot \cdot k \qquad \text{and} \qquad 2\ 3 \cdot \cdot \cdot \cdot n.$$

A term T of M will present the same number of inversions in letters and subscripts with respect to these orders as the corresponding term, a_1T, of Δ presents with respect to the orders

$$a\ b\ c \cdot \cdot \cdot \cdot k \qquad \text{and} \qquad 1\ 2\ 3 \cdot \cdot \cdot \cdot \cdot n.$$

Hence it is $+mM$ which gives terms of Δ and, since $i+j=2$, this is in accordance with the theorem.

Consider, now, the general case: m in the ith row and jth column. Carry the ith row over $i-1$ rows to the top of Δ and then carry the jth column over $j-1$ columns to the extreme left of Δ. By Th. 7, the resulting determinant is

$$(-1)^{i+j-2}\Delta, \tag{9}$$

and in it m is in the upper left-hand corner. It follows, then, from the case first considered, that $+mM$ gives terms of (9).

Hence $(-1)^{i+j}mM$ gives terms of

$$(-1)^{2i+2j-2}\Delta,$$

and therefore of Δ, since $2i+2j-2$ is even, q. e. d.*

EXERCISES

1. Prove Lemma 2.

2. Determine, by each of the three methods above described, the signs to be given to the following products:

(*a*) $b_3c_1a_2$; (*b*) $c_2a_4d_3b_1$; (*c*) $d_5b_3e_1a_4c_2$.

8. Cramer's Rule. In § 2, we stated that the three simultaneous equations in three unknowns,

$$(1)\qquad \begin{aligned} a_1x+b_1y+c_1z&=k_1,\\ a_2x+b_2y+c_2z&=k_2,\\ a_3x+b_3y+c_3z&=k_3,\end{aligned}$$

have the solution

$$(2)\qquad x=\frac{|k\ b\ c|}{|a\ b\ c|},\qquad y=\frac{|a\ k\ c|}{|a\ b\ c|},\qquad z=\frac{|a\ b\ k|}{|a\ b\ c|},$$

provided $\quad |a\ b\ c|\neq 0.$

This rule for finding the solution of (1) is due to Gabriel Cramer (1760). We proceed to prove it.

Assuming that equations (1) have a solution, we begin by multiplying them respectively by $+A_1$, $-A_2$, $+A_3$, *i.e.* by the signed minors of a_1, a_2, a_3 in the determinant $|a\ b\ c|$. Adding the resulting equations, we have

$$(a_1A_1-a_2A_2+a_3A_3)x+(b_1A_1-b_2A_2+b_3A_3)y \\ +(c_1A_1-c_2A_2+c_3A_3)z=k_1A_1-k_2A_2+k_3A_3.$$

The coefficient of x is the evaluation of $|a\ b\ c|$ by the minors of the a's. Similarly, the constant term is $|k\ b\ c|$; cf. § 4,

* All the theorems of this paragraph have been proved directly from the definition of a determinant, without the use of any of the preceding theorems, of §§ 4, 5, 6. So the paragraph could be inserted immediately after § 3. Its importance, in comparison with that of §§ 4, 5, 6, is not, however, sufficient to justify this.

Ex. 1. The coefficients of y and z are $|\,b\ b\ c\,|$ and $|\,c\ b\ c\,|$, and these determinants are zero (Th. 4). Consequently,

$$|\,a\ b\ c\,|\,x = |\,k\ b\ c\,|.$$

Since we are assuming that $|\,a\ b\ c\,| \neq 0$,

$$x = \frac{|\,k\ b\ c\,|}{|\,a\ b\ c\,|}.$$

This is the value of x, as given by (2). Multiplying the equations (1) respectively by $-B_1$, $+B_2$, $-B_3$ and adding, we obtain the value of y. That of z is arrived at in a similar manner.

What we have proved is this: If the equations (1) *have* a solution, it is given by formulas (2). It follows, then, that equations (1) have *at most one* solution, since formulas (2) give unique values for x, y, z.

It remains to show that these values of x, y, z actually are a solution, *i.e.* actually satisfy equations (1) in all cases. This can be done by direct substitution. Setting the values into the first of equations (1) and multiplying through by $|\,a\ b\ c\,|$, we have

$$(3)\quad a_1\,|\,k\ b\ c\,| + b_1\,|\,a\ k\ c\,| + c_1\,|\,a\ b\ k\,| - k_1\,|\,a\ b\ c\,| = 0.$$

By proper rearrangement of columns in the first two determinants (cf. Ths. 6–8), this becomes

$$a_1\,|\,b\ c\ k\,| - b_1\,|\,a\ c\ k\,| + c_1\,|\,a\ b\ k\,| - k_1\,|\,a\ b\ c\,| = 0.$$

The left-hand side here is the evaluation of the determinant

$$\begin{vmatrix} a_1 & b_1 & c_1 & k_1 \\ a_1 & b_1 & c_1 & k_1 \\ a_2 & b_2 & c_2 & k_2 \\ a_3 & b_3 & c_3 & k_3 \end{vmatrix}$$

by the minors of the first row. But this determinant is zero, because the first two rows are identical. Consequently, (3) is a true equation and the values of x, y, z given by (2) satisfy the first of the equations (1). In like manner it can be shown that they satisfy the other two equations.

This completes the proof that equations (1), provided $|a\ b\ c| \neq 0$, have a *unique* solution, which is given by Cramer's rule. Both the proof and the rule can be generalized to the case of any number of simultaneous linear equations in the same number of unknowns. We state the result in general form.

THEOREM 10. *A number of simultaneous linear equations in the same number of unknowns, for which the determinant of the coefficients of the unknowns does not vanish, has one and only one solution, which is given by Cramer's rule.**

EXERCISES

1. Deduce the value of y given by (2).

2. Prove that the values of x, y, z given by (2) actually satisfy the third of equations (1).

3. Give Cramer's rule for four simultaneous linear equations in four unknowns. First write down the equations and then the formulas analogous to formulas (2). No proof is required.

Solve the following systems of simultaneous equations.

4. $$\begin{aligned} 2x - y + 3z + t &= 6, \\ -x + 2y + 4z + 3t &= -6, \\ 3x - 2y - z + 4t &= -1, \\ 4x + 3y - 5z - 4t &= 8. \end{aligned}$$

5. $$\begin{aligned} 3y - 4z + 2t - 4 &= 0, \\ 2x + 3z - 4t + 3 &= 0, \\ -4x + 2y + 3t + 11 &= 0, \\ 3x - 4y + 2z - 5 &= 0. \end{aligned}$$

9. Three Equations in Two Unknowns. Compatibility. The *three* linear equations,

$$\begin{aligned} a_1x + b_1y + c_1 &= 0, \\ a_2x + b_2y + c_2 &= 0, \\ a_3x + b_3y + c_3 &= 0, \end{aligned} \tag{1}$$

* In case the determinant of the coefficients does vanish, the facts are more complex. For two equations in two unknowns, they are given in the footnote on p. 360. For a treatment of the general case, cf. Bôcher, *Introduction to Higher Algebra*, Ch. IV.

in the *two* unknowns x, y are said to be *compatible*, or *consistent*, if they have a simultaneous solution. They will, in general, be *incompatible*, since a solution of two of them will not, in general, satisfy the third. It is important, then, to determine the condition for their compatibility.

Two cases arise, according as the minors C_1, C_2, C_3 of the elements c_1, c_2, c_3 in the determinant $|a\ b\ c|$ are not, or are, all zero. In case they are not all zero, at least one of them must be different from zero. Suppose that $C_3 = |a_1\ b_2|$ is not zero. Then the first two of the equations (1) have, by Th. 10, one and only one solution, namely:

$$(2) \qquad x = \frac{|b_1\ c_2|}{|a_1\ b_2|}, \qquad y = -\frac{|a_1\ c_2|}{|a_1\ b_2|}.$$

This will be a solution of the third equation if and only if

$$a_3|b_1\ c_2| - b_3|a_1\ c_2| + c_3|a_1\ b_2| = 0,$$

or, since the left-hand side here is the expansion of $|a\ b\ c|$ by the minors of the third row, if and only if

$$(3) \qquad |a\ b\ c| = 0.$$

Before formulating this result as a theorem, we give a definition.

Definition. *The numbers* a_1, a_2, a_3 *and* b_1, b_2, b_3 *are proportional:*

$$a_1 : a_2 : a_3 = b_1 : b_2 : b_3,$$

if and only if there exist two numbers l *and* m, *not both zero, such that*

$$(4) \qquad la_1 = mb_1, \qquad la_2 = mb_2, \qquad la_3 = mb_3.$$

If b_1, b_2, b_3 are all zero and we take $l = 0$ and m any number $\neq 0$, equations (4) are satisfied no matter what values a_1, a_2, a_3 have. In other words, three arbitrary numbers a_1, a_2, a_3, on the one hand, and 0, 0, 0, on the other, are *always* proportional. In particular, a_1, a_2, a_3 may also all be zero.

Suppose, now, that b_1, b_2, b_3 are not all zero and let b_1, for example, be not zero. If, then, l were 0, we should have, from the first of equations (4), $m=0$; but $l=0$, $m=0$ is contrary to the definition. Consequently, in this case, l cannot be 0. Hence we can divide each equation through by l. The result is the equations

$$a_1 = kb_1, \qquad a_2 = kb_2, \qquad a_3 = kb_3,$$

where k has the value m/l. Conversely, if in any given case there exists a number k, zero or not zero, such that these equations hold, then a_1, a_2, a_3 and b_1, b_2, b_3 are proportional. For, the equations are but a special case of equations (4), when $l=1$ and $m=k$. We have thus proved the following theorem:

If b_1, b_2, b_3 are not all zero, the numbers a_1, a_2, a_3 and b_1, b_2, b_3 are proportional if and only if there exists a number k, zero or not zero, such that

$$(5) \qquad a_1 = kb_1, \qquad a_2 = kb_2, \qquad a_3 = kb_3.$$

By application of the definition it is easy to show that the minors C_1, C_2, C_3 in the above discussion are not all zero when and only when a_1, a_2, a_3 and b_1, b_2, b_3 are not proportional; cf. Ex. 2. The foregoing result can be stated, then, as follows:

THEOREM 11. *If, in the equations* (1),

$$a_1 : a_2 : a_3 \neq b_1 : b_2 : b_3,$$

the three equations will be compatible when and only when the determinant of their coefficients vanishes. They then have one and only one solution.

The case in which $C_1 = C_2 = C_3 = 0$ is left to the student; cf. Ex. 3.

EXERCISES

1. Show that the equations,

$$a_1x + b_1 = 0, \qquad a_2x + b_2 = 0, \qquad a_1 \neq 0, \quad a_2 \neq 0,$$

are consistent if and only if $|a \ b| = 0$.

2. The proportion $a_1 : a_2 : a_3 = b_1 : b_2 : b_3$ is valid if and only if the three two-rowed determinants, which are formed from the array

$$\begin{matrix} a_1 & a_2 & a_3 \\ b_1 & b_2 & b_3 \end{matrix}$$

by dropping each column in turn, are all zero. Prove this theorem.

3. If in the equations (1) a_1, a_2, a_3, b_1, b_2, b_3 are not all zero and a_1, a_2, a_3 are proportional to b_1, b_2, b_3, the three equations will be compatible if and only if c_1, c_2, c_3 are proportional to a_1, a_2, a_3 and b_1, b_2, b_3. They then have infinitely many solutions. Prove this theorem.

In each of the following exercises determine whether or not the given system of simultaneous equations is compatible. If it is, find the solution.

4. $3x - y + 5 = 0, \quad 4x + 3y - 2 = 0, \quad 5x + 7y - 9 = 0.$

5. $2x + y - 1 = 0, \quad 3x - 2y + 2 = 0, \quad x - 4y + 2 = 0.$

6. $x - 2y + 3 = 0, \quad -3x + 6y - 9 = 0, \quad 2x - 4y + 6 = 0.$

7. $4x + 2y - 1 = 0, \quad 2x + y - 5 = 0, \quad -6x - 3y + 2 = 0.$

8. Theorem. *If, in the equations*

$$\begin{aligned} a_1x + b_1y + c_1z + d_1 &= 0,\\ a_2x + b_2y + c_2z + d_2 &= 0,\\ a_3x + b_3y + c_3z + d_3 &= 0,\\ a_4x + b_4y + c_4z + d_4 &= 0, \end{aligned}$$

the four minors D_1, D_2, D_3, D_4 of the elements d_1, d_2, d_3, d_4 in the determinant $|a\ b\ c\ d|$ are not all zero, the equations are compatible when and only when $|a\ b\ c\ d| = 0$. They then have one and only one solution. Prove this theorem.

Determine in each case if the four given equations are compatible. If so, what is the common solution?

9. $$\begin{aligned} 2x + 3y + 4z - 5 &= 0,\\ 5x - 2y + 3z + 4 &= 0,\\ x + 3y - 7z + 5 &= 0,\\ 3x - 4y + 5z + 2 &= 0. \end{aligned}$$

10. $$\begin{aligned} 4x - 2y + 2z - 5 &= 0,\\ 2x - y + z + 3 &= 0,\\ 6x - 3y + 3z + 4 &= 0,\\ -2x + y - z + 2 &= 0. \end{aligned}$$

11. State the generalization of Theorem 11 and the theorem of Ex. 8 for the case of $n + 1$ linear equations in n unknowns.

10. Homogeneous Linear Equations. The equations,

$$
(1) \qquad \begin{aligned} a_1x + b_1y + c_1z = 0, \\ a_2x + b_2y + c_2z = 0, \\ a_3x + b_3y + c_3z = 0, \end{aligned}
$$

form what is called a system of *homogeneous* * linear equations. In considering them, we assume that not all the coefficients $a_1, b_1, \cdots, c_3$ are zero.

Let x_0, y_0, z_0 be a simultaneous solution of the equations (1). Then kx_0, ky_0, kz_0, where k is an arbitrary constant, is also a solution of (1). For, if these values are substituted for x, y, z in (1), we have

$$
\begin{aligned} k(a_1x_0 + b_1y_0 + c_1z_0) = 0, \\ k(a_2x_0 + b_2y_0 + c_2z_0) = 0, \\ k(a_3x_0 + b_3y_0 + c_3z_0) = 0. \end{aligned}
$$

The three parentheses in these equations all have the value zero, since x_0, y_0, z_0 is a solution of (1). Hence the equations are true, q. e. d.

This proof is applicable to the general case of n homogeneous linear equations in n unknowns. Hence we can state the following theorem:

THEOREM 12. *If $x_0, y_0, z_0, \cdots, t_0$ is a simultaneous solution of n homogeneous linear equations in the n unknowns $x, y, z, \cdots, t$, then $kx_0, ky_0, kz_0, \cdots, kt_0$, where k is an arbitrary constant, is also a solution.*

An obvious solution of the equations (1) is 0, 0, 0. This is the only solution, if the determinant $|a\ b\ c|$ is not 0. For, equations (1) are a special form of equations (1), § 8, when $k_1 = k_2 = k_3 = 0$. If $|a\ b\ c| \neq 0$, the latter equations have, by Th. 10, just one solution, given by formulas (2), § 8. But this solution is 0, 0, 0, since each of the determinants in the numerators in (2) now contains a column of zeros.

This result is also general:

THEOREM 13. *If the determinant of the coefficients of n homogeneous linear equations in the n unknowns $x, y, z, \cdots, t$ does*

* Cf. p. 348.

not vanish, the only simultaneous solution of the equations is $x = 0$, $y = 0$, $z = 0$, $\cdots$, $t = 0$.

If, then, the equations are to have a solution other than the *obvious* solution 0, 0, 0, $\cdots$, 0, it is necessary that the determinant of the coefficients vanish. It can be shown, conversely, that if this determinant does vanish the equations will have solutions other than the obvious solution. That is, the following theorem is true.

THEOREM 14. *A system of n homogeneous linear equations in n unknowns has a solution other than the obvious solution,* 0, 0, 0, $\cdots$, 0, *if and only if the determinant of the coefficients vanishes.*

To complete the proof of this theorem in the case of equations (1), we must show that, if $|a\ b\ c| = 0$, the equations have solutions other than 0, 0, 0. This we shall do by actually exhibiting such solutions.

By hypothesis, $|a\ b\ c| = 0$. Then two cases arise, as follows:

Case 1. *Not all the minors in* $|a\ b\ c|$ *are zero.* In this case at least one minor in $|a\ b\ c|$ does not vanish. Suppose that the minor, $|a_1\ b_2|$, in the upper left-hand corner is not zero.* We proceed, then, to show that equations (1) have a solution x, y, z, in which $z = 1$; *i.e.* that the equations,

$$\begin{aligned} a_1x + b_1y + c_1 &= 0, \\ a_2x + b_2y + c_2 &= 0, \\ a_3x + b_3y + c_3 &= 0, \end{aligned} \tag{2}$$

obtained from the equations (1) by setting $z = 1$, have a simultaneous solution for x and y.

Since, by hypothesis, $|a\ b\ c| = 0$ and $|a_1\ b_2| \neq 0$, equations (2) have, according to Th. 11, just one solution, that given by formulas (2), § 9, namely:

$$x = \frac{|b_1\ c_2|}{|a_1\ b_2|}, \qquad y = -\frac{|a_1\ c_2|}{|a_1\ b_2|}. \tag{3}$$

* If in any particular case this minor were zero, the equations and the terms in them could be rearranged, so that the minor would not be zero.

Consequently, equations (1) have the solution

$$(4)\qquad x=\frac{|b_1\ c_2|}{|a_1\ b_2|},\qquad y=-\frac{|a_1\ c_2|}{|a_1\ b_2|},\qquad z=1,$$

and hence, by Th. 12, the solution

$$(5)\qquad x=|b_1\ c_2|,\qquad y=-|a_1\ c_2|,\qquad z=|a_1\ b_2|,$$

or, finally, again by Th. 12, the solutions

$$(6)\qquad x=k|b_1\ c_2|,\qquad y=-k|a_1\ c_2|,\qquad z=k|a_1\ b_2|.$$

There are infinitely many solutions given by (6), since k may have any value. Inasmuch as $|a_1\ b_2|\neq 0$, only one of these solutions is the solution 0, 0, 0, namely the one for which $k=0$. Hence the theorem is proved in this case.

Furthermore, (6) gives *all* the solutions of (1). To prove this, let x_0, y_0, z_0 be an arbitrary solution of (1). If $z_0=0$, then $x_0=y_0=0$, since for $z=0$ the first two of equations (1) become

$$a_1x+b_1y=0,\qquad a_2x+b_2y=0,$$

and the only solution of these equations is 0, 0, because $|a_1\ b_2|\neq 0$; cf. Th. 13. If $z_0\neq 0$, then x_0/z_0, y_0/z_0, 1 is a solution of (1) and x_0/z_0, y_0/z_0 is therefore a solution of (2). But the only solution of (2) is given by formulas (3). Hence it follows that

$$\frac{x_0}{z_0}=\frac{|b_1\ c_2|}{|a_1\ b_2|},\qquad \frac{y_0}{z_0}=-\frac{|a_1\ c_2|}{|a_1\ b_2|},$$

or that

$$x_0=k|b_1\ c_2|,\qquad y_0=-k|a_1\ c_2|,\qquad z_0=k|a_1\ b_2|,$$

where k has a definite value, not zero.

We may state the final result by saying that *every solution of* (1) *is proportional to the solution* (5), meaning, thereby, that it is given by equations of the form (6); cf. § 9, eq. (5).

Case 2. All the minors in $|a\ b\ c|$ are zero. In this case it follows, by § 9, Ex. 2, that

$$a_1:b_1:c_1=a_2:b_2:c_2=a_3:b_3:c_3.$$

This means that the left-hand sides of equations (1) are proportional to one another. Consequently, all the solutions of one of the equations are solutions of the other two, and hence are all the solutions of the system (1).

The equation thus singled out must be one in which the three coefficients are not all zero. This is true of at least one of the equations (1), since, by hypothesis, not all the coefficients in (1) are zero. Let it be true of, say, the first equation:

$$a_1x + b_1y + c_1z = 0, \tag{7}$$

and let a_1, for example, be not zero.

In solving (7), the values of y and z can be chosen at pleasure: $y = k$, $z = l$, and the value of x is then determined. Consequently, all the solutions of (7), and hence of (1), are given by

$$x = -k\frac{b_1}{a_1} - l\frac{c_1}{a_1}, \qquad y = k, \qquad z = l. \tag{8}$$

Here there are *two* arbitrary constants, k and l. We say, then, that the equations (1) have a *two-parameter family* of solutions in this case; and, in distinction, a *one-parameter family* of solutions in Case 1.

The proof of Theorem 14, for $n = 3$, is now complete. In the general case the facts and, consequently, the proof are much more complicated.* See p. 403, footnote.

EXERCISES

1. Prove Theorem 14 for the case $n = 2$.

2. Prove the Theorem: *If $x_1, y_1, \cdots, t_1$ and $x_2, y_2, \cdots, t_2$ are two simultaneous solutions of n homogeneous linear equations in the n unknowns $x, y, z, \cdots, t$, then $x_1 + x_2, y_1 + y_2, \cdots, t_1 + t_2$ is also a solution.* Take first $n = 3$.

3. (Continuation of Ex. 2.) Show, further, that *$kx_1 + lx_2, ky_1 + ly_2, \cdots, kt_1 + lt_2$ is a solution.*

* Cf. Bôcher, *Introduction to Higher Algebra*, Ch. IV.

Solve the following systems of simultaneous equations, obtaining all the solutions in each case.

4. $4x-2y=0, \quad -6x+3y=0.$

5. $3x+5y+8z=0, \quad 4x-y+z=0, \quad x+2y-2z=0.$

6. $3x+2y-2z=0, \quad 2x+3y-z=0, \quad 8x+7y-5z=0.$

7. $x-2y+3z=0, \quad -3x+6y-9z=0, \quad 2x-4y+6z=0.$

II. Applications

11. The Straight Line. *Equation of the Line through Two Points, in Determinant Form.* Let $(\bar{x}, \bar{y})$ be an arbitrary point on the line determined by the two points (x_1, y_1), (x_2, y_2), and let

$$Ax + By + C = 0 \tag{1}$$

be the equation of the line. Since $(\bar{x}, \bar{y})$, (x_1, y_1), (x_2, y_2) lie on the line, we must have

$$\begin{aligned} A\bar{x} + B\bar{y} + C &= 0, \\ Ax_1 + By_1 + C &= 0, \\ Ax_2 + By_2 + C &= 0. \end{aligned} \tag{2}$$

These equations are linear and homogeneous in the three unknowns A, B, C. They have a solution for A, B, C other than the obvious solution 0, 0, 0, inasmuch as there is a line (1) on which the three points $(\bar{x}, \bar{y})$, (x_1, y_1), (x_2, y_2) lie. Consequently, by Th. 14, the determinant of the coefficients vanishes.

In other words, every point $(\bar{x}, \bar{y})$ or, on dropping the dashes, every point (x, y) on the line satisfies the equation

$$\begin{vmatrix} x & y & 1 \\ x_1 & y_1 & 1 \\ x_2 & y_2 & 1 \end{vmatrix} = 0. \tag{3}$$

By a careful retracing of the steps, it can be shown, conversely, that every point (x, y) satisfying (3) lies on the line. It would follow, then, that (3) is the equation of the line. We shall adopt, however, a quite different method to prove this, regarding the foregoing work as primarily of value in furnishing us equation (3).

To show that equation (3) represents the line through (x_1, y_1), (x_2, y_2), develop the determinant by the minors of the first row. Equation (3) then takes on the usual form (1) of a linear equation in x and y; moreover, the values obtained for A and B:

$$A = y_1 - y_2 \qquad B = x_2 - x_1,$$

are not both zero, since the given points do not coincide. Consequently, (3) represents some straight line.

This line is the required line, if the coördinates of the given points satisfy (3). They do, for, if we replace x, y in the determinant by x_1, y_1, or by x_2, y_2, two rows of the determinant will be identical and hence the value of the determinant will be zero.

Three Points on a Line. Let the three points, which we assume are distinct, be (x_1, y_1), (x_2, y_2), (x_3, y_3). The equation of the line through the second and third is, according to (3),

$$\begin{vmatrix} x & y & 1 \\ x_2 & y_2 & 1 \\ x_3 & y_3 & 1 \end{vmatrix} = 0. \tag{4}$$

The first point lies on this line if and only if (x_1, y_1) satisfies (4), *i.e.* if and only if

$$\begin{vmatrix} x_1 & y_1 & 1 \\ x_2 & y_2 & 1 \\ x_3 & y_3 & 1 \end{vmatrix} = 0. \tag{5}$$

This result we state as follows:

THEOREM 15. *The three points (x_1, y_1), (x_2, y_2), (x_3, y_3) are collinear, if and only if the determinant in* (5) *vanishes.*

Three Lines through a Point. Consider the three distinct lines

$$\begin{aligned} A_1x + B_1y + C_1 &= 0, \\ A_2x + B_2y + C_2 &= 0, \\ A_3x + B_3y + C_3 &= 0. \end{aligned} \tag{6}$$

They are parallel, by Ch. II, § 10, Th. 3, if and only if

$$A_1 : B_1 = A_2 : B_2 = A_3 . B_3,$$

i.e. if and only if

(7) $$A_1 : A_2 : A_3 = B_1 : B_2 : B_3.$$

Suppose, now, that the three lines go through a point. This means, analytically, that the equations (6) have a common solution for x, y, *i.e.* are compatible. Hence, it follows, by Th. 11, since in this case (7) cannot hold, that $|A\ B\ C| = 0$.

The determinant $|A\ B\ C|$ vanishes also when the three lines are parallel, since then (7) is valid and the first two columns in the determinant are proportional.

Conversely, if $|A\ B\ C| = 0$, the lines (6) are parallel or concurrent. For, if the determinant vanishes by virtue of the first two columns being proportional, (7) holds and the lines are parallel. On the other hand, if (7) does not hold, equations (6), by Th. 11, are compatible and this means, geometrically, that the three lines have a point in common.

We have thus proved the theorem:

THEOREM 16. *The three lines* (6) *are concurrent or parallel if and only if the determinant of their coefficients vanishes:*

$$|A\ B\ C| = 0.$$

EXERCISES

Find the equations of the following lines in determinant form.

1. The line through (x_1, y_1) with intercept b on the axis of y.

2. The line with intercepts a and b.

Find, in determinant form, the equations of the lines required in the following exercises of Chapter II. Reduce the equation each time to the usual form.

3. Ex. 1, § 1. **4.** Ex. 2, § 1. **5.** Ex. 4, § 1. **6.** Ex. 6, § 1.

7. Ex. 7, § 1. **8.** Ex. 10, § 1. **9.** Ex. 1, § 5. **10.** Ex. 3, § 5.

By the method of this paragraph, do the following exercises at the end of Chapter III concerning three lines through a point or three points on a line.

11. Ex. 1. **12.** Ex. 2. **13.** Ex. 3.

14. Ex. 4. **15.** Ex. 5. **16.** Ex. 6.

Are the lines given in the following exercises concurrent? parallel?

17. Ex. 4, § 9. **18.** Ex. 5, § 9. **19.** Ex. 7, § 9.

12. The Circle and the Conics. *Equation of the Circle through Three Points.* If the three points (x_1, y_1), (x_2, y_2), (x_3, y_3), which we assume are not collinear, lie on the circle

$$A(x^2 + y^2) + Bx + Cy + D = 0, \tag{1}$$

it follows that

$$\begin{aligned} A(x_1^2 + y_1^2) + Bx_1 + Cy_1 + D &= 0, \\ A(x_2^2 + y_2^2) + Bx_2 + Cy_2 + D &= 0, \\ A(x_3^2 + y_3^2) + Bx_3 + Cy_3 + D &= 0. \end{aligned} \tag{2}$$

In (1) and (2) we have four homogeneous linear equations in the four unknowns A, B, C, D, which have a solution other than the obvious solution, 0, 0, 0, 0. Consequently, by Th. 14,

$$\begin{vmatrix} x^2 + y^2 & x & y & 1 \\ x_1^2 + y_1^2 & x_1 & y_1 & 1 \\ x_2^2 + y_2^2 & x_2 & y_2 & 1 \\ x_3^2 + y_3^2 & x_3 & y_3 & 1 \end{vmatrix} = 0. \tag{3}$$

Equation (3) is the equation of the circle through the three given points. For, if we develop the determinant in (3) by the minors of the first row, we obtain an equation of the form (1), where

$$A = \begin{vmatrix} x_1 & y_1 & 1 \\ x_2 & y_2 & 1 \\ x_3 & y_3 & 1 \end{vmatrix} \neq 0,$$

since the three points were assumed non-collinear (Th. 15). Consequently, equation (3) represents a circle, or a point, or it has no locus; cf. Ch. IV, § 2. That it represents a circle, and, in particular, the required circle, is clear since the coördinates of each of the three points satisfy it.

Condition that Four Points Lie on a Circle.

THEOREM 17. *The four points* (x_1, y_1), (x_2, y_2), (x_3, y_3), (x_4, y_4), *of which we assume no three collinear, lie on a circle if and only if*

$$(4) \qquad \begin{vmatrix} x_1^2 + y_1^2 & x_1 & y_1 & 1 \\ x_2^2 + y_2^2 & x_2 & y_2 & 1 \\ x_3^2 + y_3^2 & x_3 & y_3 & 1 \\ x_4^2 + y_4^2 & x_4 & y_4 & 1 \end{vmatrix} = 0.$$

The proof is left to the student.

Conic through Five Points. The general equation of the straight line (1), § 11 contains *three* constants, A, B, C, entering homogeneously — one in each term — and we can always pass just one line through *two* points. Also, the general equation (1) of the circle contains *four* homogeneous constants A, B, C, D, and through *three* points (non-collinear) we can always pass just one circle.

The general equation of a conic,

$$(5) \qquad Ax^2 + Bxy + Cy^2 + Dx + Ey + F = 0,$$

contains *six* homogeneous constants, and accordingly we should expect that *through* FIVE *points we can, in general, pass just one conic.*

We prove this by writing down the equation of the conic through the five points (x_1, y_1), (x_2, y_2), (x_3, y_3), (x_4, y_4), (x_5, y_5).

Proceeding as in the cases of the straight line and circle, we find as the probable equation:

$$(6) \qquad \begin{vmatrix} x^2 & xy & y^2 & x & y & 1 \\ x_1^2 & x_1y_1 & y_1^2 & x_1 & y_1 & 1 \\ x_2^2 & x_2y_2 & y_2^2 & x_2 & y_2 & 1 \\ x_3^2 & x_3y_3 & y_3^2 & x_3 & y_3 & 1 \\ x_4^2 & x_4y_4 & y_4^2 & x_4 & y_4 & 1 \\ x_5^2 & x_5y_5 & y_5^2 & x_5 & y_5 & 1 \end{vmatrix} = 0.$$

When the determinant is developed by the minors of the first row, equation (6) takes on the form (5). Two cases then arise, according as the values obtained for A, B, C are not, or are, all zero.

Case 1. *A, B, C not all zero.* In this case it follows that equation (6) represents some conic, in particular, a conic through the five given points, since it is clear that the coördinates of each of the points satisfy the equation.

We state, without proof, that this case occurs unless four, or all five, of the given points are collinear.

If no three of the points are collinear the conic just found must be non-degenerate. It is the only conic through the five points. For, if there were a second conic through them, the two conics (both non-degenerate) would intersect in five points, and this is impossible.*

If three of the points are collinear, the conic found must be degenerate; † in particular it must consist of two straight lines (Fig. 1). Clearly, these lines are uniquely determined by the five points and hence so is the conic.

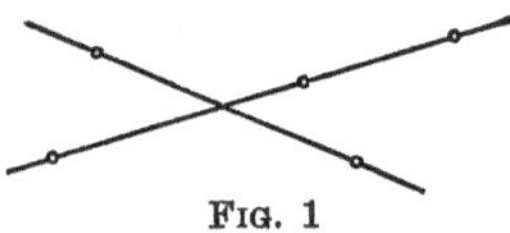

Fig. 1

The results of this case we formulate as a theorem:

Theorem 18. *Through five points, no four of which are collinear, there passes one and only one conic. If three of the points are collinear, the conic is degenerate; otherwise, it is non-degenerate.*

Case 2. $A = B = C = 0$. Then $D = E = F = 0$ also, and equation (6) reduces to the trivial equation: $0 = 0$. Stated without proof.

This case occurs if at least four of the five points are col-

* That two non-degenerate conics cannot intersect in more than four points is geometrically evident; an analytical proof is beyond the scope of this book.

† If it were non-degenerate, we should have a non-degenerate conic intersected by a line in three points — an impossibility.

linear. If just four are collinear, there are infinitely many degenerate conics through the five points, each consisting of the line of the four points and some line through the fifth.

If *all* five points are collinear, their line, taken with any line in the plane, forms a degenerate conic through them, so that here, too, there are infinitely many degenerate conics through the five points.*

Parabolas through Four Points. Demanding that the conic defined by equation (5) be a parabola puts one condition on the coefficients in (5), namely,

$$B^2 - 4AC = 0. \tag{7}$$

Consequently, we cannot prescribe more than four points through which a parabola must pass.

Let $(0, 0)$, $(1, 1)$, $(-1, 1)$, $(3, 9)$ be the four points. Then

$$\begin{array}{r} F = 0, \\ A + \quad B + \quad C + \quad D + \quad E + F = 0, \\ A - \quad B + \quad C - \quad D + \quad E + F = 0, \\ 9A + 27B + 81C + 3D + 9E + F = 0. \end{array} \tag{8}$$

To solve equations (7) and (8) simultaneously, find the values of D, E, F in terms of A, B, C from the first three of equations (8):

$$D = -B, \qquad E = -A - C, \qquad F = 0, \tag{9}$$

and substitute them in the fourth equation. The result is

$$B = -3C. \tag{10}$$

Hence (7) becomes

$$9C^2 - 4AC = 0,$$

and

$$C = 0 \quad \text{or} \quad C = \tfrac{4}{9}A.$$

From equations (9) and (10) we have, then:

$$\begin{array}{lllll} C = 0, & B = 0, & D = 0, & E = -A, & F = 0, \\ \text{or } C = \tfrac{4}{9}A, & B = -\tfrac{4}{3}A, & D = \tfrac{4}{3}A, & E = -\tfrac{13}{9}A, & F = 0. \end{array}$$

* There is a *one-parameter family* of degenerate conics in the first case, a *two-parameter family* in the second; cf. p. 390. Can the student explain why?

Setting $A = 1$ in the first case and $A = 9$ in the second, we find as the resulting equations

$$x^2 - y = 0,$$
$$9x^2 - 12xy + 4y^2 + 12x - 13y = 0.$$

There are, then, two parabolas through the four given points. We state without proof that this is, in general, true. Of course, one or both of the parabolas may be degenerate, and for special positions of the four points the two may coincide. Finally, if the four points are collinear, there are an infinite number of degenerate parabolas through them.

EXERCISES

1. State and prove the theorem giving the condition that six points, no four of which are collinear, lie on a (non-degenerate or degenerate) conic. If four or more of the points are collinear, is there a conic through the six?

Find, in determinant form, the equations of the circles required in the following exercises of Chapter IV. Reduce the equation each time to the usual form.

2. Ex. 1, § 4. **3.** Ex. 2, § 4. **4.** Ex. 3, § 4.

In each of the following exercises determine whether or not the four given points lie on a circle.

5. $(0, 0)$, $(3, 0)$, $(0, 1)$, $(2, -1)$.

6. $(2, 0)$, $(-3, 0)$, $(0, 4)$, $(-1, 4)$.

7. $(a, 0)$, $(b, 0)$, $(0, c)$, $\left(0, \dfrac{ab}{c}\right)$.

Find, in each exercise that follows, the equation of the conic through the given five points. Is the conic non-degenerate?

8. $(0, 0)$, $(2, 0)$, $(0, 2)$, $(5, 2)$, $(2, 5)$.

Ans. $2x^2 - 3xy + 2y^2 - 4x - 4y = 0$.

9. $(1, 0)$, $(-1, 0)$, $(0, 1)$, $(0, -1)$, $(1, 1)$.

10. (1, – 1), (1, 1), (3, 11), (– 3, – 11), (5, 19).

Ans. $15x^2 - y^2 = 14.$

11. (0, 0), (2, 1), (3, – 4), (0, 2), (– 2, 0).

12. (1, 2), (0, 1), (6, – 1), (– 1, – 2), (3, 0).

Find the equations of the parabolas through each of the following sets of four points. Are they degenerate or not?

13. (0, 0), (1, 1), (1, – 1), (4, 2).

Ans. $y^2 - x = 0$; $(x - y)(x - y - 2) = 0$.

14. (0, 0), (3, 1), (1, 3), (6, 3).

15. (2, 0), (0, 1), (– 1, 1), (5, – 2).

16. (2, 1), (7, 0), (4, 3), (5, – 2).

In each of the following exercises determine whether or not the six given points lie on a conic. If they do, find if the conic is degenerate.

17. (0, 0), (1, – 1), (1, 3), (5, 5), (2, 4), (6, 3).

18. (– 1, – 1), (0, 2), (– 1, 0), (5, 2), (0, – 1), (9, 5).

19. (0, 1), (1, 0), (1, – 1), (3, 1), (– 1, 3), (– 3, – 2).

20. (0, 0), (2, 0), (– 1, 1), (3, 1), (5, – 1), (– 4, 2).

EXERCISES ON CHAPTER XVI

Evaluate each of the following determinants, expressing the result, if it is different from zero, in factored form.

1. $\begin{vmatrix} 1 & a & a^2 \\ 1 & b & b^2 \\ 1 & c & c^2 \end{vmatrix}$. 2. $\begin{vmatrix} a+b & ab & c \\ b+c & bc & a \\ c+a & ca & b \end{vmatrix}$. 3. $\begin{vmatrix} 0 & -a & -b \\ a & 0 & -c \\ b & c & 0 \end{vmatrix}$.

Ans. to Ex. 1. $(a - b)(b - c)(c - a)$.

Ans. to Ex. 2. $-(a - b)(b - c)(c - a)(a + b + c)$.

4. $\begin{vmatrix} a-b & a+b & 1 \\ b-c & b+c & 1 \\ c-a & c+a & 1 \end{vmatrix}$. 5. $\begin{vmatrix} 1 & a & a^2 & a^3 \\ 1 & b & b^2 & b^3 \\ 1 & c & c^2 & c^3 \\ 1 & d & d^2 & d^3 \end{vmatrix}$.

Prove that the following determinants have the value zero.

6. $$\begin{vmatrix} a_1 & b_1 & 2a_1+3b_1 \\ a_2 & b_2 & 2a_2+3b_2 \\ a_3 & b_3 & 2a_3+3b_3 \end{vmatrix}.$$

7. $$\begin{vmatrix} 1 & 3a_1-b_1+2 & a_1 & b_1 \\ -1 & 3a_2-b_2-2 & a_2 & b_2 \\ 2 & 3a_3-b_3+4 & a_3 & b_3 \\ 1 & 3a_4-b_4+2 & a_4 & b_4 \end{vmatrix}.$$

Definition. Let $p_1, p_2, \cdots, p_n$, $q_1, q_2, \cdots, q_n$, and $r_1, r_2, \cdots, r_n$ be three columns (or rows) of a determinant. The third is said to be a *linear combination* of the first two, if two numbers, k and l, exist such that

$$r_1 = kp_1 + lq_1, \qquad r_2 = kp_2 + lq_2, \cdots\cdots, r_n = kp_n + lq_n.$$

In the determinant of Ex. 6, for example, the third column is a linear combination of the first two; and in that of Ex. 7 the second column is a linear combination of the third, fourth, and first.

8. Theorem. *If one column, or row, of a determinant is a linear combination of two others, the value of the determinant is zero.* Prove this theorem. How can it be extended?

Solve the following equations for x.

9. $$\begin{vmatrix} x+1 & 4 & 2 \\ x-9 & 5 & -3 \\ x-1 & -1 & 1 \end{vmatrix} = 0.$$

10. $$\begin{vmatrix} x-5 & 2 & -1 & 3 \\ 6 & -3x & 4 & 2 \\ 7 & x+4 & 3 & -1 \\ x & 6 & 2 & 4 \end{vmatrix} = 0.$$

Determine k so that the following equations have solutions other than 0, 0, 0; then find the solutions.

11. $\begin{aligned} kx + 3y + z &= 0, \\ x + 4y - 3z &= 0, \\ kx + y + 3z &= 0. \end{aligned}$

12. $\begin{aligned} kx - y + z &= 0, \\ 4x - 2y + kz &= 0, \\ 6x - 3y + (k+1)z &= 0. \end{aligned}$

Determine k so that the following equations are compatible. Find the common solution in each case.

13. $\begin{aligned} 2x - 3y - k &= 0, \\ kx - y - k &= 0, \\ kx + 3y - 5 &= 0. \end{aligned}$

14. $\begin{aligned} x - ky + 2k &= 0, \\ kx - 4y + 5 &= 0, \\ x + 2y - 1 &= 0. \end{aligned}$

Find all the solutions of the following equations.

15. $\begin{aligned} x + y - z &= 0, \\ 2x - y - z &= 0. \end{aligned}$

16. $\begin{aligned} 2x + 3y + 6z &= 0, \\ 3x - 6y + 2z &= 0. \end{aligned}$

17. Show that all the solutions of the equations,

$$l_1x + m_1y + n_1z = 0,$$
$$l_2x + m_2y + n_2z = 0,$$

are given by $x : y : z = |m \ n| : |n \ l| : |l \ m|$,
provided not all three of the determinants on the right are zero.

Applications

18. Show that the area of the triangle with vertices at the points (x_1, y_1), (x_2, y_2), (x_3, y_3) is

$$\pm \tfrac{1}{2}\begin{vmatrix} x_1 & y_1 & 1 \\ x_2 & y_2 & 1 \\ x_3 & y_3 & 1 \end{vmatrix}.$$

19. Prove that the equation of the line of slope λ through the point (x_1, y_1) can be written in the form

$$\begin{vmatrix} x & y & 1 \\ x_1 & y_1 & 1 \\ 1 & \lambda & 0 \end{vmatrix} = 0.$$

20. Show that every equation of the form

$$\begin{vmatrix} x & y & 1 \\ a_1 & a_2 & a_3 \\ b_1 & b_2 & b_3 \end{vmatrix} = 0,$$

where the minors of x and y are not both zero, represents a straight line.

21. Show that the points (x_1, y_1), (x_2, y_2) are collinear with the origin when and only when

$$\begin{vmatrix} x_1 & y_1 \\ x_2 & y_2 \end{vmatrix} = 0.$$

22. Prove that the distinct lines L_1, L_2 of Ch. II, § 10, are parallel if and only if

$$\begin{vmatrix} A_1 & B_1 \\ A_2 & B_2 \end{vmatrix} = 0.$$

23. Show that the lines L_1, L_2 of Ch. II, § 10, are identical, if and only if the three two-rowed determinants, which are formed from the array

$$\begin{matrix} A_1 & B_1 & C_1 \\ A_2 & B_2 & C_2 \end{matrix}$$

by dropping each column in turn, are all zero.

24. Show that the discriminant, Δ, of the quadratic equation

$$Ax^2 + Bx + C = 0, \qquad A \neq 0,$$

(cf. Ch. IX, § 5) can be written in the form

$$\Delta \equiv -\begin{vmatrix} 2A & B \\ B & 2C \end{vmatrix}.$$

25. Show that the discriminant, Δ, of the general equation of the second degree in x and y (cf. Ch. XII, § 4) can be written as

$$\Delta \equiv \frac{1}{2}\begin{vmatrix} 2A & B & D \\ B & 2C & E \\ D & E & 2F \end{vmatrix}.$$

26. Prove that the polars of all points (having polars) with respect to a degenerate conic are concurrent or parallel.

Suggestion. The conic can be represented either by

$$ax^2 + by^2 = 0 \qquad \text{or by} \qquad y^2 = c.$$

27. By applying Ex. 26, show that the general equation of the second degree represents a degenerate conic when and only when its discriminant, as given by the determinant in Ex. 25, vanishes.

Suggestion. Demand that the polars of three non-collinear points, as (0, 0), (1, 0), (0, 1), be concurrent or parallel. The equation of the polar of (x_1, y_1) is that of Ex. 2, p. 188.

28. Prove that, if the general equation of the second degree represents a non-degenerate conic, the line $ax + by + c = 0$ will be tangent to the conic if and only if

$$\begin{vmatrix} 2A & B & D & a \\ B & 2C & E & b \\ D & E & 2F & c \\ a & b & c & 0 \end{vmatrix} = 0.$$

Suggestion. Apply the second method of Ch. IX, § 5. The equation of the tangent at (x_1, y_1) is given by Ex. 2, p. 188.

Note to p. 390. Theorem 14 leads to an important result concerning the compatibility (cf. § 9) of equations (2), p. 388, namely:

Theorem. *If equations* (2) *are compatible, the determinant of their coefficients vanishes.*

For, if equations (2) have a solution, x_0, y_0, then equations (1) have a solution, x_0, y_0, 1, not the obvious solution, 0, 0, 0. Consequently, by Th. 14, $| \, a \;\, b \;\, c \, | = 0$.

The extension of the theorem and the proof to the equations of § 9, Ex. 8, and to the general case of § 9, Ex. 11, is immediate.

The determinant of the coefficients of the equations of § 9, Ex. 7, vanishes; the equations are, however, incompatible, — they represent three parallel lines. In other words, the converse of the theorem is not true; cf. Th. 11.

SOLID ANALYTIC GEOMETRY

CHAPTER XVII

PROJECTIONS. COÖRDINATES

1. Directed Line-Segments. In the Introduction to Plane Analytic Geometry directed line-segments on a line L were defined and discussed. Since L might be situated anywhere in space, the theory there developed holds equally well for the geometry of space. The student should review the details of this theory. Of the formulas, let him recall in particular the relation,

$$(1) \qquad MM_1 + M_1M_2 + \cdots + M_{n-2}M_{n-1} + M_{n-1}N = MN,$$

which holds for any $n+1$ points, $M, M_1, M_2, \cdots, M_{n-1}, N$, lying on L.

2. Projection of a Broken Line. Given a point P and a line L in space. The projection of P on L is defined as the foot, M, of the perpendicular dropped from P on L, or as the point M in which the plane p, passing through P perpendicular to L, meets L. If P lies on L, it is its own projection on L.

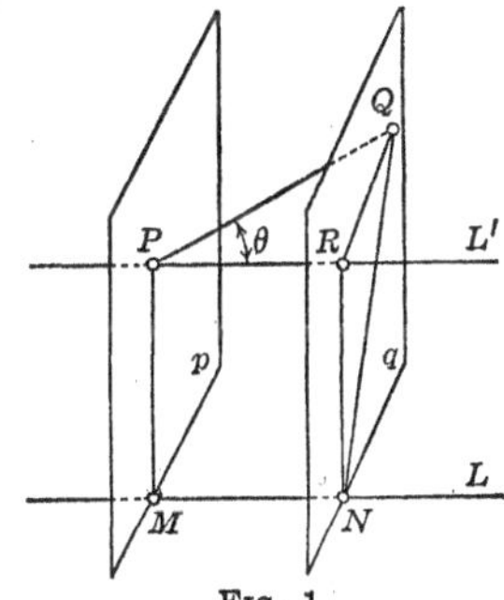

FIG. 1

Let PQ be any directed line-segment in space and let M and N be the projections of P and Q on L. The projection of the directed line-segment PQ on L is defined as the directed line-segment MN.

If p and q are the planes through P and Q perpendicular to L, the projection, MN, of PQ on L is equal to the directed line-segment intercepted by the planes p and q on any parallel to L. For example, it is equal to the directed line-segment PR in Fig. 1.*

Consider a broken line joining P to Q and consisting of the directed line-segments PP_1, P_1P_2, $\cdots$, $P_{n-1}Q$, which do not necessarily lie in a plane. The sum of the projections of these directed line-segments is

$$MM_1 + M_1M_2 + \cdots + M_{n-1}N.$$

By (1), § 1, this sum is equal to MN, *i.e.* to the projection on L of the directed line-segment PQ.

Thus Theorem 1 of the Introduction, § 3, is extended to the geometry of space:

THEOREM. *The sum of the projections, on any line L of space, of the directed line-segments, PP_1, P_1P_2, $\cdots$, $P_{n-1}Q$, of any broken line joining a point P of space with a second point Q is equal to the projection on L of the directed line-segment PQ.*

Theorem 2 of the Introduction, § 3, may be extended in a similar manner. Let the student state and prove the result.

The projection of a point P on a plane K is defined as the foot of the perpendicular dropped from P on K. If P lies in K, it is its own projection on K.

Let a plane K and a line L be given. If L is not perpendicular to K, the projection of L on K is defined as the line in which the plane through L perpendicular to K intersects K. If L is perpendicular to K, the projection of L on K is merely a point, the point in which it meets K.

3. The Angle between Two Directed Lines. Given any two indefinite straight lines in space and on each of them a sense; to define the angle between these two *directed lines.*

* In drawing this figure, we have placed ourselves in space so that the plane through L and P appears to us as a vertical plane.

If the lines meet, they lie in a plane. The angle θ between them shall be defined as the angle between the half-lines, or rays, issuing from their point of intersection in the given directions (Fig. 2).

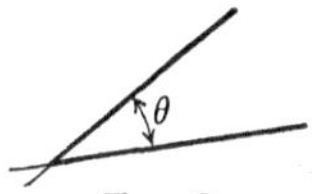

FIG. 2

If the lines do not meet, choose an arbitrary point A of space, and draw from A two rays respectively parallel to and having the same senses as the given lines. The angle between the given lines shall be defined as the angle between these rays.*

Remarks. The angle θ is the angle *between* the directed lines, not the angle *from* one *to* the other. It has always a positive or zero value, *i.e.* a numerical, and not an algebraic, value.

It is futile, in the geometry of space, to try to distinguish between positive and negative angles. For instance, suppose that, in an attempt to define the angle *from* one of two directed lines lying in a plane *to* the other, we should agree that angles measured in the counter-clockwise sense are to be considered positive and those measured in the clockwise sense, negative. Then the angle *from* the one directed line *to* the other, if viewed from a certain side of the plane, would appear positive; but, viewed from the other side of the plane, the same angle would be negative. Viewing the angle from one side of the plane is as justifiable as viewing it from the other, since the plane is immersed in space and not displayed on a blackboard or on the page of a book. Consequently, we should still be at a loss as to whether the angle is positive or negative.

There are two angles *between* the rays shown in Fig. 2, namely, θ and $360° - \theta$. One of these is necessarily less than or equal to 180°. It is this angle which we agree to take as the angle between the directed lines.

* For example, in Fig. 1, the line of PQ, directed from P to Q, and L, directed to the right, are two directed lines. The angle between them is the angle θ constructed by choosing A on the first line, at P, and by drawing through P the line L' parallel to and having the same sense as L.

EXERCISES

1. Fasten a sheet of paper to the floor with one edge against the wall, and tack a second sheet to the wall with one edge along the floor. Draw on each sheet a directed line so that the two lines meet. (It is wise to draw the lines before fixing the sheets in position.) Crease a third sheet of paper so as to form an angle which will just fit between the two directed rays. By measuring this angle with a protractor determine the angle between the two directed lines.

2. Repeat Ex. 1 with two directed lines differing widely in position from the first two chosen.

3. What is the angle between the two lines of Ex. 1, if the sense on one of them is reversed?

4. By the method of Ex. 1 find the angle between two directed lines, one on the floor and one on the wall, if the two lines do not meet.

5. Prove that, if L and L' are any two lines in space and any plane F is passed through L, there will be a plane W through L' perpendicular to F. That is, show that the above method is applicable to the problem of determining the angle between any two directed lines.

4. Value of the Projection of a Directed Line-Segment. Assign to a line L of space a sense and adopt a unit of length for all measurements in space. Then a directed line-segment AB on L is represented by an algebraic number, equal numerically to the length of AB and positive or negative according as the direction from A to B is the same as, or opposite to, the direction given to L; cf. Introduction, § 2.

In particular, to the projection MN on L of a directed line-segment PQ corresponds a certain algebraic value or number, which we can, without confusion, denote also by MN. Clearly,

$$\text{Proj. } PQ = -\text{Proj. } QP.$$

Let the length $|PQ|$ of the directed line-segment PQ be given and also the angle θ which the line of PQ, directed from P to Q, makes with the directed line L. If PQ lies in a plane with L, we know from Plane Trigonometry that

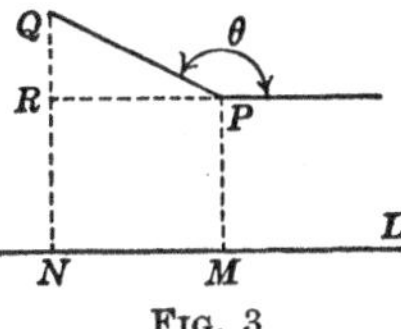

Fig. 3

(1) $MN = \text{Proj.}_L\, PQ = |PQ| \cos \theta.$

The general case, in which PQ is not in a plane with L, is shown in Fig. 1. Consider the projection, PR, of PQ on the line L' through P, parallel to and having the same sense as L. Since PQ lies in a plane with L', we have the previous case. Consequently,

$$PR = |PQ| \cos \theta.$$

But $PR = MN$ and thus formula (1) is established in the general case.

EXERCISES

1. Draw Fig. 1 for various positions of P and Q and in each case verify formula (1).

2. By application of (1) verify that Proj. $PQ = -$ Proj. QP.

3. If P and Q lie in a plane perpendicular to L, M and N coincide and $MN = 0$. Prove this by applying formula (1).

4. Prove that the directed line-segments MN and $M'N'$, which are the projections on L of two directed line-segments PQ and $P'Q'$ on the same line, are proportional to PQ and $P'Q'$:

$$\frac{MN}{M'N'} = \frac{PQ}{P'Q'}.$$

5. Prove that the theorem of the preceding exercise is true if PQ and $P'Q'$ are on parallel lines.

5. Coördinates. Three *directed* lines drawn through a point O of space, so that each is perpendicular to the other two, form a *system of rectangular coördinate axes.* The coördinates of a point P of space with respect to the system of axes are defined as the numbers which represent algebraically the projections of the directed line-segment OP on the three directed lines. Thus, if Ox, Oy, Oz denote the three directed lines and x, y, z stand for the coördinates of P,

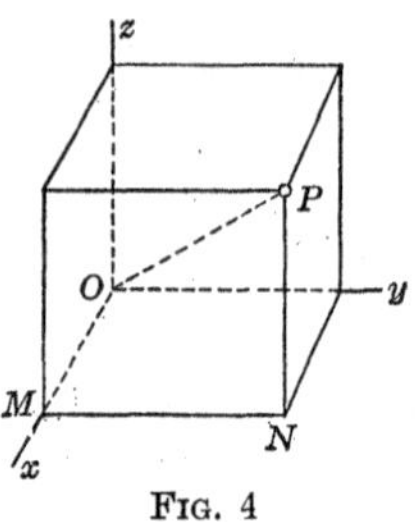

FIG. 4

$$x = \text{Proj.}_{Ox}\, OP, \qquad y = \text{Proj.}_{Oy}\, OP,$$
$$z = \text{Proj.}_{Oz}\, OP.$$

The projections of OP on the three directed lines can be constructed by passing planes through P perpendicular to the three lines (Fig. 4). These form with the planes of the lines a rectangular parallelepiped, or box. The directed edges of the box which issue from O are the projections of OP.

Every point P of space has unique coördinates (x, y, z). Conversely, if any three numbers x, y, z are given, there is a unique point P having these numbers as its coördinates. This point can be located, either by constructing a box or, more simply, by laying off $OM = x$ on the axis of x, then $MN = y$ on a parallel through M to the axis of y, and finally $NP = z$ on a parallel through N to the axis of z, as shown in Fig. 4. It is to be remembered that OM, MN, and NP are *directed* line-segments. The direction of OM, for example, is the same as, or opposite to, that of Ox, according as the number x is positive or negative. The figure is drawn for the case that x, y, z are all positive.

The point O is the *origin* of coördinates, the directed lines Ox, Oy, Oz are the *coördinate axes*, and the planes xOy, yOz, zOx are the *coördinate planes*. The origin has the coördinates $(0, 0, 0)$, a point on a coördinate axis always has two of its coördinates zero, and a point in a coördinate plane always has one zero coördinate. Thus the point on the axis of y three units distant from O in the positive direction has the coördinates $(0, 3, 0)$, and the point in the (y, z)-plane, whose coördinates in that plane are $y = 2$, $z = 3$, has the coördinates $(0, 2, 3)$.

Octants. Bounded by the coördinate planes there are eight regions, called *octants.* It is clear that, if the x-coördinate of one point of an octant is, for example, negative, the x-coördinates of all points of the octant are negative; similarly, for the y- and z-coördinates. Thus we can speak of the octant $(-, +, +)$, and mean thereby that octant in which the x-coördinate of every point is negative and the y- and z-coördinates are both positive. The octant $(+, +, +)$ is known as the *first* octant; we make no attempt to number the others.

Figures. In drawing a figure in a plane to represent a figure in space, we make use of what is known as a parallel projection. The axes of y and z are represented by two perpendicular lines and the axis of x by a line drawn in a convenient direction. All distances in the (y, z)-plane or in any parallel plane are drawn to scale, so that a figure in such a plane appears as it actually is in space. Distances on or parallel to the axis of x are foreshortened a convenient amount.

The direction of the line representing the axis of x and the amount of foreshortening along this axis depend largely on the figure in space which is to be represented. In general, however, we shall draw the line representing the axis of x at an angle of 120° with that representing the axis of y and take as the unit distance on the axis of x three-fourths the unit distance on the other axes.

Right-Handed and Left-Handed Coördinate Systems. The system of axes in Fig. 4 is the one we shall employ. Another system in common use is shown in Fig. 5. The essential difference between the two is this: If, from a point on the *negative* axis of x, we view the (y, z)-plane, the direction of rotation from the positive y-axis to the positive z-axis is that of a right-handed screw in case of the first system and that of a left-handed screw in case of the second. Accordingly, the system we are using is called a *right-handed* system; the other, a *left-handed* system.

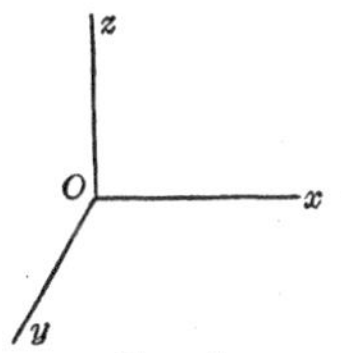

z
O x
y

FIG. 5

Any other rectangular system of axes is essentially the same as one or the other of these; that is, it is either right-handed or left-handed, by the above test (cf. Ex. 7).

EXERCISES

1. Plot the following points, drawing the line representing the axis of x at an angle of 120° with that representing the axis of y, and taking $\frac{1}{2}$ in. as the unit on the axes of y and z and $\frac{3}{8}$ in. as the unit on the axis of x.

(*a*) $(0, 3, 0)$; (*b*) $(0, 1, 3)$; (*c*) $(2, 5, 0)$;
(*d*) $(4, 0, 0)$; (*e*) $(0, -2, 0)$; (*f*) $(4, 1, 3)$;
(*g*) $(5, -2, 4)$; (*h*) $(3, 2, -5)$; (*i*) $(-2, 3, 1\frac{1}{2})$;
(*j*) $(1, -1, -3)$; (*k*) $(-2, 4, -3)$; (*l*) $(-1, -1, -2)$.

2. Determine the coördinates of the point P in Fig. 4 when the units on the axes are taken as in Ex. 1.

3. The same for the point marked by the period in "Fig. 4," if this point is $\frac{1}{2}$ a unit above the (x, y)-plane.

4. What are the coördinates of the projections of each of the following points on the coördinate axes? On the coördinate planes?

(*a*) $(3, 5, 2)$; (*b*) $(-3, 2, -1)$; (*c*) (x, y, z).

5. What equation is satisfied by the coördinates of those points and only those points which lie in the (y, z)-plane? In the (z, x)-plane? In the (x, y)-plane?

6. What two equations are satisfied by the coördinates of those points and only those points which lie on the x-axis? On the y-axis? On the z-axis?

7. Through a point O draw three mutually perpendicular lines, which, when directed, are to serve respectively as the axes of x, y, and z. Show that there are eight possible combinations of the directions which can be given to the lines and that, of the eight resulting systems of axes, four are right-handed and four, left-handed.

6. Projections of a Directed Line-Segment on the Axes. Given the points P_1 and P_2 with the coördinates (x_1, y_1, z_1) and (x_2, y_2, z_2). To determine the projections of the directed line-segment P_1P_2 on the coördinate axes, project the broken line P_1OP_2 on each of the axes in turn. Since always

$$\text{Proj. } P_1P_2 = \text{Proj. } P_1O + \text{Proj. } OP_2,$$

it follows that

$$\text{Proj. } P_1P_2 = \text{Proj. } OP_2 - \text{Proj. } OP_1.$$

But the projections of OP_2 and OP_1 on the three axes are, by definition, the coördinates of P_2 and P_1. Consequently, the projections of the directed line-segment P_1P_2 on the three axes are, respectively,

(1) $$x_2 - x_1 \qquad y_2 - y_1 \qquad z_2 - z_1.$$

By passing planes through the points P_1 and P_2 perpendicular to the three axes, we obtain on the axes the actual projections, X_1X_2, Y_1Y_2, Z_1Z_2, of the directed line-segment P_1P_2.* The planes also determine a rectangular parallelepiped, or box, whose three dimensions are equal to the numerical values of the three projections. Accordingly, the edges of the box, when properly directed, are precisely equal to the projections. In particular, the three edges emanating from P_1, *i.e.* the directed line-segments P_1R, P_1S, P_1T, are equal respectively to the three projections X_1X_2, Y_1Y_2, Z_1Z_2.

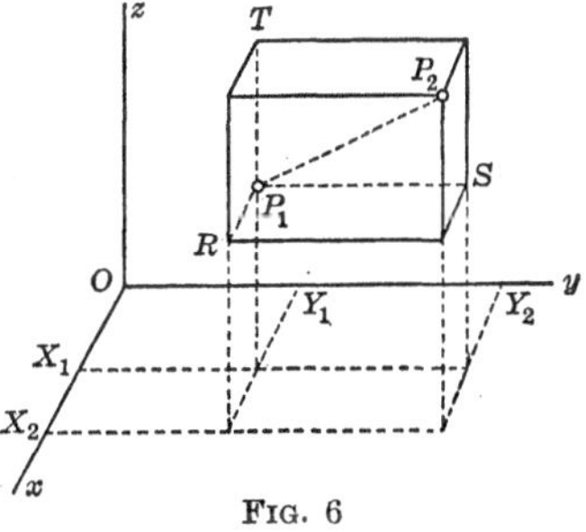

FIG. 6

EXERCISES

1. Plot P_1P_2 when P_1 is the point (*b*) of Ex. 1, § 5, and P_2 is (*c*). Determine the projections from the figure and verify by applying formulas (1).

* To keep the figure simple, only X_1X_2 and Y_1Y_2 are shown.

2. The same when

i) P_1 is (*c*) and P_2 is (*f*); ii) P_1 is (*f*) and P_2 is (*g*).

3. The projections of P_1P_2 on the axes are 2, -5, 3 and those of P_2P_3 are $-3, 2, 1$. What are the projections of P_1P_3? Justify your answer.

4. If the points P_1, R, S, T of Fig. 6 have, respectively, the coördinates (2, 1, 3), (5, 1, 3), (2, 4, 3), (2, 1, 6), what are the coördinates of P_2?

5. If the projections of P_1P_2 on the axes are 3, -5, -2 and P_1 has the coördinates (2, 1, 3), what are the coördinates of P_2?

7. Distance between Two Points. Let the two points be the points P_1, P_2 of § 6. Then the segment P_1P_2 is a diagonal of the box in Fig. 6. It is a simple matter to show that the square of the length of a diagonal equals the sum of the squares of the lengths of the edges:

$$P_1P_2^2 = P_1R^2 + P_1S^2 + P_1T^2.$$

Hence $$D^2 = (x_2 - x_1)^2 + (y_2 - y_1)^2 + (z_2 - z_1)^2,$$

and $$D = \sqrt{(x_2 - x_1)^2 + (y_2 - y_1)^2 + (z_2 - z_1)^2}.$$

Inasmuch as it is the squares of the quantities (1), § 6, which appear here under the radical, it is immaterial that these quantities have algebraic values, *i.e.* may in some cases be positive and in others, negative; cf. Ch. I, § 3.

EXERCISES

1. Find the distances between the following pairs of points, expressing the results correct to three significant figures.

(*a*) (5, 1, 4), (4, 3, 2); (*b*) (2, -1, 3), (-1, 1, -3);

(*c*) (2, -1, 8), (-2, -3, 5); (*d*) (3, 6, -2), (5, -1, 4);

(*e*) (2, -3, 5), (-1, 4, 5); (*f*) (1, 2, 4), (1, -3, 4).

2. Find the distances of each of the following points from the origin:

(*a*) (4, 2, 8); (*b*) (3, -5, -2); (*c*) (x, y, z).

3. Find the distances of each of the points of Ex. 2 from the coördinate axes.

4. Find the lengths of the projections on the coördinate planes of the line-segment joining the points (2, 3, 5) and (5, 6, 7). Draw a figure showing the projections, or line-segments equal to them.

5. What equation is satisfied by the coördinates of those points and only those points which lie on the *unit sphere,* — the sphere whose center is at the origin and whose radius is one unit?

8. Mid-Point of a Line-Segment. Let $P_1:(x_1, y_1, z_1)$ and $P_2:(x_2, y_2, z_2)$ be the extremities of the line-segment P_1P_2. If $P:(x, y, z)$ is the mid-point of P_1P_2, the directed line-segments P_1P and PP_2 are equal and have, therefore, equal projections on the coördinate axes. Thus we have, by (1), § 6,

$$x - x_1 = x_2 - x,$$

and similar equations in the y- and z-coördinates. Hence,

$$(1) \qquad x = \frac{x_1 + x_2}{2}, \qquad y = \frac{y_1 + y_2}{2}, \qquad z = \frac{z_1 + z_2}{2}.$$

This result can be stated in words as follows: *The coördinates of the mid-point of a line-segment are, respectively, the averages of the corresponding coördinates of the end-points of the segment.*

EXERCISES

1. Determine the coördinates of the mid-point of each of the line-segments given by the pairs of points in Ex. 1, § 7. Draw figures and check your answers.

2. Show that the sum of the squares of the diagonals of the quadrilateral whose successive vertices are at the four points (5, 0, 0), (0, 6, 0), (1, 2, 3), (3, − 2, 8) is double the sum of the squares of the line-segments joining the mid-points of the opposite sides. *N. B.* The four points do not lie in a plane.

3. Show that the line-segments joining the mid-points of the opposite sides of the quadrilateral of Ex. 2 intersect and bisect each other.

9. Division of a Line-Segment in a Given Ratio. Let it be required to find the coördinates (x, y, z) of the point P dividing the line-segment P_1P_2 in the given ratio m_1/m_2. Since the directed line-segments P_1P and PP_2 are to be in the ratio m_1/m_2, this must also be the ratio of their projections on any one of the axes (§ 4, Ex. 4). Accordingly, we obtain the equation

$$\frac{x - x_1}{x_2 - x} = \frac{m_1}{m_2},$$

and similar equations involving the y- and z-coördinates.

Solving these equations respectively for x, y, and z gives, as the required coördinates of the point P,

$$(1)\quad x = \frac{m_2x_1 + m_1x_2}{m_2 + m_1}, \quad y = \frac{m_2y_1 + m_1y_2}{m_2 + m_1}, \quad z = \frac{m_2z_1 + m_1z_2}{m_2 + m_1}.$$

External Division. It is sometimes of value to have at hand formulas giving the coördinates (x, y, z) of a point P which lies on the line of P_1 and P_2, but exterior to the segment P_1P_2, and whose distances to P_1 and P_2 are in a given ratio m_1/m_2, not equal to unity. Let the student show that in this case it is the directed line-segments P_1P and P_2P which are in the given ratio m_1/m_2, and that the type of equation now obtained is

$$\frac{x - x_1}{x - x_2} = \frac{m_1}{m_2},$$

so that the required coördinates of P are

$$(2)\quad x = \frac{m_2x_1 - m_1x_2}{m_2 - m_1}, \quad y = \frac{m_2y_1 - m_1y_2}{m_2 - m_1}, \quad z = \frac{m_2z_1 - m_1z_2}{m_2 - m_1}.$$

The point P is said to divide the segment P_1P_2 *internally*, in the first case; *externally*, in the second. The numbers m_1 and m_2 entering into the ratio of division do not have to be the actual lengths of the corresponding line-segments, but may be any numbers proportional to these lengths.*

EXERCISES

1. Find the coördinates of the point on the line-segment joining (2, – 3, 6) with (5, 4, – 2), which is twice as far from the first point as from the second. *Ans.* $(4, \frac{5}{3}, \frac{2}{3})$.

2. Find the point on the line through the two points of Ex. 1, which is outside the line-segment bounded by them and is twice as far from the first point as from the second.

3. Find the point which divides internally the line-segment from (2, 3, 4) to (5, – 3, 0) in the ratio 3 : 4.

4. The preceding exercise for external division.

EXERCISES ON CHAPTER XVII †

1. Show that the points (2, 4, 3), (4, 1, 9), (10, – 1, 6) are the vertices of an isosceles right triangle.

2. Prove that the tetrahedron with vertices at the points (0, 0, 0), (0, 1, 1), (1, 0, 1), (1, 1, 0) is a regular tetrahedron.

3. Show that the points $(0, 0, \sqrt{2})$, (1, 1, 0), $(0, 0, -\sqrt{2})$, (– 1, – 1, 0), (2, – 2, 0) are the vertices of a regular pyramid with a square base.

4. Given the points A, B, C with coördinates (2, – 3, 5), (4, 2, 3), (6, 7, 1). By proving that $AB + BC = AC$, show that the three points lie on a line.

5. Show that the three points of Ex. 4 lie on a line by proving that their projections on each of two coördinate planes lie on a line. Justify this method of proof.

6. Determine the point on the axis of y which is equidistant from the two points (3, – 2, 4), (– 2, 6, 5).

* Thus, in the case of internal division, if $P_1P = 100$ cm. and $PP_2 =$ 25 cm., m_1 and m_2 might be properly and wisely chosen as 4 and 1.

† Exercises 1–6, 14–18 of Ch. XIX, § 1, and Exercises 1–8, 18–22 of Ch. XX, § 1, may be introduced here, if it seems desirable.

7. Determine the point in the (y, z)-plane which is equidistant from the three points (3, 0, 2), (2, 3, 0), (1, 0, 0).

8. Two vertices of a regular tetrahedron are at the points $(0, 0, 2\sqrt{2})$, (0, 2, 0). If the other two vertices lie in the (x, y)-plane, find their coördinates.

9. A regular pyramid, of altitude h, has a square base whose vertices lie on the axes of x and y and whose edges are of length a. What are the coördinates of the vertices of the pyramid?

10. If P is the mid-point of the line-segment P_1P_2, and P and P_2 have the coördinates (3, -2, 5) and (-2, 4, 3) respectively, what are the coördinates of P_1?

11. If P divides the line-segment P_1P_2 internally in the ratio 2 : 3, and P_1 and P have respectively the coördinates (1, 4, 3) and (3, 2, -1), determine the coördinates of P_2.
Ans. (6, -1, -7).

12. Find the ratio in which the point B of Ex. 4 divides the segment AC of that exercise. *Ans.* 1 : 1.

13. A point with x-coördinate 6 lies on the line joining the two points (2, -3, 4), (8, 0, 10). Find its other two coördinates.

Suggestion. Determine the ratio in which the point divides the line-segment bounded by the two given points.

14. Find the point in which the line joining the two points (2, -3, 1), (5, 4, 6) meets the (z, x)-plane.
Ans. $(3\frac{2}{7}, 0, 3\frac{1}{7})$.

15. If the length of the line-segment P_1P_2 is D and the lengths of its projections on the coördinate planes are D_1, D_2, D_3, show that

$$2D^2 = D_1^2 + D_2^2 + D_3^2.$$

16. Show that the lines joining the mid-points of the opposite sides of any quadrilateral $ABCD$ intersect and bisect each other. *N.B.* The points A, B, C, D do not necessarily lie in a plane.

17. Show that the sum of the squares of the diagonals of any quadrilateral is twice the sum of the squares of the line-segments joining the mid-points of the opposite sides.

18. Prove that the center of gravity (intersection of the medians) of the triangle with vertices at (x_1, y_1, z_1), (x_2, y_2, z_2), (x_3, y_3, z_3) has the coördinates

$$\tfrac{1}{3}(x_1 + x_2 + x_3), \qquad \tfrac{1}{3}(y_1 + y_2 + y_3), \qquad \tfrac{1}{3}(z_1 + z_2 + z_3).$$

19. Prove that the lines joining the vertices of a tetrahedron with the centers of gravity of the opposite faces all go through a point P, which divides each of them in the ratio $3:1$.

20. Prove that the lines joining the mid-points of opposite edges of a tetrahedron all go through a point, which bisects each of them. Show that this point is identical with the point P of Ex. 19.

CHAPTER XVIII

DIRECTION COSINES. DIRECTION COMPONENTS

1. Direction Cosines of a Directed Line. Given a directed line L in space; to find a means of determining or fixing its direction.

The directed line L makes definite angles, α, β, γ, with the positive axes of x, y, z, respectively. If L does not go through the origin, O, draw L' through O parallel to L and agreeing with it in sense. Then α, β, γ are equal respectively to the angles which L' makes with the axes (Ch. XVII, § 3). The angles α, β, γ are called the *direction angles* of the directed line L.

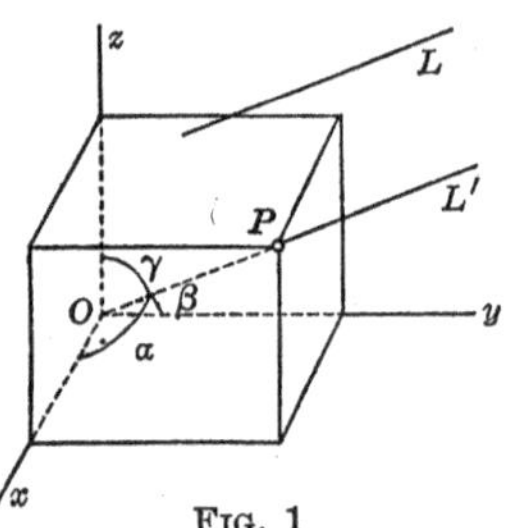

FIG. 1

Direction Cosines. The cosines of the angles α, β, γ, namely $\cos\alpha$, $\cos\beta$, $\cos\gamma$, are known as the *direction cosines* of L. Since α, β, γ are, by definition (Ch. XVII, § 3), angles between 0° and 180° inclusive, they are uniquely determined when their cosines are given, and conversely. Accordingly, we can use either the direction angles or the direction cosines to fix the direction of L. We choose the direction cosines.

Evidently, two directed lines which are parallel and have the same sense have the same direction angles and the same direction cosines.

Exercise. If two lines are parallel but have opposite senses, show that the direction angles of one are the supplements of the direction angles of the other, and that the direction cosines of one are the negatives of the direction cosines of the other.

Example 1. What are the direction cosines of the positive axis of y?

Here, $\alpha = 90°, \quad \beta = 0°, \quad \gamma = 90°;$

and $\cos\alpha = 0, \quad \cos\beta = 1, \quad \cos\gamma = 0.$

Example 2. Find the direction cosines of the line bisecting the angle between the negative axis of y and the positive axis of z, and directed upward.

In this case,

$$\alpha = 90°, \quad \beta = 135°, \quad \gamma = 45°;$$

$$\cos\alpha = 0, \quad \cos\beta = -\tfrac{1}{2}\sqrt{2}, \quad \cos\gamma = \tfrac{1}{2}\sqrt{2}.$$

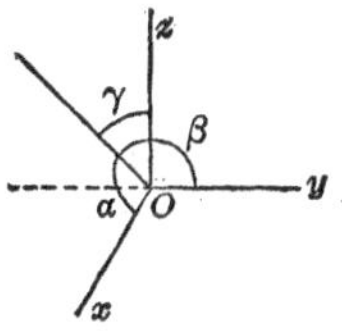

FIG. 2

THEOREM 1. *The sum of the squares of the direction cosines of a directed line is equal to unity:*

$$(1) \qquad \cos^2\alpha + \cos^2\beta + \cos^2\gamma = 1.$$

To prove this theorem, take a point $P: (x_0, y_0, z_0)$ on L' (Fig. 1) so that the direction from O to P will be the direction of L', and consider the projections of the directed line-segment OP on the axes. These are equal, on the one hand, to the coördinates x_0, y_0, z_0 of P (Ch. XVII, § 5), and on the other, to the quantities $OP\cos\alpha$, $OP\cos\beta$, $OP\cos\gamma$ (Ch. XVII, § 4). Hence

$$(2) \qquad x_0 = OP\cos\alpha, \qquad y_0 = OP\cos\beta, \qquad z_0 = OP\cos\gamma;$$

$$(3) \qquad \cos\alpha = \frac{x_0}{OP}, \qquad \cos\beta = \frac{y_0}{OP}, \qquad \cos\gamma = \frac{z_0}{OP}.$$

Thus $$\cos^2\alpha + \cos^2\beta + \cos^2\gamma = \frac{x_0^2 + y_0^2 + z_0^2}{OP^2}.$$

But $$OP^2 = x_0^2 + y_0^2 + z_0^2,$$

and the theorem is proved.

We have shown, then, that every directed line has definite direction cosines, the sum of whose squares is unity. The converse is also true: *Any three numbers, the sum of whose squares is unity, are the direction cosines of some directed line.*

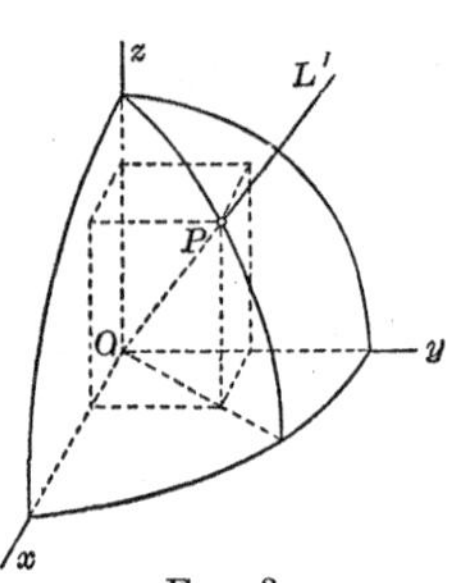

FIG. 3

Preliminary to proving this, we revert to the proof of Theorem 1 and choose the point P in particular as the point in which the ray issuing from O in the direction of L' meets the unit sphere (Fig. 3). Then $OP = 1$, or

$$x_0^2 + y_0^2 + z_0^2 = 1,$$

and (2) becomes

$$(4) \quad x_0 = \cos\alpha, \quad y_0 = \cos\beta, \quad z_0 = \cos\gamma.$$

That is, *the direction cosines of a ray issuing from O are equal to the coördinates of the point in which the ray pierces the unit sphere.*

The desired proof is now simple. If there are given any three numbers, x_0, y_0, z_0, the sum of whose squares is unity, they will be the coördinates of some point P of the unit sphere, and hence they will also be the direction cosines of a certain directed line, namely, the line L' passing through O and P and directed from O to P, q. e. d.

Example 3. The three numbers $\frac{2}{3}$, $\frac{1}{3}$, $-\frac{2}{3}$ are the direction cosines of some directed line, for

$$(\tfrac{2}{3})^2 + (\tfrac{1}{3})^2 + (-\tfrac{2}{3})^2 = \tfrac{4}{9} + \tfrac{1}{9} + \tfrac{4}{9} = 1.$$

The direction angles of the line are, respectively, 48° 11′, 70° 32′, 131° 49′.

Example 4. A directed line makes angles of 60° and 45° with the axes of x and y, respectively. What angle does it make with the axis of z?

Here,

$$\cos\alpha = \cos 60° = \tfrac{1}{2}, \qquad \cos\beta = \cos 45° = \tfrac{1}{2}\sqrt{2}.$$

Hence

$$(\tfrac{1}{2})^2 + (\tfrac{1}{2}\sqrt{2})^2 + \cos^2\gamma = 1 \qquad \text{and} \qquad \cos\gamma = \pm\tfrac{1}{2}.$$

Thus $\gamma = 60°$ or $120°$. There are, then, two directed lines making the given angles with the x- and y-axes. The one makes an angle of $60°$ with the z-axis; the other, an angle of $120°$.

Direction Cosines of the Line through Two Points. Let P_1P_2 be a directed line-segment lying on the directed line L and having the same sense as L. If X_1X_2, Y_1Y_2, Z_1Z_2 are the projections on the axes of P_1P_2 (Ch. XVII, Fig. 6), we have, by Ch. XVII, § 4,

$$X_1X_2 = D\cos\alpha, \qquad Y_1Y_2 = D\cos\beta, \qquad Z_1Z_2 = D\cos\gamma,$$

where D is the length of the segment P_1P_2.
Hence

$$(5)\quad \cos\alpha = \frac{X_1X_2}{D}, \qquad \cos\beta = \frac{Y_1Y_2}{D}, \qquad \cos\gamma = \frac{Z_1Z_2}{D}.$$

The content of these equations can be stated in words, as follows:

THEOREM 2. *If a directed line L is given and on L any directed line-segment P_1P_2 having the same sense as L is chosen, the direction cosines of L are equal to the projections of P_1P_2 on the axes, each divided by the length of P_1P_2.*

If P_1 and P_2 have the coördinates (x_1, y_1, z_1) and (x_2, y_2, z_2), formulas (5) become, by Ch. XVII, §§ 6, 7,

$$(6)\quad \cos\alpha = \frac{x_2 - x_1}{D}, \qquad \cos\beta = \frac{y_2 - y_1}{D}, \qquad \cos\gamma = \frac{z_2 - z_1}{D},$$

where $\qquad D = \sqrt{(x_2 - x_1)^2 + (y_2 - y_1)^2 + (z_2 - z_1)^2}.$

These are the formulas giving the direction cosines of the line passing through $P_1: (x_1, y_1, z_1)$ and $P_2: (x_2, y_2, z_2)$, and directed from P_1 to P_2.

EXERCISES

1. What are the direction cosines of a line parallel to the axis of z and having the same sense? Having the opposite sense?

2. A line bisects the angle between the positive axes of y and z and is directed upward. What are its direction cosines?

3. What are the direction cosines of a directed line which lies in the (z, x)-plane and makes an angle of 30° with the positive z-axis? Two answers.

4. A line in the (x, y)-plane has the slope $\sqrt{3}$. What are its direction angles and direction cosines, if it is directed forwards?

5. Construct the *directed* lines through the origin having the following direction cosines. What are the direction angles?

(*a*) $-1, 0, 0$; (*b*) $\frac{1}{2}, \frac{1}{2}\sqrt{3}, 0$; (*c*) $-\frac{1}{2}\sqrt{3}, 0, \frac{1}{2}$;
(*d*) $0, -\frac{1}{2}\sqrt{2}, \frac{1}{2}\sqrt{2}$; (*e*) $-\frac{3}{5}, -\frac{4}{5}, 0$; (*f*) $\frac{1}{3}\sqrt{3}, -\frac{1}{3}\sqrt{3}, \frac{1}{3}\sqrt{3}$.

6. Find the direction angles and the direction cosines of a line if

(*a*) $\cos\alpha = 1$; (*b*) $\cos\beta = -\frac{1}{2}$, $\cos\gamma = \frac{1}{2}\sqrt{3}$;
(*c*) $\cos\alpha = \frac{1}{2}$, $\cos\beta = -\frac{1}{2}\sqrt{2}$.

7. Find the direction angles and the direction cosines of a directed line if

(*a*) $\alpha = 120°, \beta = 60°$; (*b*) $\alpha = 135°, \gamma = 120°$;
(*c*) $\beta = \frac{\pi}{4}, \gamma = \frac{\pi}{3}$; (*d*) $\alpha = 45°, \beta = \gamma$;
(*e*) $\alpha = \beta = \gamma$; (*f*) $\alpha = \gamma = 180° - \beta$.

8. Find the direction cosines of the line passing through the origin and each of the following points, and directed from the origin to the point:

(*a*) $(2, 3, 6)$; (*b*) $(4, -1, 8)$; (*c*) $(3, -4, 0)$; (*d*) $(5, 8, -1)$.

9. Find the direction cosines of the lines determined by the pairs of points in Ex. 1 of Ch. XVII, § 7, if each line is directed from the first of the given points to the second.

10. A line-segment P_1P_2 has the length 6 and the line of P_1 and P_2, directed from P_1 to P_2, has the direction cosines $-\frac{2}{3}$, $\frac{1}{3}, \frac{2}{3}$. If the coördinates of P_1 are $(-3, 2, 5)$, what are those of P_2?

2. Angle between Two Directed Lines. Let it be required to find the angle θ between the two directed lines, L_1 and L_2, whose direction angles are $\alpha_1, \beta_1, \gamma_1$ and $\alpha_2, \beta_2, \gamma_2$.

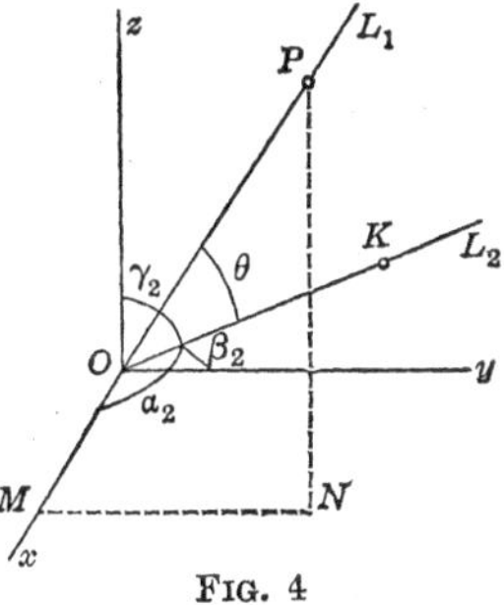

FIG. 4

We can assume without loss of generality that L_1 and L_2 pass through the origin.* Take any point $P: (x_0, y_0, z_0)$ on L_1, so that the direction from O to P is that of L_1, and draw the broken line $OMNP$, whose directed segments OM, MN, NP are, respectively, the coördinates x_0, y_0, z_0 of P. The projection of this broken line on L_2 equals the projection of OP on L_2:

(1) $\text{Proj.}_{L_2}\, OP = \text{Proj.}_{L_2}\, OM + \text{Proj.}_{L_2}\, MN + \text{Proj.}_{L_2}\, NP.$

By Ch. XVII, § 4,

$$\text{Proj.}_{L_2}\, OP = OP \cos\theta.$$

Similarly, $\quad \text{Proj.}_{L_2}\, OM = |OM| \cos \measuredangle KOM,$

where K is a point on L_2, such that OK has the direction of L_2. If the directed line-segment OM has the direction of the positive axis of x, as is the case in the figure, we have

$$|OM| = OM, \qquad \measuredangle KOM = \alpha_2,$$

and therefore,

$$\text{Proj.}_{L_2}\, OM = OM \cos\alpha_2.$$

If OM has the direction of the negative axis of x,

$$|OM| = -OM \quad \text{and} \quad \measuredangle KOM = 180^\circ - \alpha_2;$$

in this case, then,

$$\text{Proj.}_{L_2}\, OM = -OM \cos(180^\circ - \alpha_2) = OM \cos\alpha_2.$$

Consequently, in either case, we have, since $OM = x_0$,

$$\text{Proj.}_{L_2}\, OM = x_0 \cos\alpha_2.$$

* For, if they did not, we could consider equally well the angle between the two parallel lines through the origin having respectively the same senses as the given lines.

Similarly,

$$\text{Proj.}_{L_2}\, MN = y_0 \cos \beta_2, \qquad \text{Proj.}_{L_2}\, NP = z_0 \cos \gamma_2.$$

Thus (1) becomes

(2) $$OP \cos \theta = x_0 \cos \alpha_2 + y_0 \cos \beta_2 + z_0 \cos \gamma_2.$$

By (2), § 1,

$$x_0 = OP \cos \alpha_1, \qquad y_0 = OP \cos \beta_1, \qquad z_0 = OP \cos \gamma_1.$$

Substituting these values in (2) and dividing through by OP, we obtain, finally,

(3) $$\cos \theta = \cos \alpha_1 \cos \alpha_2 + \cos \beta_1 \cos \beta_2 + \cos \gamma_1 \cos \gamma_2.$$

We have, then, the result: *The cosine of the angle between two directed lines equals the sum of the products of the corresponding direction cosines of the lines.*

Example. Find the angle between the two directed lines whose direction cosines are, respectively, $\frac{2}{3}, -\frac{1}{3}, -\frac{2}{3}$ and $\frac{3}{7}, \frac{2}{7}, \frac{6}{7}$.

Here $$\cos \theta = \frac{2 \cdot 3 + (-1) \cdot 2 + (-2) \cdot 6}{3 \cdot 7} = -\frac{8}{21},$$

whence θ is found to be 112° 24′.*

Parallel and Perpendicular Directed Lines. L_1 and L_2 are perpendicular if and only if $\theta = 90°$ or $\cos \theta = 0$, that is, if and only if

(4) $$\cos \alpha_1 \cos \alpha_2 + \cos \beta_1 \cos \beta_2 + \cos \gamma_1 \cos \gamma_2 = 0.$$

In words: *Two directed lines are perpendicular, if and only if the sum of the products of the corresponding direction cosines of the lines is equal to zero.* Thus the directed lines which have the direction cosines $\frac{2}{3}, -\frac{1}{3}, \frac{2}{3}$ and $\frac{2}{15}, \frac{14}{15}, \frac{5}{15}$ are perpendicular, since

$$\frac{2 \cdot 2 - 1 \cdot 14 + 2 \cdot 5}{3 \cdot 15} = 0.$$

* It is to be remembered that the angle between two directed angles is an angle between 0° and 180° inclusive; cf. Ch. XVII, § 3.

We repeat here the results concerning parallelism obtained in § 1. The directed lines L_1 and L_2 are parallel and have the same sense if and only if they have equal direction cosines:

$$(5)\quad \cos\alpha_1 = \cos\alpha_2, \qquad \cos\beta_1 = \cos\beta_2, \qquad \cos\gamma_1 = \cos\gamma_2.$$

On the other hand, they are parallel, but with opposite senses, when and only when the corresponding direction cosines are negatives of each other:

$$(6)\quad \cos\alpha_1 = -\cos\alpha_2, \qquad \cos\beta_1 = -\cos\beta_2, \qquad \cos\gamma_1 = -\cos\gamma_2.$$

EXERCISES

In each of the following exercises find the angle between two directed lines with the given direction cosines.

1. $\frac{2}{7}, \frac{3}{7}, \frac{6}{7}; \ \frac{6}{7}, \frac{2}{7}, -\frac{3}{7}.$ 2. $\frac{2}{3}, -\frac{1}{3}, -\frac{2}{3}; \ -\frac{2}{3}, \frac{1}{3}, \frac{2}{3}.$

3. $\frac{2}{3}, -\frac{2}{3}, \frac{1}{3}; \ \frac{8}{9}, \frac{4}{9}, \frac{1}{9}.$ 4. $\frac{3}{13}, -\frac{12}{13}, \frac{4}{13}; \ -\frac{9}{11}, \frac{6}{11}, \frac{2}{11}.$

5. $\dfrac{2}{\sqrt{14}}, -\dfrac{1}{\sqrt{14}}, \dfrac{3}{\sqrt{14}}; \ \dfrac{2}{\sqrt{6}}, \dfrac{1}{\sqrt{6}}, -\dfrac{1}{\sqrt{6}}.$

6. $\dfrac{4}{\sqrt{21}}, \dfrac{2}{\sqrt{21}}, \dfrac{1}{\sqrt{21}}; \ \dfrac{2}{\sqrt{14}}, \dfrac{1}{\sqrt{14}}, -\dfrac{3}{\sqrt{14}}.$

7. Show that three directed lines with the direction cosines

$$\tfrac{12}{13}, -\tfrac{3}{13}, -\tfrac{4}{13}, \qquad \tfrac{4}{13}, \tfrac{12}{13}, \tfrac{3}{13}, \qquad \tfrac{3}{13}, -\tfrac{4}{13}, \tfrac{12}{13},$$

are mutually perpendicular.

8. Find the angle subtended at the point (5, 2, 3) by the points (2, 0, – 3), (– 9, 7, 5). *Ans.* 79° 1′.

9. Determine the angles of the triangle with vertices at the points (1, 0, 0), (0, 2, 0), (0, 0, 3).

3. Direction Components of an Undirected Line. The quantities $\frac{2}{7}, -\frac{3}{7}, \frac{6}{7}$ are the direction cosines of some line, properly directed, and the quantities $-\frac{2}{7}, \frac{3}{7}, -\frac{6}{7}$ are the direction cosines of this line, oppositely directed.

Both sets of direction cosines are proportional to the quantities 2, -3, 6. Consequently, these quantities pertain, not to the line directed in the one sense or the other, but to the line bare of sense, *i.e.* to the *undirected* line. We call them *direction components* of the *undirected* line.

It is clear that instead of 2, -3, 6 we might have taken equally well -2, 3, -6, or 4, -6, 12, or 200, -300, 600, since the two sets of direction cosines are proportional to the quantities in any one of these triples. In other words, the direction components of the undirected line are not uniquely determined. There are infinitely many sets of direction components; if one set is 2, -3, 6, all are given by the quantities 2ρ, -3ρ, 6ρ, where ρ is an arbitrary number, not zero.

Conversely, if we have given the set of direction components, 2, -3, 6, of the undirected line, and divide each by the square root of the sum of the squares of the three, *i.e.* by

$$\sqrt{(2)^2 + (-3)^2 + (6)^2} = 7,$$

we obtain the direction cosines, $\frac{2}{7}$, $-\frac{3}{7}$, $\frac{6}{7}$, of the line directed in one sense. Those of the line directed in the opposite sense are the negatives, -2, 3, -6, of the given direction components, each divided by the above square root.

The General Case. Let a line L be given and on it the arbitrary directed line-segment P_1P_2 whose projections on the axes are X_1X_2, Y_1Y_2, Z_1Z_2 (Ch. XVII, Fig. 6). The direction cosines of L, when directed in the sense of P_1P_2, are, by (5), § 1,

$$(1) \qquad \frac{X_1X_2}{D}, \quad \frac{Y_1Y_2}{D}, \quad \frac{Z_1Z_2}{D};$$

if, however, L is oppositely directed, in the sense of P_2P_1, they are

$$(2) \qquad \frac{X_2X_1}{D}, \quad \frac{Y_2Y_1}{D}, \quad \frac{Z_2Z_1}{D},$$

where, in each case,

$$(3) \qquad D = \sqrt{\overline{X_1X_2}^2 + \overline{Y_1Y_2}^2 + \overline{Z_1Z_2}^2}.$$

The two sets of direction cosines are proportional to the quantities *

$$(4) \qquad X_1X_2, \qquad Y_1Y_2, \qquad Z_1Z_2.$$

These quantities pertain merely to the *undirected* line L. We call them a set of *direction components* of L.

Since the quantities (4) are the projections of P_1P_2 on the axes, this definition can be stated as follows.

DEFINITION. *A set of direction components of an* UNDIRECTED *line L are the projections on the axes of a directed line-segment on L.*

Instead of X_1X_2, Y_1Y_2, Z_1Z_2, we might have taken, as direction components of L, X_2X_1, Y_2Y_1, Z_2Z_1, *i.e.* the projections of P_2P_1 on the axes; or $3\,X_1X_2$, $3\,Y_1Y_2$, $3\,Z_1Z_2$, *i.e.* the projections of a directed line-segment on L having the same sense as P_1P_2 but three times the length.

There are, then, *infinitely many* sets of direction components for L. *Any two sets are*, however, *proportional*. For, two arbitrary sets consist of the projections on the axes, X_1X_2, Y_1Y_2, Z_1Z_2 and $X_1'X_2'$, $Y_1'Y_2'$, $Z_1'Z_2'$, of two arbitrary directed line-segments, P_1P_2 and $P_1'P_2'$, on L. But the projections of $P_1'P_2'$ and P_1P_2 on any line are in the same ratio as $P_1'P_2'$ and P_1P_2 (Ch. XVII, § 4, Ex. 4), and, therefore,

$$\frac{X_1'X_2'}{X_1X_2} = \frac{P_1'P_2'}{P_1P_2}, \qquad \frac{Y_1'Y_2'}{Y_1Y_2} = \frac{P_1'P_2'}{P_1P_2}, \qquad \frac{Z_1'Z_2'}{Z_1Z_2} = \frac{P_1'P_2'}{P_1P_2},$$

or

$$(5) \quad X_1'X_2' = \rho\, X_1X_2, \qquad Y_1'Y_2' = \rho\, Y_1Y_2, \qquad Z_1'Z_2' = \rho\, Z_1Z_2,$$

where the factor of proportionality, ρ, is $P_1'P_2'/P_1P_2$, q. e. d.

Not all three direction components can be zero. For, if X_1X_2, Y_1Y_2, Z_1Z_2 were all zero, then, by (3), $D = |\,P_1P_2\,|$ would be zero. But this is absurd, since P_1 and P_2 are distinct points.

We summarize our results in the form of a theorem.

* The factor of proportionality is, in the first case, $1/D$; in the second, $-1/D$.

THEOREM 1. *An undirected line L has infinitely many sets of direction components; if l, m, n is one set, all the sets are given by $\rho l, \rho m, \rho n$, where ρ is an arbitrary number, not zero; moreover, l, m, n are not all zero. Any line parallel to L has the same sets of direction components as L.*

The last statement has not been explicitly proved. We leave the proof to the student; cf. Ch. XVII, § 4, Ex. 5.

Example 1. Find the direction components of a line parallel to the axis of z.

Take any directed line-segment P_1P_2 on the line. Its projections on the axes are $0, 0, P_1P_2$, or, if a is the number representing P_1P_2, they are $0, 0, a$. One set of direction components is, then, $0, 0, a$ $(a \neq 0)$; a simpler set, and the one generally used, is $0, 0, 1$.

Example 2. A line bisects the angle between the positive axes of y and z. What are its direction components?

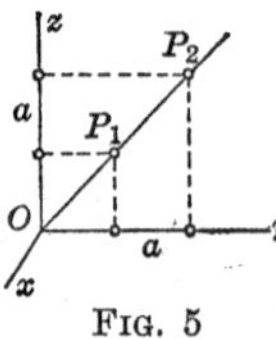

FIG. 5

The projection on the x-axis of any directed line-segment P_1P_2 on the given line is zero, and the projections on the axes of y and z are equal. If the number representing both the latter projections is a, a set of direction components for the line is $0, a, a$. A simpler set is $0, 1, 1$.

Geometrical Representation of Direction Components. The directed line-segments X_1X_2, Y_1Y_2, Z_1Z_2, which are the projections of P_1P_2 on the axes, represent geometrically the set of direction components (4) of L. Instead of them we prefer to use the equal directed line-segments P_1R, P_1S, P_1T, issuing from P_1 (Fig. 6). These form what we shall call a *directed trihedral*; P_1R, P_1S, P_1T are its *directed edges*, and P_1, its *vertex*.

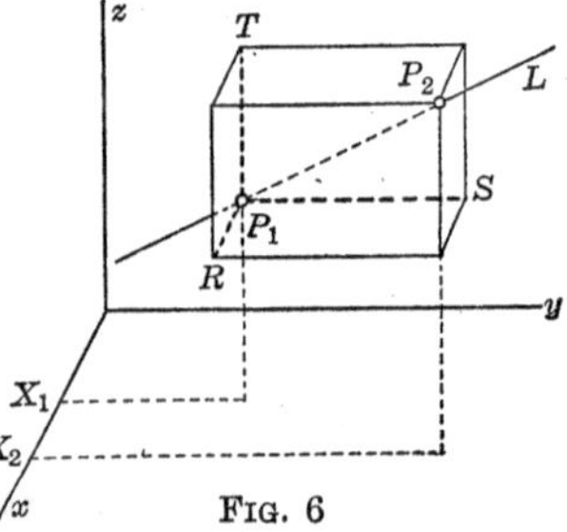

FIG. 6

The directed trihedral P_1–RST represents the set of direction components (4). Any second set, consisting of the projections on the axes, $X_1'X_2'$, $Y_1'Y_2'$, $Z_1'Z_2'$, of any second directed line-segment $P_1'P_2'$ on L is represented by the directed trihedral P_1'–$R'S'T'$. For the two directed trihedrals we have, from (5),

$$(6) \qquad P_1'R' = \rho P_1R, \qquad P_1'S' = \rho P_1S, \qquad P_1'T' = \rho P_1T.$$

Because of this relationship we call them *similar*. That is, *two directed trihedrals are similar, if homologous directed edges are proportional, i.e.* if the directions of the three edges of one trihedral are *all* the same as, or *all* opposite to, the directions of the three edges of the other, and if the lengths of homologous edges are proportional.

Since any two sets of direction components of L are in the relation (5), the directed trihedrals representing them are in the relation (6) and are, therefore, similar. Consequently, the directed trihedrals representing the infinitely many sets of direction components of L are all similar.

Construction of a Line with Given Direction Components. Let it be required to construct the line L passing through a given point P_1 in space and having the direction components 4, – 3, 2.

Construct a directed trihedral P_1–RST with P_1 as vertex and with edges P_1R, P_1S, P_1T defined, both in length and direction, by the numbers 4, – 3, 2. Complete the box determined by the trihedral and draw the diagonal P_1P_2 issuing from P_1. The line of this diagonal is the required line L. For, the projections of P_1P_2 on the axes have the values 4, – 3, 2.

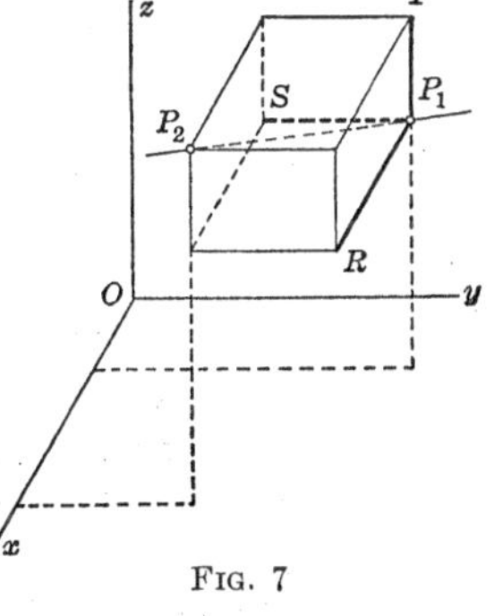

FIG. 7

Incidentally, we have shown that the triple 4, – 3, 2 is actually a set of direction components of some line, L. We proceed to show, further, that L and the lines parallel to it are the *only* lines having this triple or, more generally, the triples 4ρ, -3ρ, 2ρ, $\rho \neq 0$, as direction components.

Evidently L is the only line through P_1 with the direction components 4, -3, 2. For, these components determine the trihedral at P_1 *uniquely*, the trihedral determines the box *uniquely*, and the box the line.

If we took 8, -6, 4 instead of 4, -3, 2 as the given direction components, the resulting trihedral would have edges with the same directions, but twice as long, as the edges of the original trihedral, *i.e.* it would be similar to the original trihedral. The diagonal of the new box which issues from P_1 would be on a line with the diagonal P_1P_2 of the old box and so the same line L would be determined. Similarly, if any multiple, 4ρ, -3ρ, 2ρ, of 4, -3, 2, where ρ is any positive or negative number, were taken as the direction components.

Finally, if we start from a new point P_1', it is clear that the line L' through it with the given direction components will be parallel to L or, in case P_1' lies on L, the same as L.

The reasoning here is perfectly general, applying to any triple of numbers, l, m, n, not all zero. The result is the following converse of Theorem 1.

THEOREM 2. *If l, m, n are any three numbers, not all zero, the triples of numbers ρl, ρm, ρn, where ρ is arbitrary but not zero, are sets of direction components of some undirected line L and of the lines parallel to L, and of these lines only.*

Remark. If one direction component is zero, the corresponding edge of the directed trihedral disappears, and the box becomes a rectangle, with L along its diagonal. If two direction components are zero, the directed trihedral becomes a directed line and L lies along this line.

EXERCISES

In each of the following exercises find all the sets of direction components of the given line and then choose from them a simple set.

1. A line parallel to the axis of x.

2. The line bisecting the angle between the positive axis of x and the negative axis of z.

3. A line in the (x, y)-plane having the slope 2.

Ans. $\rho, 2\rho, 0$; $1, 2, 0$.

4. A line in the (y, z)-plane making an angle of $60°$ with the y-axis. Two answers.

5. A line making equal angles with the three coördinate axes.

6. The line through the origin and the point $(2, 1, 3)$.

7. The line through the points $(2, 3, 5)$, $(4, 7, 8)$.

8. What can you say of the position of a line if one of its direction components is zero? If two are zero?

In each of the following exercises construct the line through the given point with the given direction components.

	Point	*Direction Components*
9.	Origin,	$3, 5, 2$.
10.	Origin,	$2, -3, 6$.
11.	$(2, 4, 3)$,	$\rho, \rho, 2\rho$ $(\rho \neq 0)$.
12.	$(5, -4, 6)$,	$3, 0, -1$.
13.	$(2, 5, -3)$,	$0, 1, 0$.

4. Formulas for the Use of Direction Components. *Direction Components of the Line through Two Points.* Let the two points $P_1 : (x_1, y_1, z_1)$, $P_2 : (x_2, y_2, z_2)$ be given. Since the projections of P_1P_2 on the axes are (Ch. XVII, § 6),

$$(1) \qquad x_2 - x_1, \qquad y_2 - y_1, \qquad z_2 - z_1,$$

these three quantities are a set of direction components of the undirected line passing through P_1 and P_2.

Relationship between Direction Components and Direction Cosines. We saw in § 3 that the direction components of an undirected line are any three numbers, not all zero, proportional to the direction cosines of the line, directed in one sense or the other.

Conversely, starting with the arbitrary set of direction components, X_1X_2, Y_1Y_2, Z_1Z_2, of an undirected line, and dividing each component by the square root of the sum of the squares of the three, *i.e.* by the quantity D given by (3), § 3, we obtain the direction cosines (1), § 3 of the line, directed in one sense. And dividing the negatives of the direction components by the same square root, we get the direction cosines (2), § 3 of the line, directed oppositely.

We state this result as a theorem.

THEOREM. *If l, m, n are a set of direction components of an undirected line L, the direction cosines of L, when given a sense, are*

$$(2) \qquad \cos\alpha = \frac{\pm l}{\sqrt{l^2+m^2+n^2}}, \quad \cos\beta = \frac{\pm m}{\sqrt{l^2+m^2+n^2}}, \quad \cos\gamma = \frac{\pm n}{\sqrt{l^2+m^2+n^2}},$$

where either all the upper signs, or all the lower signs, are to be chosen according to the sense which has been given to L.

If we had used, as the direction components of L, the arbitrary set ρl, ρm, ρn ($\rho \neq 0$) instead of the particular set l, m, n, the same formulas (2) would have resulted. For if in (1) l, m, n are replaced by ρl, ρm, ρn, ρ comes out as a factor from the square root in the denominator of each fraction and cancels the ρ in the numerator, so that the fractions are left unchanged.*

Example 1. A directed line has the direction cosines $\frac{2}{3}$, $-\frac{1}{3}$, $-\frac{2}{3}$. What are the direction components of the line, undirected?

Obviously, 2, -1, -2 are one set of direction components, the one generally used; all sets are given by 2ρ, $-\rho$, -2ρ, where $\rho \neq 0$.

Example 2. An undirected line has the direction components 4, -3, 12. What are the direction cosines of the line when directed?

The sum of the squares of the given direction components is 169. Hence the direction cosines of the directed line are either $\frac{4}{13}$, $-\frac{3}{13}$, $\frac{12}{13}$, or $-\frac{4}{13}$, $\frac{3}{13}$, $-\frac{12}{13}$, depending on the sense in which the line is directed.

Angles between Two Undirected Lines. Between two directed lines there is but one angle θ such that $0 < \theta \leq 180°$. Between two undirected lines L_1 and L_2 there are, in general, two such angles, as can be seen readily from a figure. The two angles are supplementary and have, therefore, cosines which are negatives of each other.

If the direction components of L_1 and L_2 are l_1, m_1, n_1 and l_2, m_2, n_2, these cosines are given by the formula

$$\cos\theta = \pm\frac{l_1 l_2 + m_1 m_2 + n_1 n_2}{\sqrt{l_1^2 + m_1^2 + n_1^2}\sqrt{l_2^2 + m_2^2 + n_2^2}}. \tag{3}$$

To establish this formula, write down by use of (2) the direction cosines of each line, directed in either sense, and then apply formula (3), § 2.

It follows from (3) that the two lines L_1 and L_2 are perpendicular when and only when

$$l_1 l_2 + m_1 m_2 + n_1 n_2 = 0. \tag{4}$$

* If ρ is negative, it is $-\rho$ which comes out as a factor from each square root and hence each sign $\pm$ becomes $\mp$.

The lines L_1 and L_2 are parallel if and only if their direction components are proportional, *i.e.* if and only if

$$(5) \qquad l_2 = \rho l_1, \qquad m_2 = \rho m_1, \qquad n_2 = \rho n_1,$$

where ρ is a number not 0. This follows directly from Theorems 1, 2, § 3.

EXERCISES

1. From the direction cosines of the directed lines of Exs. 1, 4, 6, § 2, find the direction components of the lines, undirected.

2. From the direction components of the undirected lines of Exs. 9–13, § 3, find the direction cosines of the lines, directed first in one sense and then in the other.

3. A line has the direction components 2, 8, 9. What are its direction cosines, if it is directed upwards?

4. Find the direction components of the lines joining the origin with the points (*c*), (*f*), (*i*), (*l*) of Ex. 1 of Ch. XVII, § 5.

5. Find the direction components of the lines determined by the pairs of points in Ex. 1 of Ch. XVII, § 7.

In each of the following cases determine the angles between two lines with the given direction components. First test the lines for parallelism or perpendicularity.

6. 3, 4, -1; 5, -2, 7.

7. 4, -2, 6; -6, 3, -9.

8. 2, -1, 3; 2, 1, -1.

9. -3, 4, 2; 5, 8, 1.

10. Show that the line joining the origin to the point (2, 1, 1) is perpendicular to the line determined by the points (3, 5, -1), (4, 3, -1).

5. Line Perpendicular to Two Given Lines. If two lines, intersecting in a point P, are given, there is a single line through P perpendicular to each of them, namely, the line through P perpendicular to the plane determined by them.

More generally, let any two non-parallel lines,* L_1 and L_2, with the direction components l_1, m_1, n_1 and l_2, m_2, n_2 be given. Let L be a line perpendicular to each of them† and let it be required to find for it a set of direction components, l, m, n.

We begin with a special case. Let 2, 3, 1 and 1, 4, 2 be the direction components of L_1 and L_2, respectively. Since L, with the direction components l, m, n, is perpendicular to L_1 and also to L_2, we have, by (4), § 4,

$$(1) \qquad \begin{aligned} 2l + 3m + n &= 0, \\ l + 4m + 2n &= 0. \end{aligned}$$

From these *two* homogeneous linear equations it is impossible to determine uniquely the *three* unknowns l, m, n. But this was to be expected. For, there is not a unique set of direction components, l, m, n, of L, but infinitely many sets.

In general, then, there will be a set for which $n = 1$. To determine the values of m and n for this set, we must solve simultaneously the equations

$$(2) \qquad \begin{aligned} 2l + 3m + 1 &= 0, \\ l + 4m + 2 &= 0. \end{aligned}$$

The solutions are $l = \frac{2}{5}$, $m = -\frac{3}{5}$. Consequently, one set of direction components of L is $\frac{2}{5}$, $-\frac{3}{5}$, 1. A simpler set is 2, -3, 5.

In the general case, since L is perpendicular to both L_1 and L_2, it follows that

$$(3) \qquad \begin{aligned} l_1 l + m_1 m + n_1 n &= 0, \\ l_2 l + m_2 m + n_2 n &= 0. \end{aligned}$$

* From now on we drop the qualifying adjective "undirected," and speak merely of lines and directed lines, as usual.

† There are infinitely many common perpendiculars to L_1 and L_2. They are, however, all parallel to one another and hence the direction components of any one of them will be the direction components of all the others.

Here, too, we try to find a set of direction components l, m, n of L, for which $n = 1$. Then equations (3) become

$$(4)\qquad \begin{aligned} l_1 l + m_1 m &= -n_1,\\ l_2 l + m_2 m &= -n_2. \end{aligned}$$

The solutions of these equations are

$$(5)\qquad l = \frac{m_1 n_2 - m_2 n_1}{l_1 m_2 - l_2 m_1}, \qquad m = \frac{n_1 l_2 - n_2 l_1}{l_1 m_2 - l_2 m_1}.$$

These values for l and m, together with $n = 1$, form a set of direction components of L. A simpler set is

$$m_1 n_2 - m_2 n_1, \qquad n_1 l_2 - n_2 l_1, \qquad l_1 m_2 - l_2 m_1,$$

or, in determinant form,

$$(6)\qquad \begin{vmatrix} m_1 & n_1 \\ m_2 & n_2 \end{vmatrix}, \qquad \begin{vmatrix} n_1 & l_1 \\ n_2 & l_2 \end{vmatrix}, \qquad \begin{vmatrix} l_1 & m_1 \\ l_2 & m_2 \end{vmatrix}.$$

For the special case first treated, these determinants have the values

$$\begin{vmatrix} 3 & 1 \\ 4 & 2 \end{vmatrix}, \qquad \begin{vmatrix} 1 & 2 \\ 2 & 1 \end{vmatrix}, \qquad \begin{vmatrix} 2 & 3 \\ 1 & 4 \end{vmatrix},$$

or $6 - 4 = 2$, $1 - 4 = -3$, $8 - 3 = 5$. But $2, -3, 5$ were the direction components found, and thus the work in the special case is checked.

Rule of Thumb. To obtain the determinants in (6) easily, write the two given sets of direction components under one another:

$$\begin{matrix} l_1 & m_1 & n_1, \\ l_2 & m_2 & n_2. \end{matrix}$$

The first determinant in (6) is formed from the *second* and *third* columns of this array; the second is formed from the *third* and *first* columns — not the first and third — and the third from the *first* and *second* columns. Thus the sets of numbers, 2 3, 3 1, 1 2, represent the columns used in the three determinants. The first set, 2 3, is all that need

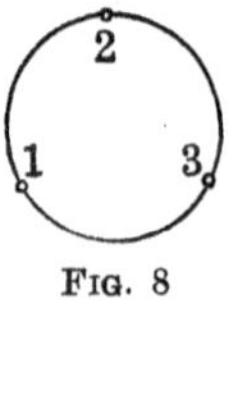

FIG. 8

be remembered. For, by advancing the numbers of this set according to the *cyclic order* 1 2 3 1 2 · · ·, this set 2 3 becomes 3 1, *i.e.* the second set; and advancing the second set 3 1 cyclicly, we get the third set, 1 2.

Critique. Not all the determinants (6) are zero, for if

$$l_1m_2 - l_2m_1 = m_1n_2 - m_2n_1 = n_1l_2 - n_2l_1 = 0,$$

then
$$l_1 : l_2 = m_1 : m_2 = n_1 : n_2$$

or
$$l_1 : m_1 : n_1 = l_2 : m_2 : n_2$$

and hence the lines L_1 and L_2 would be the same or parallel, which is contrary to hypothesis.

In obtaining the solution (5) of equations (4) we assumed, tacitly, that $l_1m_2 - l_2m_1 \neq 0$; there was, then, a set of components l, m, n, for which $n = 1$. If $l_1m_2 - l_2m_1 = 0$, at least one of the two remaining determinants cannot be zero. If, for example, $n_1l_2 - n_2l_1 \neq 0$, there will be a set of components l, m, n, for which $m = 1$, and we can find this set by putting $m = 1$ in (3) and solving the resulting equations for l and n.

EXERCISES

In each of the following exercises determine the direction components of a line which is perpendicular to each of two lines having the given direction components. Actually solve the equations and then check the result by the rule of thumb.

1. $\begin{cases} 3, & 4, & 2; \\ 1, & 2, & 3. \end{cases}$ **2.** $\begin{cases} 5, & 6, & -3; \\ 2, & -4, & -1. \end{cases}$ **3.** $\begin{cases} 1, & 2, & -1; \\ 3, & -1, & 2. \end{cases}$

4. $\begin{cases} 2, & 1, & -1; \\ 4, & 2, & 3. \end{cases}$ **5.** $\begin{cases} 0, & 1, & 0; \\ 0, & 0, & 1. \end{cases}$ **6.** $\begin{cases} 3, & 0, & 2; \\ 1, & 0, & -1. \end{cases}$

7. The two directed lines L_1 and L_2 passing through the origin and having respectively the direction cosines $\frac{1}{2}\sqrt{2}$, $\frac{1}{2}\sqrt{2}$, 0 and $-\frac{1}{2}\sqrt{2}$, $\frac{1}{2}\sqrt{2}$, 0 are perpendicular to each other. Find the direction cosines of a third line L_3 through the origin, perpendicular to both L_1 and L_2, and so directed that L_1, L_2, L_3

form a directed trihedral (with edges of indefinite length) which is right-handed by the test of Ch. XVII, § 5.

Ans. 0, 0, 1.

8. The above problem, if the direction cosines of L_1 and L_2 are, respectively, $\frac{6}{7}, \frac{2}{7}, -\frac{3}{7}$, and $\frac{2}{7}, \frac{3}{7}, \frac{6}{7}$. *Ans.* $\frac{3}{7}, -\frac{6}{7}, \frac{2}{7}$.

6. Three Lines Parallel to a Plane. Given three lines, with the direction components $l_1, m_1, n_1, l_2, m_2, n_2$, and l_3, m_3, n_3. The lines will be parallel to a plane or will lie in a plane, if and only if there is a line, with the direction components l, m, n, which is perpendicular to each of them, *i.e.* if and only if

$$\begin{aligned} l_1l + m_1m + n_1n &= 0, \\ l_2l + m_2m + n_2n &= 0, \\ l_3l + m_3m + n_3n &= 0. \end{aligned} \tag{1}$$

But, by Ch. XVI, § 10, these three homogeneous linear equations have a solution for l, m, n, other than the obvious solution 0, 0, 0, when and only when the determinant of their coefficients vanishes:

$$\begin{vmatrix} l_1 & m_1 & n_1 \\ l_2 & m_2 & n_2 \\ l_3 & m_3 & n_3 \end{vmatrix} = 0. \tag{2}$$

We have, therefore, the theorem:*

THEOREM. *Three lines are parallel to a plane or lie in a plane, if and only if the determinant of their direction components has the value zero.*

EXERCISES

1. Show that three lines through the origin with the direction components 2, −1, 5, 3, 2, −4, 7, 0, 6 lie in a plane.

* The proof covers not only the general case, when the given lines have but one common perpendicular direction and the equations (1) a one-parameter family of solutions, but also the special case in which the given lines are parallel, when the lines have infinitely many common perpendicular directions and the equations (1) a two-parameter family of solutions.

In each of the following exercises show that three lines with the given direction components are parallel to a plane.

2. 3, 1, 2; 5, – 4, 3; 1, 6, 1.

3. –1, 1, 2; 2, –1, 1; 1, 1, 8.

EXERCISES ON CHAPTER XVIII

1. Show that the triangle with vertices at the points (1, 3, – 5), (3, 4, – 7), (2, 5, – 3) is a right triangle.

2. Prove that the points (2, – 1, 5), (3, 4, – 2), (6, 2, 2), (5, –3, 9) are the vertices of a parallelogram.

3. Show that (2, 3, 0), (4, 5, – 1), (3, 7, 1), (1, 5, 2) are the vertices of a square.

4. Prove that the three points A, B, C with the coördinates (5, –2, 3), (2, 0, 2), (11, –6, 5) lie on a line by showing that the line AB has the same direction as the line AC.

5. Show that the two points (4, – 2, – 6), (– 6, 3, 9) lie on a line with the origin.

6. Show that two points (x_1, y_1, z_1), (x_2, y_2, z_2) lie on a line with the origin when and only when their coördinates are proportional: $x_1 : y_1 : z_1 = x_2 : y_2 : z_2$.

7. Show that the four points A, B, C, D, with the coördinates (3, 4, 2), (1, 6, 2), (3, 5, 1), (4, 5, 0), lie in a plane by proving that the sum of the angles which BC and CD subtend at angle A equals the angle which BD subtends at A.

8. Find the projection, on a directed line having $\frac{2}{7}$, $-\frac{3}{7}$, $\frac{6}{7}$ as its direction cosines, of the directed line-segment joining the origin to the point (5, 2, 4). *Ans.* 4.

Suggestion. Use the method of § 2 or employ formula (2) of Ch. XVII, § 4.

9. Show that the projection, on a directed line having $\cos\alpha$, $\cos\beta$, $\cos\gamma$ as its direction cosines, of the directed line-segment joining the origin to the point (x, y, z) is

$$x \cos\alpha + y \cos\beta + z \cos\gamma.$$

10. Find the projection, on a directed line with the direction cosines $\frac{8}{9}$, $\frac{4}{9}$, $-\frac{1}{9}$, of the directed line-segment joining the point $(3, -2, -5)$ to the point $(8, 0, -2)$. *Ans.* 5.

11. The previous problem, in the general case.

12. Two lines, L_1 and L_2, have the direction components 1, 1, 0 and 0, 1, -1, respectively. Find the direction components of a line which is perpendicular to L_1 and makes an angle of 30° with L_2. *Ans.* $1, -1, 2$.

13. Prove that each two *opposite* edges of the tetrahedron, with vertices at the points $(0, 0, 0)$, $(1, 1, 0)$, $(0, 1, -1)$, $(1, 0, -1)$, are perpendicular.

14. A tetrahedron has three pairs of opposite edges. Prove that, if the edges of each of two pairs are perpendicular, the edges of the third pair are also perpendicular. Choose the coördinate axes skillfully.

15. Prove the identity

$$(\mu_1\nu_2 - \mu_2\nu_1)^2 + (\nu_1\lambda_2 - \nu_2\lambda_1)^2 + (\lambda_1\mu_2 - \lambda_2\mu_1)^2$$
$$\equiv (\lambda_1^2 + \mu_1^2 + \nu_1^2)(\lambda_2^2 + \mu_2^2 + \nu_2^2) - (\lambda_1\lambda_2 + \mu_1\mu_2 + \nu_1\nu_2)^2.$$

In the following exercises λ_1, μ_1, ν_1, λ_2, μ_2, ν_2 (and λ_3, μ_3, ν_3) denote the direction cosines of directed lines, L_1, L_2 (and L_3), which we can assume go through the origin. In solving the exercises, the identity of Ex. 15 will be found useful.

16. If θ is the angle between L_1 and L_2, show that

$$\sin^2\theta = (\mu_1\nu_2 - \mu_2\nu_1)^2 + (\nu_1\lambda_2 - \nu_2\lambda_1)^2 + (\lambda_1\mu_2 - \lambda_2\mu_1)^2.$$

17. Prove that if L_1 and L_2 are perpendicular, the direction cosines of their common perpendicular L_3 are

$$\pm(\mu_1\nu_2 - \mu_2\nu_1), \qquad (\nu_1\lambda_2 - \nu_2\lambda_1), \qquad \pm(\lambda_1\mu_2 - \lambda_2\mu_1).$$

18. Show that, if the plus signs are taken in the above formulas, L_3 will be so directed that the lines L_1, L_2, L_3 will form a directed trihedral which is right-handed by the test of Ch. XVII, § 5.

19. Prove that, if L_1, L_2, and L_3 are mutually perpendicular, the determinant, $|\lambda \ \mu \ \nu|$, of their direction cosines has the value $+1$ or -1, according as the directed trihedral consisting of L_1, L_2, L_3 is right-handed or left-handed.

CHAPTER XIX

THE PLANE

1. Surfaces and Equations. *Example* 1. The equation

$$x = 5$$

is satisfied by the coördinates of those points and only those points which lie in the plane parallel to the (y, z)-plane and 5 units in front of it. We say that the equation *represents* this plane.

Example 2. Consider the equation

$$x = y.$$

The points *in the* (x, y)*-plane* whose coördinates satisfy it are the points of the line L bisecting the angle between the positive x- and y-axes. Since z is unrestricted by the equation, the points *in space* whose coördinates satisfy it are the points which lie directly above or below L, or are on L, *i.e.* the points of the vertical plane through L. The equation, then, represents this plane.

Example 3. The equation

$$x^2 + z^2 = 25$$

represents, *in the* (z, x)*-plane,* a circle, C, with its center at the origin and of radius 5. But the equation does not restrict in any way the value of y. Consequently, it represents *in space* the circular cylinder formed by drawing through each point of the circle C a line parallel to the axis of y and extending indefinitely in both directions.

Surfaces. The planes and the cylinder represented by the three equations considered are known as *surfaces*; the cylinder is a *curved* surface and the planes are *plane* surfaces. In generalizing the foregoing discussions we should, then, say:

An equation in x, y, z represents, usually, a surface. The surface consists of all those points and only those points whose coördinates, when substituted for x, y, z in the equation, satisfy it.*

Shifting the point of view, we assume now that it is a surface, and not an equation, which is given. Then we should say:

The equation of a given surface is an equation in x, y, z which is satisfied by the coördinates of every point of the surface and by the coördinates of no other point.

Problem 1. Find the equation of the sphere whose center is at the origin and whose radius is a.

A point (x, y, z) lies on this sphere if and only if the square of its distance from the origin is equal to a^2:

$$x^2 + y^2 + z^2 = a^2.$$

Therefore, this is the required equation.

Problem 2. Find the equation of the plane which passes through the axis of x and makes an angle of 30° with the (x, y)-plane, as shown in Fig. 1.

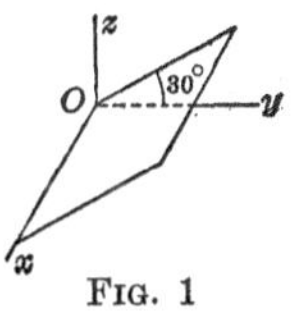

This plane intersects the (y, z)-plane in the line whose equation *in the (y, z)-plane* is

$$z = \tan 30^\circ\, y \qquad \text{or} \qquad z = \tfrac{1}{3}\sqrt{3}\, y.$$

FIG. 1

But this equation, considered as an equation in x, y, z, leaves x unrestricted; consequently, it represents *in space* the given plane, *i.e.* it is the equation of the given plane.

* An equation in x, y, z does not always represent a surface. For example, the equation $x^2 + y^2 = 0$ represents a line, namely, the z-axis; the equation $x^2 + y^2 + z^2 = 0$ represents just one point, the origin; and the equation $x^2 + y^2 + z^2 + 1 = 0$ represents no point whatsoever.

EXERCISES

What does each of the following equations represent? Draw a figure in each case.

1. $z = 0$.	5. $2y + 3z = 6$.	9. $x^2 + 2y^2 = 4$.
2. $y + 3 = 0$.	6. $y = \lambda x + b$.	10. $y^2 - z^2 = 9$.
3. $x + y = 0$.	7. $x^2 + y^2 = a^2$.	11. $x^2 - 9z^2 = -9$.
4. $z - 2x = 0$.	8. $z^2 = 2x$.	12. $x^2 + y^2 + z^2 = 4$.

13. Which of the surfaces represented by the above equations pass through the origin? Which contain a coördinate axis?

Find the equations of the following surfaces.

14. The (y, z)-plane.

15. The plane parallel to the (x, y)-plane and 3 units above it.

16. The plane parallel to the (z, x)-plane and 2 units to the left of it.

17. The plane bisecting the angle between the (x, y)- and (y, z)-planes and passing through the first octant.

18. The plane perpendicular to the (x, y)-plane whose trace* on that plane has the slope 3 and the intercept 2 on the axis of y.

19. The circular cylinder whose radius is 3 and whose axis is parallel to the x-axis and passes through the point $(0, 1, 2)$.

20. The parabolic cylinder whose rulings are parallel to the y-axis and whose trace on the (z, x)-plane is a parabola with its vertex at the origin and its focus at the point $(2, 0, 0)$.

21. The elliptic cylinder whose rulings are parallel to the z-axis and whose trace on the (x, y)-plane is an ellipse which has its center at the origin, its foci on the x-axis, and axes of lengths 6 and 4.

* The trace of a surface on a plane is the line, or curve, of intersection of the surface with the plane.

22. The hyperbolic cylinder whose rulings are parallel to the z-axis and whose trace on the (x, y)-plane is a rectangular hyperbola with its center at the origin and foci at the points $(0, \pm 2, 0)$.

23. The sphere whose center is at the point (1, 0, 0) and whose radius is unity.

24. The sphere whose center is at the point (1, 2, − 3) and whose radius is 5.

2. Plane through a Point with Given Direction of its Normals. Let there be given a point P_0 with the coördinates (x_0, y_0, z_0) and a line L with the direction components l, m, n. Through P_0 perpendicular to L there is just one plane.* We propose to find its equation.

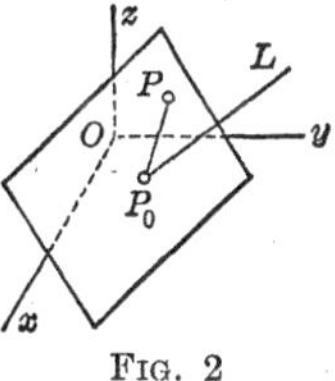

FIG. 2

Let $P : (x, y, z)$ be any point of the plane other than P_0. Then it determines with P_0 a line, P_0P, which is perpendicular to the line L. Since, by Ch. XVIII, § 4,

$$x - x_0, \quad y - y_0, \quad z - z_0$$

are the direction components of P_0P and l, m, n are the direction components of L, it follows, by (4), Ch. XVIII, § 4, that

$$l(x - x_0) + m(y - y_0) + n(z - z_0) = 0. \tag{1}$$

Conversely, if $P: (x, y, z)$ be any point other than P_0 whose coördinates satisfy equation (1), this equation says that the line P_0P is perpendicular to L and hence that P lies in the plane.

The coördinates x_0, y_0, z_0 of the excepted point, P_0, obviously satisfy equation (1). We have shown, then, that this equation is satisfied by the coördinates of those points and only those points which lie in the plane. Hence it is the equation of the plane.

There are infinitely many lines L perpendicular or, as we say, *normal* to the plane, and they are all parallel to one an-

* The figure is drawn for the special case in which L passes through P_0.

other. It is their common direction or, analytically, their common direction components, which are essential. Accordingly, we speak of (1) as *the equation of a plane through a given point with a given direction of its normals.*

EXERCISES

In each of the following exercises find the equation of the plane through the given point with the given direction of its normals.

	Point	Direction		Point	Direction
1.	(2, 1, 3),	1, 1, −2.	5.	Origin,	3, −2, 0.
2.	(−5, 3, 4),	−2, 2, 1.	6.	(5, −8, 2),	0, 1, 0.
3.	(4, −3, 2),	5, 0, 3.	7.	(3, 1, 0),	0, 0, 1.
4.	Origin,	2, −3, 5.	8.	Origin,	l, m, n.

9. Find the equations of the three planes which pass through the point (5, 6, −3) and are parallel respectively to the coördinate planes.

10. How is a plane situated if one of the direction components of its normals is zero? If two are zero?

3. The General Equation of the First Degree. Since any plane can be determined by one of its points and the direction components of a normal, the result of the preceding paragraph embraces all planes. Moreover, equation (1) of that paragraph is of the first degree in x, y, z. We have thus proved the theorem: *Every plane can be represented analytically by a linear equation in x, y, z.*

Given, conversely, *the general equation of the first degree* in x, y, z, namely,

$$Ax + By + Cz + D = 0, \tag{1}$$

where A, B, C, D are any four constants, of which A, B, C are not all zero.*

* In dealing with equation (1), here and henceforth, we shall always assume that A, B, C are not all zero.

Let (x_0, y_0, z_0) be a point whose coördinates satisfy equation (1):

(2) $$Ax_0 + By_0 + Cz_0 + D = 0.$$

Subtracting equation (2) from equation (1) we obtain

$$A(x - x_0) + B(y - y_0) + C(z - z_0) = 0.$$

But this equation is of the form (1), § 2, where

$$l : m : n = A : B : C.$$

Therefore it represents a plane which has A, B, C as the direction components of its normals. This result we state as a theorem.

THEOREM. *The general linear equation* (1) *always represents a plane. The coefficients A, B, C are the direction components of the normals to the plane.*

Example 1. The equation

$$2x - 3y + 4z - 6 = 0$$

represents a plane whose normals have $2, -3, 4$ as direction components. The point $(2, 2, 2)$, for example, lies in the plane, since when we set $x = 2$, $y = 2$ in the equation, we find $z = 2$ as the value of z.

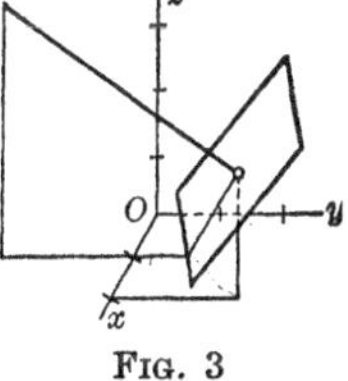

FIG. 3

We obtain a rough plot of the plane by constructing the point $(2, 2, 2)$ and the line through it with the direction components $2, -3, 4$, and by drawing then, as accurately as possible, the plane through the point perpendicular to the line.

Example 2. The equation

$$x = 2, \qquad \text{or} \qquad x + 0y + 0z = 2,$$

is the equation of a plane having the lines with the direction components $1, 0, 0$ as normals. But these lines are parallel to the axis of x and hence the plane is parallel to the (y, z)-plane. In particular, it is two units in front of that plane.

Remark. In the proof of the theorem it was assumed that there is always a point whose coördinates satisfy equation (1). This assumption is easily justified. By hypothesis, at least one of the three coefficients A, B, C is not zero. Suppose that $C \neq 0$. Then equation (1) can be solved for z:

$$z = -\frac{Ax + By + D}{C}.$$

Giving to x and y definite values, x_0 and y_0, we obtain for z from this equation a definite value, z_0. Then the point (x_0, y_0, z_0) has coördinates which satisfy (1). For example, if $x_0 = 0$ and $y_0 = 0$, then $z_0 = -D/C$, and the point is $(0, 0, -D/C)$.

EXERCISES

In each of the following exercises determine the direction components of the normals to the given plane and the coördinates of a point lying in it. Construct the plane by the method of Example 1.

1. $3x + 5y + 6z - 5 = 0.$
2. $2x - y + z - 3 = 0.$
3. $4x + 2y - 3z + 6 = 0.$
4. $5x - 2y - 3z + 4 = 0.$
5. $2x + 3y - 5 = 0.$
6. $3x - 2z - 4 = 0.$
7. $5y + 8 = 0.$
8. $2z - 7 = 0.$

4. Intercepts. Let a plane be given by means of its equation. A simple method of plotting the plane, in case it cuts the axes in three distinct points (one on each axis), consists in determining from the equation the coördinates of these three points and then in plotting the points and constructing the plane through them.

The point of intersection of a plane, for example,

$$(1) \qquad 2x - 3y + 4z - 6 = 0,$$

with the axis of x has its y- and z-coördinates both equal to zero. Consequently, to find the x-coördinate of the point, we have merely to set $y = 0$ and $z = 0$ in the equation of the plane and to solve for x. Thus, in this case, we have

$$2x - 6 = 0, \qquad \text{or} \qquad x = 3.$$

The point of intersection of the plane (1) with the axis of x is, then, (3, 0, 0). In a similar manner we find (0, -2, 0) and (0, 0, $\frac{3}{2}$) as the points of intersection of the plane with the axes of y and z respectively. By plotting these three points and joining them by lines, we obtain a good representation of the plane.

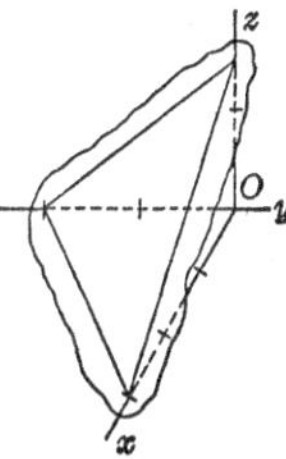

FIG. 4

The numbers 3, -2, $\frac{3}{2}$ are known as the *intercepts* of the plane (1) on the axes of x, y, z, respectively. That is, the intercept of a plane on the axis of x is the x-coördinate of the point in which the plane meets the axis of x. The intercepts on the axes of y and z are similarly defined.

A plane which passes through an axis or is parallel to an axis has no intercept on that axis. Every other plane has definite intercepts on all three axes and these intercepts determine the position of the plane unless they are all zero, that is, unless the plane goes through the origin.

EXERCISES

Determine the intercepts of the following planes on the coördinate axes, so far as they exist, and construct the planes.

1. $2x + 3y + 4z - 12 = 0.$
2. $3x - 2y + z - 6 = 0.$
3. $x + y - z - 2 = 0.$
4. $2x + 5y - 3z + 8 = 0.$
5. $x + 2y + z + 3 = 0.$
6. $x + 3y - z = 0.$
7. $2x - 3y + 12 = 0.$
8. $3y + 4z - 6 = 0.$
9. $5x + 2z = 0.$
10. $3x + 5 = 0.$

5. Intercept Form of the Equation of a Plane. Given a plane whose position is determined by its intercepts. Let these intercepts, on the axes of x, y, z, be respectively a, b, c. To find the equation of the plane in terms of a, b, c.

We have the problem of finding the equation of a plane

through the three points $(a, 0, 0)$, $(0, b, 0)$, $(0, 0, c)$. Let this equation be

(1) $$Ax + By + Cz + D = 0,$$

where the values of A, B, C, D are to be determined. Since the plane does not go through the origin, $D \neq 0$. Since it contains each of the given points, the following equations must hold:

$$Aa + D = 0, \qquad Bb + D = 0, \qquad Cc + D = 0.$$

Hence $A = -D/a$, $B = -D/b$, $C = -D/c$. Substituting these values for A, B, C in (1) and dividing through by D, we obtain

(2) $$\frac{x}{a} + \frac{y}{b} + \frac{z}{c} = 1.$$

That this is the desired equation can easily be checked by substituting successively in it the coördinates of the three points in question.

Only planes which intersect the axes in three points that are distinct can have their equations written in the form (2). A plane through the origin is an exception, because at least one of its intercepts is zero and division by zero is impossible. A plane parallel to an axis is also an exception, since it has no intercept on that axis.

EXERCISES

Find the equations of the planes with the following intercepts.

1. $2, 3, 4$.
2. $2, -3, -1$.
3. $-2, 4, 5$.
4. $-5, -3, 2$.
5. $-4, -6, -2$.
6. $2, -8, -6$.

Find the equations of the following planes.

7. With intercepts on the x- and y-axes equal to 2 and 3 and parallel to the axis of z.

8. With intercept -3 on the z-axis and parallel to the (x, y)-plane.

9. A regular quadrangular pyramid has its vertices at the points (0, 0, 6), (2, 0, 0), (0, 2, 0), (– 2, 0, 0), (0, – 2, 0). Find the equation of its faces.

10. The same, if the vertices are at the points $(0, 0, c)$, $(a, 0, 0)$, $(0, a, 0)$, $(-a, 0, 0)$, $(0, -a, 0)$.

6. Plane through Three Points. Three points, not lying in a straight line, determine a plane. In any particular case the equation of the plane can be found by the method of the preceding paragraph. In the general case, when the points are arbitrary and have the coördinates (x_1, y_1, z_1), (x_2, y_2, z_2), (x_3, y_3, z_3), this method could still be applied. It is, however, simpler to write the equation in determinant form, by analogy to the equation in that form of the straight line through two points (Ch. XVI, § 11). We have, namely,

$$(1) \qquad \begin{vmatrix} x & y & z & 1 \\ x_1 & y_1 & z_1 & 1 \\ x_2 & y_2 & z_2 & 1 \\ x_3 & y_3 & z_3 & 1 \end{vmatrix} = 0.$$

To show that this equation actually represents the plane through the three points, develop the determinant by the minors of the first row. The equation then takes on the usual form,

$$Ax + By + Cz + D = 0,$$

of a linear equation; moreover, the values obtained for A, B, C:

$$A = \begin{vmatrix} y_1 & z_1 & 1 \\ y_2 & z_2 & 1 \\ y_3 & z_3 & 1 \end{vmatrix}, \quad B = -\begin{vmatrix} x_1 & z_1 & 1 \\ x_2 & z_2 & 1 \\ x_3 & z_3 & 1 \end{vmatrix}, \quad C = \begin{vmatrix} x_1 & y_1 & 1 \\ x_2 & y_2 & 1 \\ x_3 & y_3 & 1 \end{vmatrix},$$

are not all zero, since otherwise the projections of the three points on each of the coördinate planes would lie on a line (Ch. XVI, § 11, Th. 15) * and hence so would the three points themselves. Consequently, (1) represents some plane.

* If C, for example, were zero, the three points $(x_1, y_1, 0)$, $(x_2, y_2, 0)$, $(x_3, y_3, 0)$ in the (x, y)-plane would lie on a line. But these points are the projections of the given points on the (x, y)-plane.

This is the plane through the three points, since the substitution of the coördinates of any one of the points for x, y, z in (1) makes two rows of the determinant identical and hence causes the determinant to vanish and the equation to be satisfied.

EXERCISES

Find the equations of the following planes by applying formula (1) and simplifying the result.

1. Through $(1, 2, 0)$, $(-2, 3, 3)$, $(3, -1, -3)$.
2. Through $(2, 5, -3)$, $(-2, -3, 5)$, $(5, 3, -3)$.
3. Through $(1, 1, 0)$, $(0, 1, 1)$, $(1, 0, 1)$.
4. Through $(1, 1, -1)$, $(6, 4, -5)$, $(-4, -2, 3)$.
5. Through $(4, 5, 2)$, $(-3, -2, -5)$, the origin.
6. Through (x_1, y_1, z_1), (x_2, y_2, z_2), the origin. Keep the equation in determinant form, but simplify it.
7. Establish the intercept form of the equation of a plane by applying formula (1).
8. By the method of the preceding paragraph find the equation of the plane of

   (*a*) Exercise 1; (*b*) Exercise 3; (*c*) Exercise 5.
9. Find the equations of the faces of the tetrahedron whose vertices are at the points $(0, 0, 0)$, $(0, 3, 0)$, $(2, 1, 0)$, $(1, 1, 2)$.

   Ans. $z = 0,\ 2x = z,\ 2x - 4y + z = 0,\ 2x + 2y + z = 6.$

7. Perpendicular, Parallel, and Identical Planes. Angle between Two Planes. The normals to the two planes,

$$A_1x + B_1y + C_1z + D_1 = 0, \tag{1}$$

$$A_2x + B_2y + C_2z + D_2 = 0, \tag{2}$$

have A_1, B_1, C_1 and A_2, B_2, C_2, respectively, as direction components.

The planes are perpendicular if and only if their normals are perpendicular; and parallel (or identical), if and only if the normals of one are also the normals of the other. Consequently, we have, by Ch. XVIII, § 4, the following theorems.

THEOREM 1. *The planes* (1) *and* (2) *are perpendicular when and only when*

(3) $$A_1A_2 + B_1B_2 + C_1C_2 = 0.$$

THEOREM 2. *The planes* (1) *and* (2) *are parallel* when and only when*

(4) $$A_1 : B_1 : C_1 = A_2 : B_2 : C_2.$$

The condition that the two planes be identical is analogous to the condition that two straight lines be identical; cf. Ch. II, § 10. We can state, then, the theorem:

THEOREM 3. *The planes* (1) *and* (2) *are identical when and only when*

(5) $$A_1 : B_1 : C_1 : D_1 = A_2 : B_2 : C_2 : D_2.$$

The proof of the theorem is left to the student.

Angle between Two Planes. Between the two planes (1) and (2) there are, in general, two different angles having values between 0° and 180° inclusive, and these angles are supplementary. They are equal to the angles between the normals to the two planes. Since A_1, B_1, C_1 and A_2, B_2, C_2 are the direction components of the normals, the cosines of the angles are given, according to (3), Ch. XVIII, § 4, by

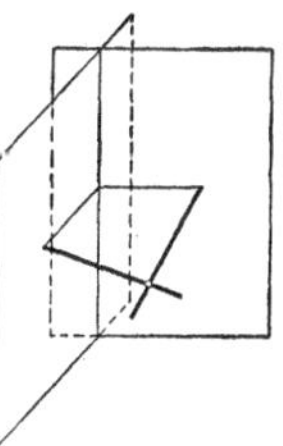

FIG. 5

(6) $$\cos\theta = \pm\frac{A_1A_2 + B_1B_2 + C_1C_2}{\sqrt{A_1^2 + B_1^2 + C_1^2}\sqrt{A_2^2 + B_2^2 + C_2^2}}.$$

EXERCISES

In each of the following exercises determine whether the given planes are parallel or perpendicular, and in case they are neither, find the angles between them.

* Or, in a single case, identical. Cf. Th. 3.

1. $5x - 3y + 2z + 4 = 0,\quad 3x - y - 9z + 2 = 0.$
2. $2x - y + 3z - 1 = 0,\quad 2x - y + 3z + 3 = 0.$
3. $x + y + z = 0,\quad 3x + 6y - 2z + 12 = 0.$
4. $2x - 2y + 4z + 5 = 0,\quad 3x - 3y + 6z - 1 = 0.$
5. $7x + 5y + 6z + 30 = 0,\quad 3x - y - 10z + 4 = 0.$
6. $2x + y + z - 3 = 0,\quad x + y - 3z + 4 = 0.$
7. $x + y + 2z = 0,\quad x + z = 0.$
8. $4x + 8y + z - 8 = 0,\quad y + z - 3 = 0.$
9. $2x + y + 3z - 2 = 0,\quad x - 2y + 5 = 0.$

10. Show that two planes are parallel when and only when their equations can be written in the forms

$$Ax + By + Cz = D, \quad Ax + By + Cz = D',$$
$$D \neq D'.$$

8. Planes Parallel or Perpendicular to a Given Plane.

Example 1. Find the equation of the plane which passes through the point $(5, 2, -4)$ and is parallel to the plane

$$2x + 4y - 6z - 7 = 0. \tag{1}$$

The normals to the plane (1) have the direction components $2, 4, -6$ or, more simply, $1, 2, -3$. The required plane has the same normals and passes through the point $(5, 2, -4)$. By (1), § 2, its equation is

$$1(x - 5) + 2(y - 2) - 3(z + 4) = 0,$$

or

$$x + 2y - 3z - 21 = 0.$$

Through a given point and parallel to a given plane there is but one plane; its equation can always be found by the above method. But through a given point and perpendicular to a given plane there is not just one plane, but infinitely many, namely, all the planes which pass through that normal to the given plane which goes through the given point. To single out one of these planes we must impose a further condition. We might demand, for instance, that the required plane pass through a second given point or, again, we might specify that it be perpendicular to a second given plane. We proceed to consider illustrative examples of these two types.

Example 2. Find the equation of the plane passing through the two points $(3, -2, 9)$, $(-6, 0, -4)$ and perpendicular to the plane

$$2x - y + 4z - 8 = 0. \tag{2}$$

First Method. Let the equation of the plane in question be

$$Ax + By + Cz + D = 0. \tag{3}$$

Since the plane contains the two given points, we must have

$$3A - 2B + 9C + D = 0, \tag{4}$$
$$-6A \qquad - 4C + D = 0. \tag{5}$$

Since it is perpendicular to the plane (1), it is necessary that

$$2A - B + 4C = 0. \tag{6}$$

In (4), (5), (6) we have *three* simultaneous linear equations in the *four* unknowns A, B, C, D. But from three linear equations it is impossible to determine uniquely the values of four unknowns. It may be possible, however, to determine the values of three of the unknowns in terms of the fourth, say the values of A, B, C in terms of D.

Accordingly, we rewrite the equations in the form

$$\begin{aligned} 3A - 2B + 9C &= -D, \\ 6A \qquad + 4C &= D, \\ 2A - B + 4C &= 0, \end{aligned}$$

and solve for A, B, C, either directly or by determinants. We *do* obtain a solution, namely,

$$A = \tfrac{1}{2}D, \qquad B = -D, \qquad C = -\tfrac{1}{2}D.$$

Hence (3) becomes

$$\tfrac{1}{2}Dx - Dy - \tfrac{1}{2}Dz + D = 0.$$

The plane represented by this equation is always the same, no matter what value, other than zero, is given to D. A simple choice is: $D = 2$. We obtain, then, as the equation of the required plane,

$$x - 2y - z + 2 = 0.$$

Second Method. Let N be any normal to the required plane. Since the plane contains the two given points, N is perpendicular to the line L_1 joining these points. Since the plane is perpendicular to the given plane (2), N is perpendicular to any line L_2 normal to (2). Thus N is a common perpendicular to the lines L_1 and L_2.

The direction components of L_1 are, by Ch. XVIII, § 4, $3-(-6), \quad -2-0, \quad 9-(-4)$, *i.e.*

$$9, \quad -2, \quad 13;$$

those of L_2 are

$$2, \quad -1, \quad 4.$$

Consequently, by (6), Ch. XVIII, § 5, the direction components of N are

$$\begin{vmatrix} -2 & 13 \\ -1 & 4 \end{vmatrix}, \quad \begin{vmatrix} 13 & 9 \\ 4 & 2 \end{vmatrix}, \quad \begin{vmatrix} 9 & -2 \\ 2 & -1 \end{vmatrix},$$

i.e. $5, -10, -5$, or $1, -2, -1$.

Our problem is now reduced to that of finding the equation of the plane which passes through one of the given points, say $(3, -2, 9)$, and has $1, -2, -1$ as the direction components of its normals. This equation is

$$1(x-3)-2(y+2)-1(z-9)=0,$$

or

$$x-2y-z+2=0.$$

Example 3. Find the equation of the plane passing through the point $(2, 5, -8)$ and perpendicular to each of the planes:

$$2x-3y+4z+1=0, \quad 4x+y-2z+6=0.$$

Either of the methods employed in the previous example is applicable. We choose the latter. A normal N to the required plane is perpendicular to the normals to both the given planes. These have, respectively, the direction components $2, -3, 4$ and $4, 1, -2$. Consequently, the direction components of N are $2, 20, 14$ or $1, 10, 7$.

The equation of the plane through $(2, 5, -8)$ with $1, 10, 7$ as the direction components of its normals is

$$(x-2)+10(y-5)+7(z+8)=0,$$

or
$$x+10y+7z+4=0.$$

This is the required equation.

EXERCISES

In each of the following exercises find the equation of the plane which is parallel to the given plane and passes through the given point. In Exs. 5, 6 find the equation directly by inspection of a figure.

	Plane	*Point*
1.	$5x-2y+3z-4=0$,	$(2, 4, 3)$.
2.	$3x+4y-8z-2=0$,	$(0, 0, 0)$.
3.	$4x-2y-6z=9$,	$(2, -1, 0)$.
4.	$3x-4z=0$,	$(5, 2, -3)$.
5.	$3x+8=0$,	$(1, -2, 5)$.
6.	$2y-5=0$,	$(4, 0, 3)$.

7. Find the equation of the plane passing through the points $(3, 1, 2)$, $(3, 4, 4)$ and perpendicular to the plane $5x+y+4z=0$. Apply both methods, checking the result of one by that of the other. *Ans.* $2x+2y-3z-2=0$.

The previous problem, if the given points and the given plane are as specified. Use either method in Exs. 8–10; in Exs. 11, 12 solve the problem directly by inspection of a figure.

	Points	*Plane*
8.	$(3, 4, 1)$, $(2, 6, -2)$,	$2x-3y+4z-2=0$.
9.	$(0, 0, 0)$, $(4, 3, 2)$,	$x+y+z=0$.
10.	$(3, 2, -4)$, $(5, -1, 3)$,	$4x-5y=8$.
11.	$(1, 0, 0)$, $(1, 2, 5)$,	$3y-7=0$.
12.	$(0, 2, 0)$, $(2, 0, 0)$,	$2z+5=0$.

13. There are infinitely many planes which pass through the two points $(2, -3, 4)$, $(-2, 3, -6)$ and are perpendicular

to the plane whose equation is $2x - 3y + 5z - 10 = 0$. Why? Justify your answer.

14. What is the equation of the plane which passes through the point $(1, -2, 1)$ and is perpendicular to each of the planes:

$$3x + y + z - 2 = 0, \qquad x - 2y + z + 4 = 0?$$

Apply both methods, checking the result of one by that of the other. *Ans.* $3x - 2y - 7z = 0$.

The previous problem for the following given planes and given point. In Exs. 15–17 use either method; in Exs. 18, 19 obtain the result directly from a figure.

	Planes	*Point*
15.	$\begin{cases} 2x + 3y + 6z - 2 = 0, \\ 6x + 2y - 3z + 4 = 0, \end{cases}$	$(1, 5, 2)$.
16.	$\begin{cases} x + y + z = 0, \\ x - y + z + 2 = 0, \end{cases}$	$(1, -1, 1)$.
17.	$\begin{cases} x + 3y + z - 4 = 0, \\ 2x + y - 6 = 0, \end{cases}$	$(0, 0, 0)$.
18.	$x = 2,\ y = 3,$	$(2, -5, 3)$.
19.	$2x + z = 0,\ 3x - z = 6,$	$(2, 1, -3)$.

20. There are infinitely many planes which pass through the point $(2, -5, 0)$ and are perpendicular to each of the planes:

$$4x - 2y - 6z + 3 = 0, \qquad -6x + 3y + 9z + 10 = 0.$$

Why? Justify your answer.

9. Distance of a Point from a Plane. To find the distance Δ of the point $P: (x_0, y_0, z_0)$ from the plane

$$Ax + By + Cz + D = 0,$$

draw a line through P perpendicular to the (x, y)-plane and mark the point Q in which this line cuts the given plane.

Then, as the figure shows, Δ is the numerical value of the product $QP \cos\theta$, *i.e.*

$$\Delta = |\, QP \cos\theta \,|,$$

where θ is the acute angle between the line QP and the normal $P'P$ to the plane.

The normal $P'P$ has the direction components A, B, C and the line QP has the direction components 0, 0, 1. Consequently,

$$\cos\theta = \pm \frac{C}{\sqrt{A^2 + B^2 + C^2}}.$$

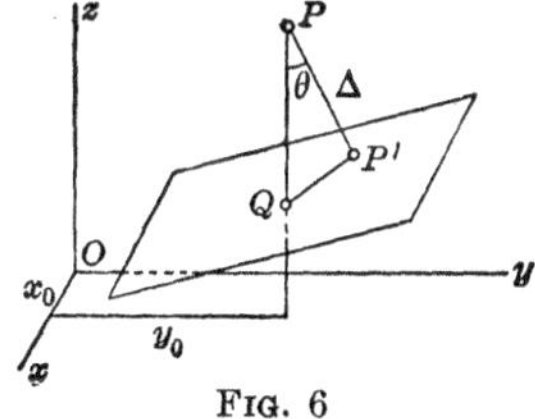

FIG. 6

It is immaterial to us which of the two signs is the proper one, for we are interested only in numerical values.

It remains to find QP. The x- and y-coördinates of Q are the same as those of P, namely, x_0, y_0; denote the z-coördinate of Q by z_Q. Since $Q:(x_0, y_0, z_Q)$ lies in the given plane, it follows that

$$Ax_0 + By_0 + Cz_Q + D = 0,$$

and hence that

$$z_Q = -\frac{Ax_0 + By_0 + D}{C}.$$

Then
$$QP = z_0 - z_Q = z_0 + \frac{Ax_0 + By_0 + D}{C},$$

or
$$QP = \frac{Ax_0 + By_0 + Cz_0 + D}{C}.$$

Multiplying the values obtained for QP and $\cos\theta$ together and taking the numerical value of the product, we obtain the desired formula:

(1)
$$\begin{cases} \Delta = \dfrac{|Ax_0 + By_0 + Cz_0 + D|}{\sqrt{A^2 + B^2 + C^2}}, \\ \text{or} \\ \Delta = \pm \dfrac{Ax_0 + By_0 + Cz_0 + D}{\sqrt{A^2 + B^2 + C^2}}, \end{cases}$$

where, in the second formula, the sign to be taken is that which gives a positive result.

In the above deduction we assumed that the line through P perpendicular to the (x, y)-plane meets the given plane, *i.e.* that the given plane is not perpendicular to the (x, y)-plane. This means, analytically, that we have assumed that $C \neq 0$.

Since we know that at least one of the three coefficients A, B, C is not zero and since the final formula (1) bears equally on A, B, and C, it is immaterial which one of the three coefficients we assume not zero. The result would have been the same if we had assumed, say, $A \neq 0$, instead of $C \neq 0$.

EXERCISES

1. Establish formula (1) on the assumption that $A \neq 0$.

Find the distance of each of the given points from the corresponding given plane. In Exs. 6, 7 check the result by inspection of a figure.

	Point	*Plane*	
2.	$(3, -2, 1)$,	$2x - y + 2z + 3 = 0$.	*Ans.* $4\frac{1}{3}$.
3.	$(2, 5, -3)$,	$6x - 3y + 2z - 4 = 0$.	
4.	$(0, 2, 1)$,	$4x + 3y + 9 = 0$.	
5.	Origin,	$8x + y - 4z - 6 = 0$.	
6.	$(3, 8, -6)$,	$y - 5 = 0$.	
7.	$(-2, 3, 4)$,	$2z + 7 = 0$.	

8. Find the lengths of the altitudes of the tetrahedron of Ex. 9, § 6.

10. Point of Intersection of Three Planes. Let there be given three planes which intersect in a point, *i.e.* three planes which have just one point in common, as, for example, the planes of the ceiling and two intersecting walls of a room, or the planes of three faces of a tetrahedron.

The point of intersection of the planes is that point whose coördinates satisfy each of the three equations of the planes.

In other words, it is the point whose coördinates form the simultaneous solution of the three equations. Consequently, to find its coördinates we have but to solve the three equations simultaneously.

Consider, for example, the three planes represented by the equations

$$\begin{aligned} 3x + 4y - 5z &= -11, \\ 2x + y + 6z &= 13, \\ x - 3y + z &= 6. \end{aligned}$$

The simultaneous solution of these three equations is most simply effected by the use of determinants (Ch. XVI, §§ 2, 8). The result is $x = 1$, $y = -1$, $z = 2$. Accordingly, the point of intersection of the three planes is $(1, -1, 2)$.

Intersections of Three Surfaces. The method to be used in finding the point (or points) of intersection of any three surfaces, given by their equations, is now obvious. The equations are to be regarded as *simultaneous equations in the unknown quantities, x, y, and z*, and solved as such.

Three Arbitrary Planes. Let the equations,

$$(1) \qquad \begin{aligned} A_1x + B_1y + C_1z + D_1 &= 0, \\ A_2x + B_2y + C_2z + D_2 &= 0, \\ A_3x + B_3y + C_3z + D_3 &= 0, \end{aligned}$$

represent three distinct planes. If the determinant, $|A\ B\ C|$, of the coefficients of x, y, and z is not zero, the three equations have a *unique* solution (Ch. XVI, § 8, Th. 10) and hence the three planes intersect in a *single* point.

Conversely, if the three planes have just one point in common, $|A\ B\ C| \neq 0$. For, if $|A\ B\ C|$ vanished, the normals to the three planes would all be parallel to a plane, M, by Ch. XVIII, § 6. Consequently, the lines of intersection of the three planes, taken in pairs, would be perpendicular to M. If there were just one such line of intersection, the three planes would have all the points of this line in common; if there were no line of intersection or more than one, the three planes would have no point in common. In either case the hypothesis is contradicted. Hence we may state the theorem:

THEOREM. *The three planes* (1) *intersect in a single point, if and only if* $|A\ B\ C| \neq 0$.*

If $|A\ B\ C| = 0$, two or all three of the planes may be parallel; these cases are easily detected by inspection of the equations. Or, the three planes, taken in pairs, may intersect in three distinct parallel lines. Or, finally, they may have a line in common. We shall learn later, Ch. XXI, § 2, how to distinguish, from the equations of the planes, between these last two cases.

EXERCISES

In each of the following exercises show that the three given planes intersect in a single point, and find the coördinates of the point.

1. The planes of Ch. XVI, § 2, Ex. 10.
2. The planes of Ch. XVI, § 2, Ex. 11.
3. The planes of Ch. XVI, § 2, Ex. 12.
4. The planes of Ch. XVI, § 2, Ex. 13.

5. Find the coördinates of the vertices of the tetrahedron whose faces lie in the planes

$$z = 0, \quad 2y - 3z = 0, \quad x - y + 3 = 0, \quad 5x - 2y + 3z = 0.$$

Find the points of intersections of the following surfaces. Draw a figure in each case.

6. $x = 4, \quad z = -2, \quad x^2 + y^2 = 25.$
7. $x + y = 2, \quad x - y = 0, \quad x^2 + z^2 - 1 = 0.$
8. $x^2 + y^2 + z^2 = 9, \quad 5x + y - 3z = 5, \quad x = z.$
Ans. $(2, 1, 2), (\frac{4}{3}, \frac{7}{3}, \frac{4}{3})$.

* Or, *the three equations* (1) *have a unique solution, if and only if* $|A\ B\ C| \neq 0$. This is the converse of Th. 10, Ch. XVI, § 8. We have thus completed, by geometric methods, the proof of an important fact in the theory of linear equations.

In each of the exercises that follow give all the information you can concerning the relative positions of the three given planes.

9. $8x - 4y - 4z + 1 = 0,$
$-2x + y + z + 5 = 0,$
$6x - 3y - 3z - 2 = 0.$

10. $9x + 6y - 3z + 7 = 0,$
$x - 2y + z + 3 = 0,$
$6x + 4y - 2z - 1 = 0.$

11. $2x - 3y + 12 = 0,\quad 3x + 5y - 1 = 0,\quad 5x + 2y + 11 = 0.$

12. $4x - 3z - 5 = 0,\quad 3x + 5z - 11 = 0,\quad 7x + 2z + 3 = 0.$

EXERCISES ON CHAPTER XIX

1. When will the plane $Ax + By + Cz + D = 0$ pass through a coördinate axis, *e.g.* the axis of z? When will it be parallel to a coördinate axis, *e.g.* the axis of x?

2. Find the equation of the plane through the axis of z and the point (1, 2, 0).

3. Find the equation of the plane through the axis of y and the point (2, 3, 1).

4. What is the equation of the plane whose intercepts are one half those of the plane $2x - 3y + 4z - 12 = 0$?

5. A perpendicular from the origin meets a plane in the point (2, − 3, 4). What is the equation of the plane?

6. A line through the point (2, 3, 7) meets a plane in the point (5, − 1, 2). Find the equation of the plane.

7. Find the equation of the plane which bisects perpendicularly the line joining the points (4, 3, − 1), (2, 5, 3).

8. Determine the point on the axis of y which is equidistant from the points (3, 7, 4), (− 1, 1, 2).

9. One vertex O of a box is at the origin and the edges issuing from O lie along the positive coördinate axes. Prove that the intercepts of the plane which bisects perpendicularly the diagonal through O are inversely proportional to the lengths of the edges.

10. For what value of m will the two planes,

$$2x + my - z = 4, \qquad 6x - 5y - 3z = 8,$$

(*a*) be perpendicular? (*b*) be parallel?

11. For what value of m will the two equations

$$mx - y + z + 3 = 0, \qquad 4x - my + mz + 6 = 0$$

represent the same plane?

12. Find the angle which the line through the points $(3, 2, -1)$, $(0, 4, 1)$ makes with the plane $2x - y - z + 3 = 0$.

Suggestion. Find first the angle between the line and a normal to the plane.

13. What angle does the plane $3x - y - z = 5$ make (*a*) with the (x, y)-plane? (*b*) with the y-axis?

14. Find the distance between the two parallel planes

$$2x - 3y + 6z + 6 = 0, \qquad 2x - 3y + 6z - 1 = 0.$$

Suggestion. Find the distance of a chosen point of the first plane from the second.

15. Show that the distance between the two parallel planes

$$Ax + By + Cz + D = 0, \qquad Ax + By + Cz + D' = 0$$

is

$$\frac{|D' - D|}{\sqrt{A^2 + B^2 + C^2}}.$$

16. There are two points on the axis of z which are distant four units from the plane $2x - y + 2z + 3 = 0$. Find their coördinates. *Ans.* $(0, 0, 4\tfrac{1}{2})$, $(0, 0, -7\tfrac{1}{2})$.

17. Show that the equation of any plane parallel to the plane

$$Ax + By + Cz + D = 0$$

can be written in the form

$$Ax + By + Cz = k.$$

18. Using the method of Ch. II, § 11, work Exs. 1–4, § 8, of the present chapter.

19. There are two planes parallel to the plane $2x - 6y + 3z = 4$ and distant 3 units from the origin. Find their equations.

20. Find the equation of the plane parallel to the plane given in Ex. 19 and so located that the point (3, 2, 8) is midway between the two planes. *Ans.* $2x - 6y + 3z - 32 = 0$.

21. Three faces of a box lie in the planes $2x - y = 6$, $x + 2y = 8$, $z = 8$ and a vertex is at the point (9, 5, 2). Find the equations of the planes of the other three faces.

22. Find the equation of the plane which passes through the point (2, − 1, 8) and is parallel to each of two lines having 2, − 3, 4 and 5, − 7, 8 as their direction components.

Ans. $4x + 4y + z - 12 = 0$.

23. Show that the equation of the plane passing through the points (x_1, y_1, z_1), (x_2, y_2, z_2) and perpendicular to the plane $Ax + By + Cz + D = 0$ can be written in the form

$$\begin{vmatrix} x & y & z & 1 \\ x_1 & y_1 & z_1 & 1 \\ x_2 & y_2 & z_2 & 1 \\ A & B & C & 0 \end{vmatrix} = 0.$$

24. Show that the four planes,

$$2x - y - z - 3 = 0, \qquad x - y + 2z - 3 = 0,$$
$$x - 2y + z = 0, \qquad x + y + z - 6 = 0,$$

meet in a point.

25. The six planes, each of which passes through the midpoint of an edge of a tetrahedron and is perpendicular to the opposite edge, go through a point.

Prove this theorem for the tetrahedron of Ex. 9, § 6.

26. Prove the theorem of Ex. 25 for the general tetrahedron, choosing the coördinate axes skillfully.

27. Show that the plane

$$\tfrac{2}{7}x + \tfrac{3}{7}y + \tfrac{6}{7}z = 5$$

is 5 units distant from the origin and that $\frac{2}{7}$, $\frac{3}{7}$, $\frac{6}{7}$ are the direction cosines of a normal to it, directed away from the origin.

28. State and prove for the plane

$$(1) \qquad x \cos \alpha + y \cos \beta + z \cos \gamma = p, \qquad p \geq 0,$$

the results corresponding to those given in the preceding exercise. Show that the equation of every plane can be written in the form (1). Prove, also, that the distance of the point (x_0, y_0, z_0) from the plane is

$$| x_0 \cos \alpha + y_0 \cos \beta + z_0 \cos \gamma - p |.$$

29. Prove that, if a plane has the intercepts a, b, c and is distant p units from the origin,

$$\frac{1}{a^2} + \frac{1}{b^2} + \frac{1}{c^2} = \frac{1}{p^2}.$$

Symmetry

30. *A surface is symmetric in the (x, y)-plane if the substitution of $-z$ for z in its equation leaves the equation essentially unchanged.* Prove this theorem and state the corresponding theorems for symmetry in the (y, z)- and (z, x)-planes.

31. *A surface is symmetric in the axis of z if the substitution of $-x$ for x, and of $-y$ for y, leaves the equation essentially unchanged.* Prove this theorem and state the corresponding theorems for symmetry in the axes of x and y.

32. Prove that *a surface is symmetric in the origin if the substitution of $-x$ for x, of $-y$ for y, and of $-z$ for z, leaves the equation essentially unchanged.*

33. Test the surfaces of the following exercises of § 1 for symmetry in each coördinate plane, in each coördinate axis, and in the origin.

(*a*) Ex. 8; (*b*) Ex. 9; (*c*) Ex. 10; (*d*) Ex. 12.

34. Prove the following theorems:

(*a*) If a surface is symmetric in each of two coördinate planes, it is symmetric in the coördinate axis in which the two planes meet.

(*b*) If a surface is symmetric in each coördinate plane, it is symmetric in the origin.

(*c*) If a surface is symmetric in a coördinate plane and in the coördinate axis perpendicular to this plane, it is symmetric in the origin.

CHAPTER XX

THE STRAIGHT LINE

1. Equations of a Curve. *Example* 1. Given the two equations

(1) $$x = 0, \qquad y = 0.$$

The points whose coördinates satisfy *both equations simultaneously* are the points on the axis of z, and no other points. We say that the two equations *represent the axis of* z.

Example 2. Consider the two equations

(2) $$3x - 4y - z + 6 = 0, \qquad 5x + 3y + 2z - 8 = 0.$$

A point whose coördinates satisfy *both* equations at once must lie in *each* of the two planes represented by the equations, *i.e.* it must be a point on the line of intersection of these planes. Conversely, the coördinates of any point on this line satisfy both equations. Thus the two equations, *considered simultaneously*, represent a line, the line of intersection of the two planes which the two equations, taken individually, define.

Example 3. Take, now, the pair of equations

(3) $$x^2 + y^2 + z^2 = 4, \qquad x - y = 0.$$

By reasoning similar to that of Example 2, it follows that these equations, taken together, represent the curve of intersection of the two surfaces which are defined by the two equations considered individually. The first equation is that of the sphere whose center is at the origin and whose radius is two units long. The second equation represents the plane through the axis of z bisecting the angle between the positive x- and y-

axes. Consequently, the two equations, considered simultaneously, represent the circle in which the plane intersects the sphere.

Example 4. Consider, lastly, the pair of equations

$$(4) \qquad x^2+y^2+z^2=4, \qquad x^2+(y-1)^2=1.$$

The first represents the sphere of Example 3. The second is the equation of the circular cylinder erected vertically on the circle in the (x, y)-plane whose center is at the point (0, 1, 0) and whose radius is unity. The curve of intersection of the two surfaces, — *i.e.* the curve represented by the two equations taken simultaneously, — is shown in Fig. 1. It does not lie in a plane; to distinguish it from curves which do, we call it a *twisted* curve.

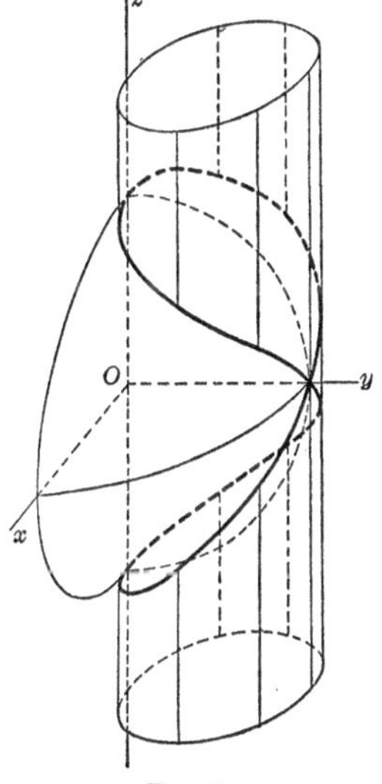

Fig. 1

Space Curves. The straight lines (1) and (2), the circle (3), and the twisted curve (4) are all called *space curves.* There are, then, three types of space curves: *straight lines, plane curves other than straight lines,* and *twisted curves.*

We now put into definitive form what we have learned from the foregoing examples:

Two equations in x, y, z, considered simultaneously, represent usually a space curve. The curve consists of all those points and only those points whose coördinates satisfy simultaneously both equations. It is the total intersection of the two surfaces which are defined by the two equations when taken individually.*

* Two equations do not always represent a curve. For example, the pair of equations $x^2+y^2=0$, $x^2+z^2=0$ represents just one point, the origin; and the equations of two parallel planes, as $x-y+z=2$ and $x-y+z=3$, represent no point at all when considered simultaneously.

If on the other hand it is a curve, and not a pair of equations, which is given, we should say:

A space curve can be represented by two simultaneous equations. These can be ANY *two equations which are satisfied simultaneously by the coördinates of every point of the curve and by those of no other point.* This means, geometrically, that the curve can be considered as the intersection of *any* two surfaces on which it lies, provided the two surfaces have no other point in common.

For example, a straight line can be considered as the intersection of any two planes through it. Consequently, it can be represented by the equations of any two of these planes. In other words, *the two equations of the line are not unique.* Thus, as equations of the axis of z we might take the equations (1) or we might take, equally well, any other two equations representing planes through the axis of z, as

$$x - 2y = 0, \qquad 17x + 56y = 0.$$

Equations (1) are, however, the simplest choice, and naturally we shall find it of aid in analytical work to choose always that pair of equations representing the curve under consideration which seems most simple.

Problem 1. What are the equations of the straight line L through the origin with the direction components 2, 3, -1?

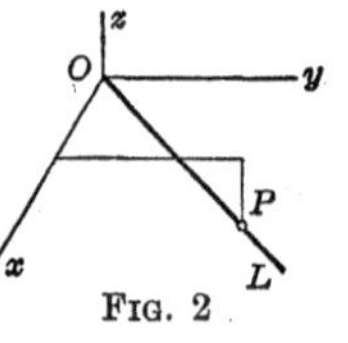

FIG. 2

If P: (x, y, z) is any point on the line, other than the origin O, then x, y, z, considered as the projections of OP on the axes, are also direction components of L. Hence, x, y, z are proportional to 2, 3, -1:

$$(5) \qquad x = 2\rho, \qquad y = 3\rho, \qquad z = -\rho,$$

where the factor of proportionality, ρ, is not constant, but variable, depending for its value on the position of P on the line.*

* From (5),

$$OP^2 = x^2 + y^2 + z^2 = 14\rho^2 \qquad \text{or} \qquad \rho = \pm\frac{OP}{\sqrt{14}}.$$

That is, ρ is equal numerically to the distance of P from O, divided by $\sqrt{14}$. Its sign depends on the side of O on which the point P lies.

Conversely, if $P:(x, y, z)$ is any point other than O, whose coördinates satisfy equations (5) for some value of ρ, these equations say that the direction of OP is that of L and hence that P lies on L.

The coördinates (0, 0, 0) of the excepted point, O, obviously satisfy (5), when ρ is given the value 0. Consequently, equations (5) represent those points and only those points which lie on L, *i.e.* they represent L.

Instead of equations (5) it is more convenient to write:

(6) $$\frac{x}{2}=\frac{y}{3}=\frac{z}{-1}.$$

This continued equality yields the three equations:

(7) $$3x-2y=0, \qquad y+3z=0, \qquad x+2z=0.$$

One of these equations must be superfluous, since we know that two equations are all that are necessary to represent a line. As a matter of fact, the three planes defined by the three individual equations all pass through L and hence one of them *is* superfluous in determining L. We prove this analytically by showing that a simultaneous solution of any two of the three equations always satisfies the third. Thus, if x_0, y_0, z_0 are any values of x, y, z which satisfy the first two equations, *i.e.* if

$$3x_0-2y_0=0, \qquad y_0+3z_0=0,$$

elimination of y_0 gives the relation

$$x_0+2z_0=0,$$

which says that these values also satisfy the third equation, q. e. d.

Since one of the equations (7) is superfluous, we might take any two of these equations, as

$$3x-2y=0, \qquad y+3z=0,$$

to represent L. It is more convenient, however, to consider the continued inequality (6) as defining L, and to call this continued inequality *the equations of L*, remembering always, that one of the equations which follow from it is superfluous.

Problem 2. Find the equations of the curve of intersection C of two circular cylinders of the same radius a, whose axes are respectively the axes of x and y.

The equations of the two cylinders are

$$y^2 + z^2 = a^2, \qquad x^2 + z^2 = a^2, \tag{8}$$

and these two equations, taken together, represent the curve C.

They are not, however, the simplest pair of equations possible. If x_0, y_0, z_0 are any set of values of x, y, z satisfying them simultaneously, *i.e.* if

$$y_0^2 + z_0^2 = a^2, \qquad x_0^2 + z_0^2 = a^2,$$

elimination of z_0 gives the relation

$$x_0^2 - y_0^2 = 0,$$

which says that x_0, y_0, z_0 also satisfy the equation

$$x^2 - y^2 = 0. \tag{9}$$

That is, the curve of intersection C of the cylinders (8) lies on the surface (9).

Conversely, the surface (9) intersects each of the cylinders (8) in the curve C *and in no other points.* For, if x_0, y_0, z_0 are any values of x, y, z satisfying equation (9) and the first, say, of equations (8), we have:

$$x_0^2 - y_0^2 = 0, \qquad y_0^2 + z_0^2 = a^2.$$

Elimination of y_0 gives the relation

$$x_0^2 + z_0^2 = a^2,$$

which says that x_0, y_0, z_0 satisfy also the second equation of (8), q. e. d.

We have proved, then, that the total intersection of any two of the surfaces (8) and (9) is the curve C. Hence any two of the equations (8) and (9) define C. A simpler pair than the pair (8) is the combination of one of the equations (8) with the equation (9), for example

$$x^2 + z^2 = a^2 \qquad x^2 - y^2 = 0. \tag{10}$$

The surface (9) consists of the two planes,

$$x - y = 0, \qquad x + y = 0,$$

passing through the z-axis and bisecting the angles between the $(y,\ z)$- and $(z,\ x)$-planes. Each of these planes intersects either cylinder in an ellipse; cf. Ch. XII, § 6. Consequently, the curve C consists of two ellipses. We have, then, not only obtained the simpler equations (10) to represent the curve, but we have also, in the process, succeeded in determining the nature of the curve.

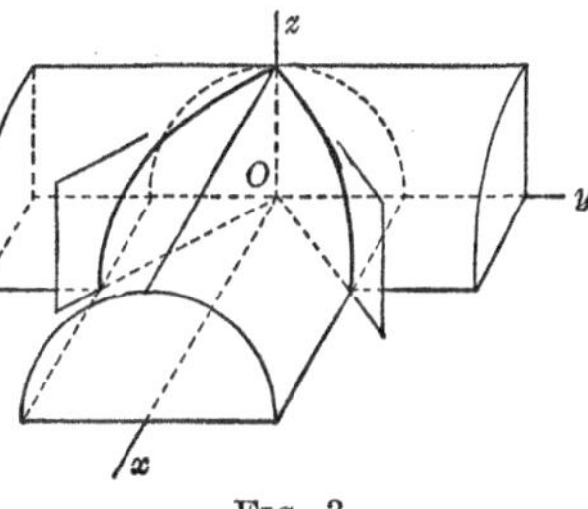

FIG. 3

EXERCISES

In each of the following exercises determine what the given equations, considered simultaneously, represent. Draw a figure.

1. $y = 0,\ z = 0.$
2. $x = 4,\ y = 0.$
3. $x + 4 = 0,\ z - 3 = 0.$
4. $2\,y + 3 = 0,\ 3\,z - 5 = 0.$
5. $y - z = 0,\ x = 3.$
6. $3\,x + 2\,y = 0,\ z - 4 = 0.$
7. $y + z = 2,\ 2\,x = 5.$
8. $x - y = 0,\ x - z = 0.$
9. $\dfrac{x}{3} = \dfrac{y}{2} = \dfrac{z}{6}.$
10. $x = -y = z.$
11. $x^2 + y^2 + z^2 - 9 = 0,\ y = 2.$
12. $x^2 + y^2 + z^2 - 16 = 0,\ x + z = 4.$
13. $x^2 + y^2 = 4,\ 2\,y + z = 3.$
14. $x^2 + y^2 + z^2 - 25 = 0,\ x^2 + y^2 = 16.$
15. $x^2 + y^2 + z^2 - 16 = 0,\ 4\,x^2 + (y - 2)^2 = 4.$
16. $x^2 + y^2 = a^2,\ z^2 = ay.$
17. Which of the curves represented by the above pairs of equations pass through the origin?

Find the equations of the following curves:

18. The axis of y.

19. The line in the (x, y)-plane 3 units in front of the (y, z)-plane.

20. The line 2 units to the left of the (z, x)-plane and 3 units above the (x, y)-plane.

21. The line $\frac{5}{2}$ units behind the (y, z)-plane and $\frac{7}{3}$ units to the right of the (z, x)-plane.

22. The line which lies in the plane passing through the x-axis and bisecting the angle between the positive y- and z-axes and is 4 units above the (x, y)-plane.

23. The line through the origin with the direction components 1, -1, -1.

24. The line through the origin with the direction components 2, 0, 3.

25. The circle of radius 3 whose center is on the axis of z and whose plane is 4 units above the (x, y)-plane.

26. The circle of radius 2 whose center is at the origin and whose plane passes through the y-axis and bisects the angle between the positive axis of x and the negative axis of z.

Find a simpler pair of equations to represent the curve given in each of the following exercises and then identify the curve. Draw a figure.

27. $2y - 3z = 0,\ 5y + 16z = 0.$

28. $x + y + z = 2,\ x + y - 5 = 0.$

29. The curve of Ex. 14.

30. $x^2 + y^2 + z^2 = a^2,\ 2y^2 + z^2 = a^2.$

2. Line of Intersection of Two Planes. Let the plane

$$A_1x + B_1y + C_1z + D_1 = 0$$

and the plane

$$A_2x + B_2y + C_2z + D_2 = 0$$

be two planes which meet. According to the theory of the preceding paragraph, their line of intersection is represented by the two *simultaneous* equations

$$(1) \qquad \begin{aligned} A_1x + B_1y + C_1z + D_1 = 0, \\ A_2x + B_2y + C_2z + D_2 = 0, \end{aligned}$$

i.e. by these equations, where the only sets of values of x, y, z considered are those which satisfy *both* equations.

Since the line lies in each plane, it is perpendicular to the normals to each plane. That is, it is a common perpendicular to the normals to the two planes. From this fact its direction components can easily be determined, by the method of Ch. XVIII, § 5.

Consider, for example, the line (2) of § 1. The normals to the two planes determining this line have 3, -4, -1 and 5, 3, 2 as direction components. Consequently,

$$\begin{vmatrix} -4 & -1 \\ 3 & 2 \end{vmatrix}, \quad \begin{vmatrix} -1 & 3 \\ 2 & 5 \end{vmatrix}, \quad \begin{vmatrix} 3 & -4 \\ 5 & 3 \end{vmatrix},$$

or -5, -11, 29 are the direction components of the line.

The planes determining the line (1) have normals with A_1, B_1, C_1 and A_2, B_2, C_2 as direction components. Hence, *the direction components of the line* (1) *are*

$$(2) \qquad \begin{vmatrix} B_1 & C_1 \\ B_2 & C_2 \end{vmatrix}, \quad \begin{vmatrix} C_1 & A_1 \\ C_2 & A_2 \end{vmatrix}, \quad \begin{vmatrix} A_1 & B_1 \\ A_2 & B_2 \end{vmatrix}.$$

Parallel Planes. If two planes are parallel and so have no line of intersection, their equations are *incompatible, i.e.* they have no *simultaneous* solution. For example, the equations

$$2x + y - z - 3 = 0, \qquad 4x + 2y - 2z + 5 = 0,$$

which represent two parallel planes, are obviously incompatible, since the first says that the quantity $2x + y - z$ has the value 3, whereas the second says that it has the value $-\frac{5}{2}$.

EXERCISES

Find the direction components of each of the following lines. Determine also a point on the line and hence construct the line.

1. $2x + y + 3z = 0, \quad x - 2y + 4z = 0.$
2. $3x - 4y - 6z + 7 = 0, \quad 2x - y - 2z + 1 = 0.$
3. $x + 3y - 3z + 5 = 0, \quad 3x + 4y + 6z - 5 = 0.$
4. $2x - y - z = 0, \quad x + y = 5.$
5. $3x - 5y - z - 1 = 0, \quad x - 4 = 0.$
6. $x + 2y = 0, \quad y - 3z = 0.$

7. What are the equations of the edges of the tetrahedron of Ch. XIX, § 6, Ex. 9? Draw a figure and label the edges and the pairs of equations to correspond.

8. The same for the tetrahedron of Ch. XIX, § 10, Ex. 5.

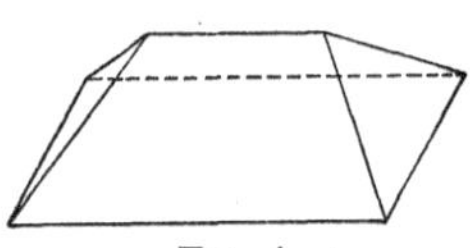

FIG. 4

9. Show that the lines of Exs. 1 and 4 are perpendicular.

10. Show that the lines of Exs. 3 and 6 are parallel.

11. Each side of a hip roof (Fig. 4) makes an angle with the horizontal whose tangent is $\frac{1}{2}$. What angle do the edges of the roof make with the horizontal?

3. Line through a Point with Given Direction Components. Let it be required to find the equations of the line L which goes through the point $P_0: (x_0, y_0, z_0)$ and has the direction components l, m, n.

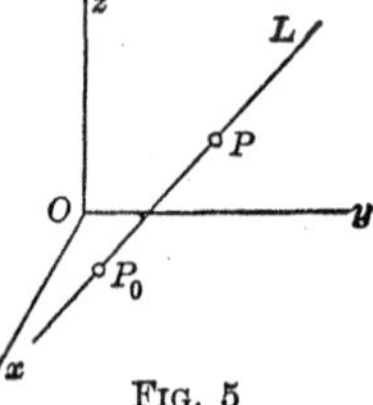

FIG. 5

Let $P: (x, y, z)$ be any point on L, other than P_0. Then the direction components of P_0P, namely,

$$(1) \qquad x - x_0, \qquad y - y_0, \qquad z - z_0,$$

are direction components of L. They are, then, proportional to l, m, n:

$$(2) \qquad x - x_0 = \rho l, \qquad y - y_0 = \rho m, \qquad z - z_0 = \rho n,$$

where the factor of proportionality, ρ, varies in value as P changes position.*

Conversely, if $P:(x, y, z)$ is any point other than P_0, for which equations (2) hold for some value of ρ, *i.e.* for which the quantities (1) are proportional to l, m, n, it follows that the direction of P_0P is that of L and hence that P lies on L.

The coördinates (x_0, y_0, z_0) of the excepted point P_0 obviously satisfy (2), when $\rho = 0$. Consequently, equations (2) are satisfied by the coördinates of those points and only those points which lie on L and so they represent L.

Instead of equations (2), we can write

$$(3) \qquad \frac{x - x_0}{l} = \frac{y - y_0}{m} = \frac{z - z_0}{n}.$$

We call this continued equality *the equations of the line*, remembering from § 1 that, of the three equations which in general result from it, one is superfluous.

If one of the denominators in (3) is zero, so is the corresponding numerator; thus, if $n = 0$, then $z - z_0 = 0$. For, equations (3) are but an abbreviated form of the equations of proportionality † (2), and if $n = 0$ the last equation in (2) reduces immediately to $z - z_0 = 0$.

Suppose, for example, that the line is to go through the point $(3, 4, -6)$ and have 0, 5, 3 as its direction components. Here $l = 0$, and hence $x - x_0 = 0$, *i.e.* $x - 3 = 0$. Thus,

* From (2), $$P_0P^2 = \rho^2(l^2 + m^2 + n^2),$$
or $$\rho = \pm\frac{P_0P}{\sqrt{l^2 + m^2 + n^2}}.$$
That is, the numerical value of ρ is proportional always to the distance of P from P_0. The sign of ρ depends on the side of P_0 on which the point P is situated.

† Cf. Ch. XVI, § 9, eq. (5).

$x-3=0$ is one of the equations of the line. The other, obtained from the equality of the last two members of (3), is

$$\frac{y-4}{5}=\frac{z+6}{3}, \qquad \text{or} \qquad 3x-5z-42=0.$$

Hence the desired equations are

$$x-3=0, \qquad 3x-5z-42=0.$$

Again, let the point be $(1, 2, -3)$ and let the direction components be $0, 0, 1$. In this case, $l=0$, and $m=0$; therefore $x-x_0=0$ and $y-y_0=0$, *i.e.*

$$x-1=0, \qquad y-2=0,$$

and these are the equations of the line. This result might have been obtained directly, by inspection, since it is clear that the line is parallel to the axis of z.

Reduction of a Continued Equality to the Form (3). *Example* 1. What are the direction components of the line

$$\frac{x-5}{6}=-\frac{y+4}{3}=\frac{z+3}{2}?$$

This continued equality will be of the form (3), if the minus sign before the second fraction is associated with the denominator. Accordingly, the direction components are $6, -3, 2$.

Example 2. Consider a more complicated case:

$$\frac{2x-1}{3}=\frac{2-5y}{4}=\frac{3z}{2}.$$

To put this continued equality in the form (3), divide the numerator and denominator of each fraction by the coefficient of the variable in the numerator:

$$\frac{x-\frac{1}{2}}{\frac{3}{2}}=\frac{y-\frac{2}{5}}{-\frac{4}{5}}=\frac{z}{\frac{2}{3}}.$$

Here, each variable x, y, z, as it appears in the numerator, has the coefficient unity, and there is complete conformity to (3).

The direction components of the line are, then, $\frac{3}{2}$, $-\frac{4}{5}$, $\frac{2}{3}$ or 45, -24, 20. Furthermore, it is clear that the line goes through the point $(\frac{1}{2}, \frac{2}{5}, 0)$.

EXERCISES

In each of the following exercises find the equations of the line through the given point with the given direction components.

	Point	*Components*
1.	$(2, -3, 1)$,	$5, 2, -4$.
2.	$(0, 0, 0)$,	$3, -1, 2$.
3.	$(4, -1, -2)$,	$-6, 5, 8$.
4.	$(2, 0, -3)$,	$1, 1, 1$.
5.	$(3, 2, -8)$,	$1, 3, 0$.
6.	$(2, 0, 1)$,	$4, 0, 1$.
7.	$(-3, 4, 6)$,	$0, 1, 0$.

In each of the exercises which follow, find the direction components of the line represented by the given equations and the coördinates of a point on the line. Construct the line.

8. $\dfrac{x-1}{2} = -\dfrac{y+2}{3} = \dfrac{z-1}{5}$.

9. $\dfrac{3x+2}{9} = \dfrac{y}{-6} = \dfrac{2z-1}{4}$.

10. $1 - x = y - 2 = z - 6$.

11. $3x + 4 = 2 - 5y = 4z - 7$.

12. $2x = 1 - y = 3z$.

13. Show that the lines of Exs. 8 and 10 are perpendicular.

14. Show that the lines of Exs. 9 and 12 are parallel.

4. Line through Two Points. The line through the two points (x_1, y_1, z_1), (x_2, y_2, z_2) has the direction components

$$x_2 - x_1, \qquad y_2 - y_1, \qquad z_2 - z_1.$$

It can be considered, then, as the line which has these direction components and goes through the point (x_1, y_1, z_1). Consequently, by (2), § 3, it is represented by the equations

$$(1) \quad x - x_1 = \rho(x_2 - x_1), \quad y - y_1 = \rho(y_2 - y_1), \quad z - z_1 = \rho(z_2 - z_1).$$

Instead of (1) we can write, as the equations of the line, the continued equality

$$(2) \qquad \frac{x - x_1}{x_2 - x_1} = \frac{y - y_1}{y_2 - y_1} = \frac{z - z_1}{z_2 - z_1}.$$

Since (2) is an abbreviated form of (1), it follows that, if a denominator in (2) is zero, the corresponding numerator is also zero. Thus, if the two points are (3, 5, −4), (8, 5, −4), so that $y_2 - y_1 = 0$ and $z_2 - z_1 = 0$, we have $y - y_1 = 0$ and $z - z_1 = 0$, that is,

$$y - 5 = 0, \qquad z + 4 = 0.$$

These are, then, the equations of the line. The result might have been obtained directly by noting in the beginning that the y-coördinates, and also the z-coördinates, of the two points are equal and by concluding, then, that the line is parallel to the axis of x.

EXERCISES

Find the equations of each of the following lines.

1. Through (2, 5, 8), (− 1, 6, 3).
2. Through (− 1, 0, 2), (3, 4, 6).
3. Through the origin and (5, − 2, 3).
4. Through (2, 0, 3), (0, 3, 2).
5. Through (2, − 5, 8), (2, 3, 7).
6. Through (3, − 2, − 5), (3, − 2, 6).
7. The edges of the tetrahedron of Ch. XIX, § 6, Ex. 9.
8. The edge of the quadrangular pyramid of Ch. XIX, § 5, Ex. 10.

5. Line or Plane in Given Relationship to Given Lines or Planes. *Problem* 1. To find the equations of the line through a given point perpendicular to a given plane.

The direction components of the line are those of a normal to the given plane and hence can easily be found. The problem then becomes that of finding the equations of a line through a given point with given direction components.

For example, if the given point is the origin and the given plane is

$$3x - 4y + 5z + 6 = 0,$$

the required line goes through (0, 0, 0) and has the direction components 3, -4, 5. Hence its equations are

$$\frac{x-0}{3} = \frac{y-0}{-4} = \frac{z-0}{5},$$

or

$$20x = -15y = 12z.$$

Problem 2. To find the equation of the plane through a given point perpendicular to a given line.

The direction of a normal to the plane is that of the given line and therefore the direction components of the normal are easily written down. We then have the problem of finding the equation of a plane through a given point with given direction components of its normals.

Thus, if the point is (3, -2, 1) and the line is given, as the intersection of two planes, by the equations,

$$3x - 5y - 2z + 6 = 0, \qquad 4x + y + 3z - 7 = 0,$$

the direction components of the line are, by (2), § 2,

$$\begin{vmatrix} -5 & -2 \\ 1 & 3 \end{vmatrix}, \quad \begin{vmatrix} -2 & 3 \\ 3 & 4 \end{vmatrix}, \quad \begin{vmatrix} 3 & -5 \\ 4 & 1 \end{vmatrix},$$

i.e. -13, -17, 23, or 13, 17, -23. The required plane has these direction components for its normals and passes through the point (3, -2, 1). Consequently, its equation is

$$13(x-3) + 17(y+2) - 23(z-1) = 0,$$

or

$$13x + 17y - 23z + 18 = 0.$$

Problem 3. Let two non-parallel lines, L_1 and L_2, and a point P be given. Through P parallel to each of the lines there is a unique plane. To determine this plane.

A normal to the plane is a common perpendicular to the lines L_1 and L_2 and so its direction components can be found by the method of Ch. XVIII, § 5. The problem is then that of finding the equation of a plane through a given point with given direction components of its normals.

If L_1 and L_2 have the equations

$$\frac{x-1}{5}=\frac{y-2}{3}=-\frac{z}{2}, \qquad \frac{x+3}{4}=\frac{y}{2}=\frac{z-1}{3},$$

their direction components are, respectively, 5, 3, -2 and 4, 2, 3. The direction components of a common perpendicular to them are, by Ch. XVIII, § 5, (6), 13, -23, -2. Thus the plane parallel to L_1 and L_2 and passing through a given point, say $(3, 2, -4)$, has the equation

$$13(x-3)-23(y-2)-2(z+4)=0,$$

or

$$13x-23y-2z-1=0.$$

Problem 4. Given two intersecting planes, M_1 and M_2, and a point P. Through P parallel to each of the planes there is a unique line. To find this line.

Since the line is parallel to each of the planes, it is parallel to their line of intersection. It is, therefore, itself the line of intersection of the two planes which pass through P and are parallel respectively to M_1 and M_2.

For example, if P is $(2, 0, -1)$ and M_1 and M_2 are

$$2x-3y+z-6=0, \qquad 4x-2y+3z+9=0,$$

the planes through P parallel respectively to M_1 and M_2 have the equations

$$2(x-2)-3y+(z+1)=0, \quad 4(x-2)-2y+3(z+1)=0,$$

or

$$2x-3y+z-3=0, \qquad 4x-2y+3z-5=0.$$

These equations, considered simultaneously, represent the required line.

The problem might have been solved by determining the direction components of the line of intersection of M_1 and M_2, as given by (2), § 2, and by finding the equations of the line which has these direction components and passes through P.

Problem 5. To find the equation of a line which passes through a given point and is parallel to a given line.

If the line is given as the intersection of two planes, this is the previous problem. If its equations are given in the form of a continued equality, the solution is simple. We leave it to the student.

Problem 6. To find the equations of a line which passes through a given point and is perpendicular to each of two given non-parallel lines.

The solution of this problem we also leave to the student.

EXERCISES

In each one of the following exercises in which it is possible, solve the given problem directly, by inspection of a figure.

Find the equations of the line passing through the given point and perpendicular to the given plane.

	Point	*Plane*
1.	$(2, -8, 3)$,	$x + 2y - 3z - 2 = 0.$
2.	$(0, 0, 0)$,	$6x - 2y + 5z + 3 = 0.$
3.	$(3, -4, 0)$,	$5x - 3z + 4 = 0.$
4.	$(-1, 2, 5)$,	$2z + 3 = 0.$

Find the equation of the plane passing through the given point and perpendicular to the given line.

	Point	*Line*
5.	$(-1, 2, 5)$,	$\begin{cases} 2x - 3y + 6z - 4 = 0, \\ 4x - y + 5z + 2 = 0. \end{cases}$
6.	$(0, 0, 0)$,	$\begin{cases} 3x + 2y + z - 3 = 0, \\ x + 2y + 3z + 2 = 0. \end{cases}$

	Point	*Line*
7.	$(3, -2, -8)$,	$\frac{x-1}{3} = -\frac{y}{5} = \frac{z-2}{6}$.
8.	$(5, 0, -1)$,	$1 - 3x = 2y = 4z - 3$.
9.	$(4, -\frac{1}{2}, 3)$,	$2y + 5 = 0, 3z - 2 = 0$.

Find the equation of the plane passing through the given point and parallel to each of the given lines.

	Point	*Lines*
10.	$(3, 5, 1)$	The lines given in Exs. 7, 8.
11.	$(0, 2, -3)$	The lines given in Exs. 6, 7.
12.	$(0, 0, 0)$	The lines given in Exs. 5, 6.
13.	$(2, -\frac{5}{2}, 3)$	$y = 4, z = 3$; $2x = 5, 3y = 7$.

Find the equations of the line passing through the given point and parallel to each of the given planes.

	Point	*Planes*
14.	$(0, 0, 0)$	The planes given in Exs. 1, 2.
15.	$(2, -1, -3)$	The planes given in Exs. 1, 3.
16.	$(0, 2, -\frac{3}{2})$	The planes given in Exs. 2, 3.
17.	$(\frac{1}{2}, 0, -\frac{2}{3})$	The planes given in Exs. 3, 4.

Find the equations of the line passing through the given point and parallel to the given line.

18. The point and line given in Ex. 5.
19. The point and line given in Ex. 7.
20. The point and line given in Ex. 8.
21. The point and line given in Ex. 9.

Find the equations of the line passing through the given point and perpendicular to the given lines.

	Point	*Lines*
22.	$(0, 0, 0)$	The lines given in Exs. 7, 8.
23.	$(3, -2, 5)$	The lines given in Exs. 6, 7.

	Point	*Lines*
24.	(2, 4, 0)	The lines given in Exs. 5, 6.
25.	(0, 0, 3)	The lines given in Ex. 13.

6. Angle between a Line and a Plane. Given a plane M and a line L which is not perpendicular to the plane. Project the line on the plane. The acute angle, ϕ, which the line makes with this projection is the angle between the line and the plane. It may be determined by finding the acute angle, θ, which the line makes with a normal N to the plane. For, it is clear that θ and ϕ are complementary angles.

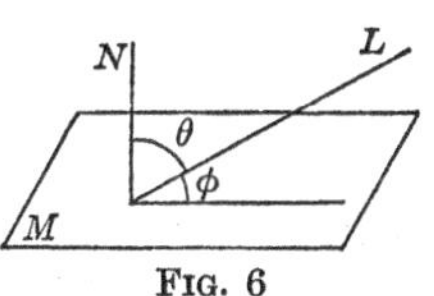

FIG. 6

Example. Find the angle between the line

$$2x + 2 = y + 1 = -4z + 28$$

and the plane

$$2x + 3y - 2z + 1 = 0.$$

The direction components of the line are 2, 4, -1; those of a normal to the plane are 2, 3, -2. Hence

$$\cos\theta = \pm\frac{2 \cdot 2 + 4 \cdot 3 + (-1)(-2)}{\sqrt{4 + 16 + 1}\sqrt{4 + 9 + 4}},$$

where we are to take that sign which makes the right-hand side positive.

Thus, $\cos\theta = \frac{18}{\sqrt{357}} = 0.9527$, and $\theta = 17° \, 42'$.

Finally, $\phi = 90° - \theta = 72° \, 18'$.

EXERCISES

Find the angle between the given line and the given plane.

	Line	*Plane*
1.	$3x + 3 = 2y + 2 = -6z - 12,$	$3x + y + 2z + 1 = 0.$
2.	$\begin{cases} 3x - 4y + 2z = 0, \\ 4x - 3y + z = 5, \end{cases}$	$3x - 2z - 12 = 0.$

3. Through (0, 0, 0), (1, 2, − 2), $2x - 6y + 3z - 4 = 0$.

4. Through (2, 5, 3), (4, − 1, 6), $y + 2z + 4 = 0$.

5. Find the angles which the plane given in Ex. 1 makes with the coördinate axes.

6. Find the angles which the line given in Ex. 2 makes with the coördinate planes.

7. Point of Intersection of a Line and a Plane. Given a plane,

$$(1) \qquad Ax + By + Cz + D = 0,$$

and a line,

$$(2) \quad A_1x + B_1y + C_1z + D_1 = 0, \quad A_2x + B_2y + C_2z + D_2 = 0,$$

which intersects the plane. The coördinates of the point of intersection satisfy the equation of the plane, since the point lies in the plane. They also satisfy *each* of the equations of the line, for the point lies on the line. They are, then, the *simultaneous solution* of the three equations

$$(3) \qquad \begin{aligned} Ax + By + Cz + D &= 0, \\ A_1x + B_1y + C_1z + D_1 &= 0, \\ A_2x + B_2y + C_2z + D_2 &= 0. \end{aligned}$$

Example. Find the coördinates of the point of intersection of the line and the plane of § 6. Take, as the equations of the line, those obtained by equating the first and second members, and then the first and third members, of the continued equality which represents the line. Then the three equations, which are to be solved simultaneously, are

$$2x - y = -1, \qquad x + 2z = 13, \qquad 2x + 3y - 2z = -1.$$

The solution is found to be $x = 1$, $y = 3$, $z = 6$. Thus, the line meets the plane in the point (1, 3, 6).

Intersection of a Curve and a Surface. The above method applies also to the problem of finding the point (or points) of intersection of a curve and a surface which are given by their equations. The two equations of the curve and the equation of the surface are to be considered as *simultaneous equations in the unknowns*, x, y, z, and solved as such.

Plane and Line Arbitrary. The line (2) has one and just one point in common with the plane (1), if and only if the three planes (3) intersect in a single point, *i.e.* by the theorem of Ch. XXIX, § 10, if and only if the determinant of the coefficients of x, y, z in equations (3) does not vanish.

If this determinant vanishes, it follows, either directly or from the discussion in Ch. XXIX, § 10, that the line either is parallel to the plane or lies in it.

EXERCISES

Show that the given line has just one point in common with the given plane and find the coördinates of the point.

1. The line and plane of Ex. 1, § 6.
2. The line and plane of Ex. 2, § 6.
3. The line and plane of Ex. 3, § 6.

Find the points of intersection of the given curve with the given surface. Draw a figure for each exercise.

	Curve	*Surface*
4.	$x = y = z,$	$x^2 + y^2 + z^2 = 1.$
5.	$x^2 + y^2 + z^2 = 29,\ z = 2,$	$4x - 3y = 0.$
6.	$12 - 6x = 2y + 2 = 3z - 9,$	$x^2 + y^2 = 5.$
7.	$x^2 + y^2 + z^2 = 12,\ x = y,$	$x^2 + y^2 = 8.$

Find out all you can about the relative positions of the given line and the given plane.

	Line	*Plane*
8.	$\begin{cases} 2x - 3y + z = 0, \\ 4x - 5y + 6 = 0, \end{cases}$	$4x - 6y + 2z + 1 = 0.$
9.	$x = y = z.$	$5x + 3y - 8z - 3 = 0.$
10.	$\begin{cases} 2x + 3y - 8 = 0, \\ 3x - 2y + 1 = 0, \end{cases}$	$8x - y - 6 = 0.$

8. Parametric Representation of a Curve. *The Straight Line.* Given a *directed* straight line passing through the point $P_0:(x_0, y_0, z_0)$ and having the direction cosines $\cos\alpha$, $\cos\beta$, $\cos\gamma$. Let $P:(x, y, z)$ be an arbitrary point of the line other than P_0 and let r be the algebraic distance from P_0 to P, positive if the direction from P_0 to P is that of the line and negative if this direction is opposite to that of the line.

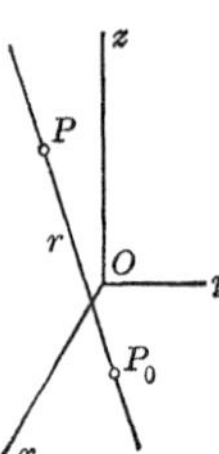

FIG. 7

The projections of P_0P, each divided by r, are equal respectively to the direction cosines of the line, by Ch. XVIII, § 1, Th. 2. Thus

$$\frac{x - x_0}{r} = \cos\alpha, \qquad \frac{y - y_0}{r} = \cos\beta, \qquad \frac{z - z_0}{r} = \cos\gamma.$$

These equations can be put into the form

$$(1) \quad x = x_0 + r\cos\alpha, \qquad y = y_0 + r\cos\beta, \qquad z = z_0 + r\cos\gamma.$$

Equations (1) give the coördinates (x, y, z) of the point P on the given line at the arbitrary distance r from P_0. If r is allowed to vary through all values, positive, zero, and negative, P takes on all positions on the line, and always its coördinates are given by equations (1). These equations, then, represent the line. Since they express the coördinates of the point $P:(x, y, z)$ tracing the line in terms of the *auxiliary variable*, or *parameter*, r, we call them a *parametric representation* of the line.

If the line is determined by the point $P_0:(x_0, y_0, z_0)$ and its direction components l, m, n, we have, according to (2), § 3, the following parametric representation

$$(2) \qquad x = x_0 + \rho l, \qquad y = y_0 + \rho m, \qquad z = z_0 + \rho n.$$

The parameter ρ is not, in general, equal to the distance from P_0 to $P:(x, y, z)$, but is merely proportional to this distance.*

* Cf. footnote, p. 479.

The Helix. Given the cylinder

$$x^2 + y^2 = a^2$$

with the axis of z as axis. The circle in the (x, y)-plane on which the cylinder is erected has the parametric representation (Ch. VII, § 10):

$$x = a \cos \theta, \qquad y = a \sin \theta, \qquad z = 0,$$

where θ is the angle which the radius, OP, to the point $P: (x, y, 0)$ makes with the positive axis of x. On the ruling of the cylinder through P mark the point P' at a height above P* equal to a constant multiple, $k\theta$, of the angle θ. The coördinates of P' are

(3) $\quad x = a \cos \theta, \qquad y = a \sin \theta, \qquad z = k\theta.$

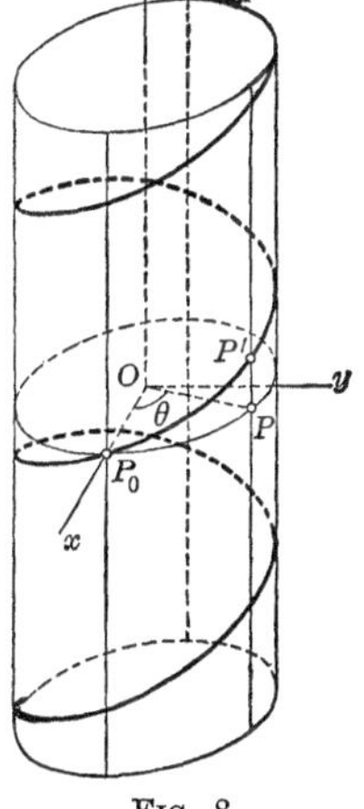

FIG. 8

When $\theta = 0$, P and P' coincide in the point P_0 on the axis of x. As θ increases from 0 to 2π, the point P traces the circle, and the point P', always directly above P, traces a locus on the cylinder, encircling it just once. When θ increases from 2π to 4π, P retraces the circle, whereas P' continues on its rising path, encircling the cylinder a second time. Consequently, when θ increases through all positive values, the locus traced by P' encircles the cylinder infinitely many times.

As θ decreases from zero through all negative values, P', starting from P_0, encircles the lower half of the cylinder infinitely many times. The complete locus of P' is, then, an unbroken curve continuously winding about the cylinder in both directions. It is this curve which is represented parametrically by the equations (3).

Since the height of P' above (or below) P is always proportional to the angle θ through which the radius OP of the circle has turned, the curve (3) which P' traces is mounting on the

* Above P, if θ is positive; below P, if θ is negative.

cylinder with a uniform steepness. It is the curve of the thread of a machine screw and is called a *circular screw* or a *circular helix.*

The Twisted Cubic. Consider the curve represented parametrically by the equations

$$(4) \qquad x = at, \qquad y = bt^2, \qquad z = ct^3,$$

where t is the parameter, and a, b, c are constants, not zero. Like the helix, this curve is a twisted curve. It is known as a *twisted cubic.*

Points of Intersection of a Curve and a Surface. Example 1. The straight line of § 6 can be represented parametrically by setting each of the members of the continued equality

$$2x + 2 = y + 1 = -4z + 28$$

equal to a parameter t and by solving the three resulting equations for x, y, z:

$$(5) \qquad x = \tfrac{1}{2}t - 1, \qquad y = t - 1, \qquad z = 7 - \tfrac{1}{4}t.$$

A point of this line lies in the plane of § 6,

$$2x + 3y - 2z + 1 = 0,$$

if and only if its coördinates, as given by (5), satisfy the equation of the plane; *i.e.* if and only if t is a solution of the equation

$$2(\tfrac{1}{2}t - 1) + 3(t - 1) - 2(7 - \tfrac{1}{4}t) + 1 = 0,$$

or

$$\tfrac{9}{2}t - 18 = 0.$$

Hence $t = 4$. But the point $t = 4$ of the line (5), *i.e.* the point corresponding to the value 4 for the parameter t, has the coördinates:

$$x = 2 - 1 = 1, \qquad y = 4 - 1 = 3, \qquad z = 7 - 1 = 6.$$

Hence the line intersects the plane in the point (1, 3, 6).

Example 2. Find the points of intersection of the twisted cubic

(6) $$x = t, \quad y = t^2, \quad z = t^3$$

with the plane $$2x + y - z = 0.$$

A point of the cubic lies in the given plane when and only when its coördinates, as given by (6), satisfy the equation of the plane. Hence the solutions of the equation

$$2t + t^2 - t^3 = 0$$

for t determine all the points of intersection.

One solution is $t = 0$; the others are $t = 2$, $t = -1$. The points of the cubic (6) corresponding to these values of t are respectively $(0, 0, 0)$, $(2, 4, 8)$, $(-1, 1, -1)$. Thus the cubic meets the plane in these three points and in no further point.

The above method may be used to find the points of intersection of any given curve with a given surface, provided the curve is defined by a parametric representation. The simplicity and effectiveness of the method is one of the advantages of representing a curve parametrically.

EXERCISES

Find a parametric representation for each of the following straight lines.

1. Through $(2, -3, 5)$ with the direction cosines $\frac{2}{7}, -\frac{6}{7}, \frac{3}{7}$.

2. Through $(2, -3, 5)$ with the direction components $2, -6, 3$.

3. Through (x_1, y_1, z_1), (x_2, y_2, z_2); cf. (1), § 4.

4. $3x - 4y - 6z + 7 = 0$, $2x - y - 2z + 1 = 0$.

Suggestion. First find the direction components of the line and the coördinates of a point on it.

The same for the lines of the following exercises of previous paragraphs.

5. Ex. 1, § 3.

6. Ex. 2, § 3.

7. Ex. 1, § 4.

8. Ex. 2, § 4.

9. Ex. 1, § 2.

10. Ex. 3, § 2.

Find equations for the lines with the following parametric representations.

11. $x = 3 - 2t, \quad y = t - 4, \quad z = 3t + 2.$

12. $x = 3t, \quad y = 5t, \quad z = -8t.$

13. If from the point (3, 2, − 6) one proceeds 12 units in the direction whose cosines are $\frac{2}{3}, -\frac{1}{3}, \frac{2}{3}$, what are the coördinates of the point reached? *Ans.* (11, − 2, 2).

14. Draw to scale the circular helix for which $a = 4$, $k = 2$.

15. Show that the twisted cubic (6) is the total intersection of the parabolic cylinder $y = x^2$ with the cylinder $z = x^3$. Hence construct the cubic.

16. Find a parametric representation of the curve

$$y^2 = 2x, \qquad z = 3y^3.$$

17. The same for the curve of § 1, Example 4.

Suggestion. Let $x = \sin 2\theta$.

By the method of this paragraph find, in each of the following exercises, the point (or points) of intersection of the given curve and the given surface.

18. The line and plane of Ex. 1, § 6.

19. The line and plane of Ex. 3, § 6.

20. The line and cylinder of Ex. 6, § 7.

EXERCISES ON CHAPTER XX

1. Find the equations of the line which passes through the point (1, −2, 3) and intersects the axis of z at right angles.

Find the coördinates of the points in which the given line meets the coördinate planes and hence construct the line.

2. $\begin{cases} x + 5y - z - 7 = 0, \\ 2x - 5y + 3z + 1 = 0. \end{cases}$ 3. $x - 1 = \dfrac{y+2}{3} = 2 - z.$

Show that the first of the two following lines intersects the axis of z and that the second intersects the axis of x.

4. $\begin{cases} 2x + 3y - z + 2 = 0, \\ x - 2y + 2z - 4 = 0. \end{cases}$ 5. $\dfrac{x+4}{4} = \dfrac{2y+6}{3} = \dfrac{3z+4}{2}.$

6. What is the condition that the line

$$A_1x + B_1y + C_1z + D_1 = 0, \qquad A_2x + B_2y + C_2z + D_2 = 0,$$

where $C_1C_2 \neq 0$, meet the axis of z? *Ans.* $C_1D_2 - C_2D_1 = 0$.

Show that the following lines are identical.

7. $\begin{cases} 2x + y - 5z + 2 = 0, \\ 6x - 2y - 5z + 1 = 0; \end{cases}$ $\begin{cases} 4x + y - 8z + 3 = 0, \\ 2x - 2y + z - 1 = 0. \end{cases}$

8. $\begin{cases} x + y - z + 1 = 0, \\ 2x + 3y - 1 = 0; \end{cases}$ $\dfrac{x+1}{3} = \dfrac{1-y}{2} = z - 1.$

9. Find the equations of the altitudes of the tetrahedron of Ch. XIX, § 6, Ex. 9.

10. Find the equation of the plane which contains the point $(2, -1, 5)$, is perpendicular to the plane $2x - y + 3z = 4$, and is parallel to the line

$$5x + 2y + 3z = 0, \qquad 4x + y + 2z - 8 = 0.$$

Ans. $3x - 9y - 5z + 10 = 0$.

11. Find the equation of the plane which passes through the points $(2, -1, 3)$, $(5, 0, 2)$ and is parallel to the line

$$2x - 5 = 1 - y = 2 - 3z.$$

12. Find the equations of the line which contains the point $(4, 2, -3)$, is parallel to the plane $x + y + z = 0$, and is perpendicular to the line whose equations are $x + 2y - z = 5$, $z = 4$. *Ans.* $6x - 24 = 3y - 6 = -2z - 6$.

13. A line is parallel to the plane $2x - 3y + 4 = 0$. If the perpendicular from the origin on the line meets it in the point $(2, 5, -3)$, what are the equations of the line?

14. Show that the equation of the plane which passes through the point (x_0, y_0, z_0) and is parallel to two (non-parallel)

lines having l_1, m_1, n_1 and l_2, m_2, n_2 as direction components can be written in the form

$$\begin{vmatrix} x - x_0 & y - y_0 & z - z_0 \\ l_1 & m_1 & n_1 \\ l_2 & m_2 & n_2 \end{vmatrix} = 0.$$

15. Two lines with the direction components l_1, m_1, n_1 and l_2, m_2, n_2 intersect in the point (x_0, y_0, z_0). Find the equation of the plane containing them.

16. Find the equations of the line determined by the point $(2, -1, 0)$ and the point of intersection of the three planes

$$x - y + z = 0, \quad 3x - 2y + 4z = 0, \quad 2x + y + z - 4 = 0.$$

17. Find the equations of the line through the origin and the point of intersection of the plane and the line whose equations are $x + 2y - 3z + 4 = 0$ and $3x + 1 = 2 - 2y = z + 3$.

18. Determine the equations of the line which lies in the plane $2x - y + z - 3 = 0$ and is perpendicular to the line

$$3x + 2y - z - 8 = 0, \qquad x - 3y + 1 = 0$$

in the point in which this line meets the plane.

Ans. $\dfrac{x-2}{12} = \dfrac{y-1}{19} = -\dfrac{z}{5}.$

19. A line through the origin with the direction cosines $\frac{2}{7}$, $-\frac{3}{7}$, $\frac{6}{7}$ intersects the plane $3x + 5y + 2z - 6 = 0$ in the point P. Find the length of OP. *Ans.* 14.

Suggestion. Represent the line parametrically.

20. A line through the point A: $(3, -2, 5)$ with the direction cosines $\frac{4}{9}$, $\frac{1}{9}$, $\frac{8}{9}$ meets the plane $2x + 3y - z + 7 = 0$ in the point P. What is the length of AP?

Loci

21. Find the locus of a point which is always equidistant from the three points $(2, 0, 3)$, $(0, -2, 1)$, $(4, 2, 0)$.

22. Determine the point in the plane $x - y - 2z = 0$ which is equidistant from the three points (2, 1, 5), (4, −3, 1), (−2, −1, 3). *Ans.* $(\frac{7}{5}, 1, \frac{1}{5})$.

23. Show that the locus of a point moving so that it is always equidistant from three given non-collinear points is a line perpendicular to the plane of the three points. In what point does it intersect this plane?

Suggestion. Choose the coördinate axes skillfully.

24. Find the locus of a point which is equidistant from the points (2, 3, 0), (4, −1, 2) and also equidistant from the points (5, 2, −3), (3, 0, 1).

25. The previous problem for any four non-coplanar points P_1, P_2 and P_3, P_4. Show that the locus is a line which is perpendicular to each of the lines P_1P_2, P_3P_4 and goes through the center of the sphere determined by the four points.

26. Find the locus of a point which moves so that the difference of the squares of its distances from two given points is constant.

CHAPTER XXI

THE PLANE AND THE STRAIGHT LINE. ADVANCED METHODS

1. Linear Combination of Two Planes. A *linear combination* of two planes,

$$(1) \qquad A_1x + B_1y + C_1z + D_1 = 0,$$

$$(2) \qquad A_2x + B_2y + C_2z + D_2 = 0,$$

shall be defined as any plane

$$(3) \quad \lambda_1(A_1x + B_1y + C_1z + D_1) + \lambda_2(A_2x + B_2y + C_2z + D_2) = 0,$$

whose equation is obtained by multiplying the equations of the two planes by constants, λ_1, λ_2, and adding the results. The constants λ_1, λ_2 can be chosen at pleasure, provided merely that the coefficients of x, y, z in (3), namely $\lambda_1A_1 + \lambda_2A_2$, $\lambda_1B_1 + \lambda_2B_2$, $\lambda_1C_1 + \lambda_2C_2$, do not all vanish. In particular, the case $\lambda_1 = \lambda_2 = 0$ is thus excluded.

For example, if the given planes are

$$(4) \qquad x - 2y - 2z + 9 = 0,$$

$$(5) \qquad 2x - 3y - 2z + 8 = 0,$$

and we multiply the equation of the first by -3 and the equation of the second by 2 and add, the plane defined by the resulting equation,

$$(6) \qquad x + 2z - 11 = 0,$$

is a linear combination of the given planes.

If the planes (1) and (2) meet, the plane (3) passes through their line of intersection. For, if (x_0, y_0, z_0) is an arbitrary point of this line, then

$$
(7) \qquad \begin{aligned} A_1x_0 + B_1y_0 + C_1z_0 + D_1 = 0, \\ A_2x_0 + B_2y_0 + C_2z_0 + D_2 = 0, \end{aligned}
$$

since the point lies in both the planes (1) and (2). If it is also to lie in the plane (3), the equation

$$
\lambda_1(A_1x_0 + B_1y_0 + C_1z_0 + D_1) + \lambda_2(A_2x_0 + B_2y_0 + C_2z_0 + D_2) = 0
$$

must be a true equation; and this it is, since the two parentheses on the left-hand side both vanish by virtue of equations (7). Thus the plane (3) contains the arbitrary point (x_0, y_0, z_0) on the line of intersection of the planes (1) and (2) and hence contains the whole line, q. e. d.

We have thus proved the theorem:

THEOREM 1. *A linear combination of two intersecting planes is a plane through their line of intersection.*

For example, the plane (6) passes through the line of intersection of the planes (4) and (5).

If the planes (1) and (2) are parallel, it follows, by Ch. XIX, § 7, Th. 2, that

$$
A_2 = \rho A_1, \qquad B_2 = \rho B_1, \qquad C_2 = \rho C_1, \qquad \rho \neq 0.
$$

The direction components of the normals to the plane (3), namely,

$$
\lambda_1 A_1 + \lambda_2 A_2, \qquad \lambda_1 B_1 + \lambda_2 B_2, \qquad \lambda_1 C_1 + \lambda_2 C_2,
$$

become, then,

$$
(\lambda_1 + \rho\lambda_2) A_1, \qquad (\lambda_1 + \rho\lambda_2) B_1, \qquad (\lambda_1 + \rho\lambda_2) C_1.
$$

Now $\lambda_1 + \rho\lambda_2 \neq 0$, since otherwise the coefficients of x, y, z in (3) would all be zero; that is, in this case we must exclude, according to the definition, not only the values $\lambda_1 = \lambda_2 = 0$ but also the values of λ_1 and λ_2 for which $\lambda_1/\lambda_2 = -\rho$. It follows, then, that the plane (3) is parallel to or identical with the plane (1). Thus we have the theorem:

THEOREM 2. *A linear combination of two parallel planes is a plane parallel to them or coincident with one of them.*

Plane through a Line and a Point. It is now a simple matter to find the equation of a plane determined by a line and a point.

For example, let the line be the line of intersection of the planes (4) and (5) and let the point be $(-5, -1, 2)$. By Th. 1, the plane

$$(8) \qquad \lambda_1(x-2y-2z+9)+\lambda_2(2x-3y-2z+8)=0$$

passes through the given line. If it is also to contain the given point, $(-5, -1, 2)$, we must have

$$\lambda_1(-5+2-4+9)+\lambda_2(-10+3-4+8)=0,$$
$$2\lambda_1-3\lambda_2=0.$$

This equation determines the ratio λ_1/λ_2. It will be satisfied if, in particular, we take $\lambda_1=3$ and $\lambda_2=2$. Then (8) becomes

$$3(x-2y-2z+9)+2(2x-3y-2z+8)=0,$$

or

$$7x-12y-10z+43=0.$$

This is the equation of the required plane.

In the general case, when the given line is the line common to two intersecting planes, (1) and (2), and (x_1, y_1, z_1) is the given point, not on the line, the procedure is quite the same. The plane (3) passes through the given line. Demanding, further, that it contain the point (x_1, y_1, z_1) leads to an equation for the determination of the ratio λ_1/λ_2, and any values for λ_1 and λ_2 which have this ratio yield, when substituted in (3), the equation of the required plane.

Converses of Theorems 1, 2. Since every plane through the line of intersection of the given planes (1) and (2) can be thought of as determined by this line and a point (x_1, y_1, z_1) external to it, we have proved that every plane through the line of intersection of the given planes is a linear combination of them. This is the converse of Theorem 1.

The converse of Theorem 2 can be proved in a similar manner. The details are left to the student. Both converses can be stated in a single theorem.

THEOREM 3. *Any plane through the line of intersection of two intersecting planes, or parallel to two parallel planes, is a linear combination of the two planes.*

Projecting Planes of a Line. Consider the line L:

$$(9) \qquad \begin{cases} x - 2y - 2z + 9 = 0, \\ 2x - 3y - 2z + 8 = 0, \end{cases}$$

in which the planes (4) and (5) intersect, and also the plane (6):

$$x + 2z - 11 = 0.$$

This plane passes through L, since it is a linear combination of the planes (4) and (5). In particular, it is the plane through L which is perpendicular to the (z, x)-plane, for equation (6) contains no term in y. It is, then, the plane which *projects* L on the (z, x)-plane. Accordingly, it is known as a *projecting plane* of L.

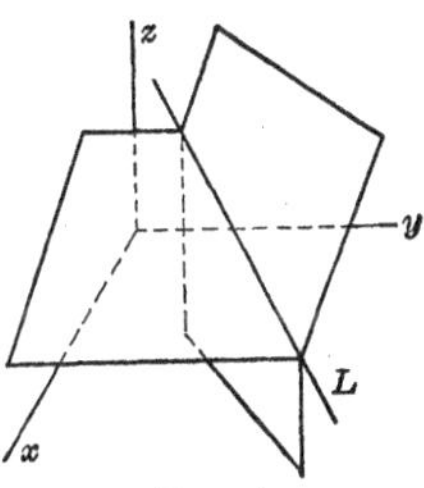

FIG. 1

Equation (6) represents, *in space*, this projecting plane. Considered merely in the (z, x)-plane, it defines the line which is the actual projection of L on the (z, x)-plane.

By combining equations (4) and (5) linearly so that the resulting equation contains no term in x, *e.g.* by multiplying the first of the equations by -2 and adding it to the second, we obtain the equation of the plane which projects L on the (y, z)-plane, namely,

$$(10) \qquad y + 2z - 10 = 0.$$

In a similar manner the equation of the plane which projects L on the (x, y)-plane is found to be

$$(11) \qquad x - y - 1 = 0.$$

The planes (6), (10), and (11) are the *three projecting planes* of L. Any two of the three projecting planes of a line will, in general,* determine the line. For example, the pair of equations

$$(12) \qquad x + 2z - 11 = 0, \qquad y + 2z - 10 = 0$$

* Exceptions occur when the line is parallel to, or lies in, a coördinate plane; cf. Ex. 13.

is as proper an analytic representation of L as the pair (9), and much more simple.

Equations (6), (10), and (11) are equivalent to the continued equality

$$(13) \qquad x - 11 = y - 10 = -2z,$$

and conversely. In other words, the representation of a line by means of its projecting planes is essentially the same as the representation of it by means of a continued equality of the usual form. The three equations which result from equating the members of the equality are the equations of the projecting planes.

Furthermore, we now have a method of finding, from the representation of a line as the intersection of two planes, a representation of it by a continued equality. Thus, in the case of L, we passed from the equations (9) to the continued equality (13).

EXERCISES

Find the equation of the plane determined by the given line and the given point.

	Line	*Point*
1.	$\begin{cases} 2x - 3y + 4z - 2 = 0, \\ 5x + 2y - z + 5 = 0, \end{cases}$	(0, 0, 0).
2.	$\begin{cases} 4x + 2y + 5z = 0, \\ 3x - 5y - 2z = 8, \end{cases}$	(3, − 1, 2).
3.	$3x - 5y = 6,\ 2x + 3z = 9,$	(4, 3, − 5).
4.	$\dfrac{x+1}{5} = \dfrac{y-1}{-2} = \dfrac{z-3}{4},$	(1, − 1, 2).

5. Find the equation of the plane containing the line of Ex. 1 and having the intercept 2 on the axis of y.

6. Find the equation of the plane passing through the line of Ex. 2 and having equal intercepts, not zero, on the axes of x and y. *Ans.* $13x + 13y + 22z + 8 = 0.$

7. What is the equation of the plane which contains the line of Ex. 4 and is perpendicular to the plane $3x - y + 4z = 0$?

Ans. $4x + 8y - z - 1 = 0$.

8. Find the equations of the line which is the projection of the line of Ex. 3 on the plane $2x + y - 3z + 5 = 0$.

9. Find the equations of the projecting planes of the line of Ex. 1 and from them determine a continued equality which represents the line.

10. The preceding exercise for the line of Ex. 2.

11. The line of Ex. 3 is defined by two of its projecting planes. What is the equation of the third?

12. What are the equations of the projecting planes of the line of Ex. 4?

13. A line which is not parallel to or in a coördinate plane has three projecting planes, which are distinct; a line parallel to or in *one* coördinate plane has three projecting planes, just two of which are identical; a line parallel to or in *two* coördinate planes — *i.e.* parallel to or coincident with an axis — has but two projecting planes, which are distinct. Consequently there are always at least two distinct projecting planes of a line and the line is determined by them. Prove these statements.

14. A line not parallel to or in the (x, y)-plane can be represented by equations of the form

$$x = az + b, \qquad y = cz + d;$$

a line parallel to or in the (x, y)-plane can have its equations put into the form

$$y = ax + b, \qquad z = c,$$

unless it is parallel to or coincident with the y-axis; in this case, its equations can be written as

$$x = a, \qquad z = b.$$

Prove these statements.

15. Prove that the plane determined by the point (x_2, y_2, z_2) and the line through the point (x_1, y_1, z_1) with the direction

components l, m, n can have its equation written in the form

$$\begin{vmatrix} x - x_1 & y - y_1 & z - z_1 \\ x_2 - x_1 & y_2 - y_1 & z_2 - z_1 \\ l & m & n \end{vmatrix} = 0.$$

Suggestion. Determine the direction components of the normals to the plane.

2. Three Planes through a Line. Three Points on a Line.* By means of the results of the preceding paragraph we can prove the following theorem.

THEOREM 1. *Three planes pass through a line or are parallel, when and only when any one of them is a linear combination of the other two.*

If the three planes pass through a line, or are parallel, any one of them passes through the line of intersection of the other two, or is parallel to the other two. Consequently, by Th. 3, § 1, this plane is a linear combination of the other two.

Conversely, if a *particular* one of the planes is a linear combination of the other two, it goes through the line of intersection of these two, if they intersect, or is parallel to them, if they are parallel (Ths. 1, 2, § 1). It follows then, furthermore, by the first part of the proof, that *any* one of the three planes is a linear combination of the other two, q. e. d.

For example, the three planes,

$$\begin{aligned} 3x - 2y + z + 6 &= 0, \\ 2x + 5y - 3z - 2 &= 0, \\ 4x - 9y + 5z + 14 &= 0, \end{aligned} \tag{1}$$

pass through a line, inasmuch as the equation of the third can be obtained by multiplying that of the first by 2, that of the second by -1, and by adding the results.

Three Points on a Line. The three points $P_1: (x_1, y_1, z_1)$, $P_2: (x_2, y_2, z_2)$, $P_3: (x_3, y_3, z_3)$ lie on a line if and only if the

* It is assumed, here and in § 4, that the given planes, or the given points, are distinct.

direction components of P_1P_3 are proportional to those of P_1P_2:

$$x_3 - x_1 = \rho(x_2 - x_1), \quad y_3 - y_1 = \rho(y_2 - y_1), \quad z_3 - z_1 = \rho(z_2 - z_1).$$

These equations can be rewritten as

$$\begin{aligned} x_3 &= (1-\rho)\,x_1 + \rho x_2, \\ y_3 &= (1-\rho)\,y_1 + \rho y_2, \\ z_3 &= (1-\rho)\,z_1 + \rho\, z_2. \end{aligned} \tag{2}$$

They then say that the coördinates of P_3 are a linear combination of the coördinates of P_1 and P_2 with constants of combination, $1 - \rho$ and ρ, *whose sum is unity.* Since *any* one of the points might have been called P_3, this result can be stated more generally.

THEOREM 2. *Three points lie on a line when and only when the coördinates of any one of them can be expressed as a linear combination of those of the other two, with constants of combination whose sum is unity.*

This theorem is of importance because of the analogy between it and Theorem 1, and because of its theoretical value in later work. It has not the practical value of Theorem 1, since testing three points for collinearity can be more easily done directly.

Thus, if the three points are (2, −1, 5), (4, 2, 6), (−2, −7, 3), the direction components of P_1P_2 and P_1P_3 are, respectively, 2, 3, 1 and − 4, − 6, − 2. Since these triples are proportional, the three points lie on a line.

EXERCISES

What can you say of the three planes in each of the following exercises?

1. $2x - y - z = 2,\ 3x + y + 2z = 1,\ 5x - 5y - 6z = 7.$
2. $x + 3y - z = 1,\ 3x - 5y + 7z = 3,\ 3x + 2y + 2z = 3.$
3. $6x - 3y + 9z = 2,\ 2x - y + 3z = 0,\ -4x + 2y - 6z = 3.$

Are the three given points collinear?

4. (5, 3, 4), (1, 5, 10), (11, 0, – 5).

5. (– 13, 12, – 15), (– 5, 6, – 11), (7, – 3, – 5).

6. (2, – 3, 8), (5, 4, 7), (8, 10, 6).

7. Determine k so that the three planes

$$kx - 3y + z = 2, \quad 3x + 2y + 4z = 1, \quad x - 8y - 2z = 3$$

will pass through a line.

8. Determine k so that the three points (2, 3, k), (5, 5, 1), (–1, 1, 9) will be collinear.

3. Line in a Plane. From Theorem 1 of the preceding paragraph follows immediately the theorem:

THEOREM. *A line lies in a plane, if and only if the plane is a linear combination of any two planes which determine the line.*

For example, the line of intersection of the first two of the planes (1), § 2 lies in the third plane, since the third plane was shown to be a linear combination of the first two.

A second method of testing whether or not a given line lies in a given plane presents itself if the line is represented parametrically. The line through the point (2, –1, 3) with the direction components 3, 2, – 4 has the parametric representation (Ch. XX, § 8, (2)):

$$x = 3t + 2, \qquad y = 2t - 1, \qquad z = -4t + 3.$$

It will lie in the plane

$$2x + 5y + 4z - 11 = 0,$$

if and only if the coördinates of every one of its points satisfy the equation of the plane, *i.e.* if and only if

$$2(3t + 2) + 5(2t - 1) + 4(-4t + 3) - 11 = 0$$

is a true equation for all values of t. But the equation reduces to

$$0 \cdot t + 0 = 0,$$

and so is satisfied by all values of t. Consequently, the line lies in the plane.

It is clear that this method can also be applied to test whether or not a given curve, represented parametrically, lies on a given surface.

EXERCISES

In each of the following exercises determine whether or not the given line lies in the given plane. Apply both methods.

	Line	*Plane*
1.	$\frac{x-1}{2}=\frac{y+2}{-3}=\frac{z+3}{-1}$,	$x+y-z-2=0$.
2.	$2x-3=1-5y=3z$,	$4x+5y-3z-7=0$.
3.	$4x-1=-3y=1-2z$,	$2x-y+3z+5=0$.

4. Find the conditions under which the line through the point (x_0, y_0, z_0) with the direction components l, m, n will lie in the plane $Ax+By+Cz+D=0$.

Ans. $Al+Bm+Cn=0,\ Ax_0+By_0+Cz_0+D=0$.

5. Does the twisted cubic $x=t,\ y=t^2,\ z=2\cdot t^3$ lie on the surface $2x^3-z=0$?

6. Show that the curve

$$x=a\sin^2 t, \qquad y=a\sin t\cos t, \qquad z=a\cos t$$

lies on the sphere $x^2+y^2+z^2=a^2$.

4. Four Points in a Plane. Four Planes through a Point. Given the four points P_1, P_2, P_3, P_4, with the coördinates (x_1, y_1, z_1), (x_2, y_2, z_2), (x_3, y_3, z_3), (x_4, y_4, z_4). To determine when they lie in a plane.

Unless all four points lie on a line, there will be three of them which determine a plane. Let these three be P_2, P_3, P_4. The equation of the plane through them is, by Ch. XIX, § 6,

$$\begin{vmatrix} x & y & z & 1 \\ x_2 & y_2 & z_2 & 1 \\ x_3 & y_3 & z_3 & 1 \\ x_4 & y_4 & z_4 & 1 \end{vmatrix} = 0.$$

The four points will be coplanar if and only if $P_1: (x_1, y_1, z_1)$ lies in this plane, *i.e.* if and only if

$$(1) \qquad \begin{vmatrix} x_1 & y_1 & z_1 & 1 \\ x_2 & y_2 & z_2 & 1 \\ x_3 & y_3 & z_3 & 1 \\ x_4 & y_4 & z_4 & 1 \end{vmatrix} = 0.$$

If the four points do lie on a line, they are certainly coplanar. On the other hand, equation (1) is satisfied in this case also. For, equations (2), § 2, must hold for some value of ρ, since the points P_1, P_2, P_3 are collinear. Consequently, if in the determinant in (1) we subtract from the third row the first row multiplied by $1-\rho$ and the second row multiplied by ρ, the new third row will consist exclusively of zeros,* and hence the determinant will vanish.

We have proved, then, the following theorem.

THEOREM 1. *The four points (x_1, y_1, z_1), (x_2, y_2, z_2), (x_3, y_3, z_3), (x_4, y_4, z_4) lie in a plane if and only if the determinant in* (1) *vanishes.*

Four Planes through a Point. Let the four planes be

$$(2) \qquad \begin{aligned} A_1x + B_1y + C_1z + D_1 &= 0, \\ A_2x + B_2y + C_2z + D_2 &= 0, \\ A_3x + B_3y + C_3z + D_3 &= 0, \\ A_4x + B_4y + C_4z + D_4 &= 0. \end{aligned}$$

Form the determinant of the coefficients in these equations, namely, $|A_1B_2C_3D_4|$ or, more simply, $|ABCD|$. In this determinant let Δ_1, Δ_2, Δ_3, Δ_4 be the minors of the elements D_1, D_2, D_3, D_4; for example, $\Delta_4 = |A_1B_2C_3|$.†

We first prove the following Theorem:

THEOREM 2. *The normals to the planes* (2) *will all be parallel to a plane if and only if*

$$\Delta_1 = \Delta_2 = \Delta_3 = \Delta_4 = 0.$$

* The fourth element in the new third row is $1 - (1-\rho) - \rho = 0$.

† Throughout this paragraph we denote determinants by writing the elements of their principal diagonals between two parallel bars.

For, if the normals to the four planes are parallel to a plane, the normals of *any* three are parallel to this plane, and hence the determinant of their direction components vanishes, by Ch. XVIII, § 6. But this determinant is one of the four determinants Δ in question. Hence all four determinants Δ vanish.

Conversely, if all four determinants Δ vanish, the normals of each set of three planes are parallel to a plane, and this plane can be taken as the same plane in all four cases.* Hence the normals to all four planes are parallel to it.

Suppose, now, that the planes have one and only one point in common. Then the four determinants Δ are not all zero. For, if they were all zero, the normals to the four planes would all be parallel to a plane. Consequently, either the four planes would be all parallel or all the lines of intersection obtained by taking them in pairs would be parallel or identical (cf. Ch. XIX, § 10), and hence the planes would have either no point in common or a whole line of points in common. But this contradicts the hypothesis that they meet in just one point. At least one of the determinants Δ, then, does not vanish. Let us assume, say, that $\Delta_4 = |A_1B_2C_3|$ is not zero. Then the first three planes meet in a single point (Ch. XIX, § 10), whose coördinates, found by Cramer's rule, are

$$x = -\frac{|D_1B_2C_3|}{\Delta_4}, \qquad y = -\frac{|A_1D_2C_3|}{\Delta_4}, \qquad z = -\frac{|A_1B_2D_3|}{\Delta_4}$$

or

$$(3) \quad x = -\frac{|B_1C_2D_3|}{\Delta_4}, \qquad y = \frac{|A_1C_2D_3|}{\Delta_4}, \qquad z = -\frac{|A_1B_2D_3|}{\Delta_4}.$$

Since this point lies in the fourth plane, we must have

$$-A_4|B_1C_2D_3| + B_4|A_1C_2D_3| - C_4|A_1B_2D_3| + D_4|A_1B_2C_3| = 0$$

* This is obvious if the four planes are parallel. In the contrary case, when at least two of the planes, say the first two, are not parallel, the statement is substantiated as follows. The normals to the first two planes and the third are parallel to a plane M_1, and the normals to the first two planes and the fourth are parallel to a plane M_2. But M_1 and M_2, since they are both parallel to the normals of the first two planes are, in any case, parallel to each other and hence *can* always be taken as the same plane.

or, since the expression on the left is the development of the determinant $|ABCD|$ by the minors of the last row,

$$(4) \qquad |ABCD| = 0.$$

Conversely, if $|ABCD| = 0$ and not all four of the determinants Δ are zero, the planes (2) meet in a single point. For, we can assume that $\Delta_4 \neq 0$. Then the first three planes meet in a single point (3) and this point lies in the fourth plane, since by hypothesis (4) holds.

Thus we have proved the theorem:

THEOREM 3. *The four planes* (2) *meet in a single point if and only if the determinant of their coefficients vanishes and not all four minors* Δ_1, Δ_2, Δ_3, Δ_4 *are zero.**

If the normals to the planes (2) are all parallel to a plane, the determinant $|ABCD|$ obviously vanishes, for then $\Delta_1 = \Delta_2 = \Delta_3 = \Delta_4 = 0$ and the expansion of $|ABCD|$ by the minors of the fourth column, namely:

$$-D_1\Delta_1 + D_2\Delta_2 - D_3\Delta_3 + D_4\Delta_4$$

has the value zero.

Conversely, if $|ABCD|$ vanishes by virtue of the vanishing of Δ_1, Δ_2, Δ_3, Δ_4, the normals of the planes (2) are, by Theorem 2, all parallel to a plane.

Consequently, we can combine Theorems 2 and 3 in the more general, though less useful, theorem:

THEOREM 4. *The four planes* (2) *meet in a single point or, their normals are all parallel to a plane, if and only if the determinant of their coefficients vanishes.*

Finally, we enumerate the cases which can occur when the normals to the four planes are parallel to a plane. First, the

* Stated algebraically this theorem reads: *The four equations* (2) *are compatible and have, moreover, a unique solution, if and only if* $|ABCD| = 0$ *and* Δ_1, Δ_2, Δ_3, Δ_4 *are not all zero.* This theorem includes the theorem of Ch. XVI, § 9, Ex. 8, and also its converse. It is to be noted that it was geometric considerations which led us here to a proof which covered the converse as well as the theorem.

planes can be all parallel; this case can easily be detected by inspection. Secondly, the planes can all go through a line; the test of § 3 reveals this case. Finally, whatever lines of intersection the planes have, when taken in pairs, are all parallel; this case will make itself known by exclusion of the others.

Example. Consider the four planes

$$\begin{aligned} x+\ y+3z-\ 6&=0,\\ 4x+2y-5z+\ 8&=0,\\ 8y-7z+22&=0,\\ 2x-3y+\ z-\ 7&=0. \end{aligned}$$

Here $|ABCD|=0$ and $\Delta_4\neq 0$, as can easily be verified. Hence the four planes meet in a single point. Let the student show further that the last three planes pass through a line, which is intersected by the first plane in the point in question.

EXERCISES

Do the four given points lie in a plane? If so, do three, or do all four, lie on a line?

1. $(2, 3, 1)$, $(1, 5, 2)$, $(-3, 4, -1)$, $(-2, 2, -2)$.
2. $(2, 5, 3)$, $(0, 2, -3)$, $(1, 3, 7)$, $(-1, -1, 15)$.
3. $(1, 2, -1)$, $(3, 1, 2)$, $(-1, 3, -4)$, $(7, -1, 8)$.
4. $(0, 2, 1)$, $(1, 0, 2)$, $(-1, -1, 1)$, $(4, 2, 3)$.

5. For what values of k will the four points $(k, -5, 6)$, $(4, -4, k)$, $(5, 1, 2)$, $(2, 0, 7)$ lie in a plane?

What can you say of the relative positions of the four given planes?

6. The planes of Ch. XVI, § 9, Ex. 9.
7. The planes of Ch. XVI, § 9, Ex. 10.

8. $$\begin{aligned} 3x-5y+2z+3&=0,\\ 2x+4y-3z-1&=0,\\ 5x-\ y-\ z+2&=0,\\ 8x-6y+\ z+5&=0. \end{aligned}$$

9. $$\begin{aligned} 2x+3y-\ z+4&=0,\\ x-2y+3z-2&=0,\\ 3x+\ y+2z+3&=0,\\ 5x+4y+\ z-5&=0. \end{aligned}$$

10. For what value of k will the four planes,

$$kx + y - z - 6 = 0, \qquad x - y + z = 0,$$
$$x + ky + z - 3 = 0, \qquad 2x + y + 4z - 1 = 0,$$

go through a point?

5. Two Intersecting Lines. Given the two distinct lines

$$\begin{cases} A_1x + B_1y + C_1z + D_1 = 0, \\ A_2x + B_2y + C_2z + D_2 = 0; \end{cases} \qquad \begin{cases} A_1'x + B_1'y + C_1'z + D_1' = 0, \\ A_2'x + B_2'y + C_2'z + D_2' = 0. \end{cases}$$

The two lines intersect in a point, when and only when the four planes which in pairs determine them meet in a single point. The condition for this is given in Theorem 3 of the preceding paragraph.

The two lines are parallel if and only if the normals to the four planes are all parallel to a plane. Theorem 2 of § 4 tells when this occurs.

These results can be combined in the general theorem:

THEOREM 1. *Two lines intersect or are parallel when and only when the determinant of the coefficients in the equations of the four planes which in pairs determine the two lines vanishes.*

The simplest way to decide in any case whether the two lines intersect or are parallel is to compute the direction components of the lines and compare them.

The above proof assumes tacitly that the four planes in question are distinct; otherwise the theorems of § 4 could not be applied. Theorem 1 still holds, however, in the exceptional case when one of the planes determining one line is identical with one of the planes determining the other line. For, the two lines lie, then, in a plane and hence intersect or are parallel; on the other hand, the determinant in question contains two rows which are proportional and hence it vanishes.

Lines Given by Continued Equalities. Let the first line be determined by the point $P_1 : (x_1, y_1, z_1)$ and the direction components l_1, m_1, n_1 and the second by the point $P_2 : (x_2, y_2, z_2)$ and the direction components l_2, m_2, n_2.

The lines are parallel if and only if l_1, m_1, n_1 are proportional to l_2, m_2, n_2.

If the lines are not parallel, there is a unique plane which contains the first line and is parallel to or contains the second. A normal to this plane is perpendicular to both lines and hence has the direction components $|m_1 n_2|, |n_1 l_2|, |l_1 m_2|$; cf. Ch. XVIII, § 5. The equation of the plane is, then,

$$(1) \qquad |m_1 n_2|(x - x_1) + |n_1 l_2|(y - y_1) + |l_1 m_2|(z - z_1) = 0,$$

or

$$(2) \qquad \begin{vmatrix} x - x_1 & y - y_1 & z - z_1 \\ l_1 & m_1 & n_1 \\ l_2 & m_2 & n_2 \end{vmatrix} = 0.$$

If this plane contains the point P_2 it will contain the entire second line, and conversely. Consequently, *the two non-parallel lines intersect if and only if*

$$(3) \qquad \begin{vmatrix} x_2 - x_1 & y_2 - y_1 & z_2 - z_1 \\ l_1 & m_1 & n_1 \\ l_2 & m_2 & n_2 \end{vmatrix} = 0.$$

Equation of the Plane Determined by Two Intersecting or Parallel Lines. If two given lines intersect or are parallel, the plane in which they lie can be determined by one of the lines, say the first, and a point of the second which does not lie on the first. Its equation, then, can be found by the method of § 1. This method, though always applicable, is designed primarily for the case when at least one of the lines is given as the intersection of two planes.

If both lines are represented by continued equalities, the plane which they determine has (2) as its equation, in case the lines intersect. If the lines are parallel, a similar equation for their plane can be found; cf. § 1, Ex. 15.

EXERCISES

Show that the given lines intersect or are parallel. In each case find the equation of the plane which they determine.

1. $\begin{cases} 2x - 3y + 1 = 0, \\ 3x - y - 2z = 0; \end{cases}$ $\quad \begin{cases} 2x - y - z = 0, \\ x - 2y + 1 = 0. \end{cases}$

Ans. The plane is $5x - 4y - 2z + 1 = 0$.

2. $\begin{cases} x + 2y + 5 = 0, \\ x - y - z = 0; \end{cases}$ $\quad \begin{cases} 2x + 7y + z + 1 = 0, \\ 3x + 3y - z + 4 = 0. \end{cases}$

3. $\dfrac{y-3}{2} = \dfrac{y-2}{-5} = \dfrac{z-1}{3},$ $\quad \dfrac{x-1}{-4} = y + 2 = \dfrac{z-6}{2}.$

4. $\dfrac{4-3x}{2} = \dfrac{y}{2} = 1 - z,$ $\quad 3x + 1 = 4 - y = 2z - 3.$

5. $\begin{cases} 4x - y + 3z + 1 = 0, \\ 2x + 3y + 5z - 3 = 0; \end{cases}$ $\quad 1 - x = \dfrac{y-2}{2} = \dfrac{z+1}{3}.$

6. Distance of a Point from a Line. Distance between Two Lines. Let it be required to find the distance D of the point $P_2 : (x_2, y_2, z_2)$ from the line L which passes through the point $P_1 : (x_1, y_1, z_1)$ and has the direction cosines $\cos\alpha$, $\cos\beta$, $\cos\gamma$.

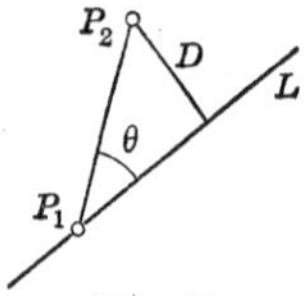

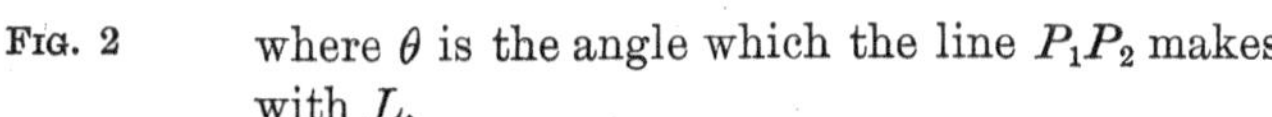

FIG. 2

It is clear from the figure that

$$D = P_1P_2 \sin\theta,$$

where θ is the angle which the line P_1P_2 makes with L.

By Ex. 16 at the end of Ch. XVIII,

$$\sin^2\theta = |\mu_1\nu_2|^2 + |\nu_1\lambda_2|^2 + |\lambda_1\mu_2|^2, \tag{1}$$

where λ_1, μ_1, ν_1 and λ_2, μ_2, ν_2 are the direction cosines of P_1P_2 and L.

Now,

$$\lambda_1 = \frac{x_2 - x_1}{P_1P_2}, \qquad \mu_1 = \frac{y_2 - y_1}{P_1P_2}, \qquad \nu_1 = \frac{z_2 - z_1}{P_1P_2};$$

$$\lambda_2 = \cos\alpha, \qquad \mu_2 = \cos\beta, \qquad \nu_2 = \cos\gamma.$$

Hence

$$|\mu_1\nu_2| = \begin{vmatrix} \dfrac{y_2 - y_1}{P_1P_2} & \dfrac{z_2 - z_1}{P_1P_2} \\ \cos\beta & \cos\gamma \end{vmatrix} = \frac{1}{P_1P_2}\begin{vmatrix} y_2 - y_1 & z_2 - z_1 \\ \cos\beta & \cos\gamma \end{vmatrix}.$$

Similar values are found for $|\nu_1\lambda_2|$ and $|\lambda_1\mu_2|$.

Substituting these values in (1), multiplying the resulting equation through by $P_1P_2^2$ and extracting the square root of both sides, we obtain, as the final result:

$$D=\sqrt{\begin{vmatrix} y_2-y_1 & z_2-z_1 \\ \cos\beta & \cos\gamma \end{vmatrix}^2+\begin{vmatrix} z_2-z_1 & x_2-x_1 \\ \cos\gamma & \cos\alpha \end{vmatrix}^2+\begin{vmatrix} x_2-x_1 & y_2-y_1 \\ \cos\alpha & \cos\beta \end{vmatrix}^2}.$$

Distance between Two Skew Lines. Let the line through the point $P_1:(x_1, y_1, z_1)$ with the direction components l_1, m_1, n_1 and the line through the point $P_2:(x_2, y_2, z_2)$ with the direction components l_2, m_2, n_2 be two *skew* lines, *i.e.* two lines which neither intersect nor are parallel. To find the distance D between them measured along their common perpendicular.

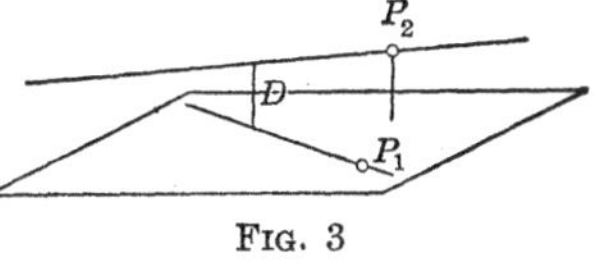

FIG. 3

The plane through the first line parallel to the second has the equation (1), § 5, namely,

$$|m_1n_2|(x-x_1)+|n_1l_2|(y-y_1)+|l_1m_2|(z-z_1)=0.$$

The required distance D is the uniform distance of the second line from this plane, or it is the distance of the point $P_2:(x_2, y_2, z_2)$ from this plane. Thus,

$$D=\pm\frac{|m_1n_2|(x_2-x_1)+|n_1l_2|(y_2-y_1)+|l_1m_2|(z_2-z_1)}{\sqrt{|m_1n_2|^2+|n_1l_2|^2+|l_1m_2|^2}}$$

or

$$D=\pm\frac{\Delta}{\sqrt{|m_1n_2|^2+|n_1l_2|^2+|l_1m_2|^2}},$$

where Δ is the determinant in formula (3), § 5, and where that sign is to be chosen which will make the right-hand side positive.

Distance between Two Parallel Lines. The distance between two parallel lines can be found as the distance of a point on one of the lines from the other line.

EXERCISES

Find the distance of the given point from the given line.

	Point	*Line*
1.	(2, 3, 4),	$\frac{4-x}{2}=\frac{y}{6}=\frac{1-z}{3}$. *Ans.* $\frac{3}{7}\sqrt{101}=4.31$.
2.	(0, 0, 0),	$\frac{x-3}{2}=\frac{y+2}{-5}=\frac{z-2}{4}$.
3.	(−1, 2, −3),	$3x+1=4-y=2z-3$.
4.	(2, −1, 5),	$\begin{cases}3x+2y+2z+2=0,\\6x+5y+6z+2=0.\end{cases}$
5.	(3, 1, −1),	$\begin{cases}2x-3y+1=0,\\3x-y-2z=0.\end{cases}$

Find the distance between the two given lines.

6. The lines of Exs. 1, 2. *Ans.* $\frac{15}{89}\sqrt{89}=1.59$.

7. The lines of Exs. 2, 3.

8. The lines of Exs. 1, 3.

9. The lines of Exs. 2, 4.

10. The lines of Exs. 3, 4.

11. The lines of Exs. 4, 5.

12. A cube has edges of length a. Find the distance between a diagonal and an edge skew to it. *Ans.* $\frac{1}{2}\sqrt{2}\,a$.

7. Area of a Triangle. Volume of a Tetrahedron. We first prove the following theorem.

THEOREM 1. *If a region of area A in a plane M is projected on a second plane M', the area of the projected region equals $A\cos\theta$, where θ is the acute angle between M and M'.**

In the case of a rectangle whose sides are respectively parallel and perpendicular to the line of intersection of M and M', the proof is immediate. For, when the rectangle is projected on M', one dimension remains the same and the other is multiplied by $\cos\theta$.

* The theorem is trivial if M is parallel or perpendicular to M'; we exclude these cases.

The area A of an arbitrary region in M is the limit approached by the sum B of the areas of rectangles, of the type just described, which are inscribed in the region:

$$A = \lim B.$$

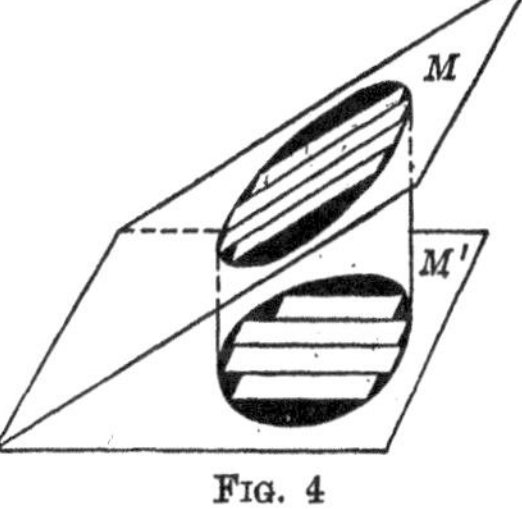

FIG. 4

If A' is the area of the projected region and B' is the sum of the areas of the projected rectangles, evidently

$$A' = \lim B'.$$

Since the area of each projected rectangle is $\cos\theta$ times the area of the original rectangle,

$$B' = B\cos\theta.$$

Hence $$A' = \lim B\cos\theta = \cos\theta \lim B$$

or $$A' = A\cos\theta, \qquad \text{q. e. d.}$$

Let the areas of the projections of the given region on the coördinate planes be denoted by A_{yz}, A_{zx}, A_{xy}, and let the normals to M have the direction angles α, β, γ. By Th. 1,*

$$A_{yz} = |A\cos\alpha|, \qquad A_{zx} = |A\cos\beta|, \qquad A_{xy} = |A\cos\gamma|.$$

Hence

(1) $$A^2 = A_{yz}^2 + A_{zx}^2 + A_{xy}^2.$$

Thus we have proved the theorem:

THEOREM 2. *The sum of the squares of the areas of the projections of a region on the three coördinate planes equals the square of the area of the region.*

Area of a Triangle. It is now easy to write down a formula for the area A of the triangle whose vertices are at the points $P_1: (x_1, y_1, z_1)$, $P_2: (x_2, y_2, z_2)$, $P_3: (x_3, y_3, z_3)$. For, the areas of the projections of the triangle on the coördinate planes are, by Ex. 18 at the end of Ch. XVI,

$$A_{yz} = \pm\tfrac{1}{2}|y_1 z_2 1|, \qquad A_{zx} = \pm\tfrac{1}{2}|x_1 z_2 1|, \qquad A_{xy} = \pm\tfrac{1}{2}|x_1 y_2 1|,$$

* The absolute value signs are necessary since α, β, γ are not necessarily acute angles.

where $|y_1z_2 1|$, for example, is the determinant whose three columns are y_1, y_2, y_3; z_1, z_2, z_3; 1, 1, 1. Hence, by (1),

$$(2) \qquad A = \tfrac{1}{2}\sqrt{|y_1z_2 1|^2 + |x_1z_2 1|^2 + |x_1y_2 1|^2}.$$

Volume of a Tetrahedron. Let the above triangle be the base of the tetrahedron and let the fourth vertex be at the point $P_0 : (x_0, y_0, z_0)$. The volume V of the tetrahedron is known from Solid Geometry to be equal to one third the area A of the base times the length D of the altitude:

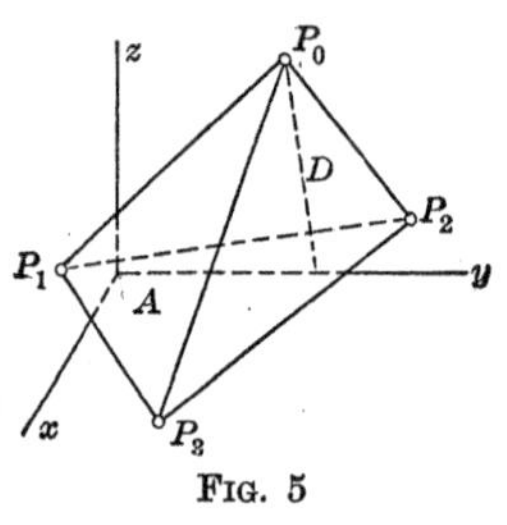

FIG. 5

$$(3) \qquad V = \tfrac{1}{3} AD.$$

The equation of the plane of the base, the plane of P_1, P_2, P_3, is given in determinant form in Ch. XIX, § 6. This equation, when the determinant is developed by the minors of the first row, becomes

$$|y_1z_2 1|x - |x_1z_2 1|y + |x_1y_2 1|z - |x_1y_2z_3| = 0.$$

The distance of the point $P_0 : (x_0, y_0, z_0)$ from this plane, *i.e.* the length D of the altitude of the tetrahedron, is, by Ch. XIX, § 9,

$$(4) \quad D = \pm \frac{|y_1z_2 1|x_0 - |x_1z_2 1|y_0 + |x_1y_2 1|z_0 - |x_1y_2z_3|}{\sqrt{|y_1z_2 1|^2 + |x_1z_2 1|^2 + |x_1y_2 1|^2}}.$$

Substituting in (3) the values of A and D as given by (2) and (4), and at the same time writing the numerator in (4) in determinant form, we obtain, as the value of V,

$$(5) \qquad V = \pm \tfrac{1}{6} \begin{vmatrix} x_0 & y_0 & z_0 & 1 \\ x_1 & y_1 & z_1 & 1 \\ x_2 & y_2 & z_2 & 1 \\ x_3 & y_3 & z_3 & 1 \end{vmatrix},$$

where that sign is to be taken which yields a positive result.

EXERCISES

Find the areas of the following triangles.

1. With vertices at $(2, -1, 3)$, $(4, 3, -2)$, $(3, 0, -1)$.
Ans. $\frac{1}{2}\sqrt{134} = 5.79$.

2. With vertices at $(0, 0, 0)$, (x_1, y_1, z_1), (x_2, y_2, z_2).

3. Cut from the plane $2x - 3y + 4z - 12 = 0$ by the coördinate planes.

4. With vertices at $(a, 0, 0)$, $(0, b, 0)$, $(0, 0, c)$.

Find the volumes of the following tetrahedra.

5. That of Ch. XIX, § 6, Ex. 9.

6. That of Ch. XIX, § 10, Ex. 5.

7. Included between the plane $2x - 3y + 4z = 12$ and the coördinate planes.

8. Included between the coördinate planes and the plane with intercepts a, b, c on the axes.

9. With vertices at $(0, 0, 0)$, (x_1, y_1, z_1), (x_2, y_2, z_2), (x_3, y_3, z_3).

EXERCISES ON CHAPTER XXI

1. Find a parametric representation of the line
$$x - y - z + 1 = 0, \qquad 2x - y + z - 8 = 0.$$

2. What are the equations of the projecting planes of an arbitrary line passing through the origin?

3. Find the equation of the plane which contains the line of Ex. 1 and is parallel to the line $2x - 3 = y - 3 = 2z - 1$.

4. Find the equation of the plane which is perpendicular to the plane $2x + 5y - 3z - 2 = 0$ and meets it in the line in which it intersects the (x, y)-plane.
Ans. $6x + 15y + 29z - 6 = 0$.

5. Do Ex. 16 at the end of Ch. XX without finding the coördinates of the point of intersection of the three given planes.

Suggestion. Find two planes through the given point, each containing the line of intersection of a pair of given planes.

6. Do Ex. 17 at the end of Ch. XX without finding the coördinates of the point of intersection of the given line and the given plane.

7. Find the equations of the line which contains the point $(2, 0, -1)$ and intersects each of the lines

$$\begin{cases} 2x - y + 3z = 0, \\ 3x + y - 2z = 2; \end{cases} \qquad \begin{cases} x + y + 2z - 5 = 0, \\ 3x + 4y - z + 1 = 0. \end{cases}$$

8. Find the equations of the line which intersects each of the lines given in Ex. 7 and is parallel to the line $4 - 6x = y + 9 = 2z$.

9. A plane intersects the (x, y)-plane in the line whose equation in the (x, y)-plane is $2x + 3y = 12$. If the plane cuts from the first octant a tetrahedron whose volume is 12, find its equation. *Ans.* $2x + 3y + 4z - 12 = 0$.

10. There are two planes which contain the line

$$x + 2y + z + 1 = 0, \qquad 2x + y - z - 7 = 0$$

and make angles of 30° with the plane $x - z + 2 = 0$. Find their equations. *Ans.* $x - y - 2z - 8 = 0,\ 2x + y - z - 7 = 0$.

11. Find the equations of the planes which contain the line given in Ex. 10 and are $\sqrt{2}$ units distant from the point $(2, 2, -3)$.

12. The planes through the edges of a trihedral angle perpendicular to the opposite faces pass through a line. Prove this theorem in the case that the faces lie in the planes

$$2x - y + z = 0, \qquad 4x - y + 3z = 0, \qquad 3x - 2y - z = 0.$$

13. Prove the theorem of Ex. 12 in the general case.

The Equations $\lambda u + \mu v = 0$, $uv = 0$ *

14. Theorem. *If* $u = 0$, $v = 0$ *are the equations of two surfaces, the equation* $\lambda u + \mu v = 0$, $\lambda\mu \neq 0$, *represents in general* † *a*

* Cf. Ch. IX, §§ 3, 4.

† In particular, it may represent a curve or a point; cf. footnotes, pp. 445, 167.

surface which contains the total intersection of the two surfaces, if they intersect, and has no other point in common with either of them. If the given surfaces do not intersect, the equation represents in general a surface not meeting either of them or it has no locus.* Prove this theorem.

15. Find the equation of the sphere which contains the circle

$$x^2 + y^2 + z^2 - 4 = 0, \qquad z - 5 = 0$$

and passes through the point (3, 0, 2).

Ans. $x^2 + y^2 + z^2 + 3z = 19$.

16. Prove that the curve of intersection of the cylinders $x^2 + y^2 = 4$, $y^2 + z^2 = 5$ lies on the surface $x^2 - z^2 + 1 = 0$ and is the total intersection of this surface with each cylinder.

17. THEOREM. *If $u = 0$, $v = 0$, $w = 0$ are the equations of three planes which meet in a single point, the equation $\lambda u + \mu v + \nu w = 0$ represents a plane through this point. Conversely, every plane which contains this point is a linear combination of the three given planes.* Prove this theorem.

18. Find the equation of the plane which is determined by the points (0, 0, 0), (1, 2, 0) and the point of intersection of the three planes

$$2x + y + 3z + 1 = 0, \quad x + y - 4z + 2 = 0, \quad 2x - 2y + 5z - 3 = 0.$$

Ans. $2x - y - 6z = 0$.

19. THEOREM. *The equation $uv = 0$ represents those points and only those points which lie on the surfaces $u = 0$ and $v = 0$.* Prove this theorem.

20. What do the following equations represent?

(*a*) $x^2 - y^2 = 0$; (*c*) $x^2 - xy - xz + yz = 0$;
(*b*) $x^4 - y^4 = 0$; (*d*) $xy - xz - 2y + 2z = 0$.

* In particular, it may represent a curve or a point; cf. footnotes, pp. 445, 168.

BISECTORS OF THE ANGLES BETWEEN TWO PLANES *

21. What is the locus of the inequality

$$2x - y + 2z - 4 > 0?$$

22. There are four regions lying between the planes

$$2x - y + 2z - 4 = 0, \qquad 8x + 4y + z - 8 = 0.$$

Find the pairs of simultaneous inequalities representing these regions, specifying the region which each pair represents.

23. Find the equations of the planes bisecting the angles between the two planes of Ex. 22.

Ans. $2x + 7y - 5z + 4 = 0, \quad 14x + y + 7z - 20 = 0.$

24. The same for the following pairs of planes

$$(a)\ \begin{matrix} x - y + z - 2 = 0, \\ x + y - z + 3 = 0; \end{matrix} \qquad (b)\ \begin{matrix} 3x - 6y + 2z - 4 = 0, \\ 6x + 2y - 9z - 5 = 0. \end{matrix}$$

25. Find the equation of that bisector of the angle between the two planes of Ex. 24 (*b*) which passes through the region between the two planes which contains the origin.

26. The planes which bisect the dihedral angles of any trihedral angle meet in a line. Prove this theorem when the faces lie in the planes

$$x + y + z - 1 = 0, \qquad x - y + z - 1 = 0, \qquad 2x + y - z + 1 = 0.$$

27. Prove the theorem of Ex. 26 in the general case.

Suggestion. Cf. Exs. 28, 29 at the end of Ch. XIII and Ex. 28 at the end of Ch. XIX.

28. Show that by a proper choice of axes two arbitrarily chosen skew lines, L_1 and L_2, can have their equations written as

$$x = c, \quad z = my; \qquad x = -c, \quad z = -my; \qquad cm \neq 0.$$

29. Prove that, if the line L_1 of Ex. 28 is taken as the z-axis, the x- and y-axes can be so chosen that L_2 has the equations, $x = c$, $z = my$, where $c \neq 0$.

* Cf. Ch. XIII, §§ 6, 7, 8.

CHAPTER XXII

SPHERES, CYLINDERS, CONES. SURFACES OF REVOLUTION

1. Equation of the Sphere. The equation of the sphere whose center is at the origin and whose radius is ρ is, according to Ch. XIX, § 1,

$$(1) \qquad x^2 + y^2 + z^2 = \rho^2.$$

It can be shown in a similar manner that, if the center is at the point (α, β, γ) and the radius is ρ, the equation of the sphere is

$$(2) \qquad (x - \alpha)^2 + (y - \beta)^2 + (z - \gamma)^2 = \rho^2.$$

Thus the sphere whose center is at the point $(2, -3, 4)$ and whose radius is 6 has the equation

$$(x - 2)^2 + (y + 3)^2 + (z - 4)^2 = 36,$$

or

$$x^2 + y^2 + z^2 - 4x + 6y - 8z - 7 = 0.$$

EXERCISES

Find the equations of the following spheres and reduce the results to their simplest form.

1. Center at $(3, 1, 2)$; radius, 5.

   Ans. $x^2 + y^2 + z^2 - 6x - 2y - 4z - 11 = 0.$

2. Center at $(-2, 3, -6)$; radius, 7.
3. Center at $(4, 0, 0)$; radius, 4.
4. Center at $(0, -5, 0)$; radius, 2.
5. Center at $(0, -4, 3)$; radius, 5.
6. Center at $(\frac{2}{3}, -\frac{1}{3}, 0)$; radius, 1.

7. Center at $(\frac{1}{2}, -\frac{2}{3}, \frac{7}{6})$; radius, $\frac{4}{3}$.
8. Center at $(0, 0, a)$; radius, a.
9. Center at $(a, 0, a)$; radius, $a\sqrt{2}$.
10. Center at (a, a, a); radius, $a\sqrt{3}$.

2. General Form of the Equation. The equation of a sphere can always be written in the form

$$(1) \qquad x^2 + y^2 + z^2 + Ax + By + Cz + D = 0,$$

as is seen by expanding equation (2), § 1.

Let us investigate whether, conversely, every equation of the form (1) represents a sphere.

Consider, first, the particular equation

$$(2) \qquad x^2 + y^2 + z^2 - 2x + 6y + 4z - 35 = 0.$$

If we complete the square of the terms in x, and do the same for the terms in y and in z, the equation becomes

$$(3) \quad (x - 1)^2 + (y + 3)^2 + (z + 2)^2 = 35 + 1 + 9 + 4 = 49.$$

This equation is of the form (2), § 1, where $\alpha = 1$, $\beta = -3$, $\gamma = -2$, $\rho = 7$, and hence represents a sphere whose center is at the point $(1, -3, -2)$ and whose radius is 7.

If the constant term, -35, in (2) is replaced by 14, the right-hand member of (3) becomes $-14 + 1 + 9 + 4 = 0$. In this case, then, we have

$$(x - 1)^2 + (y + 3)^2 + (z + 2)^2 = 0.$$

The point $(1, -3, -2)$ has coördinates satisfying this equation. For the coördinates (x, y, z) of any other point at least one of the parentheses is not zero and the left-hand side of the equation is positive. Consequently, the equation represents the single point $(1, -3, -2)$ or, if we define a *null sphere* (a sphere of zero radius) as a point, it represents a null sphere.

If the constant term, -35, in (2) is replaced by 15, the equation becomes

$$(x - 1)^2 + (y + 3)^2 + (z + 2)^2 = -1.$$

Since the left-hand member of this equation can never be negative, no matter what values are assigned to x, y, z, the equation represents no point whatever in space.

These three examples indicate what to expect of the general equation (1). On completing the squares for the pairs of terms in x, y, and z, respectively, in (1), the equation takes on the form (2), § 1, where

$$\alpha = -\frac{A}{2}, \qquad \beta = -\frac{B}{2}, \qquad \gamma = -\frac{C}{2}, \tag{4}$$
$$\rho^2 = \frac{A^2 + B^2 + C^2 - 4D}{4}.$$

Hence, we have the following

THEOREM. *Equation* (1) *represents a sphere, a single point, or no point whatever, according as the quantity*

$$A^2 + B^2 + C^2 - 4D$$

is positive, zero, or negative. In case it represents a sphere, the coördinates of the center and the square of the radius are given by formulas (4).

Consider, more generally, the equation

$$a(x^2 + y^2 + z^2) + bx + cy + dz + e = 0. \tag{5}$$

If $a = 0$, but b, c, and d are not all zero, the equation represents a plane.

If $a \neq 0$, the equation can be divided through by a, and it then becomes

$$x^2 + y^2 + z^2 + \frac{b}{a}x + \frac{c}{a}y + \frac{d}{a}z + \frac{e}{a} = 0.$$

This equation is of the form (1) and hence the foregoing considerations apply to it.

EXERCISES

Determine what the following equations represent. Apply in each case the *method* of completing the square. Do *not* merely substitute numerical values in formulas (4).

1. $x^2 + y^2 + z^2 + 4x - 6y - 2z + 5 = 0.$

Ans. A sphere, radius 3, with center at $(-2, 3, 1)$.

2. $x^2 + y^2 + z^2 - 6x + 8y + 4z + 29 = 0$.

Ans. The point $(3, -4, -2)$.

3. $x^2 + y^2 + z^2 - 2x + 4y + 2z + 9 = 0$. *Ans.* No point.

4. $x^2 + y^2 + z^2 - 4x + 2y + 4z = 0$.

5. $x^2 + y^2 + z^2 + 6x - 8y + 16 = 0$.

6. $x^2 + y^2 + z^2 + 3z - 4 = 0$.

7. $x^2 + y^2 + z^2 - 2ay - 2bz = 0$.

8. $x^2 + y^2 + z^2 = 2ax$.

9. $x^2 + y^2 + z^2 - 4x - 6y + 13 = 0$.

10. $x^2 + y^2 + z^2 + 9 = 0$.

11. $3x^2 + 3y^2 + 3z^2 - 2x - 12y + 6z + 14 = 0$.

12. $2x^2 + 2y^2 + 2z^2 + 2x - 6y - 2z - 7 = 0$.

13. $3x^2 + 3y^2 + 3z^2 - 4x + 2y + 4z + 6 = 0$.

14. $5x^2 + 5y^2 + 5z^2 - 5x + 6y - 8z - 5 = 0$.

3. Sphere through Four Points. Through four points which do not lie in a plane there passes a single sphere. If the points are (x_1, y_1, z_1), (x_2, y_2, z_2), (x_3, y_3, z_3), (x_4, y_4, z_4), the equation of the sphere, in determinant form, is

$$\begin{vmatrix} x^2 + y^2 + z^2 & x & y & z & 1 \\ x_1^2 + y_1^2 + z_1^2 & x_1 & y_1 & z_1 & 1 \\ x_2^2 + y_2^2 + z_2^2 & x_2 & y_2 & z_2 & 1 \\ x_3^2 + y_3^2 + z_3^2 & x_3 & y_3 & z_3 & 1 \\ x_4^2 + y_4^2 + z_4^2 & x_4 & y_4 & z_4 & 1 \end{vmatrix} = 0.$$

To prove this, develop the determinant by the minors of the elements of the first row. The equation takes on the form (5), § 2. The coefficient of $x^2 + y^2 + z^2$ is the determinant $| x_1 y_2 z_3 1 |$, and is different from zero, since the four points do not lie in a plane (Ch. XXI, § 4). Consequently, the equation represents a sphere, a point, or no point whatever.

But the coördinates of the four given points satisfy the equation, since the substitution for x, y, z of the coördinates of any one of these points makes the first row of the determinant identical with a later row, so that the determinant vanishes. Therefore the equation actually represents a sphere and this sphere is the one through the four given points.

EXERCISES

Find the equations of the spheres through the following sets of four points.

1. (1, 0, 0), (0, 1, 0), (0, 0, 1), the origin.

Ans. $x^2 + y^2 + z^2 - x - y - z = 0.$

2. (1, 1, 1), (– 1, 1, 1), (1, – 1, 1), (1, 1, – 1).

3. (2, 3, 1), (5, – 1, 2), (4, 3, – 1), (2, 5, 3).

4. The vertices of the tetrahedron of Ch. XIX, § 6, Ex. 9.

5. The vertices of the tetrahedron formed by the coördinate planes and the plane $2x - 3y + 4z - 12 = 0$.

6. The vertices of the tetrahedron of Ch. XIX, § 10, Ex. 5.

7. When will five points, no four of which are coplanar, lie on a sphere?

Do the five given points lie on a sphere?

8. (0, 0, 0), (– 1, 0, – 1), (3, 1, 0), (2, 4, – 4), (3, 3, – 4).

9. (0, 2, 3), (4, 1, 0), (– 4, 5, 0), (1, 5, – 1), (4, 2, – 5).

4. Tangent Plane to a Sphere. Let P be a point of a sphere and let the radius to P be drawn. The plane through P perpendicular to the radius is the tangent plane to the sphere at P.

If the sphere is

(1) $$x^2 + y^2 + z^2 = \rho^2$$

and P has the coördinates (x_1, y_1, z_1), the radius to P has the direction components x_1, y_1, z_1. The tangent plane at P is the plane through (x_1, y_1, z_1), whose normals have these direction components. Consequently, its equation is

$$x_1(x - x_1) + y_1(y - y_1) + z_1(z - z_1) = 0,$$

or $$x_1x + y_1y + z_1z = x_1^2 + y_1^2 + z_1^2.$$

Since the point (x_1, y_1, z_1) lies on the sphere (1),

$$x_1^2 + y_1^2 + z_1^2 = \rho^2,$$

and hence the equation of the tangent plane, in final form, is

$$(2) \qquad x_1x + y_1y + z_1z = \rho^2.$$

In a similar manner the equation of the tangent plane to the sphere

$$(3) \qquad (x-\alpha)^2 + (y-\beta)^2 + (z-\gamma)^2 = \rho^2$$

at the point (x_1, y_1, z_1) of the sphere can be shown to be

$$(4) \quad (x_1-\alpha)(x-\alpha)+(y_1-\beta)(y-\beta)+(z_1-\gamma)(z-\gamma) = \rho^2.$$

The use of (4) to find the equation of the tangent plane to a sphere whose equation is in the form

$$(5) \qquad x^2 + y^2 + z^2 + Ax + By + Cz + D = 0$$

involves the reduction of the equation of the sphere to the form (3). Thus, if the sphere is

$$(6) \qquad x^2 + y^2 + z^2 - 2x + 6y + 4z - 35 = 0,$$

the equation must first be rewritten as

$$(x-1)^2 + (y+3)^2 + (z+2)^2 = 49.$$

The equation of the tangent plane at the point $(3, -6, 4)$, for example, is then, according to (4),

$$(3-1)(x-1)+(-6+3)(y+3)+(4+2)(z+2)=49,$$

or

$$(7) \qquad 2x - 3y + 6z - 48 = 0.$$

The coördinates of the center (α, β, γ) and the square of the radius, ρ^2, of a sphere whose equation is in the form (5) are given by formulas (4), § 2. If these values for α, β, γ, ρ^2 are substituted in (4) and the equation obtained is simplified, the result is

$$(8) \quad x_1x + y_1y + z_1z + \frac{A}{2}(x + x_1) + \frac{B}{2}(y + y_1) + \frac{C}{2}(z + z_1) + D = 0.$$

This is the equation of the tangent plane at the point (x_1, y_1, z_1) to a sphere whose equation is in the form (5). By means of it the equation of the tangent plane to (6) at the point (3, – 6, 4) can be written down directly. We have, namely,

$$3x - 6y + 4z - (x+3) + 3(y-6) + 2(z+4) - 35 = 0,$$

and this reduces to the equation (7) obtained by the indirect method.

EXERCISES

Find the equation of the tangent plane to each of the following spheres at the given point.

1. $x^2 + y^2 + z^2 = 9$ at $(2, -2, -1)$.
2. $x^2 + y^2 + z^2 = 49$ at $(3, -6, 2)$.
3. $(x-1)^2 + (y-2)^2 + (z+3)^2 = 81$ at $(2, 6, 5)$.
4. $x^2 + (y+5)^2 + (z-4)^2 = 9$ at $(1, -3, 2)$.
5. $x^2 + y^2 + z^2 - 2x - 4y + 4z = 0$ at the origin.
6. $x^2 + y^2 + z^2 - 6x + 4y + 10z - 11 = 0$ at $(1, 1, 1)$.

7. Find the volume of the tetrahedron cut from the first octant by the tangent plane at (1, 2, 3) to the sphere

$$2x^2 + 2y^2 + 2z^2 + 2x - 3y - 4z - 12 = 0.$$

8. The coördinates of one of the points of intersection of the plane $2x - y - 2 = 0$ with the sphere of Ex. 1 are (2, 2, 1). Find the angle between the plane and the sphere.

Ans. 72° 39′.

9. Find the angle which the line $x = y = z$ makes with the sphere of Ex. 5. *Ans.* 11° 6′.

5. The Circle. A plane intersects a sphere in a circle, is tangent to it, or fails to meet it, according as the distance D of the center of the sphere from the plane is less than, equal to, or greater than, the radius ρ of the sphere.

In other words, the equations of a sphere and a plane, considered simultaneously, represent a circle, a point, or no point whatever, according as $D < \rho$, $D = \rho$, or $D > \rho$.

Consider, for example, the sphere and the plane

$$(1)\qquad \begin{aligned} x^2+y^2+z^2-2x+6y+4z-35=0,\\ x-2y-2z+7=0. \end{aligned}$$

The center of the sphere is at the point $(1, -3, -2)$ and its radius is $\rho=7$; cf. § 2. The distance of the center from the plane is

$$D=\frac{1+(-2)(-3)+(-2)(-2)+7}{\sqrt{1+4+4}}=\frac{18}{3}=6.$$

Consequently, the plane meets the sphere in a circle, and equations (1), considered simultaneously, are the equations of the circle.

It is readily seen that ρ^2-D^2, *i.e.* $7^2-6^2=13$, is the square of the radius of the circle. Hence the radius of the circle is $\sqrt{13}$. The center of the circle is the point of intersection of the plane and the line through the center of the sphere perpendicular to the plane. Its coördinates are thus found to be $(-1, 1, 2)$.

Radical Plane of Two Spheres. Given the two spheres

$$(2)\qquad \begin{aligned} x^2+y^2+z^2+2x-2z-7=0,\\ x^2+y^2+z^2+x+4y-10z-9=0. \end{aligned}$$

Subtract the equation of the second from that of the first. The resulting equation,

$$(3)\qquad x-4y+8z+2=0,$$

represents a plane. This plane is known as the *radical plane* of the two spheres. Since its equation is a linear combination of the two spheres, we conclude, by Ex. 14 at the end of Ch. XXI, the following:

If the spheres intersect, the radical plane is the plane of their common circle; if the spheres are tangent, it is their tangent plane at the point of tangency; and if the spheres fail to meet,* the radical plane intersects neither of them.

* Two concentric spheres have no radical plane; this is the only exceptional case.

Conversely, the spheres intersect in a circle, are tangent, or fail to meet, according as their radical plane intersects one of them in a circle, is tangent to it, or fails to meet it. Thus the question of the relationship between two spheres is reduced to that of the relationship between a plane and a sphere, and this we have already discussed.

The center of the first of the spheres (2) is at the point $(-1, 0, 1)$ and its radius is 3. The distance of the center from the plane (3) is found to be 1. The radical plane and the first sphere intersect, then, in a circle, and consequently this is true of the two spheres.

Equations (2), considered simultaneously, are a pair of equations of the circle. A simpler pair consists in one of the equations (2) and the equation (3).

EXERCISES

In each of the following exercises, determine what the given equations represent. If they represent a circle, find its center and radius; if they represent a point, find its coördinates.

1. $x^2 + y^2 + z^2 - 25 = 0,\ z = 4.$
2. $x^2 + y^2 + z^2 - 6x - 4y = 0,\ 2x + y + 2z - 1 = 0.$
3. $x^2 + y^2 + z^2 - 4x - 2y + 6z - 2 = 0,\ 6x + 2y - 3z + 5 = 0.$
4. $x^2 + y^2 + z^2 - 2x = 0,\ 8x - y - 3 = 0.$

Find the radical plane of the spheres given in the following exercises. If the spheres intersect in a circle, find its center and radius. If the spheres are tangent, find the coördinates of the point of contact.

5. $x^2 + y^2 + z^2 = 13,\ x^2 + y^2 + z^2 + 3x - 4 = 0.$
6. $\begin{cases} x^2 + y^2 + z^2 - 2x - 4y - 8z - 4 = 0, \\ x^2 + y^2 + z^2 + 2y - 5z - 5 = 0. \end{cases}$
7. $x^2 + y^2 + z^2 - 2 = 0,\ 2x^2 + 2y^2 + 2z^2 + 3x - 4y + z + 2 = 0.$

Find the equation of the sphere determined by the given circle and the given point; cf. Ex. 15 at the end of Ch. XXI.

8. The circle of Ex. 1 and the origin.

9. The circle of Ex. 2 and the point (1, 1, 1).

10. The circle of Ex. 5 and the point (1, −2, 3).

6. Cylinders. Given a plane curve, not a straight line, and through each point of the curve draw an indefinite straight line perpendicular to the plane. The surface generated by these lines is called a *cylinder*. The lines are its *rulings*, or *generators*, and the given curve its *directrix*.

We shall consider here only cylinders whose rulings are parallel to a coördinate axis.

If the rulings of a cylinder are parallel, for example, to the axis of z, the equation of the cylinder does not contain z. For, the directrix can be thought of as lying in the (x, y)-plane, and its equation in this plane will represent *in space* the cylinder, inasmuch as the points whose coördinates satisfy the equation are those points and only those points which lie on the directrix, or directly above or below it, *i.e.* which lie on the cylinder. But this equation does not contain z, q. e. d.

Conversely, a curved surface represented by an equation in which z does not appear is a cylinder whose rulings are parallel to the axis of z. For, the equation defines *in the* (x, y)*-plane* a curve, and *in space* it represents those points and only those points which lie on, or directly above or below, this curve, *i.e.* which lie on the cylinder erected on the curve.

We have, then, the following theorem: *A curved surface is a cylinder with rulings parallel to a coördinate axis when and only when its equation does not contain the variable corresponding to that axis.*

Quadric Cylinders. A cylinder whose directrix is a conic is known as a *quadric cylinder*. In particular, it is called *elliptic* (or *circular*), *hyperbolic*, or *parabolic*, according as the directrix is an ellipse (or circle), a hyperbola, or a parabola.

Figure 1 shows the quadric cylinders of the three types, whose equations are

$$(1)\qquad \frac{x^2}{a^2}+\frac{y^2}{b^2}=1, \qquad \frac{x^2}{a^2}-\frac{y^2}{b^2}=\pm 1, \qquad x^2=2\,my.$$

The two hyperbolic cylinders represented by the second equation are known as *conjugate* hyperbolic cylinders, and the planes

$$(2)\qquad \frac{x^2}{a^2}-\frac{y^2}{b^2}=0,$$

shown also in Fig. 1, as their common *asymptotic planes.*

The elliptic cylinder, or either hyperbolic cylinder, of Fig. 1 is symmetric in each point of the axis of z. That is, every

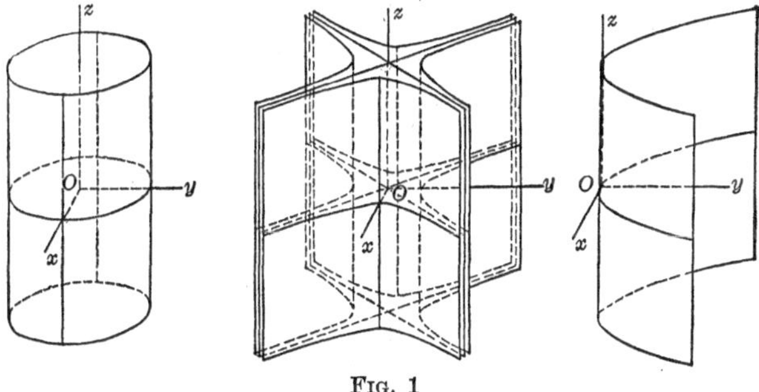

FIG. 1

quadric cylinder whose directrix is a central conic is symmetric in each point of the line drawn through the center of the conic parallel to the rulings. This line is called the *axis* of the cylinder.

Sections of Quadric Cylinders. The curve in which a quadric cylinder is met by a plane, M, which is not parallel to the rulings, we shall call a (plane) *section* of the cylinder.

THEOREM 1. *A section of a quadric cylinder is a conic of the same type as the directrix.*

We give the proof in the case in which the directrix, D, is a central conic. Let M intersect the plane K of D in the line L.* As coördinate axes in K take Ox parallel to L and Oy perpendicular to L, as shown. The equation of D, referred to these axes, is of the form (Ch. XII, § 3):

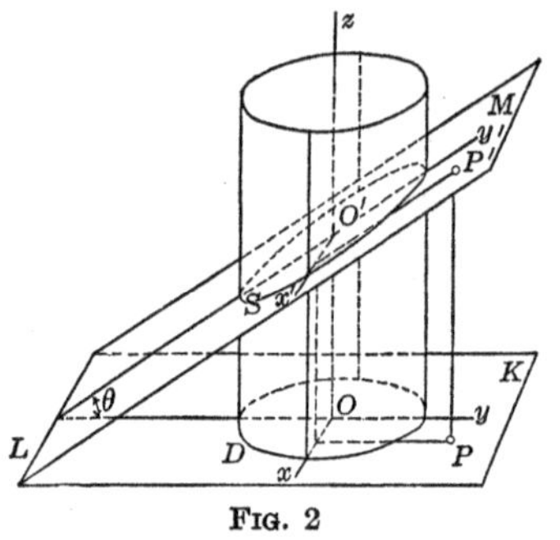

FIG. 2

(3) $Ax^2 + Bxy + Cy^2 + F' = 0.$

Draw in M the rectangular axes, $O'x'$, $O'y'$, whose projections on K are respectively Ox, Oy. Let P': (x', y') be an arbitrary point of M and let P: (x, y) be its projection on K. Then

(4) $$x = x', \qquad y = y' \cos \theta,$$

where θ is the acute angle between M and K.

Since D is the curve in K into which the section, S, of the cylinder by M projects, the equation of S is obtained from equation (3) of D by substituting for x and y in (3) their values as given by (4). Thus the equation of S is

(5) $$Ax'^2 + Bx'y' \cos \theta + Cy'^2 \cos^2 \theta + F'' = 0.$$

This equation represents a conic and, furthermore, a conic of the same type as D, since the discriminant of the quadratic terms:

$$B^2 \cos^2 \theta - 4\,AC \cos^2 \theta = (B^2 - 4\,AC) \cos^2 \theta$$

is of the same sign as the discriminant, $B^2 - 4\,AC$, of the quadratic terms in (3), q. e. d.

It is clear that (5) is independent of the height OO' at which M cuts the axis of the cylinder. In other words, *the sections by two parallel planes are congruent conics.*

Suppose, now, that $B^2 - 4\,AC > 0$ and that to F'' is given in turn the values 1, -1, 0. Then (3) represents in turn a hyper-

* If M is parallel to K, the section by M is congruent to the directrix; in this case, then, no further proof is required.

bola, the conjugate hyperbola, and the common asymptotic lines; but this is true, also, of (5). We have, then, the following result.

THEOREM 2. *The sections of two conjugate hyperbolic cylinders by a plane M are two conjugate hyperbolas whose common asymptotes are the lines in which M cuts the common asymptotic planes of the cylinders.*

Returning to the general case, we assume that there is given a second cylinder with vertical rulings, whose directrix, $\bar{D}$, is similar and similarly placed to D, or, if D is a hyperbola, is similar and similarly placed to D or to the conjugate of D. The equation of $\bar{D}$, as a curve in K, can be written, according to Ex. 40, p. 260, in the form:

$$(6) \qquad Ax^2 + Bxy + Cy^2 + Dx + Ey + F = 0.$$

The equation of the section $\bar{S}$ of the second cylinder by the plane M is, then,

$$(7) \quad Ax'^2 + Bx'y' \cos\theta + Cy'^2 \cos^2\theta + Dx' + Ey' \cos\theta + F = 0.$$

Since equations (5) and (7) fulfill the conditions of Ex. 40, p. 260, it follows that S and $\bar{S}$ are in the same relation as D and $\bar{D}$. We have thus proved, in the case in which D and $\bar{D}$ are central conics, the following theorem.

THEOREM 3. *If the directrices of two cylinders (with parallel rulings) are similar and similarly placed conics, or if, in the case of hyperbolas, each is similar and similarly placed either to the other or to the conjugate of the other, then the sections of the cylinders by the same plane or by two parallel planes stand in like relationship.*

The converses of Theorems 1, 2, 3 are true, as is readily seen. The three theorems and their converses can be stated equally well in terms of projections. Thus Theorem 1 and its converse are equivalent to the theorem: *A plane curve is a conic of a certain type if and only if its projection on a plane not perpendicular to its plane is a conic of this type.*

EXERCISES

1. Do Exs. 9, 10, 11 of Ch. XIX, § 1.

2. Do Exs. 20, 21, 22 of Ch. XIX, § 1.

What does each of the following equations represent?

3. $4x^2 + y^2 - 8x + 4y - 4 = 0.$

4. $3x^2 + 6x - 2y + 1 = 0.$

5. $xy + 2x - y - 6 = 0.$

6. Prove Theorem 1 when D is a parabola.

7. Prove Theorem 3 when D is a parabola.

8. State Theorems 2 and 3 and their converses in terms of projections.

9. Show that if a central conic S in a plane M projects into the central conic D in the plane K, then the center O' of S projects into the center O of D.

7. Cones. Let a plane curve, not a straight line, and a point O, not in the plane of the curve, be given. Draw an indefinite straight line through O and each point of the curve. The surface formed by these lines is known as a cone. The lines are its *rulings*, or *generators*, and the point O is its *vertex*.

If the given curve is a circle and O lies on the line L through its center perpendicular to its plane, the cone can be generated by the rotation about L of any ruling. Accordingly, it is known as a *cone of revolution* or a *circular cone*. The line L is its *axis* and the constant angle between L and a ruling is the *generating angle*.

Problem. To find the equation of a cone of revolution whose vertex is at the origin, whose axis is the axis of z, and whose generating angle is ϕ.

Let $P: (x, y, z)$ be any point of the cone other than O. The ruling R on which P lies determines with the axis of z a plane M which cuts the (x, y)-plane in a line L. Direct the line L as shown in the figure and denote the projection of OP

on L, thus directed, by r. Then the directed line L and the axis of z form in the plane M a system of coördinate axes, with respect to which the point P has the coördinates (r, z).

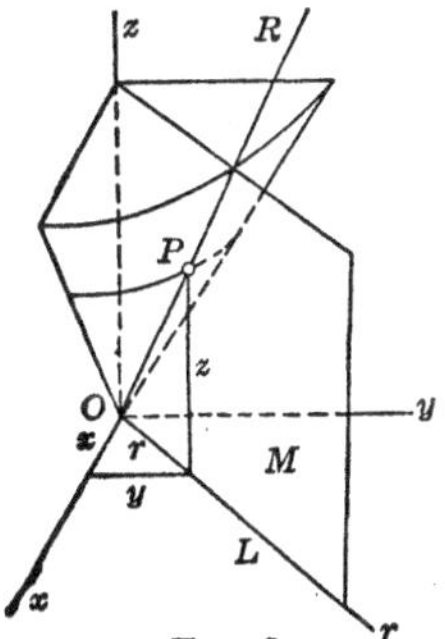

The equation of the ruling R, as a line of M, is

$$r = z \tan \phi.$$

Since P lies on R, its coördinates (r, z) satisfy this equation. But, clearly,

$$r = \pm \sqrt{x^2 + y^2},$$

where the plus sign or the minus sign is to be taken, according as P lies on the upper nappe of the cone (as shown) or on the lower nappe. Consequently, the coördinates (x, y, z) of P satisfy the equation*

$$\pm \sqrt{x^2 + y^2} = z \tan \phi,$$

or

(1) $$x^2 + y^2 - z^2 \tan^2 \phi = 0.$$

FIG. 3

Conversely, every point whose coördinates satisfy this equation lies on the cone, for the steps can be retraced. Hence this is the equation of the cone.

Equation (1) is homogeneous in x, y, z. This is characteristic of the equation of a cone with its vertex at the origin. In fact, we can state the theorem: *A curved surface is a cone with its vertex at the origin, when and only when its equation is homogeneous in* x, y, z.

Before giving the proof we consider a particular homogeneous equation

(2) $$4x^2 - 3y^2 + 12z^2 = 0.$$

If x_1, y_1, z_1 is a solution of this equation, *i.e.* if

(3) $$4x_1^2 - 3y_1^2 + 12z_1^2 = 0,$$

* The coördinates of the origin, originally ruled out, clearly satisfy the equation.

kx_1, ky_1, kz_1, where k is any constant, is also a solution. For, the equation

$$4(kx_1)^2 - 3(ky_1)^2 + 12(kz_1)^2 = 0 \quad \text{or} \quad k^2(4x_1^2 - 3y_1^2 + 12z_1^2) = 0$$

is true, inasmuch as the parenthesis has, by (3), the value zero.

It follows that, if $P:(x_1, y_1, z_1)$ is an arbitrary point, not the origin, on the surface represented by (2),* any point with coördinates of the form (kx_1, ky_1, kz_1) is also on the surface. But the points (kx_1, ky_1, kz_1), where k is an arbitrary constant, are all the points of the line OP passing through the origin and P (Ex. 6 at the end of Ch. XVIII). That is, the line OP through the origin lies wholly on the surface. But P was an arbitrary point on the surface, other than O, and therefore the surface is formed by lines through the origin, *i.e.* it is a cone with its vertex at the origin.

The cone can be constructed by drawing the curve in which it intersects a plane not passing through O and by joining O with the points of this curve by straight lines. If $y = 1$ is the plane taken, the curve of intersection is the ellipse

$$4x^2 + 12z^2 = 3,$$

and thus the cone is as shown in Fig. 4.

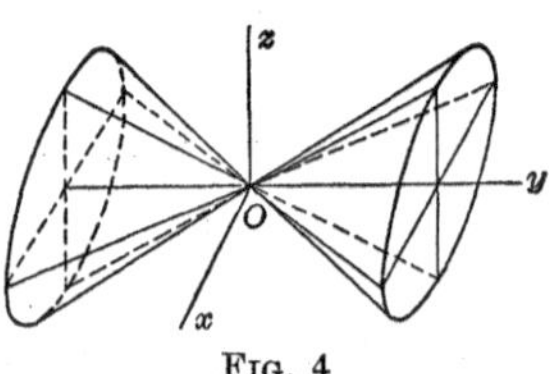

FIG. 4

This example suggests the following proof for the theorem.

An equation in x, y, z is homogeneous if and only if, when x_1, y_1, z_1 (not all zero) is a solution, kx_1, ky_1, kz_1, where k is an arbitrary number, is also a solution. On the other hand, a curved surface is a cone with its vertex at the origin if and only if, when $P:(x_1, y_1, z_1)$ is a point of the surface other than O, an arbitrary point, (kx_1, ky_1, kz_1), of OP is also a point of the surface. Thus the algebraic condition that an equation be homogeneous is equivalent to the geometrical condition that a

* The locus of (2) is actually a surface (and not a curve or a point), since all pairs of values (x, z) lead to points on it.

curved surface be a cone with its vertex at the origin, and the theorem is proved.

Quadric Cones. A cone represented by a *homogeneous* equation of the *second* degree in x, y, z, *i.e.* by an equation of the form,

(4) $$ax^2 + by^2 + cz^2 + dxy + eyz + fzx = 0,$$

is called a *quadric* cone.

We state, without proof, that equation (4), if it represents a cone,* can always be transformed by a rotation of axes (cf. Ch. XXIV, § 6) into the equation,

(5) $$Ax^2 + By^2 - Cz^2 = 0,$$

where A, B, and C are positive constants.

The quadric cone (5) is, in particular, a cone of revolution, when and only when it can be written in the form (1), *i.e.* when and only when $A = B$.

All quadric cones are of *one* general type. They cannot be classified into three types, corresponding to those for quadric cylinders. For ellipses, hyperbolas, and parabolas can all be obtained as plane sections of any one of them.†

EXERCISES

Construct the cones represented by the following equations. If the cone is circular, determine its axis and the generating angle.

1. $x^2 + y^2 - z^2 = 0.$
2. $x^2 - 3y^2 + z^2 = 0.$
3. $4x^2 - y^2 - z^2 = 0.$
4. $4x^2 + y^2 - 4z^2 = 0.$
5. $6x^2 - 3y^2 - 2z^2 = 0.$
6. $y^2 - 2xz = 0.$

* Equation (4) represents in general a *curved* surface, and hence, by the theorem, a cone. Under special conditions it may, however, represent two planes, a single plane, a line, or merely the origin. For example, $x^2 - y^2 = 0$ represents two planes; $(x - y)^2 = 0$, a single plane; $x^2 + y^2 = 0$, a line; and $x^2 + y^2 + z^2 = 0$, only the origin. These cases are here excluded.

† This was proved geometrically for the case of a cone of revolution in Ch. VIII, § 10. An analytical proof covering all cases will be given later, Ch. XXIII, § 5.

Find the equations of the following cones.

7. The cone of revolution whose vertex is at the origin, whose axis is the axis of y, and whose generating angle is 30°.

8. The cone of revolution whose vertex is at $(0, 0, a)$, whose axis is the axis of z, and whose generating angle is 45°.

Ans. $x^2 + y^2 - (z - a)^2 = 0.$

9. The quadric cone which has its vertex at the origin and intersects the plane $z = 1$ in the ellipse whose center is on the z-axis, whose transverse axis is parallel to the axis of y, and whose major and minor axes are 6 and 4.

10. The quadric cone which has its vertex at $(0, 1, 0)$ and intersects the (z, x)-plane in the hyperbola $2x^2 - z^2 = 4$.

11. The cone of revolution which has the line bisecting the angle between the positive y- and z-axes as axis and which contains the y- and z-axes. *Ans.* $x^2 = 2yz.$

12. The cone of revolution which has the line $x = y = z$ as axis and passes through the coördinate axes.

Ans. $xy + yz + zx = 0.$

8. Surfaces of Revolution. Let a plane curve and a line L in the plane of the curve be given. The surface generated by the curve when the plane is rotated about L through 360° is known as a *surface of revolution.* The line L is its *axis.*

It is clear that spheres are surfaces of revolution. So also are circular cylinders and circular cones; in the one case the generating curve is a line parallel to L, and in the other, it is a line which intersects L.

Quadric Surfaces of Revolution. The surfaces obtained by rotating the conics about their *axes,* together with the spheres and the circular cylinders and cones, are known as *quadric surfaces of revolution.*

Problem. To find the equation of the *ellipsoid of revolution* generated when the ellipse in the (y, z)-plane, whose equation in that plane is

$$\frac{y^2}{a^2}+\frac{z^2}{b^2}=1, \qquad a>b,$$

is rotated about its conjugate axis, the axis of z.

Let M be the rotating plane in an arbitrary position, and let $P:(x, y, z)$ be an arbitrary point on the ellipse in M. Establish in M the same system of axes as was set up in the plane M of Fig. 3 in finding the equation of a cone of revolution. The equation of the ellipse in M, referred to these axes, is

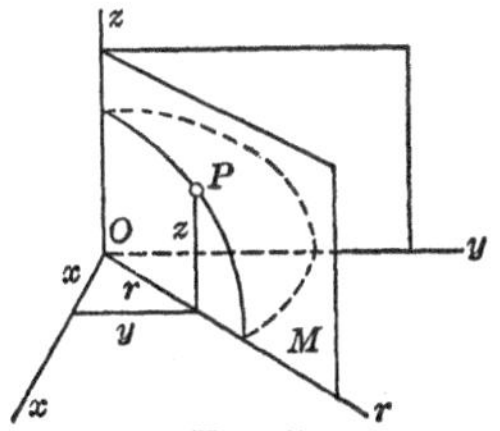

$$\frac{r^2}{a^2}+\frac{z^2}{b^2}=1.$$

FIG. 5

Since P lies on the ellipse, its coördinates (r, z) satisfy this equation. Consequently, inasmuch as

$$r^2=x^2+y^2,$$

the coördinates (x, y, z) of P satisfy the equation

$$(1) \qquad \frac{x^2}{a^2}+\frac{y^2}{a^2}+\frac{z^2}{b^2}=1, \qquad a>b,$$

and this is the equation of the ellipsoid of revolution.

In a similar manner, the equation of the ellipsoid of revolution obtained by rotating the ellipse about its transverse axis, the axis of y, is found to be

$$(2) \qquad \frac{x^2}{b^2}+\frac{y^2}{a^2}+\frac{z^2}{b^2}=1, \qquad a>b.$$

The first of the two ellipsoids of revolution (Fig. 6) is often called an *oblate spheroid*; and the second (Fig. 7), a *prolate spheroid*. Both approach as their limits the sphere, whose center is at the origin and whose radius is a, when b is made to approach a as its limit.

The *hyperboloids of revolution* generated when the hyperbolas

$$(3) \qquad \frac{y^2}{a^2}-\frac{z^2}{b^2}=1, \qquad \frac{y^2}{a^2}-\frac{z^2}{b^2}=-1,$$

situated in the (y, z)-plane, are rotated about the axis of z, — the conjugate axis of the first hyperbola, and the transverse axis of the second, — have respectively the equations:

$$\frac{x^2}{a^2}+\frac{y^2}{a^2}-\frac{z^2}{b^2}=1; \tag{4}$$

$$\frac{x^2}{a^2}+\frac{y^2}{a^2}-\frac{z^2}{b^2}=-1, \quad \text{or} \quad -\frac{x^2}{a^2}-\frac{y^2}{a^2}+\frac{z^2}{b^2}=1. \tag{5}$$

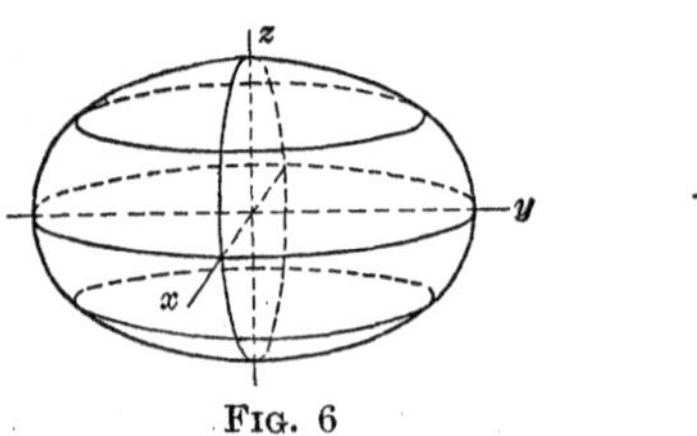

FIG. 6

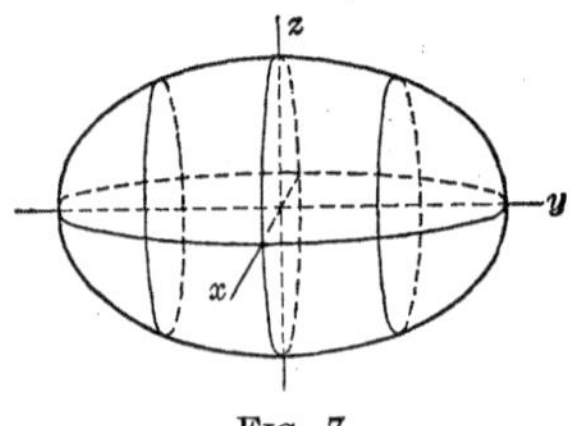

FIG. 7

The two hyperboloids of revolution are shown in Figs. 8 and 9. The first is known as a *hyperboloid of one sheet* or an *unparted hyperboloid*; the second, as a *hyperboloid of two sheets*, or a *biparted hyperboloid.*

Taken together, the hyperboloids (4) and (5) are known as *conjugate* hyperboloids of revolution. They are generated by

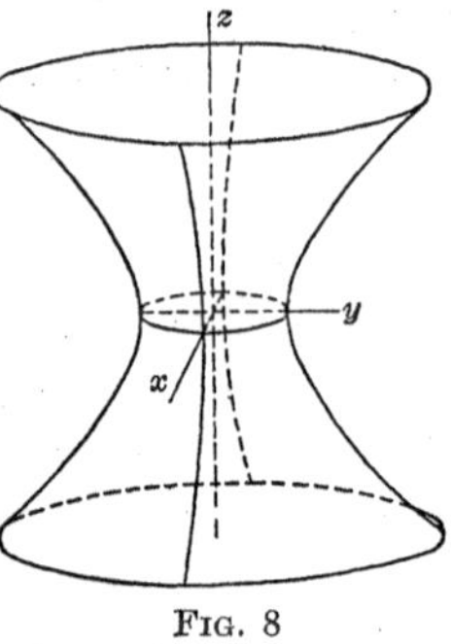

FIG. 8

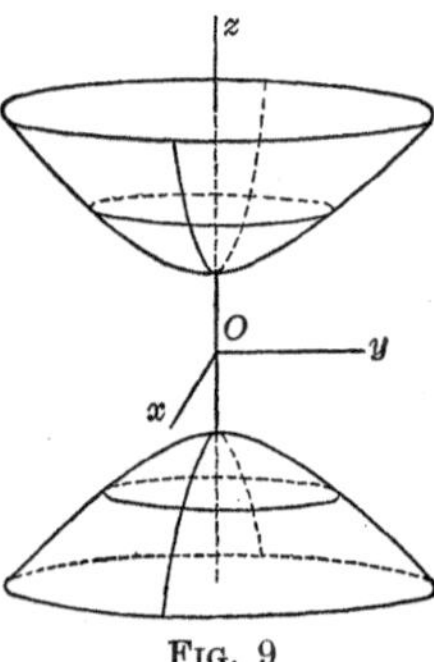

FIG. 9

the conjugate hyperbolas (3) revolving about the same axis, the axis of z. The cone which results from the rotation about this axis of the common asymptotes of the hyperbolas, namely, the cone

(6) $$\frac{x^2}{a^2}+\frac{y^2}{a^2}-\frac{z^2}{b^2}=0,$$

is called the common *asymptotic cone* of the conjugate hyperboloids. See Ch. XXIII, Fig. 5.

Since a parabola has but one axis, there can be obtained from it but one quadric surface, or paraboloid, of revolution, namely, that which results from rotating it about its axis. If the equation of the parabola, in the (y, z)-plane, is

$$y^2 = 2\,mz,$$

the equation of the *paraboloid of revolution* is

(7) $$x^2 + y^2 = 2\,mz.$$

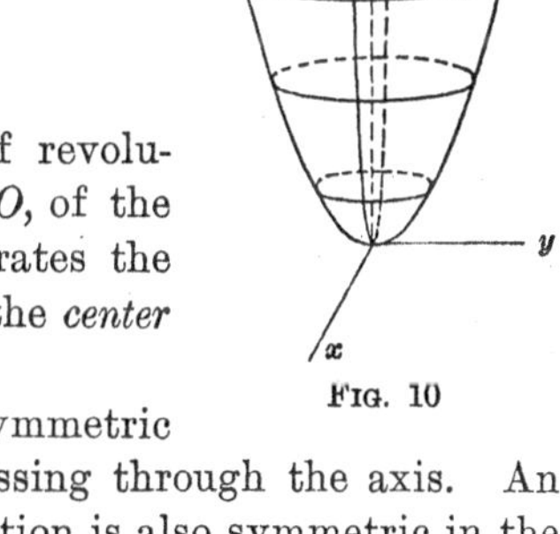

FIG. 10

An ellipsoid or hyperboloid of revolution is symmetric in the center, O, of the ellipse or hyperbola which generates the surface. Accordingly, we call O the *center* of the surface.

Every surface of revolution is symmetric in its axis and in every plane passing through the axis. An ellipsoid or hyperboloid of revolution is also symmetric in the plane through the center perpendicular to the axis, and in every line through the center lying in this plane. Thus the surface (1) is symmetric, not only in the axis of z and in all planes through this axis, but also in the (x, y)-plane and all lines lying in this plane and passing through O.

EXERCISES

Establish each of the following equations.

1. Equation (2). 2. Equation (4). 3. Equation (5).

What surface does each of the following equations represent? Construct the surface.

4. $x^2 + y^2 + 4z^2 = 4.$

5. $4x^2 + 9y^2 + 9z^2 = 36.$

6. $2x^2 + 3y^2 + 2z^2 = 6.$

7. $x^2 + y^2 - 2z^2 = 2.$

8. $3x^2 - 4y^2 - 4z^2 = 12.$

9. $3x^2 - 5y^2 + 3z^2 = 15.$

10. $x^2 + y^2 = 8z.$

11. $y^2 + z^2 = -4x.$

12. The parabola of the text is rotated about the axis of y. Find the equation of the surface generated and construct it.

13. The surface generated by the rotation of a circle of radius a about a line L in the plane of the circle at the distance $b > a$ from its center is called an *anchor ring*, or *torus.* Find its equation, if L is the axis of z and the circle is in the (y, z)-plane with its center on the axis of y.

Ans. $(\sqrt{x^2 + y^2} - b)^2 + z^2 = a^2.$

EXERCISES ON CHAPTER XXII

1. Find the equation of the sphere having the line-segment joining the two points $(3, 2, -1)$, $(5, 4, 3)$ as a diameter.

2. Find the equation of the sphere which has its center at the point $(5, -2, 3)$ and is tangent to the plane $3x+2y+z=0.$

3. A sphere has its center in the plane $x + y + 3z - 2 = 0$ and passes through the three points $(2, 3, 1)$, $(2, -1, 5)$, $(-2, -3, 3)$. Find its equation.

4. Find the equation of the sphere which has its center on the line $4x + 8 = 3y + 7 = 4z$ and passes through the points $(4, 3, -1)$, $(3, 2, 3)$.

5. A sphere is tangent to the plane $x - 2y - 2z = 7$ in the point $(3, -1, -1)$ and goes through the point $(1, 1, -3)$. Find its equation. *Ans.* $x^2 + y^2 + z^2 - 10y - 10z - 31 = 0.$

6. There are two spheres passing through the points $(4, 0, 3)$, $(5, 4, 0)$, $(5, 1, 3)$ and having the radius 3. Find their equations.

Ans. $$\begin{cases} x^2 + y^2 + z^2 - 6x - 4y - 2z + 5 = 0. \\ 3x^2 + 3y^2 + 3z^2 - 22x - 8y - 2z + 19 = 0. \end{cases}$$

7. Find the equations of the spheres which are tangent to the plane $x + 2y - 2z - 12 = 0$ and pass through the three points $(3, -2, 0)$, $(2, -3, 0)$, $(3, 1, -3)$.

8. Find the equations of the spheres which are tangent to the planes $2x - y + 2z + 1 = 0$, $6x + 3y - 2z + 5 = 0$ and have their centers on the line $1 - x = y + 1 = 2z$.

9. Find the equation of the sphere inscribed in the tetrahedron formed by the plane $2x + 2y + z - 4 = 0$ and the coördinate planes.

10. A sphere goes through the point (4, 6, 3) and meets the (x, y)-plane in a circle whose center is at the point (1, 2, 0) and whose radius is 5. Find its equation.

Ans. $x^2 + y^2 + z^2 - 2x - 4y - 3z - 20 = 0.$

11. Find the equations of the tangent line to the circle

$$x^2 + y^2 + z^2 - x + 4z = 0, \quad 3x - 2y + 4z + 1 = 0$$

at the point $(1, -2, -2)$.

12. Does the line $2x - 1 = y + 3 = 4 - z$ intersect the sphere

$$x^2 + y^2 + z^2 - 6x + 8y - 4z + 4 = 0?$$

13. Show that the radical planes of three spheres, taken in pairs,* pass through a line or are parallel.

Orthogonality †

14. Prove that the plane $x + y + z - 1 = 0$ intersects the sphere of Ex. 12 orthogonally.

15. When does the plane $ax + by + cz + d = 0$ intersect the sphere

$$x^2 + y^2 + z^2 + Ax + By + Cz + D = 0$$

orthogonally?

Ans. When and only when $aA + bB + cC = 2d$.

* It is assumed that no two of the three spheres are concentric; cf. footnote, p. 530.

† Cf. Exs. 18–26 at the end of Ch. IV.

16. Show that the line $3x + 8 = -6y - 7 = 2z + 13$ intersects the sphere of Ex. 12 orthogonally.

17. Find the condition that the line

$$a_1x + b_1y + c_1z + d_1 = 0, \qquad a_2x + b_2y + c_2z + d_2 = 0$$

intersect the sphere of Ex. 15 orthogonally.

18. Find the angle of intersection of the spheres (2) of § 5.

19. Prove that the two spheres

$$x^2 + y^2 + z^2 - 9 = 0, \qquad x^2 + y^2 + z^2 - 6x + 8y + 9 = 0$$

intersect orthogonally.

20. When does the sphere of Ex. 15 intersect the sphere

$$x^2 + y^2 + z^2 + A'x + B'y + C'z + D' = 0$$

orthogonally?

Ans. When and only when $AA' + BB' + CC' = 2D + 2D'$.

21. Find the equation of the sphere containing the circle of Ex. 11 and intersecting orthogonally the first of the spheres of Ex. 19.

Loci

22. Do Ex. 1 of Ch. V, § 1, when P is not restricted to lie in a plane.

23. The same for Ex. 4 of Ch. V, § 1.

24. A point P moves so that the ratio of its distances from two fixed points is constant. Find its locus.

25. A point P moves so that its distance from a given plane M is proportional to the square of its distance to a given point P_0, not in M. If P remains always on the same side of M as P_0, find its locus.

26. If, in the preceding exercise, P_0 lies in M and P may be on either side of M, what is the locus of P?

27. What is the locus of a point which moves so that its distance from a given line is proportional to its distance from a given plane perpendicular to the line?

28. What is the locus of a point which moves so that its distance from a given point is proportional to its distance to a given plane through the point?

29. What is the locus of a point P which moves so that the difference of the squares of its distances from a given point and a given sphere is constant, if the distance from P to the sphere is measured along a tangent line to the sphere through P?

30. A given point P_0 is distant $2a$ units from a given plane and P' is an arbitrary point in the plane. What is the locus of the point P so chosen on the line P_0P' that $P_0P \cdot P_0P' = 4a^2$?

Suggestion. Take the coördinates of P' as auxiliary variables.

31. A is a fixed point and R an arbitrary point of a given sphere whose center is O. The radius OA is produced four times its length to the point A' and the radius OR, twice its length to the point R'. What is the locus of the point of intersection of AR and $A'R'$?

CHAPTER XXIII

QUADRIC SURFACES

1. The Ellipsoid. *A quadric surface is any surface defined by an equation of the second degree in x, y, z.* The sphere and the quadric cylinders, cones and surfaces of revolution studied in the previous chapter are special types of quadric surfaces. We proceed to consider more general types.

The surface defined by the equation

$$\frac{x^2}{a^2}+\frac{y^2}{b^2}+\frac{z^2}{c^2}=1 \tag{1}$$

is known as an *ellipsoid.* If two of the three numbers a, b, c are equal, it is in particular an ellipsoid of revolution (Ch. XXII, § 8). To construct the surface in the general case when no two of the three numbers a, b, c are equal, plot first the *sections* by the coördinate planes, that is, the curves of intersection with the planes $x=0$, $y=0$, $z=0$. These are, respectively, the ellipses

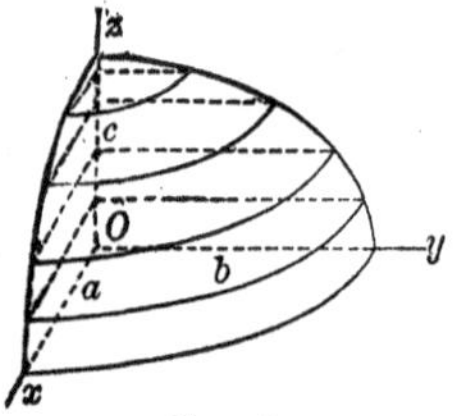

FIG. 1

$$\frac{y^2}{b^2}+\frac{z^2}{c^2}=1, \qquad \frac{x^2}{a^2}+\frac{z^2}{c^2}=1, \qquad \frac{x^2}{a^2}+\frac{y^2}{b^2}=1.$$

The parts of these ellipses which lie in the quarter-planes bounding the first octant connect the points $(a, 0, 0)$, $(0, b, 0)$, $(0, 0, c)$, as shown.

The section of (1) by a plane $z=k$ parallel to the (x, y)-plane has the equations:

$$\frac{x^2}{a^2}+\frac{y^2}{b^2}=1-\frac{k^2}{c^2}, \qquad z=k.$$

If $k^2 < c^2$, these equations represent, in the plane $z=k$, an ellipse whose center is on the axis of z and whose axes lie in the (z, x)- and (y, z)-planes and have the lengths

$$2a\sqrt{1-\frac{k^2}{c^2}}, \qquad 2b\sqrt{1-\frac{k^2}{c^2}}.$$

As k increases from 0 toward c as its limit, this ellipse, rising from the section by the (x, y)-plane, grows continuously smaller and shrinks finally to a point, — the point $(0, 0, c)$. Similarly, if k decreases from 0 toward $-c$ as its limit.

The surface generated by the changing ellipse is the ellipsoid. Fig. 2 (or Fig. 6 of Ch. XXII *) shows it in its entirety. The surface is evidently symmetric in the origin, O, and in the coördinate axes and coördinate planes. O is called the *center* of the ellipsoid; the coördinate axes, the *axes* of the ellipsoid; and the coördinate planes, the *principal planes* of the ellipsoid. The sections by the principal planes are known as the *principal sections.*

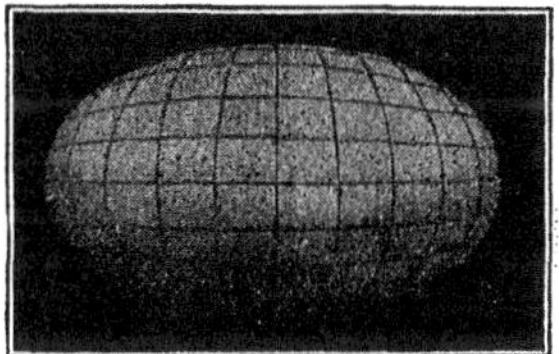

Fig. 2

The dimensions of the ellipsoid, measured along the axes, are $2a$, $2b$, $2c$. These numbers, in the order of their magnitude, are known as the *major axis, mean axis,* and *minor axis* of the ellipsoid.

Here, and throughout the chapter, we speak of a (plane) *section* of a surface only when the plane in question meets the surface in a *curved* line. Sections by parallel planes we shall call *parallel sections.*

* Figs. 6–10 of Ch. XXII, drawn originally to represent quadric surfaces of revolution, picture equally well the corresponding general quadric surfaces studied in this chapter. One has merely to imagine that a different ratio of foreshortening along the axis of x has been chosen.

EXERCISES

Construct the following ellipsoids, drawing accurately the principal sections and the sections parallel to one principal plane. What are the lengths of the axes?

1. $\frac{x^2}{16}+\frac{y^2}{25}+\frac{z^2}{9}=1.$

2. $9x^2+36y^2+4z^2=36.$

3. Discuss the generation of the ellipsoid (1) by sections parallel to the (z, x)-plane.

4. Prove that the sections of the ellipsoid (1), which are parallel to a principal plane, *e.g.* the (x, y)-plane, are similar and similarly placed ellipses; cf. p. 260.

2. The Hyperboloids. *The Hyperboloid of One Sheet.* The surface represented by the equation

$$\frac{x^2}{a^2}+\frac{y^2}{b^2}-\frac{z^2}{c^2}=1 \tag{1}$$

is called a *hyperboloid of one sheet* or an *unparted hyperboloid.* If $a=b$, it is in particular a hyperboloid of revolution of one sheet (Ch. XXII, § 8).

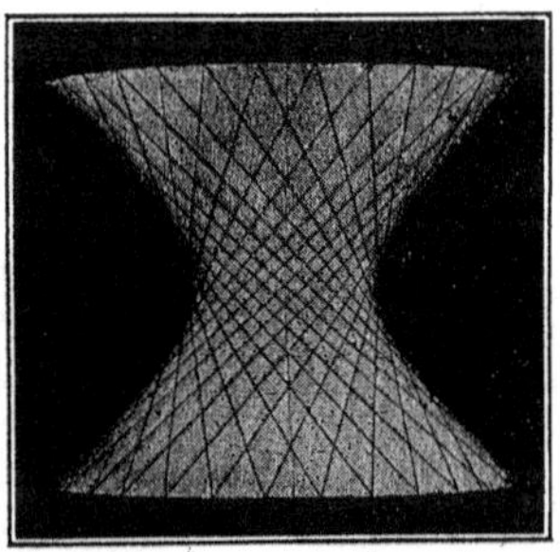

FIG. 3

In the general case, $a \neq b$, the surface can be constructed by the method of § 1. The sections by the vertical coördinate planes, $x=0$ and $y=0$, are the hyperbolas

$$\frac{y^2}{b^2}-\frac{z^2}{c^2}=1, \quad \frac{x^2}{a^2}-\frac{z^2}{c^2}=1. \tag{2}$$

The sections by the planes $z=k$ are ellipses. The smallest one is the section by $z=0$, the (x, y)-plane; it is known as the *minimum ellipse.* The general one increases in size as its distance from the (x, y)-plane increases. The surface can be thought of as generated by it; cf. Fig. 8, Ch. XXII.

The Hyperboloid of Two Sheets. This surface, also known as the *biparted hyperboloid,* is defined by the equation

$$(3) \qquad \frac{x^2}{a^2}+\frac{y^2}{b^2}-\frac{z^2}{c^2}=-1 \quad \text{or} \quad -\frac{x^2}{a^2}-\frac{y^2}{b^2}+\frac{z^2}{c^2}=1.$$

A particular case, when $a=b$, is the hyperboloid of revolution of two sheets.

In the general case, $a \neq b$, the sections by the vertical coördinate planes are the hyperbolas conjugate to the hyperbolas (2). The (x, y)-plane, $z=0$, does not intersect the surface. This is true of all the planes $z=k$, for which $k^2 < c^2$. The planes $z = \pm c$ meet the surface in the points $(0, 0, \pm c)$, and the planes $z=k$, where $k^2 > c^2$, meet it in ellipses, which increase in size as k increases in numerical value; cf. Fig. 9, Ch. XXII.

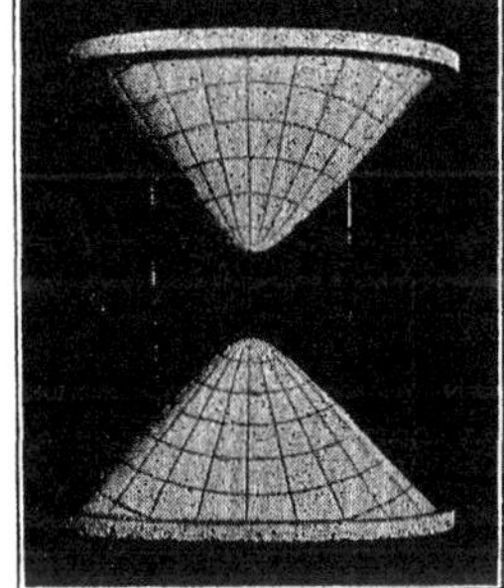

FIG. 4

Center, Axes, Principal Planes. Each hyperboloid is symmetric in the origin O and in the coördinate axes and coördinate planes; O is the *center,* the coordinate axes, the *axes,* and the coördinate planes, the *principal planes* for each surface. The sections by the principal planes are the *principal sections.*

The Asymptotic Cone. The hyperboloids (1) and (3) are called *conjugate hyperboloids.* We have seen that each vertical coördinate plane intersects them in conjugate hyperbolas whose common asymptotes pass through the origin. This is true also of any vertical plane,

$$(4) \qquad y = mx,$$

which passes through the axis of z. For, the sections of (1) and (3) by the plane (4) are also the sections by this plane of the cylinders,

$$x^2\left(\frac{1}{a^2}+\frac{m^2}{b^2}\right)-\frac{z^2}{c^2}=1, \qquad x^2\left(\frac{1}{a^2}+\frac{m^2}{b^2}\right)-\frac{z^2}{c^2}=-1,$$

whose equations are obtained by eliminating y from (4) and (1), and from (4) and (3). But these cylinders are conjugate hyperbolic cylinders whose common asymptotic planes are

$$x^2\left(\frac{1}{a^2}+\frac{m^2}{b^2}\right)-\frac{z^2}{c^2}=0. \tag{5}$$

Consequently, by Ch. XXII, § 6, Th. 2, their sections by the plane (4) are conjugate hyperbolas whose common asymptotes, defined by equations (4) and (5), pass through the origin, q. e. d.

The equation of the locus of these asymptotes, as the plane (4) rotates about the axis of z, is obtained by eliminating m (now an auxiliary variable expressing the motion of the plane) from equations (4) and (5). The result is

$$\frac{x^2}{a^2}+\frac{y^2}{b^2}-\frac{z^2}{c^2}=0. \tag{6}$$

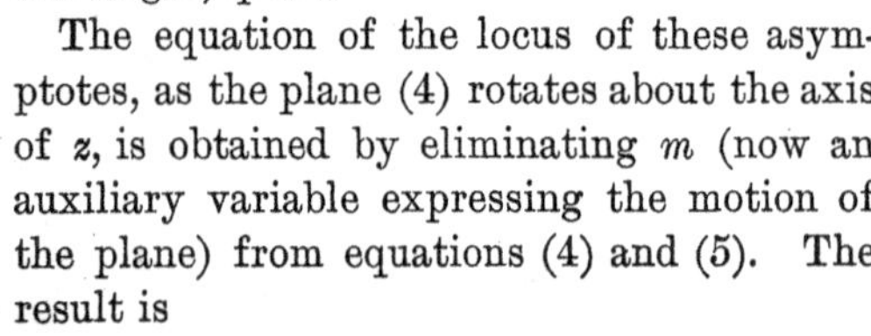

FIG. 5

The locus of the asymptotes is, therefore, a cone whose vertex is at the origin. This cone is called the *asymptotic cone* of each of the hyperboloids (1) and (3). Evidently (1) lies wholly without the cone, that is, on the convex side of it, while (3) lies wholly within it.

EXERCISES

Construct the following hyperboloids, drawing accurately the principal sections which exist and the sections parallel to one principal plane.

1. $\frac{x^2}{9}+\frac{y^2}{4}-\frac{z^2}{25}=1.$ 2. $\frac{x^2}{9}+\frac{y^2}{4}-\frac{z^2}{25}=-1.$

3. $9x^2-16y^2+36z^2=144.$ 4. $-4x^2+36y^2-9z^2=36.$

5. What is the equation of the hyperboloid conjugate to the hyperboloid of Ex. 3? of Ex. 4? Give also in each case the equation of the common asymptotic cone.

6. Show that the sections of the hyperboloid of two sheets (3), which are parallel to a vertical principal plane, are similar and similarly placed hyperbolas.

7. The sections of the hyperboloid of one sheet (1) by two planes parallel to a vertical principal plane are similar and similarly placed hyperbolas, or are hyperbolas each of which is similar and similarly placed to the conjugate of the other. Prove this theorem and determine when each of the two cases occurs.

3. The Paraboloids. *The Elliptic Paraboloid.* The surface defined by the equation

$$\frac{x^2}{a^2}+\frac{y^2}{b^2}=2z \tag{1}$$

is called an *elliptic paraboloid.* If $a=b$, it is in particular a paraboloid of revolution (Ch. XXII, § 8).

In the general case, $a \neq b$, the sections of the surface by the vertical coördinate planes, $y=0$ and $x=0$, are the parabolas

$$x^2=2\,a^2z, \qquad y^2=2\,b^2z,$$

both of which open upwards. The (x, y)-plane intersects the surface only in the origin. A plane parallel to the (x, y)-plane and below it does not meet the surface, while a plane parallel to the (x, y)-plane and above it intersects the surface in an ellipse, which increases in size as the height of the plane increases; cf. Fig. 10, Ch. XXII.

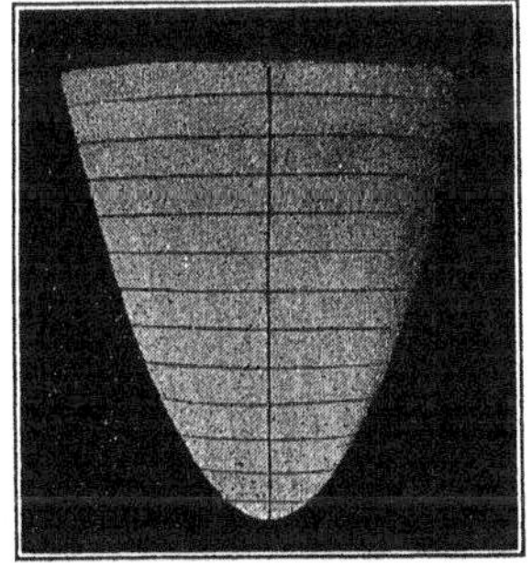

FIG. 6

The Hyperbolic Paraboloid. This is the surface defined by the equation

$$\frac{x^2}{a^2}-\frac{y^2}{b^2}=2\,z. \tag{2}$$

It is never a surface of revolution, no matter what values are assigned to a and b.

The sections of the surface by the vertical coördinate planes are the parabolas

$$x^2 = 2\,a^2z, \qquad y^2 = -\,2\,b^2z,$$

of which the first opens upwards and the second, downwards.

Fig. 7

The section by the (x, y)-plane consists of the two lines,

$$OA:\ \frac{x}{a}+\frac{y}{b}=0,$$

$$OB:\ \frac{x}{a}-\frac{y}{b}=0.$$

A section parallel to and above the (x, y)-plane is a hyperbola whose vertices are on the parabola opening upwards, whereas a section parallel to and below the (x, y)-plane is a hyperbola whose vertices are on the parabola opening downwards. It is seen, then, that the surface is saddle-shaped; it rises along the parabola which opens upwards, and falls along the parabola which opens downwards. The (z, x)-plane contains the pommel and the (y, z)-plane, the stirrups.

The surface can best be plotted by drawing the sections parallel to a vertical coördinate plane, for example, the (y, z)-plane. These sections are all parabolas opening downwards and having their vertices on the parabola in the (z, x)-plane. Figure 8 shows part of the surface constructed by means of them.

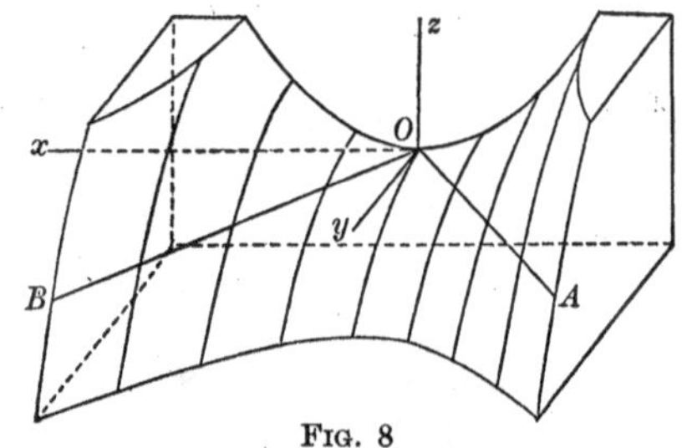

Fig. 8

Vertex, Axis, Principal Planes. Each paraboloid is symmetric in only one line, the axis of z, and in only two planes, the vertical coördinate planes. The line is known as the *axis*, and the planes as the *principal planes*. The sections by the principal planes are called the *principal sections*, and the point O, the *vertex*.

EXERCISES

Construct accurately the following paraboloids.

1. $\frac{x^2}{9}+\frac{y^2}{4}=2z.$ 2. $\frac{x^2}{9}-\frac{y^2}{4}=2z.$

3. $2x^2+3z^2=12y.$ 4. $x^2-4y^2=-8z.$

5. Prove that the sections of a paraboloid of either type, which are parallel to a principal plane, are *equal* and similarly placed parabolas.

6. Prove that the elliptic paraboloid (1) can be generated by the parabola $x^2=2a^2z$ moving so that its vertex traces the parabola $y^2=2b^2z$ in $x=0$, while its axis remains vertical and its plane parallel to the (z, x)-plane.

7. Describe and prove a method of generating the hyperbolic paraboloid, which is similar to that given in Ex. 6 for the elliptic paraboloid.

8. Show that the equation $xy=a^2z$ represents a hyperbolic paraboloid.

4. Rulings. *The Hyperboloid of One Sheet.* The equation

$$\frac{x^2}{a^2}+\frac{y^2}{b^2}-\frac{z^2}{c^2}=1 \tag{1}$$

can be written in the form

$$\left(\frac{y}{b}+\frac{z}{c}\right)\left(\frac{y}{b}-\frac{z}{c}\right)=\left(1+\frac{x}{a}\right)\left(1-\frac{x}{a}\right). \tag{2}$$

Consider the equations

$$U_1: \qquad \frac{y}{b}+\frac{z}{c}=u\left(1+\frac{x}{a}\right), \qquad u\left(\frac{y}{b}-\frac{z}{c}\right)=1-\frac{x}{a},$$

obtained from (2) by setting the first factor on the left equal to the parameter u times the first factor on the right, and then the second factor on the right equal to u times the second factor on the left.

These equations represent a one-parameter family of lines, each line being given by a particular value of the parameter u. All the lines lie on the surface (1). For, if P is an arbitrary point of the line $u = u_0$, the coördinates (x, y, z) of P satisfy equations U_1 for $u = u_0$; hence they also satisfy equation (2), since, if $u = u_0$ is eliminated from equations U_1 by multiplying them together, side for side, the result is precisely equation (2).

There will be just one line U_1 through an arbitrary point (x_0, y_0, z_0) of the surface (1) if the equations

$$(3) \qquad \frac{y_0}{b} + \frac{z_0}{c} = u\left(1 + \frac{x_0}{a}\right), \qquad u\left(\frac{y_0}{b} - \frac{z_0}{c}\right) = 1 - \frac{x_0}{a}$$

have a unique simultaneous solution for u. Let us see when this is the case.

If $1 + x_0/a \neq 0$, the first equation determines u uniquely, and the value obtained is seen to satisfy the second, since, when it is substituted in the second, this equation takes on the form (2) for $(x, y, z) = (x_0, y_0, z_0)$. In this case, then, there is just one line U_1 through (x_0, y_0, z_0).

If $1 + x_0/a = 0$, but $y_0/b - z_0/c \neq 0$, it follows from (2) that $y_0/b + z_0/c = 0$. Then the first equation of (3) is satisfied, no matter what value u has. The second equation determines u uniquely, and so in this case, too, there is just one line U_1 through (x_0, y_0, z_0).

Finally, if $1 + x_0/a = 0$ and $y_0/b - z_0/c = 0$, at least one of the equations (3) is contradictory and there is no line U_1 through (x_0, y_0, z_0). It is, however, natural to supplement the lines U_1 by the line

$$U_0: \qquad \frac{y}{b} - \frac{z}{c} = 0, \qquad 1 + \frac{x}{a} = 0,$$

for, if we divide each of the equations U_1 by u and then allow u to become infinite, the line U_1 approaches U_0 as its limit.

We have proved, then, that the lines U consisting of the family of lines U_1 and the line U_0 fill out the surface just once. They form what is called a *set of rectilinear generators* or *rulings* of the surface.

There is a second set of rulings, V, consisting of the family of lines

$$V_1: \qquad \frac{y}{b}-\frac{z}{c}=v\left(1+\frac{x}{a}\right), \qquad v\left(\frac{y}{b}+\frac{z}{c}\right)=1-\frac{x}{a},$$

and the line

$$V_0: \qquad \frac{y}{b}+\frac{z}{c}=0, \qquad 1+\frac{x}{a}=0.$$

It is readily seen that this set has the same properties as the set U. Hence we have the theorem :

THEOREM 1. *A hyperboloid of one sheet contains two sets of rulings. Through each point of the surface passes one ruling of each set.*

It is conceivable that the ruling U and the ruling V which go through the same point coincide. This is not the case, however, as will appear later ;* cf. Theorem 5.

The lines through the origin parallel to the lines U have the equations

$$(4) \qquad \begin{cases} \dfrac{y}{b}+\dfrac{z}{c}=u\dfrac{x}{a}, \\ u\left(\dfrac{y}{b}-\dfrac{z}{c}\right)=-\dfrac{x}{a}; \end{cases} \qquad \begin{cases} \dfrac{y}{b}-\dfrac{z}{c}=0, \\ x=0. \end{cases}$$

Their locus, obtained by eliminating u, is the asymptotic cone

$$\frac{x^2}{a^2}+\frac{y^2}{b^2}-\frac{z^2}{c^2}=0.$$

Moreover, the lines fill out the cone just once, as can be shown by the method used in proving Theorem 1.

Similarly, the lines through the origin parallel to the lines V, *i.e.* the lines

$$(5) \qquad \begin{cases} \dfrac{y}{b}-\dfrac{z}{c}=v\dfrac{x}{a}, \\ v\left(\dfrac{y}{b}+\dfrac{z}{c}\right)=-\dfrac{x}{a}; \end{cases} \qquad \begin{cases} \dfrac{y}{b}+\dfrac{z}{c}=0, \\ x=0\,; \end{cases}$$

* It is not difficult to give a direct proof of the fact at this point.

also fill out the asymptotic cone just once. Consequently, we have proved the theorem:

THEOREM 2. *The lines which pass through the center of a hyperboloid of one sheet and are parallel to the rulings U (or V) are precisely the elements of the asymptotic cone. In other words, there is one and only one ruling of each set which is parallel to a given element of the cone, and conversely.*

From this theorem we can draw the following conclusions.

THEOREM 3. *No three rulings of one set are parallel to a plane.*

For otherwise there would be three elements of the cone lying in a plane, and this is impossible.

THEOREM 4. *Two rulings of one set neither intersect nor are parallel; that is, they are never coplanar.*

For, they are not parallel, since no two elements of the cone are parallel; and they do not intersect, since otherwise there would be a point on the surface, through which pass two rulings of the same set.

THEOREM 5. *Two rulings of different sets either intersect or are parallel; that is, they are always coplanar.*

For, first, the rulings of the two sets are parallel in pairs, since there is just one ruling of each set which is parallel to a given element of the cone. From equations (4) and (5) it appears that $u = u_0 (\neq 0)$ and $v = v_0 (\neq 0)$ determine a pair of parallel rulings if and only if $1 + u_0 v_0 = 0$, and that the ruling $u = 0$ is parallel to V_0, and the ruling $v = 0$ to U_0.

Secondly, two non-parallel rulings of different sets intersect in just one point. For, it is easily shown that the four equations U_1 and V_1* which define in pairs two rulings which are not parallel, *i.e.* for which $1 + uv \neq 0$, have one and just one simultaneous solution for x, y, z, namely,

$$(6) \qquad x = a\frac{1 - uv}{1 + uv}, \quad y = b\frac{u + v}{1 + uv}, \quad z = c\frac{u - v}{1 + uv}, \quad 1 + uv \neq 0.$$

* The proof in the special cases, in which U_0 or V_0 or both are involved, is left to the student.

If u and v take on all possible pairs of values for which $1 + uv \neq 0$, equations (6) give the coördinates of the points of intersection of all the lines U_1 with the lines V_1, that is, the coördinates of all the points of the surface (1) except those on the lines U_0 and V_0. They constitute, then, a *parametric representation* of the surface (1) in terms of the *two* parameters u and v.

The Hyperbolic Paraboloid. The equation

$$\frac{x^2}{a^2} - \frac{y^2}{b^2} = 2z \tag{7}$$

can be written as

$$\left(\frac{x}{a} + \frac{y}{b}\right)\left(\frac{x}{a} - \frac{y}{b}\right) = 2z. \tag{8}$$

Accordingly, *there are two sets of rulings on the hyperbolic paraboloid, namely:*

U: $\qquad \frac{x}{a} + \frac{y}{b} = u, \qquad u\left(\frac{x}{a} - \frac{y}{b}\right) = 2z;$

V: $\qquad \frac{x}{a} - \frac{y}{b} = v, \qquad v\left(\frac{x}{a} + \frac{y}{b}\right) = 2z.$

The first equation of U represents a plane parallel to or coincident with the plane AOz (Fig. 8):

$$\frac{x}{a} + \frac{y}{b} = 0.$$

Consequently, the rulings U lie one each in the planes parallel to (and including) the plane AOz. Moreover, they are the total intersection of these planes with the surface; for, if the first equation of U is solved with equation (8) of the surface, the result is precisely the second equation of U. Similarly, the rulings V lie one each in the planes parallel to (and including) the plane BOz:

$$\frac{x}{a} - \frac{y}{b} = 0,$$

and are the total intersection of these planes with the surface. The planes AOz and BOz are known as the *directrix planes.*

There is just one ruling of each set through each point of the surface, for the planes parallel to (and including) a directrix plane exhaust all points of space just once, and hence their lines of intersection with the surface exhaust all points of the surface just once.

It is easily shown that the direction components of a ruling U are a, $-b$, u, and that those of a ruling V are a, b, v. Since the two triples are never proportional, two rulings of different sets are never parallel or coincident. In particular, the two rulings which pass through one and the same point of the surface are distinct.

The following theorems are now easily proved.

THEOREM 6. *Three rulings of one set are always parallel to a plane.*

For, all the rulings of a set are parallel to a directrix plane.

THEOREM 7. *Two rulings of one set are never coplanar.*

For, they do not meet since they lie in parallel planes, and they are not parallel, as inspection of the direction components just found shows.

THEOREM 8. *Two rulings of different sets always intersect.*

For, their projections on the (x, y)-plane, being lines in the direction of OA and OB respectively, intersect in a point M. Now there is but one point, P, on the surface which projects into M, since a line perpendicular to the (x, y)-plane meets the surface just once. Consequently, the two rulings in question intersect at this point P.

The coördinates of P, found by solving the four equations U and V simultaneously for x, y, z, are

$$(9) \qquad x = a\frac{u+v}{2}, \qquad y = b\frac{u-v}{2}, \qquad z = \frac{uv}{2}.$$

These equations constitute a *parametric representation* of the surface (7) in terms of the parameters u and v; there are no exceptional points.

EXERCISES

1. Find the equations of the rulings which pass through the point (3, 2, 5) of the hyperboloid of one sheet of Ex. 1, § 2.

2. The same for the hyperbolic paraboloid of Ex. 2, § 3, the point on the surface being (9, 2, 4).

Exercises 3–5. Use considerations of symmetry in the proofs.

3. The two rulings through a point P on a principal section of a hyperboloid of one sheet are equally inclined to the plane of the section and lie in a plane M which is perpendicular to the plane of the section.

4. The same for a hyperbolic paraboloid.

5. If P and P' are points of a hyperboloid of one sheet which are symmetric in the center, the rulings through P are parallel to those through P'.

6. Assuming Th. 1, § 7, prove that the plane M of Ex. 3 passes through the tangent line at P to the principal section. Hence show that the projections of the rulings of either set on a principal plane are the tangents to the principal section in that plane.

7. The same for a hyperbolic paraboloid, applying the results of Ex. 4.

8. Prove that the plane determined by two parallel rulings of a hyperboloid of one sheet is tangent to the asymptotic cone along the element which is parallel to the two rulings.

9. Prove that there are no straight lines on (*a*) an ellipsoid; (*b*) a hyperboloid of two sheets; (*c*) an elliptic paraboloid.

5. Parallel Sections. Equations (1), (3), and (6), § (2), of a hyperboloid of one sheet, H_1, of the conjugate hyperboloid of two sheets, H_2, and of the common asymptotic cone, C, can be written as the one equation

$$\frac{x^2}{a^2}+\frac{y^2}{b^2}-\frac{z^2}{c^2}=\lambda, \tag{1}$$

where λ is given the values 1, -1, and 0 in turn. The three surfaces can thus be considered simultaneously.

We propose to determine the sections of the surfaces (1) by an arbitrary plane,

$$Ax + By + Cz = k. \tag{2}$$

At least one of the coefficients A, B, C is not zero. Assume that $C \neq 0$. Then the sections in question are also the curves in which the plane (2) meets the cylinders whose equations result from the elimination of z from equations (1) and (2), namely, the cylinders

$$\frac{x^2}{a^2} + \frac{y^2}{b^2} - \frac{(k - Ax - By)^2}{c^2C^2} = \lambda. \tag{3}$$

Considering the sections from this point of view, we conclude the following theorems.

THEOREM 1. *The section of a hyperboloid or a cone is a conic.*

For, the cylinders (3) are quadric cylinders, and a section of a quadric cylinder is a conic; cf. Ch. XXII, § 6, Th. 1.

THEOREM 2. *Two parallel sections of a hyperboloid or a cone are conics of the same type. They are, moreover, similar and similarly placed, or, in the case of two hyperbolas, each is similar and similarly placed either to the other or to the conjugate of the other.*

To prove this theorem, we fix our attention on one of the surfaces (1), say the hyperboloid H_1, and give to k two arbitrarily chosen values, k_1 and k_2, thus obtaining two arbitrary parallel sections of H_1. The coefficients of the quadratic terms in the two equations (3) which result are respectively equal, since these coefficients in the general equation (3) do not contain k. Hence, by p. 260, Ex. 40, the directrices of the cylinders defined by the two equations are similar and similarly placed conics, or, in the case of two hyperbolas, each is similar or similarly placed either to the other or to the conjugate of the other. Consequently, by Ch. XXII, § 6, Th. 2, this is true also of the sections of the cylinders by the two planes, q. e. d.

THEOREM 3. *The sections of two conjugate hyperboloids and the common asymptotic cone by the same plane or by parallel planes are similar and similarly placed conics, or, in the case of hyperbolas, one of any two is similar or similarly placed either to the other or to the conjugate of the other.*

It is sufficient to prove the theorem for the sections of the three surfaces by a single plane, since its truth for sections by parallel planes will then follow from Theorem 2. Here, then, k is fixed, and λ takes on successively the values $1, -1, 0$. But the coefficients of the quadratic terms in (3) do not contain λ, and hence we reach the desired conclusion immediately, by reasoning identical with that used in the proof of Theorem 2.

The following theorem is now obvious.

THEOREM 4. *A plane which intersects a hyperboloid or a cone, but not in a non-degenerate conic, cuts it in a degenerate conic, which is of the same type as any section by a parallel plane.*

Accordingly, to ascertain the type of conic (degenerate or non-degenerate) in which a plane intersects a hyperboloid, it is necessary merely to determine the type of degenerate conic in which the parallel plane through the center meets the asymptotic cone. But the planes through the center intersect the cone in degenerate conics of all three types. Consequently, a hyperboloid has sections of all three types.

EXERCISES

1. Show that every plane section of an ellipsoid is an ellipse and that parallel sections are similar and similarly placed ellipses.

2. Prove that an elliptic paraboloid has no hyperbolic sections, that sections by parallel planes cutting the axis are similar and similarly placed ellipses, and that sections by

parallel planes parallel to the axis are *equal* and similarly placed parabolas.

3. Prove that a hyperbolic paraboloid has no elliptic sections, that sections by two parallel planes cutting the axes are hyperbolas, one of which is similar and similarly placed either to the other or to the conjugate of the other, and that sections by parallel planes parallel to the axis, but not to a directrix plane, are *equal* and similarly placed parabolas.

6. Circular Sections. Consider a section of the ellipsoid

$$(1) \qquad \frac{x^2}{a^2}+\frac{y^2}{b^2}+\frac{z^2}{c^2}=1, \qquad a > b > c,$$

by a plane M passing through the axis of y. The section is an ellipse, one of whose axes is always the mean axis, $2b$, of the ellipsoid, no matter how M is situated. When M, starting from the (x, y)-plane, rotates in either direction into coincidence with the (y, z)-plane, the second axis of the ellipse, starting from the major axis, $2a$, of the ellipsoid, decreases continuously to the minor axis, $2c$. Consequently, there must be a single position of M, in each direction of rotation, for which the second axis of the ellipse takes on the value $2b$ equal to the first. But then the ellipse is a circle.

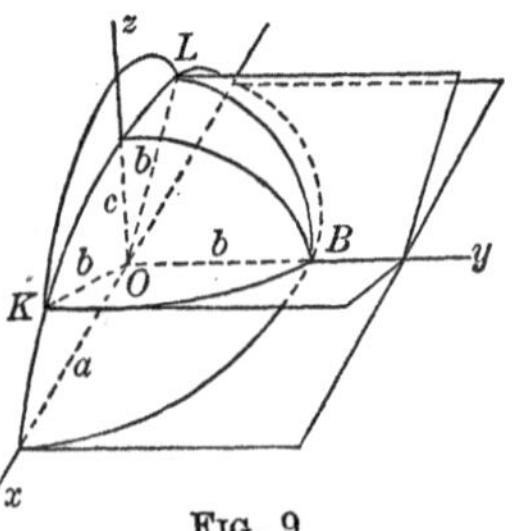

Fig. 9

These two positions, KOB and LOB, of the plane M can be constructed by describing in the upper half of the (z, x)-plane a semicircle whose center is at O and whose radius is b. The semicircle will meet the ellipsoid in the desired points K and L.

Since the sections of (1) by the planes KOB and LOB are circles, so also are the sections by planes parallel to KOB and LOB, by § 5, Ex. 1. The ellipsoid has, then, these two sets of circular sections and, as can be shown (Ex. 1), only these two.

It can be proved in the same way that the hyperboloid of one sheet

(2) $$\frac{x^2}{a^2}+\frac{y^2}{b^2}-\frac{z^2}{c^2}=1, \qquad a>b,$$

contains just two sets of circular sections. Hence it follows, by § 5, Th. 3, that this is true also of the cone and the hyperboloid of two sheets. It is to be noted, however, that there are no circular sections of the cone by planes through the vertex and none of the hyperboloid of two sheets by planes through the center.

The results obtained we now consolidate into a theorem.

THEOREM. *An ellipsoid, a hyperboloid, or a cone, which is not a surface of revolution, contains just two sets of circular sections.*

If, in Fig. 9, b approaches a as its limit, the planes KOB and LOB both approach as their limits the (x, y)-plane. Consequently, an ellipsoid of revolution has but one set of circular sections. This is true also of the hyperboloids and cones of revolution.

The circles in which the planes KOB, LOB intersect the ellipsoid evidently lie on the sphere whose center is at O and whose radius is b:

(3) $$\frac{x^2}{b^2}+\frac{y^2}{b^2}+\frac{z^2}{b^2}=1.$$

They therefore lie on the surface whose equation results from subtracting (1) from (3):

$$x^2\left(\frac{1}{b^2}-\frac{1}{a^2}\right)+z^2\left(\frac{1}{b^2}-\frac{1}{c^2}\right)=0,$$

or $$c^2(a^2-b^2)x^2-a^2(b^2-c^2)z^2=0.$$

But this surface consists of the two planes

(4) $$c\sqrt{a^2-b^2}\,x\pm a\sqrt{b^2-c^2}\,z=0.$$

Consequently, these are the equations of the planes KOB, LOB.

EXERCISES

1. To show that the ellipsoid (1) has but two sets of circular sections, prove first, using the fact that the centers of the circles of any set lie on a line (§ 8, Problem 1), that every circular section must be symmetric in a principal plane; then show that a section of (1) by a plane passing through the x-axis or the z-axis is never a circle.

2. Prove geometrically that the hyperboloid (2) has two sets of circular sections. Give a construction for the planes through the origin which yield circular sections.

3. Find the equations of the planes just mentioned.

4. Show that the elliptic paraboloid (1), § (3), where $a > b$, has two sets of circular sections, by proving first that this is true of the elliptic cylinder

$$\frac{x^2}{a^2}+\frac{y^2}{b^2}=1, \qquad a > b.$$

5. A hyperbolic paraboloid has no circular sections. Why?

7. Tangent Lines and Planes. Let the line L through the point $P_0:(x_0,\ y_0,\ z_0)$ with the direction cosines $\cos\alpha=\lambda$, $\cos\beta=\mu$, $\cos\gamma=\nu$ meet the ellipsoid

$$\text{(1)} \qquad \frac{x^2}{a^2}+\frac{y^2}{b^2}+\frac{z^2}{c^2}=1$$

in two distinct points, P_1 and P_2. To find the coördinates, $(x_1,\ y_1,\ z_1)$ and $(x_2,\ y_2,\ z_2)$, of P_1 and P_2.

The parametric representation of L is

$$\text{(2)} \qquad x=x_0+\lambda r, \qquad y=y_0+\mu r, \qquad z=z_0+\nu r,$$

where r is the algebraic distance from P_0 to $P:(x,\ y,\ z)$; cf. Ch. XX, § 8. The point P of L lies on the ellipsoid, if and only if its coördinates $(x,\ y,\ z)$, as given by (2), satisfy (1), that is, if and only if r satisfies the equation

$$(3)\qquad \left(\frac{\lambda^2}{a^2}+\frac{\mu^2}{b^2}+\frac{\nu^2}{c^2}\right)r^2+2\left(\frac{x_0\lambda}{a^2}+\frac{y_0\mu}{b^2}+\frac{z_0\nu}{c^2}\right)r$$
$$+\left(\frac{x_0^2}{a^2}+\frac{y_0^2}{b^2}+\frac{z_0^2}{c^2}-1\right)=0.$$

Since L intersects the ellipsoid in two distinct points, (3) has two distinct roots. If we denote them by r_1 and r_2, the coördinates of P_1 and P_2 are

$$x_1 = x_0 + \lambda r_1, \qquad y_1 = y_0 + \mu r_1, \qquad z_1 = z_0 + \nu r_1;$$
$$x_2 = x_0 + \lambda r_2, \qquad y_2 = y_0 + \mu r_2, \qquad z_2 = z_0 + \nu r_2.$$

Tangent Line. Suppose, now, that P_0 lies on the ellipsoid. Then one of the points of intersection of L, say P_2, coincides with P_0. Analytically, we have

$$(4)\qquad \frac{x_0^2}{a^2}+\frac{y_0^2}{b^2}+\frac{z_0^2}{c^2}=1,$$

so that the absolute term in (3) is zero; also, $r_2 = 0$ and r_1 is the distance from P_0 to P_1.

Imagine a curve drawn on the surface through P_0 and P_1, for example, an arbitrary plane section, C, through P_0 and P_1. The line L is the secant P_0P_1 of C and its limiting position, as P_1 moving along C approaches P_0 as its limit, is the tangent to C at P_0. We define this tangent as the *tangent line to the surface at P_0 in the direction of the curve C.*

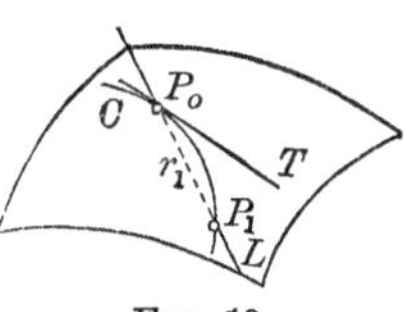

FIG. 10

When P approaches P_0, then no matter what curve C of approach is chosen r_1 approaches zero. But, when r_1 approaches zero, the coefficient of r in (3) approaches zero, and conversely. Consequently, *the line L is a tangent to the ellipsoid* (1) *at the point P_0 on the ellipsoid if and only if*

$$(5)\qquad \frac{x_0\lambda}{a^2}+\frac{y_0\mu}{b^2}+\frac{z_0\nu}{c^2}=0.$$

Tangent Plane. There are evidently infinitely many lines L tangent to the surface at P_0. For them λ, μ, ν have varying values, which, however, always satisfy (5). To obtain the locus of all the tangent lines L, we have only to eliminate the *auxiliary variables* λ, μ, ν, r from equations (2) and (5). Substituting the values of λ, μ, ν as given by (2) into (5) and suppressing the factor $1/r$, we get the equation

$$\frac{x_0(x-x_0)}{a^2}+\frac{y_0(y-y_0)}{b^2}+\frac{z_0(z-z_0)}{c^2}=0,$$

which reduces, by virtue of (4), to

$$(6) \qquad \frac{x_0x}{a^2}+\frac{y_0y}{b^2}+\frac{z_0z}{c^2}=1.$$

But this is the equation of a plane. Hence we have the theorem:

THEOREM 1. *The tangent lines at a point P_0 of an ellipsoid all lie in a plane.*

The plane of the tangent lines at P_0 we define as the *tangent plane* to the ellipsoid at P_0. Its equation is given by (6).

EXERCISES

Find for each of the following surfaces the condition that a line L through a point P_0 of the surface be tangent to the surface. Prove the analogue of Theorem 1 and deduce the equation of the tangent plane at P_0.

1. The unparted hyperboloid. 2. The biparted hyperboloid.
3. The elliptic paraboloid. 4. The hyperbolic paraboloid.
5. The cone, P_0 not being at the vertex.

6. Prove that the tangent plane to a hyperboloid of one sheet at a point P_0 is the plane determined by the rulings which pass through P_0. Hence show that the two sets of rulings found in § 4 exhaust all the straight lines on the surface.

7. The same for a hyperbolic paraboloid.

8. Let Q be a quadric surface, not a cone or a cylinder. Prove that a plane is tangent to Q if and only if it intersects

Q in a point (a degenerate ellipse) or in two intersecting lines (a degenerate hyperbola).

8. Diameters. Diametral Planes. *Problem* 1. Find the locus of the centers of parallel sections of the ellipsoid

$$(1) \qquad \frac{x^2}{a^2}+\frac{y^2}{b^2}+\frac{z^2}{c^2}=1.$$

Let the common normals to the planes of the sections have the direction components A, B, C. Let $P:(X, Y, Z)$ be the center of one of the sections and let L, with the direction cosines λ, μ, ν, be an arbitrary line through P, which lies in the plane of this section. The parametric equations of L are, then,

$$(2) \qquad x = X+\lambda r, \qquad y = Y+\mu r, \qquad z = Z+\nu r,$$

where r is the algebraic distance from P to (x, y, z).

Since the points of intersection of L with the ellipsoid are equally distant from P, their algebraic distances, r_1 and r_2, from P are negatives of each other: $r_1+r_2=0$. But r_1, r_2 are the roots of the quadratic equation

$$(3) \qquad \left(\frac{\lambda^2}{a^2}+\frac{\mu^2}{b^2}+\frac{\nu^2}{c^2}\right)r^2+2\left(\frac{\lambda X}{a^2}+\frac{\mu Y}{b^2}+\frac{\nu Z}{c^2}\right)r + \left(\frac{X^2}{a^2}+\frac{Y^2}{b^2}+\frac{Z^2}{c^2}-1\right)=0,$$

and consequently, by Ch. XIII, § 5,

$$(4) \qquad \lambda\frac{X}{a^2}+\mu\frac{Y}{b^2}+\nu\frac{Z}{c^2}=0.$$

This equation says that the direction whose components are X/a^2, Y/b^2, Z/c^2 is always perpendicular to L. But L is an arbitrary line in one of the planes of the sections and the only direction which is always perpendicular to it is that of the normals to these planes, that is, the direction whose components are A, B, C. Consequently, X/a^2, Y/b^2, Z/c^2 are proportional to A, B, C, or

$$(5) \qquad \frac{X}{a^2A}=\frac{Y}{b^2B}=\frac{Z}{c^2C}.$$

The centers of the sections lie, then, on a line through the center of the ellipsoid. Such a line is known as a *diameter.* Accordingly, we can state our result as follows.

THEOREM 1. *The locus of the centers of parallel sections of the ellipsoid* (1) *is that portion of a diameter which lies within the ellipsoid. If the direction components of the normals to the planes of the sections are A, B, C, those of the diameter are a^2A, b^2B, c^2C.*

Exercise. The tangent planes at the extremities of a diameter are parallel to the sections whose centers the diameter contains.

Problem 2. Find the locus of the mid-points of a set of parallel chords of the ellipsoid (1).

Let λ, μ, ν be the given direction cosines of one of the chords and let $P:(X, Y, Z)$ be the mid-point of the chord. Then equations (2) represent the chord parametrically.

Since P is the mid-point of the chord, the algebraic distances, r_1 and r_2, from it to the end points of the chord are negatives of each other: $r_1 + r_2 = 0$. Hence, as in Problem 1, we obtain the equation (4). But, whereas in that problem λ, μ, ν were auxiliary variables, here they are given constants, and hence (4) is the equation satisfied by the point P of the locus. But (4) represents a plane through the center of the ellipsoid. Such a plane is known as a *diametral plane.* Thus our result is:

THEOREM 2. *The locus of the mid-points of a set of parallel chords of the ellipsoid* (1) *is that portion of a diametral plane lying within the ellipsoid. If the direction components of the chords are l, m, n, the equation of the diametral plane is*

$$\frac{lx}{a^2} + \frac{my}{b^2} + \frac{nz}{c^2} = 0.$$

Conjugate Diameters and Diametral Planes.

THEOREM 3. *If a diameter D contains the centers of sections parallel to a diametral plane M, then M bisects the chords parallel to D, and conversely.*

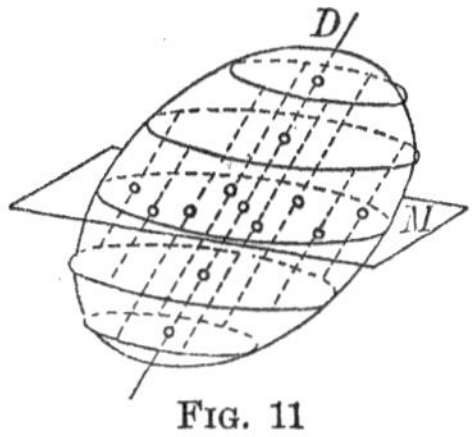

For, let D have the direction components l, m, n, and let M be the plane

$$Ax + By + Cz = 0.$$

The condition that D contain the centers of sections parallel to M is, by Th. 1, that

FIG. 11

$$l : m : n = a^2A : b^2B : c^2C.$$

The condition that M bisect the chords parallel to D is, by Th. 2, that

$$A : B : C = \frac{l}{a^2} : \frac{m}{b^2} : \frac{n}{c^2}.$$

The two conditions can both be written in the form

$$(6) \qquad \frac{l}{a^2A} = \frac{m}{b^2B} = \frac{n}{c^2C},$$

and are, therefore, equivalent, q. e. d.

A diameter D and a diametral plane M in the relationship described are said to be *conjugate*. We have, then, the following theorem.

THEOREM 4. *The diameter D with the direction components l, m, n and the diametral plane $Ax + By + Cz = 0$ are conjugate if and only if*

$$(6) \qquad \frac{l}{a^2A} = \frac{m}{b^2B} = \frac{n}{c^2C}.$$

Exercise. Show that an axis and the principal plane perpendicular to it are conjugate, and that in no other case is D perpendicular to its conjugate, M.

THEOREM 5. *If two diameters, D_1 and D_2, are conjugate in the ellipse E in which their plane meets the ellipsoid, each lies in the diametral plane of the other.*

For, since D_1 is conjugate to D_2 in E, D_1 bisects the chords of E parallel to D_2. But the diametral plane M_2 conjugate to D_2 also bisects these chords. Hence D_1 must be the line in which M_2 meets the plane of E, and so D_1 lies in M_2. Similarly, D_2 lies in the diametral plane, M_1, conjugate to D_1.

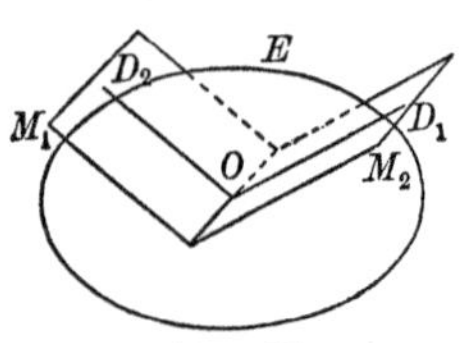

FIG. 12

THEOREM 6. *If one diameter lies in the diametral plane conjugate to a second, then the second diameter lies in the diametral plane conjugate to the first.*

Suppose that D_1 lies in the diametral plane M_2 conjugate to D_2. It will follow, then, by Th. 5, that D_2 lies in the diametral plane M_1 conjugate to D_1, if we can show that D_1 and D_2 are conjugate diameters in the ellipse E (Fig. 12). This is the case, for, since M_2 bisects all chords parallel to D_2, then D_1 bisects all chords of E parallel to D_2.

Conjugate Diameters. Conjugate Diametral Planes. Given three diameters D_1, D_2, D_3 and three diametral planes M_1, M_2, M_3 such that D_1, D_2, D_3 are the lines of intersection of M_1, M_2, M_3 or M_1, M_2, M_3 are the planes determined by D_1, D_2, D_3 (Fig. 13). Consider the following relationships:

FIG. 13

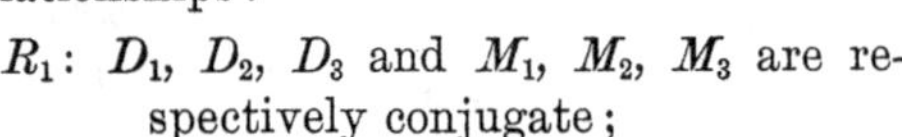

R_1: D_1, D_2, D_3 and M_1, M_2, M_3 are respectively conjugate;

R_2: Each diameter contains the centers of sections parallel to the plane of the other two;

R_3: Each diametral plane bisects the chords parallel to the line of intersection of the other two.

According to the definition of conjugacy of a diameter and a diametral plane, these relationships are equivalent:

THEOREM 7. *Any one of the relationships R is equivalent to each of the other two; that is, if any one holds, so does each of the other two.*

Three diameters in the relationship R_2 are called *conjugate diameters*, and three diametral planes in the relationship R_3 are called *conjugate diametral planes.*

There are infinitely many sets of three diameters and three diametral planes in the relationship R_1. For, let E be an arbitrary section of the ellipsoid by a plane through O, and let D_1 and D_2 be any two diameters conjugate in E. Then the diametral planes M_1 and M_2 conjugate to D_1 and D_2 will, by Th. 5, pass through D_2 and D_1 respectively. Finally, the diametral plane, M_3, conjugate to the diameter, D_3, in which M_1 and M_2 intersect will, by Th. 6, contain D_1 and D_2 and hence must be the plane of D_1 and D_2.

The following theorems follow directly by application of Theorems 2 and 1, respectively.

THEOREM 8. *The diameters with the direction components* $l_1, m_1, n_1,\quad l_2, m_2, n_2,\quad l_3, m_3, n_3$ *are conjugate if and only if*

$$\frac{l_1l_2}{a^2}+\frac{m_1m_2}{b^2}+\frac{n_1n_2}{c^2}=0,$$

$$\frac{l_2l_3}{a^2}+\frac{m_2m_3}{b^2}+\frac{n_2n_3}{c^2}=0,$$

$$\frac{l_3l_1}{a^2}+\frac{m_3m_1}{b^2}+\frac{n_3n_1}{c^2}=0.$$

THEOREM 9. *The diametral planes*

$$A_1x+B_1y+C_1z=0,\ A_2x+B_2y+C_2z=0,\ A_3x+B_3y+C_3z=0$$

are conjugate if and only if

$$a^2A_1A_2+b^2B_1B_2+c^2C_1C_2=0,$$

$$a^2A_2A_3+b^2B_2B_3+c^2C_2C_3=0,$$

$$a^2A_3A_1+b^2B_3B_1+c^2C_3C_1=0.$$

EXERCISES

State and prove for the following surfaces the theorems analogous to Theorems 1, 2.

1. The unparted hyperboloid.

2. The biparted hyperboloid.

3. The elliptic paraboloid. Show, in particular, that the diameters and diametral planes are all parallel to the axis.

4. The hyperbolic paraboloid. Describe the positions of the diameters and diametral planes.

Give reasons for the following exceptions to the theorems just proved.

5. *Hyperboloids.* There is no analogue to Theorem 1 for parallel sections parallel to an element of the asymptotic cone, and no analogue to Theorem 2 for lines parallel to an element of the asymptotic cone.

6. *Paraboloids.* There is no analogue to Theorem 1 for parallel sections parallel to the axis, and no analogue to Theorem 2 for lines parallel to the axis. In the case of a hyperbolic paraboloid there is also no analogue to Theorem 2 for a set of parallel lines parallel to a directrix plane.

State and prove for the following surfaces the theorems analogous to Theorems 3, 4.

7. The unparted hyperboloid.

8. The biparted hyperboloid.

9. There are no conjugate diameters and diametral planes for a paraboloid. Why?

10. Find for the ellipsoid (1) the equation of the diametral plane conjugate to the diameter through the point (x_0, y_0, z_0) of the surface.

11. Prove that the pairs of conjugate diameters and diametral planes are the same for two conjugate hyperboloids.

12. Discuss the conjugacy of three diameters or three diametral planes for either hyperboloid.

13. Prove Theorems 8, 9.

14. State and prove the analogues of Theorems 8, 9 for either hyperboloid.

9. Poles and Polars. Through a point $P_0: (x_0, y_0, z_0)$ not on the ellipsoid:

$$(1) \qquad \frac{x^2}{a^2}+\frac{y^2}{b^2}+\frac{z^2}{c^2}=1$$

an arbitrary line L is drawn meeting the ellipsoid in Q_1 and Q_2. What is the locus of the point P which with P_0 divides Q_1Q_2 harmonically?

By the method used in solving the corresponding problem in the plane, Ch. XIV, § 9, the locus is found to be the plane

$$(2) \qquad \frac{x_0x}{a^2}+\frac{y_0y}{b^2}+\frac{z_0z}{c^2}=1,$$

or a portion of this plane.

The point P_0 (not on the ellipsoid) and the plane (2) are said to be *pole* and *polar* in the ellipsoid: P_0 is the *pole* of (2), and (2) the *polar* of P_0. A point on the ellipsoid and the tangent plane at the point are defined to be pole and polar.

By the methods of Ch. XIV, §§ 9–11, the following theorems can now be proved.

THEOREM 1. *Let Q be a central quadric (an ellipsoid or hyperboloid). Every point in space, except the center of Q, has a polar with respect to Q.*

THEOREM 2. *Every point in space has a polar with respect to a paraboloid.*

THEOREM 3. *Let Q be a central quadric or a paraboloid. Every plane in space, which is not a diametral plane of Q, has a pole with respect to Q.*

The poles and polars considered in the following theorems are taken with respect to an arbitrarily chosen central quadric or paraboloid.

THEOREM 4*a*. *If one point lies in the polar plane of a second, the second point lies in the polar plane of the first.*

THEOREM 4*b*. *If one plane contains the pole of a second, the second plane contains the pole of the first.*

THEOREM 5*a*. *If a number of points lie on a line L, their polar planes pass through a line L' or are parallel.*

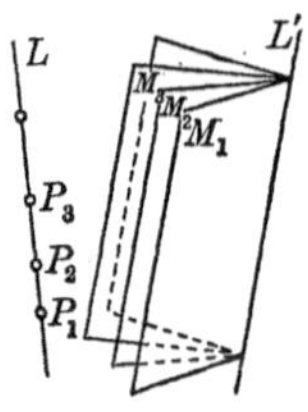

FIG. 14

THEOREM 5*b*. *If a number of planes pass through a line L' (or are parallel), their poles lie on a line L.*

Theorems 5*a*, 5*b* are peculiar to the geometry of space. We give a proof of Theorem 5*a*.

Let $P_1:(x_1,\ y_1,\ z_1)$, $P_2:(x_2,\ y_2,\ z_2)$ be two distinct points. Their polar planes are

$$u \equiv \frac{x_1x}{a^2}+\frac{y_1y}{b^2}+\frac{z_1z}{c^2}-1=0, \qquad v \equiv \frac{x_2x}{a^2}+\frac{y_2y}{b^2}+\frac{z_2z}{c^2}-1=0.$$

Let $P_3:(x_3,\ y_3,\ z_3)$ be an arbitrary point of the line P_1P_2. Then, by Ch. XXI, § 2,

$$x_3=\rho x_1+(1-\rho)x_2,\quad y_3=\rho y_1+(1-\rho)y_2,\quad z_3=\rho z_1+(1-\rho)z_2.$$

Consequently, the polar plane of P_3, namely

$$\frac{x_3x}{a^2}+\frac{y_3y}{b^2}+\frac{z_3z}{c^2}-1=0,$$

can have its equation written in the form

$$\rho u+(1-\rho)v=0,$$

and hence passes through the line of intersection of the polar planes of P_1 and P_2, if they intersect, or is parallel to them, if they are parallel, q. e. d.

Two lines L and L', in the relationship described in Theorem 5*a* or 5*b*, are each said to be *polar* or *conjugate* to the other. The following theorems concerning polar lines are readily proved; cf. Ths. 8*a*, 8*b* of Ch. XIV, § 11.

THEOREM 6*a*. *The polar of a line intersecting the quadric in two distinct points is the line of intersection of the tangent planes at these points.*

THEOREM 6*b*. *The polar of a line not meeting the quadric is the line joining the points of contact of the two planes through the line tangent to the quadric.*

THEOREM 6c. *The polar of a tangent to a quadric, not a ruling, is a second tangent line with the same point of contact. A ruling of a quadric is self-polar.*

EXERCISES

Establish formula (2) and the analogous formulas for the other quadrics. Prove the theorems stated without proof in the text.

Discuss poles and polars with respect to a sphere (cf. Ch. XIV, § 9, Exs. 9–11), showing, in particular, that two polar lines are always perpendicular.

10. One-Dimensional Strains, with Applications.* The *one-dimensional strain* which stretches all space directly away from the (y, z)-plane (or compresses all space directly towards the (y, z)-plane), so that each point is carried, along a parallel to the axis of x, to a times its original distance from the (y, z)-plane, where a is a positive constant not unity, has the equations

i) $$x' = ax, \qquad y' = y, \qquad z' = z.$$

Similarly, the equations

ii) $$x' = x, \qquad y' = by, \qquad z' = z,$$

iii) $$x' = x, \qquad y' = y, \qquad z' = cz,$$

where b and c are positive constants different from unity, represent one-dimensional strains in the directions of the axes of y and z respectively.

One-dimensional strains have the following properties:

A. Planes go into planes, and hence straight lines go into straight lines;

B. Parallel planes go into parallel planes and hence parallel straight lines go into parallel straight lines;

C. Tangent surfaces go into tangent surfaces, and tangent curves into tangent curves.

* Cf. Ch. XIV, particularly § 5.

One-dimensional strains do not in general preserve angles or areas. They never preserve volumes; for example, i) carries a portion of space of volume V into a portion of space of volume aV.

The product of the three one-dimensional strains i), ii), iii) is the transformation

$$T: \qquad x' = ax, \qquad y' = by, \qquad z' = cz.$$

It is clear from the foregoing that T carries a portion of space of volume V into a portion of space of volume $abcV$.

Applications. The sphere

$$(1) \qquad x^2 + y^2 + z^2 = 1$$

is carried by the transformation T into the ellipsoid

$$(2) \qquad \frac{x'^2}{a^2} + \frac{y'^2}{b^2} + \frac{z'^2}{c^2} = 1.$$

The volume of the sphere is $\frac{4}{3}\pi$. That of the ellipsoid is, then, $\frac{4}{3}\pi abc$.

THEOREM 1. *The volume of the ellipsoid* (2) *is*

$$V = \tfrac{4}{3}\pi abc.$$

Let the triples

$$(3) \qquad \lambda_1, \mu_1, \nu_1, \qquad \lambda_2, \mu_2, \nu_2, \qquad \lambda_3, \mu_3, \nu_3$$

be the direction cosines of three mutually perpendicular (and hence conjugate) diameters, D_1, D_2, D_3, of the sphere (1). They are, then, also the coördinates of three points, P_1, P_2, P_3, on the sphere, which are respectively extremities of D_1, D_2, D_3.

Now T carries D_1, D_2, D_3 into three diameters, D_1', D_2', D_3', of the ellipsoid (2), and carries P_1, P_2, P_3 into three points, P_1', P_2', P_3', on the ellipsoid, which are respectively extremities of D_1', D_2', D_3'. Evidently, the triples

$$(4) \qquad a\lambda_1, b\mu_1, c\nu_1, \qquad a\lambda_2, b\mu_2, c\nu_2, \qquad a\lambda_3, b\mu_3, c\nu_3$$

are both the coördinates of P_1', P_2', P_3' and the direction components of D_1', D_2', D_3'.

Considered as the direction components of D_1', D_2', D_3', the triples (4) satisfy the conditions of Th. 8, § 8. Hence we have the theorem:

THEOREM 2. *The transformation T carries three mutually perpendicular (and therefore conjugate) diameters of the sphere* (1) *into three conjugate diameters of the ellipsoid* (2).

It follows, by Th. 7, § 8, that T carries three mutually perpendicular diametral planes of (1) into three conjugate diametral planes of (2), and carries a diameter and the perpendicular diametral plane of (1) into a diameter and the conjugate diametral plane of (2).

Considered as the coördinates of P_1', P_2', P_3', the triples (4) give immediately, as the squares of the half-lengths of the diameters D_1', D_2', D_3',

$$a^2\lambda_1^2 + b^2\mu_1^2 + c^2\nu_1^2, \quad a^2\lambda_2^2 + b^2\mu_2^2 + c^2\nu_2^2, \quad a^2\lambda_3^2 + b^2\mu_3^2 + c^2\nu_3^2.$$

The sum of these squares is

$$a^2(\lambda_1^2 + \lambda_2^2 + \lambda_3^2) + b^2(\mu_1^2 + \mu_2^2 + \mu_3^2) + c^2(\nu_1^2 + \nu_2^2 + \nu_3^2).$$

Since the triples (3) are the direction cosines of three mutually perpendicular lines, so also are the triples

$$\lambda_1, \lambda_2, \lambda_3, \qquad \mu_1, \mu_2, \mu_3, \qquad \nu_1, \nu_2, \nu_3;$$

cf. Ch. XXIV, § 6. Hence the above sum has the value

$$a^2 + b^2 + c^2$$

and is therefore independent of the three conjugate diameters D_1', D_2', D_3' taken.

THEOREM 3. *The sum of the squares of the lengths of three conjugate diameters of an ellipsoid is constant.*

EXERCISES

1. Prove analytically the properties A, B, C of one-dimensional strains.

2. What angles and what areas does the transformation i) preserve?

3. Prove that i) carries a region of volume V into a region of volume aV.

4. Show that T carries a line with the direction components l, m, n into a line with the direction components al, bm, cn.

5. The plane M goes into the plane M' under T. If the direction components of the normals to M are A, B, C, what are those of the normals to M'?

6. Assuming the equation of the tangent plane to the sphere (1) at the point (x_0, y_0, z_0), deduce by means of the transformation T the equation of the tangent plane to the ellipsoid (2) at the point (x_0', y_0', z_0').

7. Show that a hyperboloid of general type can always be carried into a hyperboloid of revolution by means of a transformation of the form T.

EXERCISES ON CHAPTER XXIII

1. Find the equation of the quadric surface generated by the lines $x - \lambda z = 0$, $\lambda y - z = 0$, where λ is a parameter. Determine the equations of the second set of rulings and set up a parametric representation of the surface.

2. The same for the lines $y - \lambda - 1 = 0$, $\lambda x - z + 2 = 0$.

3. Find the equations of the planes which pass through the line $y = 2$, $x + 2z = 0$ and are tangent to the ellipsoid

$$x^2 + 3y^2 + 2z^2 = 6.$$

4. Prove that the sections of the hyperbolic paraboloid and hyperbolic cylinder:

$$\frac{x^2}{a^2} - \frac{y^2}{b^2} = 2z, \qquad \frac{x^2}{a^2} - \frac{y^2}{b^2} = 1$$

by the same plane or by parallel planes, oblique to the z-axis, are hyperbolas, each of which is similar and similarly placed to the other or to the conjugate of the other.

5. The *umbilics* of a quadric surface which has circular sections are the extremities of the diameters which contain the

centers of these sections. Find their coördinates in the case of the ellipsoid (1), § 1.

Similar Quadrics

Definition. Two central quadrics are said to be *similar* if the principal sections of one are similar, respectively, to the principal sections of the other.

6. Prove that the ellipsoids defined by the equation

$$\frac{x^2}{a^2}+\frac{y^2}{b^2}+\frac{z^2}{c^2}=\lambda, \qquad \lambda > 0, \tag{1}$$

where λ is a parameter, are similar.

7. Show that, of the hyperboloids represented by the equation

$$\frac{x^2}{a^2}+\frac{y^2}{b^2}-\frac{z^2}{c^2}=\lambda, \qquad \lambda \neq 0, \tag{2}$$

those for which λ is positive are all similar, and that this is true also of those for which λ is negative. Prove that all the hyperboloids have the same asymptotic cone.

8. Prove that all the ellipsoids (1) have the same pairs of conjugate diameters and diametral planes.

9. The same for the hyperboloids (2).

Definition. Two paraboloids of the same type are *similar*, if the principal sections of one are proportional in scale (Ch. VI, § 1) to the principal sections of the other.

10. Prove in each case that the paraboloids defined by the given equation are all similar:

$$(a)\quad \frac{x^2}{a^2}+\frac{y^2}{b^2}=2\lambda z, \quad \lambda \neq 0; \qquad (b)\quad \frac{x^2}{a^2}-\frac{y^2}{b^2}=2\lambda z, \quad \lambda \neq 0.$$

Ruled Surfaces

11. Show that the pencil of planes through the ruling V_0 of the hyperboloid (1), § 4, cuts the surface in the set of rulings U and that the pencil of planes through U_0 cuts it in the rulings V.

12. Let P be a point on the minimum ellipse of the hyperboloid (1), § 4, and let ϕ be the common angle which the two rulings through P make with the z-axis. Prove that $\tan \phi = b_1/c$, where b_1 is the half-length of the diameter of the minimum ellipse which is conjugate to the diameter through P.

13. Using the result of Ex. 12, show that a hyperboloid of revolution of one sheet can be generated by the rotation of a ruling of either set about the axis which does not meet the surface.

14. Prove that the rulings of one set on a hyperbolic paraboloid intercept proportional segments on two rulings of the other set.

Loci

15. Find the locus of a point which moves so that its distance from a fixed point bears to its distance from a fixed plane, not through the point, a constant ratio, k.

Ans. A quadric of revolution which is an ellipsoid, an elliptic paraboloid, or a hyperboloid of two sheets, according as k is less than, equal to, or greater than unity.

16. A point moves so that its distance to a fixed point bears to its distance to a fixed line, not through the point, a constant ratio. Find its locus.

Exercises 17–19. In connection with these exercises, Exs. 28, 29, p. 522 will be found useful.

17. Find the locus of a point which moves so that its distances to two skew lines are always in the same ratio, k.

18. Prove that a line which is rotated about an axis skew to it generates a hyperboloid of revolution of one sheet; cf. Ex. 13.

19. Let L and L' be two fixed skew lines and let M and M' be two planes, which pass through L and L' respectively and so move that they are always mutually perpendicular. Find the locus of their line of intersection.

20. The locus of a line which so moves that it always intersects three fixed skew lines, not parallel to a plane, is a hyperboloid of one sheet. Prove this theorem in the case that the fixed lines are

$$\begin{cases} x = c, \\ y = z\cos\theta; \end{cases} \qquad \begin{cases} x = -c, \\ y = -z\cos\theta; \end{cases} \qquad \begin{cases} x = -z\cot\theta, \\ y = c\sin\theta, \end{cases}$$

where $c \neq 0$ and $\theta \neq 0, n\frac{\pi}{2}$.

21. The locus of a line which so moves that it always intersects three fixed skew lines, parallel to a plane, is a hyperbolic paraboloid. Prove this theorem in the case that one of the fixed lines is the axis of z and the others have the equations $x = c$, $z = my$; $x = -c$, $z = -my$, where $cm \neq 0$.

22. A line moving so that it is always parallel to a fixed plane, M, and always intersects two fixed skew lines, neither of which is parallel to M, generates a hyperbolic paraboloid. Prove this theorem when M is the (x, z)-plane and the two fixed lines are the last two of the three in Ex. 21.

CHAPTER XXIV

SPHERICAL AND CYLINDRICAL COÖRDINATES. TRANSFORMATION OF COÖRDINATES

1. Spherical Coördinates. Given a point O, a ray OA issuing from O, and a half-plane m bounded by the line of the ray OA. Let P be any point of space. Join P to O and construct the half-plane, p, determined by OA and OP. Denote the distance OP by r, the angle AOP by ϕ, and the angle from the half-plane m to the half-plane p by θ. Then (r, ϕ, θ) are the *spherical coördinates* of the point P.

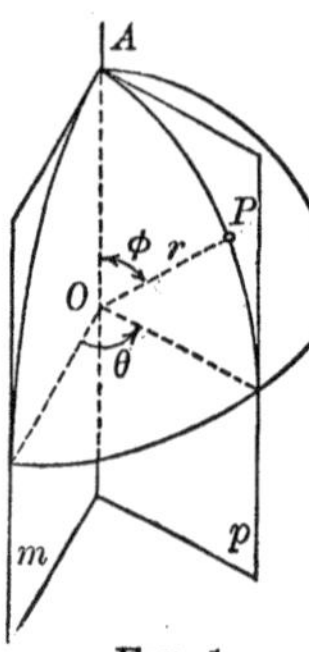

FIG. 1

For a given value, r_0, of the *radius vector* r, the point P lies on a sphere whose center is at O and whose radius is $r = r_0$. The angle θ is the *longitude* of P, measured from the *prime meridian* m, and the angle ϕ is the *colatitude* (complement of the latitude), at least for a point P on the upper half of the sphere.

The radius vector r is, by definition, positive or zero. The colatitude ϕ shall be restricted to values between 0 and π inclusive: $0 \leq \phi \leq \pi$. The longitude θ shall be unrestricted; it shall be taken as positive if measured in the direction shown, and as negative, if measured in the opposite direction.*

* It is possible to define spherical coördinates so that r or ϕ or both are also unrestricted. Systems of these extended types are not often necessary, and when exceptional need for them occurs, they can easily be introduced.

It is clear that the r- and ϕ-coördinates of a given point P are unique, while the θ-coördinate has infinitely many values, each two differing by an integral multiple of 2π.* Conversely, if r, ϕ, θ are given, such that $r \geq 0$ and $0 \leq \phi \leq \pi$, a unique point P is determined.

Let r_0, ϕ_0, θ_0 be particular values of r, ϕ, θ, such that $r_0 > 0$ and $0 < \phi_0 < \pi$. The equation $r = r_0$ represents a sphere, whose center is at O and whose radius is r_0; $\phi = \phi_0$ defines one nappe of a circular cone whose vertex is at O and whose axis lies along OA; finally $\theta = \theta_0$ represents a meridian half-plane issuing from the line of OA.

Transformation to and from Rectangular Coördinates. Let P be any point of space whose coördinates, referred to a system of rectangular axes, are (x, y, z). Let P have the spherical coördinates (r, ϕ, θ) with respect to O, OA, and m, as chosen in the figure. It is clear that

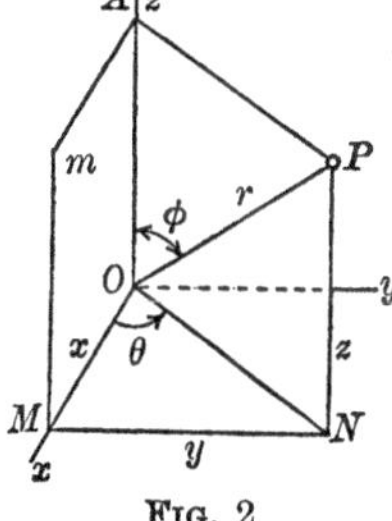

FIG. 2

$$x = ON \cos\theta, \qquad y = ON \sin\theta,$$

and $\quad ON = r \sin\phi, \qquad z = r\cos\phi.$

Hence the values of x, y, z in terms of r, ϕ, θ are

$$(1) \qquad x = r\sin\phi\cos\theta, \quad y = r\sin\phi\sin\theta, \quad z = r\cos\phi.$$

Since r is the distance from O to $P: (x, y, z)$, we have also that

$$(2) \qquad r^2 = x^2 + y^2 + z^2.$$

EXERCISES

1. Plot the points $(2, 90°, 180°)$, $(4, 60°, -30°)$, $(8, \frac{3}{4}\pi, \frac{4}{3}\pi)$.

2. Find the rectangular coördinates of the points of Ex. 1.

3. Find the spherical coördinates of the points $(0, 2, 0)$, $(3, 4, 12)$, $(2, -2, -1)$, checking each result by a figure.

* It is to be noted, however, that for every point P on the line of OA θ is undetermined and that for O in particular ϕ is also undetermined. Cf. Ch. X, § 1.

4. What do the following equations represent?

(a) $r=7$; (b) $\phi=30°$; (c) $\theta=\frac{5}{6}\pi$;
(d) $\tan\theta=1$; (e) $\tan\phi=1$; (f) $4\cos^2\phi=1$.

5. Find the equations in spherical coördinates of the following surfaces:

(a) The sphere of radius 5, center at O;
(b) The meridian half-plane of longitude 35°;
(c) The complete plane determined by this half-plane;
(d) The upper nappe of the circular cone whose vertex is at O, whose axis is along OA, and whose generating angle is 60°;
(e) The lower nappe of the cone of (d);
(f) The cone of (d).

6. What do the following pairs of equations represent?

(a) $r=3$, $\phi=120°$; (b) $r=3$, $\theta=\frac{5}{4}\pi$; (c) $\theta=30°$, $\phi=45°$;
(d) $r=3$, $\tan^2\phi=1$; (e) $\tan\theta=2$, $\phi=\frac{5}{6}\pi$; (f) $\tan\theta=-1$, $\cos^2\phi=\frac{1}{2}$.

7. Find the equations in spherical coördinates of the following curves:

(a) The small circle on the earth of colatitude 47°;
(b) The semicircle on the earth of longitude 135°;
(c) The complete circle determined by the semicircle of (b);
(d) The ray from O of colatitude 60° and longitude 25°.

8. Determine the locus of each of the following equations:
(a) $r=4\cos\phi$; (b) $r=6\sec\phi$; (c) $r=3\csc\phi$.

Find the equations in spherical coördinates of the following surfaces. Identify each surface.

9. $x^2+y^2+z^2=9$.
10. $x^2+y^2-k^2z^2=0$.
11. $x^2+y^2+z^2=4y$.
12. $4(x^2+y^2)+9z^2=36$.
13. $3x+2y=0$.
14. $3z-4=0$.
15. $2x+5=0$.
16. $x^2=2yz$.

2. Cylindrical Coördinates. Given a point O, the axis of z through O, and the plane K through O perpendicular to the axis of z. In K introduce a system of polar coördinates, as shown. Let P be any point of space and let N be its projection on K. Then the polar coördinates, r and θ, of N and the directed line-segment $NP = z$ determine the position of P. The three numbers, taken together, are known as the *cylindrical coördinates* (r, θ, z) of P.

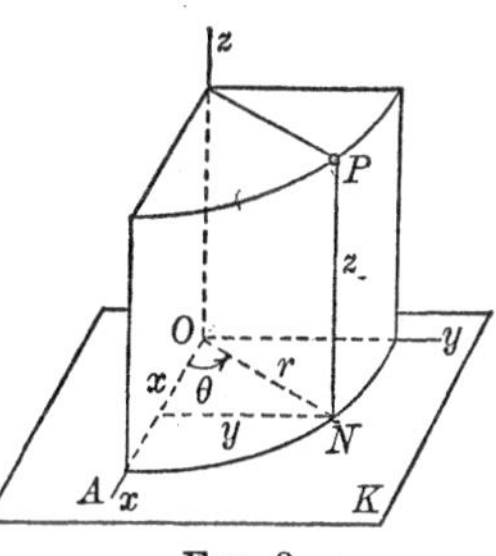

Fig. 3

As in the case of polar coördinates in the plane (Ch. X, § 1), r is restricted to be positive or zero, while θ is unrestricted. The positive direction of rotation for the measurement of θ is as indicated in the figure.

If r_0 (> 0), θ_0, z_0 are particular values of r, θ, z, the equation $r = r_0$ represents a circular cylinder whose axis is the axis of z; $\theta = \theta_0$ defines a half-plane issuing from the axis of z, and $z = z_0$ represents a plane perpendicular to the axis of z.

Transformation to and from Rectangular Coördinates. Choose in the plane K the Cartesian axes of x and y shown in Fig. 3. Referred to these axes and the axis of z, P has the rectangular coördinates (x, y, z).

It is clear that the z of the rectangular coördinates of P is precisely the z of the cylindrical coördinates of P. The formulas for x, y in terms of r, θ and for r, θ in terms of x, y are those of transformation in a plane from polar to rectangular coördinates, and *vice versa* (Ch. X, § 6). In particular,

(1) $$x = r\cos\theta, \qquad y = r\sin\theta,$$

(2) $$r^2 = x^2 + y^2.$$

EXERCISES

1. Plot the points $(2, 40°, 5)$, $(4, -\frac{3}{4}\pi, -3)$, $(0, 122°, 1)$.
2. Find the rectangular coördinates of the points of Ex. 1.

3. Find the cylindrical coördinates of the points (3, 4, 8), (12, – 5, – 3), (0, 0, – 6), checking each result by a figure.

4. What do the following equations represent?

(a) $r = 5$; (b) $\theta = 225°$; (c) $\tan\theta = 1$.

5. Find the equations in cylindrical coördinates of the following surfaces:

(a) The circular cylinder of radius 7 whose axis is the axis of z;

(b) The half-plane bounded by the axis of z, the angle from OA to it being 60°;

(c) The complete plane determined by this half-plane.

6. What do the following pairs of equations represent?

(a) $r = 3$, $\theta = -\frac{5}{6}\pi$; (c) $2z = 5$, $\theta = 120°$;
(b) $r = 5$, $z = -6$; (d) $3z - 8 = 0$, $\tan\theta = 2$.

7. Find the equations in cylindrical coördinates of the following curves:

(a) An arbitrary line parallel to the z-axis;

(b) The circle of radius 3, whose center is on the z-axis and whose plane is parallel to K and 5 units below it;

(c) An arbitrary ray perpendicular to the axis of z and issuing from a point on it;

(d) The line of this ray.

8. Determine the locus of each of the following equations:

(a) $r^2 + z^2 = 9$; (b) $r = 4\sin\theta$; (c) $r\sin\theta = 5$.

Find the equations in cylindrical coördinates of the following surfaces. Identify each surface.

9. Ex. 9, § 1. 10. Ex. 10, § 1. 11. Ex. 11, § 1.

12. Ex. 13, § 1. 13. Ex. 14, § 1. 14. Ex. 15, § 1.

15. $3(x^2 + y^2) - 2z^2 = 6$. 16. $xy + yz + zx = 0$.

17. Prove that a plane through OA (Fig. 1) or Oz (Fig. 3) is represented by the same equation in both spherical and cylindrical coördinates.

3. Triply Orthogonal Systems of Surfaces. Consider the three sets, or *families*, of planes which are parallel to the coördinate planes of a Cartesian system. These families of planes evidently have the following properties: (*a*) Through each point of space there passes just one plane of each family; (*b*) Two planes of different families intersect at right angles. We say, then, that the three families of planes form a *triply orthogonal system of planes.*

The equations of the families are, respectively,

$$(1) \qquad x = k, \qquad y = l, \qquad z = m,$$

where k, l, m are arbitrary constants, or *parameters*, each taking on any value, positive, zero, or negative.

Since through a point P: (x_0, y_0, z_0) there pass just three planes, one from each family, namely the planes $x = x_0$, $y = y_0$, $z = z_0$, and since, further, P is the only point which the three planes have in common, the position of P can be thought of as determined by the three planes. From this point of view, then, the basis of the rectangular coördinate system is seen to be the triply orthogonal system of planes (1).

In the case of a system of cylindrical coördinates, consider the three families of surfaces:

$$(2) \qquad r = k, \qquad \theta = l, \qquad z = m,$$

where, of the parameters k, l, m, k *cannot* be negative (or zero), l *may* be restricted to the range of values: $0 \leq l < 2\pi$, and m is unrestricted.

The first family of surfaces consists of the circular cylinders with the axis of z as axis; the second family is made up of the half-planes issuing from the axis of z; and the third, of the planes perpendicular to the axis of z. It is easily seen that through each point of space, with the exception of those on the z-axis, there passes just one surface of each family, and that two surfaces of different families intersect

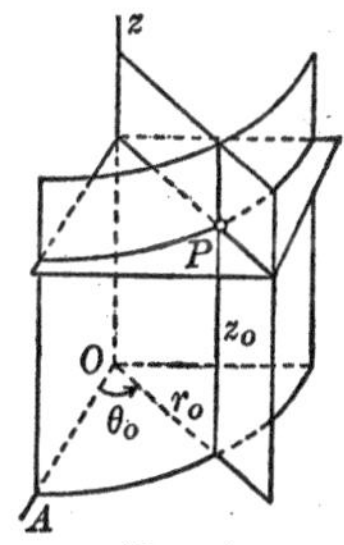

FIG. 4

orthogonally. We say, then, that the three families form a *triply orthogonal system of surfaces.*

A point $P: (r_0, \theta_0, z_0)$, not on the z-axis,* is the *single* point of intersection of the three surfaces, $r = r_0$, $\theta = \theta_0$, $z = z_0$ which pass through it. Thus the basis of cylindrical coördinates is the triply orthogonal system of surfaces (2).

EXERCISE

Write the equations of the three families of surfaces peculiar to a spherical coördinate system. Describe each family and draw a figure showing three surfaces, one from each family, their curves of intersection and their common point. Prove that the three families constitute a triply orthogonal system of surfaces and show that this system can be considered as the basis of spherical coördinates.

4. Confocal Quadrics. Consider the quadrics

$$(1) \qquad \frac{x^2}{a^2 - k} + \frac{y^2}{b^2 - k} + \frac{z^2}{c^2 - k} = 1, \qquad k < c^2,$$

$$(2) \qquad \frac{x^2}{a^2 - l} + \frac{y^2}{b^2 - l} - \frac{z^2}{l - c^2} = 1, \qquad c^2 < l < b^2,$$

$$(3) \qquad \frac{x^2}{a^2 - m} - \frac{y^2}{m - b^2} - \frac{z^2}{m - c^2} = 1, \qquad b^2 < m < a^2,$$

where a, b, c are positive constants such that $a > b > c$ and k, l, m are parameters subject to the given restrictions.

The surfaces (1) are all ellipsoids; the surfaces (2), all hyperboloids of one sheet opening out along the axis of z; and the surfaces (3), all hyperboloids of two sheets cutting the axis of x.

The coördinate planes are the principal planes of all the

* To remove this exception, add to the family of cylinders, $r = k > 0$, the axis of z, $r = 0$. Through a point P of this axis passes, then, one surface each from the first and third families, and every surface of the second. However, all these surfaces have but the *one* point P in common and hence can be considered as determining the position of P.

surfaces. It is readily shown that the sections by the (x, y)-plane of all the surfaces are *confocal* conics, the common foci being at the points $(\pm\sqrt{a^2 - b^2}, 0, 0)$. Similarly, for the sections by the (z, x)-plane of all the surfaces, and for the sections by the (y, z)-plane of the surfaces (1) and (2), — the (y, z)-plane does not cut the surfaces (3). This property of the surfaces (1), (2), (3) is expressed by calling them *confocal quadrics.*

It can be shown that through each point of space, with the exception of those in the coördinate planes, there passes just one surface of each type and that two surfaces of different types intersect orthogonally all along a curve.* Consequently, *the confocal quadrics form a triply orthogonal system of surfaces.*

This triply orthogonal system differs in one respect from those studied in § 3, in that the three surfaces, one of each type, which pass through a point P situated in a given octant intersect not only in P but also in one point of each of the other octants; this is clear since all three surfaces are symmetric in each coördinate plane. Consequently, in the so-called *ellipsoidal* coördinate system based on the confocal quadrics there are eight points with the same coördinates. This ambiguity can be avoided, however, by considering only a restricted region of space, for example, the first octant.

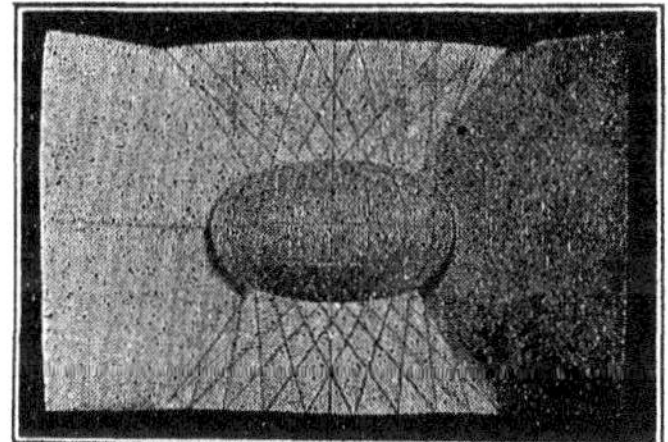

FIG. 5

The equations (1), (2), (3) can be written as the single equation

$$\frac{x^2}{a^2-\lambda}+\frac{y^2}{b^2-\lambda}+\frac{z^2}{c^2-\lambda}=1, \tag{4}$$

where λ is arbitrary except that it shall not take on the values c^2, b^2, a^2. If $\lambda < c^2$, equation (4) defines the surfaces (1), etc.; finally, if $\lambda > a^2$, (4) has no locus.

* Cf. Osgood, *Differential and Integral Calculus*, p. 326.

EXERCISE

Show that the equation

$$\frac{x^2}{a^2-\lambda}+\frac{y^2}{b^2-\lambda}=2z-\lambda, \qquad a>b,$$

where λ is a parameter not taking on the values a^2 and b^2, represents three families of paraboloids defined by the inequalities $\lambda < b^2$, $b^2 < \lambda < a^2$, $a^2 < \lambda$. Describe each family and show that the sections of all the surfaces by the common principal planes — the (y, z)- and (z, x)-planes — are confocal parabolas with the axis of z as axis. The surfaces are known as *confocal paraboloids*. They form a triply orthogonal system.

Transformation of Coördinates

5. Transformation to Parallel Axes. To transform from a system of rectangular axes to a new system of axes having the same directions as the old, but with a different origin, consider a point P whose coördinates with respect to the two systems are, respectively, (x, y, z) and (x', y', z'). Then

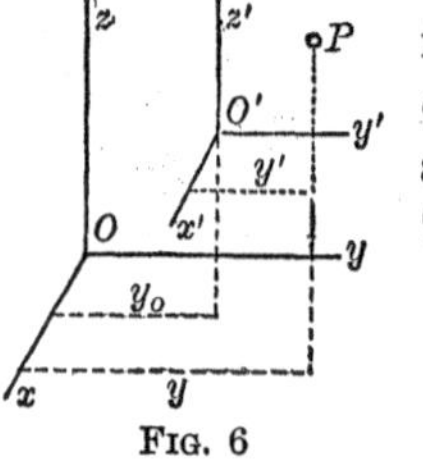

Fig. 6

$$(1) \qquad x = x' + x_0, \qquad y = y' + y_0, \qquad z = z' + z_0,$$

$$(2) \qquad x' = x - x_0, \qquad y' = y - y_0, \qquad z' = z - z_0,$$

where (x_0, y_0, z_0) are the coördinates of the new origin, O', referred to the old axes; cf. Ch. XI, § 1.

Example. What surface is represented by the equation

$$(3) \qquad 2x^2+3y^2-4z^2-4x+6y+4z-8=0?$$

Completing successively the squares of the terms in x, y, and z, we have

$$2(x-1)^2+3(y+1)^2-4(z-\tfrac{1}{2})^2=12.$$

On setting

$$x'=x-1, \qquad y'=y+1, \qquad z'=z-\tfrac{1}{2},$$

that is, on transforming to parallel axes with the new origin at the point $(1, -1, \frac{1}{2})$, the equation becomes

$$2x'^2 + 3y'^2 - 4z'^2 = 12.$$

Equation (3) is thus seen to represent a hyperboloid of one sheet whose center is at $(1, -1, \frac{1}{2})$ and whose axes are parallel to the coördinate axes. The hyperboloid opens out in the direction of the axis of z and the semi-axes of the minimum ellipse have the lengths $\sqrt{6}$ and 2.

EXERCISES

Determine and draw roughly the surface represented by each of the following equations.

1. $z^2 - 4x - 6z + 13 = 0.$
2. $9y^2 - 4z^2 + 36y + 8z - 4 = 0.$
3. $2x^2 + 2y^2 + 3z^2 - 4x + 8y - 12z + 16 = 0.$
4. $x^2 - 3y^2 + z^2 - 8x + 12y + 6z + 13 = 0.$
5. $3y^2 + 4z^2 + 4x - 6y + 16z + 27 = 0.$
6. $2x^2 + 4y^2 - 3z^2 - 8x - 24y - 30z - 19 = 0.$
7. $2x^2 - 3y^2 + z^2 + 8x + 18y - 16z - 3 = 0.$
8. $x^2 + 2y^2 + 6z^2 - 2x - 2y + 18z + 9 = 0.$
9. $2x^2 - 5y^2 + 3z^2 + 20y + 6z - 47 = 0.$
10. $y^2 - 2xz - 2x - 6y + 2z + 11 = 0.$

6. Rotation of the Axes. Through the origin O of the usual (right-handed) system of (x, y, z)-axes, choose arbitrarily three mutually perpendicular directed lines to serve as the axes of a new (right-handed) system of coördinates (x', y', z'). Let the direction angles of the axis of x', referred to the old system, be $\alpha_1, \beta_1, \gamma_1$, let those of the axis of y' be $\alpha_2, \beta_2, \gamma_2$, and those of the axis of z', $\alpha_3, \beta_3, \gamma_3$.

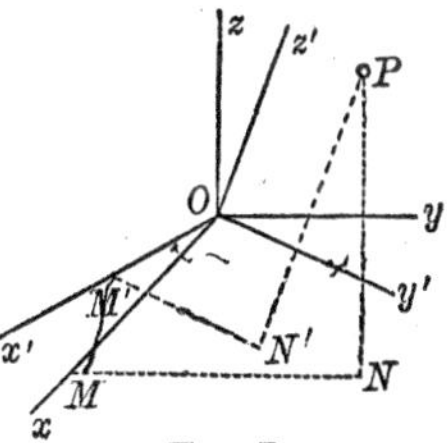

FIG. 7

Let an arbitrary point P of space

have the coördinates (x, y, z) and (x', y', z') with respect to the two systems. Join O to P by the two broken lines $OMNP$ and $OM'N'P$, where

$$OM = x,\ MN = y,\ NP = z;\quad OM' = x',\ M'N' = y',\ N'P = z'.$$

Then

$$\text{(1)}\quad \text{Proj. } OM + \text{Proj. } MN + \text{Proj. } NP = \text{Proj. } OM' + \text{Proj. } M'N' + \text{Proj. } N'P,$$

no matter on what directed line the projections are taken. Choosing the positive axes of x, y, and z in turn as this directed line, we have

$$\text{(2)}\qquad \begin{aligned} x &= x' \cos \alpha_1 + y' \cos \alpha_2 + z' \cos \alpha_3, \\ y &= x' \cos \beta_1 + y' \cos \beta_2 + z' \cos \beta_3, \\ z &= x \cos \gamma_1 + y' \cos \gamma_2 + z' \cos \gamma_3. \end{aligned}$$

Here $\cos \alpha_1$, $\cos \beta_1$, $\cos \gamma_1$ — the coefficients of x' — are the direction cosines of the axis of x'; $\cos \alpha_2$, $\cos \beta_2$, $\cos \gamma_2$, those of the axis of y'; and $\cos \alpha_3$, $\cos \beta_3$, $\cos \gamma_3$, those of the axis of z'. Let us denote these direction cosines, for the sake of brevity, by $\lambda_1, \mu_1, \nu_1,\ \ \lambda_2, \mu_2, \nu_2,\ \ \lambda_3, \mu_3, \nu_3$, respectively.

Since these triples of numbers are direction cosines and, moreover, direction cosines of three mutually perpendicular lines, we have

$$\text{(3)}\qquad \begin{aligned} \lambda_1^2 + \mu_1^2 + \nu_1^2 &= 1, &\qquad \lambda_1\lambda_2 + \mu_1\mu_2 + \nu_1\nu_2 &= 0, \\ \lambda_2^2 + \mu_2^2 + \nu_2^2 &= 1, &\qquad \lambda_2\lambda_3 + \mu_2\mu_3 + \nu_2\nu_3 &= 0, \\ \lambda_3^2 + \mu_3^2 + \nu_3^2 &= 1, &\qquad \lambda_3\lambda_1 + \mu_3\mu_1 + \nu_3\nu_1 &= 0. \end{aligned}$$

Since the three directed lines form a right-handed system, it follows by Ex. 19 at the end of Ch. XVIII that the determinant of their direction cosines has the value *plus* one:

$$\text{(4)}\qquad |\lambda\ \mu\ \nu| = 1.$$

Equations (3) and (4) express completely the fact that the three given lines through O which serve as the new axes are directed, mutually perpendicular lines forming a right-handed system.

The direction cosines of any one of the three directed lines can be expressed simply in terms of those of the other two (Exs. 17, 18 at the end of Ch. XVIII):

$$(5)\quad \begin{array}{lll} \lambda_1 = \mu_2\nu_3 - \mu_3\nu_2, & \mu_1 = \nu_2\lambda_3 - \nu_3\lambda_2, & \nu_1 = \lambda_2\mu_3 - \lambda_3\mu_2, \\ \lambda_2 = \mu_3\nu_1 - \mu_1\nu_3, & \mu_2 = \nu_3\lambda_1 - \nu_1\lambda_3, & \nu_2 = \lambda_3\mu_1 - \lambda_1\mu_3, \\ \lambda_3 = \mu_1\nu_2 - \mu_2\nu_1, & \mu_3 = \nu_1\lambda_2 - \nu_2\lambda_1, & \nu_3 = \lambda_1\mu_2 - \lambda_2\mu_1. \end{array}$$

Since α_1, α_2, α_3 are the angles which the axes of x', y', z' make with the axis of x, they are the direction angles of the axis of x with respect to the new axes. Similarly, β_1, β_2, β_3 and γ_1, γ_2, γ_3 are respectively the direction angles of the axes of y and z, referred to the new system. Consequently, the equations of transformation from the new axes to the old are

$$(6)\quad \begin{aligned} x' &= x\cos\alpha_1 + y\cos\beta_1 + z\cos\gamma_1, \\ y' &= x\cos\alpha_2 + y\cos\beta_2 + z\cos\gamma_2, \\ z' &= x\cos\alpha_3 + y\cos\beta_3 + z\cos\gamma_3. \end{aligned}$$

The direction cosines of the old axes with respect to the new are, in our notation, $\lambda_1, \lambda_2, \lambda_3,\ \ \mu_1, \mu_2, \mu_3,\ \ \nu_1, \nu_2, \nu_3$. It is clear that between these three triples there exist relations similar to the relations (3), (4), (5) for the original triples.*

The accompanying diagram gives equations (2) and (6) in skeleton. Reading across we obtain (2) and reading down we get (6). Also, the rows give the direction cosines of the old axes with respect to the new, and the columns, those of the new axes with respect to the old.

	x'	y'	z'
x	λ_1	λ_2	λ_3
y	μ_1	μ_2	μ_3
z	ν_1	ν_2	ν_3

Example 1. Transform the equation of the surface

$$(7)\qquad 13x^2 + 13y^2 + 10z^2 + 8xy - 4yz - 4xz - 36 = 0$$

to new axes through O, whose direction cosines are respectively $-\frac{1}{3}, \frac{2}{3}, \frac{2}{3};\ \ \frac{2}{3}, -\frac{1}{3}, \frac{2}{3};\ \ \frac{2}{3}, \frac{2}{3}, -\frac{1}{3}$.

* Of the new equations only those of the form (3) are different from the old. The new equation (4) is obtainable from the old by interchanging rows and columns in the determinant. Similarly, if in the present equations (5) the columns are written as rows, the result is the new equations (5).

Here
$$x = \tfrac{1}{3}(-\ x' + 2\,y' + 2\,z'),$$
$$y = \tfrac{1}{3}(\ 2\,x' - \ y' + 2\,z'),$$
$$z = \tfrac{1}{3}(\ 2\,x' + 2\,y' - \ z').$$

Substituting these values for x, y, z in (7) and simplifying the result, we obtain
$$x'^2 + y'^2 + 2\,z'^2 = 4.$$

This equation represents an ellipsoid of revolution about the z'-axis as axis. Hence (7) represents an ellipsoid of revolution whose axis is the line through O with the direction components 2, 2, -1.

Example 2. What surface is represented by the equation

$$(8)\quad 13\,x^2 + 13\,y^2 + 10\,z^2 + 8\,xy - 4\,yz - 4\,xz - 10\,x + 26\,y - 40\,z + 22 = 0\,?$$

We make the transformation to parallel axes

$$(9)\qquad x = x' + x_0, \qquad y = y' + y_0, \qquad z = z' + z_0,$$

aiming to choose the new origin (x_0, y_0, z_0) so that in the equation resulting from (8) the linear terms in x', y', z' do not appear. Substituting the values of x, y, z given by (9) into (8), collecting terms, and then setting the coefficients of x', y', and z' equal to zero, we obtain the equations:

$$13\,x_0 + 4\,y_0 - 2\,z_0 - 5 = 0,$$
$$4\,x_0 + 13\,y_0 - 2\,z_0 + 13 = 0,$$
$$x_0 + y_0 - 5\,z_0 + 10 = 0.$$

These equations have a unique solution, namely, $x_0 = 1$, $y_0 = -1$, $z_0 = 2$.

If (8) is transformed to parallel axes with the new origin at the point $(1, -1, 2)$ thus determined, it becomes

$$13\,x'^2 + 13\,y'^2 + 10\,z'^2 + 8\,x'y' - 4\,y'z' - 4\,x'z' - 36 = 0.$$

But this is the same equation in x', y', z' as (7) is in x, y, z. Hence (8) represents an ellipsoid of revolution whose center is at the point $(1, -1, 2)$ and whose axis of revolution has the direction components 2, 2, -1.

Example 3. Consider an equation in which only one of the terms in xy, yz, zx is present, for example, the equation

(10) $$2x^2 - y^2 - z^2 - 2yz - 4x + 6y + 2z + 2 = 0.$$

The term in yz in this equation can be removed by rotating the y- and z-axes about the axis of x through a suitable acute angle θ; that is, by application of the transformation

(11) $$x = x', \qquad y = y' \cos\theta - z' \sin\theta, \qquad z = y' \sin\theta + z' \cos\theta.$$

According to Ch. XII, § 5, the desired angle θ is 45°. Transforming (10) by the rotation of axes (11), where $\theta = 45°$, we obtain

$$x'^2 - y'^2 - 2x' + 2\sqrt{2}\,y' - \sqrt{2}\,z' + 1 = 0.$$

This equation can be written in the form

$$(x' - 1)^2 - (y' - \sqrt{2})^2 = \sqrt{2}(z' - \sqrt{2}),$$

and hence becomes

$$x''^2 - y''^2 = \sqrt{2}\,z'',$$

when referred to axes through the point $(1, \sqrt{2}, \sqrt{2})$ parallel to the axes of x', y', z'.

It follows, then, that (10) represents a hyperbolic paraboloid whose vertex, referred to the (x', y', z')-axes, is at the point $(1, \sqrt{2}, \sqrt{2})$ and whose axis is parallel to the axis of z'.

Remark. The general method of procedure to determine the surface represented by an equation of the second degree in x, y, z is that of Examples 1 and 2; the equation is first transformed by a change of origin to remove the linear terms in x, y, z and is then subjected to a rotation of the axes to get rid of the terms in xy, yz, zx. This method cannot be applied, however, to equation (10) of Example 3, for it is impossible to transform (10) so that the linear terms disappear, since, if this were possible, the surface would be symmetric in the new origin (Ex. 11 at the end of the chapter), whereas we know that a paraboloid has no point of symmetry. Accordingly, in Example 3 and in similar cases, the axes are first rotated to remove the terms in xy, yz, zx and then a proper change of origin, as suggested by the new equation, is made.

EXERCISES

1. Find the equations of the rotation of the axes which introduces the two directed lines through O with the direction cosines $\frac{6}{7}, \frac{2}{7}, -\frac{3}{7}$ and $\frac{2}{7}, \frac{3}{7}, \frac{6}{7}$ as the axes of x' and y' respectively.

2. Find the equations of a rotation of the axes which introduces the planes,

$$x + 2y + 2z = 0, \quad 2x - 2y + z = 0, \quad 2x + y - 2z = 0,$$

as the (x', y')-, (y', z')-, and (z, x')- planes, respectively.

3. Find the equations of a rotation of the axes which introduces the planes

$$3x + 4y + 12z = 0, \qquad 12x + 3y - 4z = 0$$

as the (x', y')- and (y', z')- planes.

4. Transform the equation of the hyperbolic paraboloid

$$x^2 - y^2 = 2mz$$

by a rotation of the x- and y-axes through an angle of $-45°$ about the axis of z. *Ans.* $x'y' = mz'$.

Determine the surface represented by each of the following equations.

5. $5x^2 + 5y^2 + 4z^2 - 6xy - 32 = 0$.

6. $3x^2 - 2y^2 - 5z^2 - 4yz + 4y + 10z + 1 = 0$.

7. $x^2 - 2y^2 + z^2 - 2xz - x + 3z - 3 = 0$.

8. Transform the equation of the surface

$$5x^2 - 2y^2 + 11z^2 + 12xy + 12yz - 14 = 0$$

by the rotation of the axes of Ex. 1. Thus identify the surface.

9. The equation

$$10x^2 + 13y^2 + 13z^2 - 4xy - 10yz - 4xz - 36y + 36z = 0$$

represents a central quadric whose axes have the direction components 2, -1, -1, 0, 1, -1, 1, 1, 1. Identify the quadric.

10. The equation

$$5x^2+5y^2+2z^2+2xy-4yz+4xz+6x-6y-12z=0$$

represents a paraboloid whose principal planes are $x+y=0$, $x-y+z=0$. Identify the paraboloid.

11. Show that the equation of a sphere whose center is at the origin is not changed by any rotation of the axes. Actually carry through the transformation.

7. The General Equation of the Second Degree. We have defined a quadric surface as any surface represented by an equation of the second degree in x, y, z, that is, by an equation of the form:

$$(1)\qquad Ax^2+By^2+Cz^2+2A'yz+2B'xz+2C'xy \\ +2A''x+2B''y+2C''z+F=0.$$

We propose now to ascertain whether there are types of quadric surfaces other than those already discussed in Chs. XXII, XXIII, and to sketch a method whereby the type of surface defined by a given equation of the form (1) can be determined.

As in the corresponding problem in the plane (Ch. XII), transformations of coördinates play an important rôle. In particular, the expressions formed from the coefficients of (1), which are invariant (Ch. XII, § 6) under any change of axes, are fundamental. Chief among these invariant expressions are the determinants

$$D=\begin{vmatrix} A & C' & B' \\ C' & B & A' \\ B' & A' & C \end{vmatrix},\qquad \Delta=\begin{vmatrix} A & C' & B' & A'' \\ C' & B & A' & B'' \\ B' & A' & C & C'' \\ A'' & B'' & C'' & F \end{vmatrix},$$

which correspond to the invariants B^2-4AC and Δ in the case of the general equation of the second degree in x and y (Ch. XII, § 6).

We state, without proof, the following theorems:

THEOREM 1. *If equation* (1) *represents a surface, and if* $D \neq 0$, *the surface is symmetric in just one point. If* $D = 0$, *there is in general no point of symmetry, and when there is one, there are infinitely many.*

THEOREM 2. *If equation* (1) *represents a surface and if* $D \neq 0$, *the surface is symmetric in three mutually perpendicular planes, and these are, in general, all the planes of symmetry. If* $D = 0$, *it is symmetric in two perpendicular planes and these are, in general, all the planes of symmetry.*

It is clear from these theorems that, in discussing equation (1), two essentially different cases arise, according as $D \neq 0$ or $D = 0$.

Case 1. $D \neq 0$. A surface defined by an equation of the form (1) for which $D \neq 0$ is symmetric in a unique point O', by Th. 1. The coördinates of O' can be found by the method of Example 2, § 6. A transformation to parallel axes with the new origin at O' removes the linear terms in (1), leaves the quadratic terms unchanged, and, as can be shown, makes the constant term into Δ/D. Thus (1) becomes

$$(2) \quad Ax'^2 + By'^2 + Cz'^2 + 2A'y'z' + 2B'x'z' + 2C'x'y' + \Delta/D = 0.$$

Since $D \neq 0$, the surface is, by Th. 2, symmetric in three mutually perpendicular planes whose common point, since it is a point of symmetry, must be O'. To determine from equations (1) or (2) the precise positions of these planes through O' is a problem of intrinsic difficulty which we shall not attempt to discuss. When once the positions are known, however, a rotation of the axes which brings the coördinate planes into coincidence with them serves, either immediately or eventually, to remove the terms in $y'z'$, $z'x'$, and $x'y'$ in (2); cf. Ex. 12 at the end of the chapter. We obtain, then, the final equation

$$(\text{I}) \qquad ax''^2 + by''^2 + cz''^2 + \Delta/D = 0.$$

For this equation, $D = abc$ and hence, since $D \neq 0$, no one of the coefficients a, b, c can be zero.

If $\Delta \neq 0$, (I) *and hence* (1) *represents a central quadric (an ellipsoid or hyperboloid) or, in case* a, b, c, Δ/D *are all of the same sign, it has no locus.**

If $\Delta = 0$, (I) *and hence* (1) *represents a cone, or, in case* a, b, c *are all of the same sign, a point.*

Case 2. $D = 0$. A surface defined by an equation of the form (1) for which $D = 0$ has in general no point of symmetry (Th. 1), and hence it is in general impossible to transform to parallel axes so that the linear terms in (1) disappear. There are, however, at least two mutually perpendicular planes of symmetry, by Th. 2. If the positions of two such planes are known, a rotation of the axes whereby two of the coördinate planes become respectively parallel to them serves, either immediately or eventually, to remove the terms in yz, zx, and xy in (1); cf. Ex. 13 at the end of the chapter. Thus (1) becomes

$$(3) \qquad ax'^2 + by'^2 + cz'^2 + 2a''x' + 2b''y' + 2c''z' + F = 0.$$

For this equation, $D = abc$, and, since $D = 0$, $abc = 0$. Now a, b, c are not all zero, since otherwise (3), and hence (1), would not be a quadratic equation. Two cases then arise, according as one or two of the coefficients a, b, c vanish.

A. *One of the coefficients* a, b, c *vanishes.* Since (3) bears equally on x', y', z' it is immaterial which one of the coefficients a, b, c we assume to be zero. Suppose that $c = 0$:

$$ax'^2 + by'^2 + 2a''x' + 2b''y' + 2c''z' + F = 0, \quad ab \neq 0.$$

By a change of origin to the point $(-a''/a, -b''/b, 0)$, this equation becomes

$$(\text{II}a) \qquad ax''^2 + by''^2 + 2c''z' + f = 0, \qquad ab \neq 0.$$

* In reducing (1) to the form (I), it was assumed that (1) represents a surface; the method of reduction is quite the same, however, if (1) has no locus. Similarly, the method of reduction in case 2 is always applicable, both when (1) represents a surface, as assumed, and when it does not.

For (IIa), $\Delta = -abc''^2$ and hence $\Delta \neq 0$ or $\Delta = 0$, according as $c'' \neq 0$ or $c'' = 0$. If $c'' \neq 0$, a change of origin to the point $(0, 0, -f/2c'')$ reduces (IIa) to

$$ax''^2 + by''^2 + 2c''z'' = 0, \qquad ab \neq 0.$$

If $c'' = 0$, (IIa) becomes

$$ax''^2 + by''^2 + f = 0, \qquad ab \neq 0.$$

Hence we conclude the following:

If $\Delta \neq 0$, (IIa) and hence (1) represents a paraboloid (elliptic or hyperbolic).

If $\Delta = 0$, (IIa) and hence (1) represents, in the case $f \neq 0$, an elliptic or hyperbolic cylinder, or it has no locus; if $f = 0$, it represents two intersecting planes or a line.

B. *Two of the coefficients a, b, c vanish.* Here again it is immaterial which two of the three coefficients we assume to be zero. Suppose that $b = c = 0$:

$$ax'^2 + 2a''x' + 2b'y' + 2c''z' + F = 0, \qquad a \neq 0.$$

By a change of origin to the point $(-a''/a, 0, 0)$ and by a proper rotation of the axes about the axis of x', this equation becomes

$$\text{(II}b\text{)} \qquad ax''^2 + 2dz'' + f = 0, \qquad a \neq 0.$$

Here Δ is always zero. *Equation* (IIb), *and hence* (1), *represents a parabolic cylinder, if $d \neq 0$; if $d = 0$, it represents two parallel planes, a single plane, or has no locus.*

Summary. The new types of loci of equations of the form (1) which have resulted from this investigation, are:

i) A point, — which is a limiting form of an ellipsoid and is frequently spoken of, in this connection, as a *null ellipsoid.* It is to be noted, from the discussion of (I), that the corresponding limiting form of a hyperboloid is a cone.

ii) Two intersecting planes, a line, two parallel planes or a single plane, all of which are limiting forms of cylinders. We shall call them *degenerate cylinders;* two intersecting planes, a degenerate hyperbolic cylinder; a line, a degenerate (or null) elliptic cylinder; two parallel planes, or a single plane, a degenerate parabolic cylinder.

We can now summarize our results:

THEOREM 3. *An equation of the form* (1), *if it has a locus, represents a central quadric, a paraboloid, a cone or a point, or a cylinder* (*non-degenerate or degenerate*).

The following table shows when each of the four cases occurs and thus furnishes a means of determining the type of surface defined by any given equation of the form (1).

	$\Delta \neq 0$	$\Delta = 0$
$D \neq 0$	Central quadric or no locus	Cone or point
$D = 0$	Paraboloid	Cylinder or no locus

Equations (IIa) and (IIb) can be written as the single equation

(II) $$ax^2 + by^2 + 2\,dz + f = 0, \qquad a \neq 0,$$

where the primes have been dropped from the variables. We have then, in conclusion, the theorem:

THEOREM 4. *An equation of the form* (1) *can be reduced by transformations of coördinates to one of the two forms:*

(I) $$ax^2 + by^2 + cz^2 + f = 0, \qquad abc \neq 0,$$

(II) $$ax^2 + by^2 + 2\,dz + f = 0, \qquad a \neq 0.$$

EXERCISES

Determine to which of the four types the quadric surface represented by each of the following equations belongs.

1. That of Ex. 7, § 6. 2. That of Ex. 8, § 6.

3. $31\,x^2 + 41\,y^2 - 23\,z^2 + 48\,yz + 72\,xz - 24\,xy$
$- 72\,x - 48\,y + 46\,z - 23 = 0.$

4. $2\,x^2 + 2\,y^2 + 2\,z^2 + yz + xz + xy - 4\,x - y - z - 4 = 0.$

5. $10x^2 + 2y^2 + 5z^2 + 6yz + 2xz + 4xy$
$$- 26x - 14y - 18z - 18 = 0.$$

6. $7x^2 + 7y^2 + 4z^2 - 8yz + 8xz - 2xy$
$$- 42x + 18y - 12z = 0.$$

7. The surface (3) is symmetric in three planes whose normals have the direction components $6, -3, 2,\ \ 2, 6, 3,\ \ -3, -2, 6$. Determine the precise nature and position of the surface.

8. The surface (4) is symmetric in three lines whose direction components are $2, -1, -1,\ \ 0, 1, -1,\ \ 1, 1, 1$. Determine its precise nature and position.

9. The surface (5) is symmetric in two planes whose normals have the direction components $1, -1, -2,\ \ 3, 1, 1$. Determine its precise nature and position.

10. Two principal planes of the surface (6) are parallel respectively to the planes

$$x - y + z = 0, \qquad x + y = 0.$$

Determine the precise nature and position of the surface.

EXERCISES ON CHAPTER XXIV

1. Prove that a curved surface whose equation in spherical coördinates does not contain r is a cone with the pole as vertex.

2. Show that a curved surface whose equation in spherical coördinates does not contain θ is a surface of revolution. What is its axis?

3. Prove that a curved surface whose equation in cylindrical coördinates does not contain z is a cylinder.

4. Show that a curved surface whose equation in cylindrical coördinates does not contain θ is a surface of revolution.

Transformation of Axes

5. Prove that the transformation to new axes through O whose direction cosines are $-\frac{1}{3}, \frac{2}{3}, \frac{2}{3},\ \ \frac{2}{3}, -\frac{1}{3}, \frac{2}{3},\ \ \frac{2}{3}, \frac{2}{3}, -\frac{1}{3}$ is

identical with a rotation of the original axes about the line $x = y = z$ through 180°.

Suggestion. Show that the equations of transformation are equivalent to those connecting the coördinates of two points symmetric in the line.

6. Find the equations of the transformation which introduces as axes the three mutually perpendicular lines through the point (x_0, y_0, z_0) with the direction angles $\alpha_1, \beta_1, \gamma_1, \alpha_2, \beta_2, \gamma_2, \alpha_3, \beta_3, \gamma_3$.

7. Set up the equations of a rotation of the axes which introduces the plane $x + y + z = 0$ as the (x, y)-plane.

8. Determine the precise nature of the curve of intersection of the plane $x + y + z = 0$ with the surface

$$x^2 - xy + yz - zx - x - y - z = 0.$$

Suggestion. Use the result of Ex. 7.

9. A line of symmetry of the surface

$$xy + yz + xz = 2$$

is the line $x = y = z$. Determine the precise nature of the surface.

10. A plane of symmetry of the surface

$$x^2 + y^2 + z^2 + xy + yz - xz - x + y - z = 0$$

is the plane $x - y - 2z = 0$. What is the exact nature of the surface?

11. Show that, if a quadric surface is symmetric in the origin, its equation contains no linear term in x, y, z, and conversely.

12. A quadric surface is symmetric in each of the coördinate planes. Prove that either the equation of the surface is of the form

$$ax^2 + by^2 + cz^2 = d,$$

or that the surface consists of two coördinate planes. Show that in the latter case the equation can be reduced to the desired form by rotating two of the axes about the third through an angle of 45°.

13. A quadric surface is symmetric in two planes which are parallel to or identical with two coördinate planes. Show that either the terms in yz, zx, xy do not appear in the equation of the surface or the surface itself consists of two planes of the type described. Prove that in the latter case the terms in yz, zx, xy can be removed from the equation by rotating two of the axes about the third through an angle of 45°.

14. Show that, if a quadric surface is symmetric in a coördinate plane or in a plane parallel to a coördinate plane, its equation contains, *in general*, at most one of the three terms in yz, zx, xy. When does the exception occur?

15. Prove that the conclusion of the previous exercise follows if the surface is symmetric in a coördinate axis or in a line parallel to a coördinate axis. When does the exception occur in this case?

INDEX

www.ingramcontent.com/pod-product-compliance
Lightning Source LLC
LaVergne TN
LVHW011244110826
845149LV00001B/44

* 9 7 8 1 4 1 8 1 8 6 0 3 6 *